MW00844050

ANALYSIS AND DESIGN OF SHALLOW AND DEEP FOUNDATIONS

ANALYSIS AND DESIGN OF SHALLOW AND DEEP FOUNDATIONS

LYMON C. REESE
WILLIAM M. ISENHOWER
SHIN-TOWER WANG

JOHN WILEY & SONS, INC.

Published by John Wiley & Sons, Inc., Hoboken, New Jersey.
Published simultaneously in Canada.

Library of Congress Cataloging-in-Publication Data:

Reese, Lymon C., 1917–
Analysis and design of shallow and deep foundations/Lymon C. Reese, William M. Isenhower, Shin-Tower Wang.
p. cm.
Includes bibliographical references and index.
ISBN-13 978-0-471-43159-6 (cloth)
ISBN-10 0-471-43159-1 (cloth)
1. Foundations—Design and construction. I. Isenhower, William M. II. Wang, Shin-Tower. III. Title.
TA775.R419 2006
624.1′5—dc22 2005013198

Printed in Mexico

10 9 8 7 6 5 4

CONTENTS

PREFACE

Advances in foundation engineering have been rapid in recent years. Of note are the maturity of the concepts of soil–structure interaction, the development of computer codes to deal with advanced topics, the advent of new methods for the support of structures, and the proliferation of technical publications and conferences that present a variety of useful information on the design and performance of foundations. This book takes advantage of these advances by presenting methods of analysis while being careful to emphasize standard methods such as site visits and the role of engineering geology.

The goals of the engineer in the design of foundations are to achieve a system that will perform according to stipulated criteria, can be constructed by established methods, is capable of being inspected, and can be built at a reasonable cost. Acceptable performance usually requires limited vertical and horizontal movement in the short and long term. Chapter 6 presents several instances where foundations have collapsed or suffered excessive movement. The purpose of presenting such failures is to emphasize that knowledge of foundation behavior must be used and that care must be taken in predicting how soil will respond to the imposition of loads. However, other kinds of failure are also possible. On occasion, the use of advanced methods for the analysis of shallow and deep foundations is omitted due to lack of knowledge or time in completing a design. The result may be an unacceptable and unfavorable design requiring more costly foundations than necessary.

This book describes methods for computing the settlement of both shallow and deep foundations. The engineer may compute a substantial safety factor for a foundation and decide that settlement is not a problem. However, modern engineering is aimed at achieving compatibility between the foundation and the superstructure, so computation of foundation settlement, even under relatively light loads, cannot be ignored.

USE OF THE BOOK IN TEACHING

The material in this book can be presented in several ways. The student is expected to have completed a basic four-semester-hour introductory course on geotechnical engineering. The book assumes that this course introduced the basic knowledge of soil behavior, one-dimensional consolidation, shear strength of soils under undrained and fully drained conditions, and laboratory testing of soils. If the course is a senior-level or a joint senior-level and introductory graduate-level course, the course should cover Chapters 1, 2, and 4 through 11 in their entirety. Portions of Chapter 3 can be presented as a review to the introductory course in geotechnical engineering. The material on the design of drilled shaft foundations should include a discussion of the basic methods of construction and should cover the methods of design for axial loading in cohesive and cohesionless soils. The design of drilled shafts in intermediate geomaterials or rock may be covered if time permits.

If this book is being used for a graduate-level course, the chapters on foundation design should be covered in their entirety, and design projects should include all foundation types in all applicable soil conditions. Chapters 2 and 3 may be omitted if students are required to complete another graduate-level course on advanced soil mechanics.

UNIQUE CONTRIBUTIONS OF THE BOOK

The movement of shallow foundations under an increasing series of loads can be computed in a relatively straightforward manner, but the same method cannot be applied to deep foundations. Detailed, comprehensive, and up-to-date methods are presented for determining the behavior of piles or drilled shafts under axial or lateral loading.

The following forms of the differential equation are presented for determining a deep foundation under axial loading:

$$E_p A_p \frac{d^2 w_z}{dz^2} = f_x C$$

and

$$f_z = \beta_z w_z$$

where $E_p A_p$ = axial stiffness of the pile, w_z = relative movement of the pile with reference to the soil at point z, C = circumference of the pile, f_z = unit load transfer at point z, and β_z = is a function depending on the depth z and the value of w_z. The differential equation is solved using the difference-equation technique.

The differential equation for the deep foundation under lateral load is

$$E_p I_p \frac{d^4y}{dx^4} + E_{py} y = 0$$

where $E_p I_p$ = lateral stiffness of the pile, y = lateral deflection of the pile with respect to the soil, x = distance along the pile, and E_{py} = lateral stiffness of the soil corresponding to y. Again, the differential equation is solved by using difference-equation techniques.

The differential equations for axial loading and lateral loading are used to produce a solution for two-dimensional pile groups under inclined and eccentric loading. The treatment of a pile or drilled shaft as a deformable body whose deformations are dependent on all of the relevant parameters is on the cutting edge of the present technology. The combination of the behavior under axial loading and lateral loading to produce a method for determining the behavior of pile groups gives the engineer a complex but rational tool to use in the solution of a problem that was beyond the scope of practice until recently.

A student version of the computer codes for the pile under an axial load, for the pile under a lateral load, and for the pile group is included with the book. The programs allow demonstration of the solution of the homework problems in this book. The professional version of the computer codes is available to the industry.

ACKNOWLEDGMENTS

The authors wish to thank their colleagues at Ensoft, Inc., Mr. José Arréllaga, Dr. Gonzalo Vásquez, Ms. Lisa Novak, and Mr. Joe Hendrix, for their assistance and contributions to the material presented in the book.

Contributions and insight by the faculty at The University of Texas at Austin were helpful to the authors. In particular, appreciation is expressed to Professor Roy E. Olson, a colleague and teacher of two of the authors. Much of the content of Chapter 3 is based on class notes and materials written by Professor Olson.

The authors also wish to acknowledge the major contributions of the late Professor Michael W. O'Neill, a close friend and colleague. His writings are the basis of much of the current practice in the design and construction of drilled shaft foundations in the United States. Chapter 11 is based on Professor O'Neill's extensive research and writing on the design of drilled shaft foundations.

Of special note are our friends in the construction industry. Over the years, we have had close involvement with Mr. Glyen Farmer of Farmer Foundation and Mr. James Parks of Parks Drilling, who allowed us to learn about the dangers and consequences of incorrect soil characterization or bad design choices on construction operations and have allowed us to have an advanced look at developments in drilled shaft construction practices. Many other contractors have been gracious in sharing their knowledge and assisting the authors in special circumstances.

The authors wish to thank Mr. Caleb Lai, Mrs. Eileen Melant, Miss Marian Wu, Miss Janie Wang, and Mrs. Georgia Reese for their diligent work in formatting, proofreading, and preparing graphics.

SYMBOLS AND NOTATIONS

A = total area (Chapter 3), area of footing (Chapter 7), cross-sectional area of pile material (Chapter 10), cross-sectional area of the socket (Chapter 11), cross-sectional area of the pile in square inches (Chapter 13)

A_b = effective area of the tip of the shaft in contact with the soil (Chapter 11)

A_f = coefficient-A at the point of failure (Chapter 3)

A_p = gross end area of the pile (Chapter 10), cross-sectional area of the tip of the pile (Chapter 10)

A_r = ram cross-sectional area (Chapter 10)

A_s = area of the wire in the spring (Chapter 3), side surface area of the pile (Chapter 10), surface area of the shaft in contact with the soil (Chapter 10)

A_w = area covered by water, water area (Chapter 3)

a_v = coefficient of compressibility (Chapters 3 and 7)

B = width of footing (Chapters 7, 9), pile diameter (Chapter 10), diameter of the base of the shaft (Chapter 11), socket diameter (Chapter 11), diameter of footing or equivalent length of a side for a square or rectangular shape (Chapter 13), pile diameter (Chapter 13)

B_b = diameter of socket (Chapter 11), diameter of drilled shaft (Chapter 11)

b = base inclination factors (Chapter 7), diameter (width) of the pile (Chapter 14)

C = parameter describing the effect of repeated loading on deformation (Chapter 14)
C_1 = coefficient to reflect arching-compression relief (Chapter 9)
C_2 = coefficient to reflect creep with time (Chapter 9)
C_α = coefficient of secondary compression (Chapter 3)
C_c = coefficient of curvature (also called the *coefficient of gradation*) (Chapter 3), compression index (Chapter 3)
C_d = effective perimeter of the pile (Chapter 10)
C_f = correction factor for K_δ when $\delta \neq \phi$ (Chapter 10)
C_u = uniformity coefficient (Chapter 3)
C_w = bulk modulus of water (Chapter 3), settlement coefficient (Chapter 13)
C_ε = strain index (Chapter 3)
c = cohesion (Chapters 6 and 7), undrained shear strength of soil (Chapter 10), undrained shear strength at the tip of the pile (Chapter 10), undrained shear strength at depth z (Chapter 14)
c_a = adhesion between the clay and the pile (Chapter 10), pile soil adhesion (Chapter 10), average undrained shear strength (Chapter 14)
c_m = mean undrained shear strength along the pile (Chapter 10)
c_n = undrained shear strength of normally consolidated clay (Chapter 10)
c_s = spacing of discontinuities (Chapter 11)
c_u = value of undrained shear strength (saturated clays) (Chapter 7), undrained shear strength at depth z (Chapter 11), average undrained shear strength of clay (Chapter 11)
c_v = coefficient of consolidation (Chapters 3 and 7)
c_w = term to adjust for the position of the water table (Chapter 9)
c_x = undrained shear strength at depth x (Chapter 10)
D = diameter of the vane (Chapter 4), depth along the piles at which the effective overburden pressure is calculated (Chapter 10), pile diameter (Chapters 10 and 12)
D_c = critical depth (Chapter 10)
D_f = ground surface (Chapter 7), depth below ground surface (base of the foundation) (Chapter 7), depth of overburden (Chapter 9)
D_i = thickness of a single segment (Chapter 10)
D_r = relative density, decimal fraction (Chapter 13)
D_s = depth of socket measured from the top of rock (not the ground surface) (Chapter 11)
D_w = depth to the water table from the ground surface (Chapter 9)
d = depth factors (Chapter 7)
dA = differential area of the perimeter along the sides of the drilled shaft over the penetration depth (Chapter 11)
E = modulus of elasticity (Chapters 9, 10, and 14), $(\sigma_1 - \sigma_3)$ (Chapter 9), efficiency factor (1 or <1) (Chapter 15), Young's modulus (Chapter 13)

E_b = Young's modulus of the granular geomaterial beneath the base of the drilled shaft, which can be different from E_{sL} (Chapter 13)
E_c = Young modulus of the shaft's cross section (Chapter 11), Young's modulus of the composite (steel and concrete) cross section of the drilled shaft (Chapter 13)
EI = nonlinear bending stiffness (Chapter 14)
E_i = Young's modulus of the recovered, intact core material (Chapter 11), the equivalent Young's modulus of the concrete in the socket, considering the stiffening effects of any steel reinforcement (Chapter 11)
E_{ir} = initial modulus of the rock (Chapter 14)
E_m = mass modulus of elasticity (Chapter 11), Young modulus of the shaft's rock mass (Chapter 11), Modulus of the in situ rock (Chapter 11)
E_p = arbitrary modulus of deformation related to the pressuremeter (Chapter 4)
E_pI_p = bending stiffness of the pile ($F - L^2$) (Chapter 12)
E_{pp} = reduced value secant to a *p-y* curve with pile deflection (Chapter 14)
E_pI_p = value of bending stiffness of a reinforced concrete pile (Chapter 14)
E_{py} = modulus (F/L) or a parameter that relates p and y (Chapter 12)
$E_{py\,\max}$ = initial slope of the *p-y* curve (Chapter 14)
E_s = stiffness of sand (Chapter 9), modulus for sand (Chapter 9), Young's modulus of the material in the softer seams within the harder weak rock (Chapter 11), soil modulus, kPa/m (Chapter 13)
E_{sL} = Young's modulus of the granular geomaterial along the sides of the socket at the base level (Chapter 13)
E_{sm} = Young's modulus of soil at the mid-depth of the socket where the decomposed rock becomes stronger with depth (Chapter 13)
$E_{s\,\max}$ = slope of the initial portion of the stress-strain curve in the laboratory (Chapter 14)
e = void ratio (Chapters 3 and 7)
F = shearing force (Chapter 3)
F_a = active force (Chapter 14)
F_h = horizontal force at the pile head (Chapter 15)
F_p = passive force (Chapter 14)
F_{pt} = total lateral force (Chapter 14)
F_s = global factor of safety (Chapter 8)
F_v = vertical force at the pile head (Chapter 15)
f = unit load transfer in skin friction (Chapter 10), load transfer in side resistance (Chapter 13), unit load transfer, kPa (Chapter 13)
f_c = cone resistance (Chapter 4)
$f_{\max}$ = peak soil friction (taken as the mean undrained shearing strength) (Chapter 10), maximum unit load transfer, kPa (Chapter 13)

f_s = average skin friction (Chapter 10), ultimate skin resistance per unit area of shaft C_d (Chapter 10), ultimate load transfer in side resistance at depth z (Chapter 11), ultimate side resistance in units of lb/in.^{2} (Chapter 11)
f_{sz} = ultimate unit side resistance in sand at depth z (Chapter 11)
f_x = unit resistance in clay at depth x measured from ground surface (Chapter 10)
G_s = specific gravity of the mineral grains (Chapter 3)
g = ground inclination factors (Chapter 7)
H = total thickness (Chapter 3), height (Chapter 4), horizontal component (Chapter 7)
H_B = horizontal force component parallel with short sides B (Chapter 7)
H_f = hydraulic load factor (Chapter 11)
H_L = horizontal force component parallel with long sides L (Chapter 7)
h = height of ram fall, in feet (Chapter 10), increment length or dx (Chapter 13)
I = inertia (Chapters 12 and 14)
I_r = rigidity index of the soil (Chapter 11)
I_z = vertical strain factor (Chapter 9)
I_ρ = influence coefficient (Chapter 11)
i = inclination factors (Chapter 7), hydraulic gradient (Chapter 7)
J = experimentally determined parameter (Chapter 14)
K = a constant in ft^3 for finding shear strength (Chapter 4), vane constant, depending on its shape (Chapter 4), coefficient of lateral earth (ratio of horizontal to vertical normal effective stress) (Chapter 10), lateral earth pressure coefficient (Chapter 10), a parameter that combines the lateral pressure coefficient and a correlation factor (Chapter 11)
K_0 = coefficient of earth pressure at rest (Chapters 4 and 14)
K_A = minimum coefficient of active earth pressure = $\tan^2(45 - \phi/2)$ (Chapter 14)
K_b = a fitting factor (Chapter 13)
K_s = fitting factor (Chapter 13)
K_{sp} = empirical coefficient that depends on the spacing of discontinuities (Chapter 11), empirical coefficient that depends on the spacing of discontinuities and includes a factor of safety of 3 (Chapter 11)
K_δ = coefficient of lateral stress at depth v (Chapter 10)
k = spring constant (Chapter 3), coefficient of permeability (Chapter 3)
k_c = constant (Chapter 10)
k_{ir} = dimensionless constant (Chapter 14)
k_{rm} = constant, ranging from 0.00005 to 0.0005 (Chapter 14)

L = flow distance (Chapter 3), limited length of the foundation (Chapter 7), penetration of the pile below ground surface (Chapter 10), length of the pile in contact with the soil (Chapter 10), penetration of the drilled shaft below ground surface (Chapter 11), penetration of the shaft (Chapter 11), depth of embedment of the drilled shaft (Chapter 11), penetration of the socket (Chapter 11), center-to-center distance between piles (Chapter 12), socket length (Chapter 13)
L_c = length of the core (Chapter 11)
L_e = embedded pile length (Chapter 10)
LI = liquidity index (Chapter 3)
LL = liquid limit (Chapter 3)
M = empirical factor that depends upon the fluidity of the concrete as indexed by the concrete slump (Chapter 11), moment (Chapter 12)
M_s = moment at the pile head (Chapter 15)
M_t = applied moment at the pile head (Chapters 12 and 14)
m = mean vertical effective stress between the ground surface and the pile tip (Chapter 10), parameter for cohesive intermediate geomaterial (Chapter 11)
m_R = mean value of resistance R (Chapter 8)
m_S = mean value of loads S (Chapter 8)
m_v = coefficient of volume compressibility (Chapter 7)
N = vertical force (Chapter 3), number of cycles of load application (Chapter 14)
N_{60} = Standard Penetration Test (SPT) penetration resistance, in blows per foot, for the condition in which the energy transferred to the top of the drive string is 60% of the drop energy of the SPT hammer (Chapter 11)
N_c = bearing capacity factor (Chapters 7, 9, and 13)
N'_c = bearing capacity factor (Chapter 7)
N_q = bearing capacity factor (Chapters 7 and 10)
N'_q = bearing capacity factor (Chapters 7 and 10)
N_{SPT} = corrected blow count from the SPT (Chapter 9)
N-value = sum of the number of blows needed to drive the sampler through the second and third intervals (Chapter 4).
N_γ = bearing capacity factor (Chapter 7)
N'_γ = bearing capacity factor (Chapter 7)
n = number of drainage boundaries (Chapter 3), number of segments (Chapter 10), number of piles (Chapter 10), characteristic parameter (Chapter 11), number of piles in a group (Chapter 15)
OCR = overconsolidation ratio (Chapter 11)
P = force applied (Chapter 3), compressive force (Chapter 13), axial force in the pile, in pounds (downward position) (Chapter 13)
P_a = allowable pile capacity (Chapter 10)

PI = plasticity index (Chapters 3 and 10)
P_s = force in the spring (Chapter 3)
P_t = lateral load (shear, horizontal) (Chapters 12 and 14), lateral force at the pile head (Chapter 15), axial force (Chapter 13)
P_u = ultimate pile capacity (Chapter 10)
P_{ult} = failure load (Chapter 10)
P_w = force in the water (Chapter 3)
P_x = axial force (Chapter 15), axial load on the pile (Chapter 14)
$P - \Delta$ = effect of axial loading on bending moment (Chapter 14)
p = total normal stress on the plane of failure (Chapter 6), intensity of the load on a homogeneous, isotropic, elastic half-space (Chapter 9), unit load (Chapter 9), steam (or air) pressure (Chapter 10), reaction from the soil due to deflection of the pile (F/L) (Chapter 12), soil resistance (Chapter 14)
p_0 = overburden pressure (Chapter 9)
p_a = atmospheric pressure (Chapter 11)
p_{c2} = ultimate soil resistance (Chapter 14)
$\bar{p}_d$ = effective overburden pressure (Chapter 10)
p_l = limit pressure (Chapter 4)
p_u = ultimate soil resistance (Chapter 14)
p_{u1} = ultimate resistance near the ground surface per unit of length along the pile (Chapter 14)
p_{u2} = ultimate resistance (Chapter 14)
p_{ur} = ultimate strength of rock (Chapter 14)
p_1^* = net limit pressure (Chapter 4)
$\bar{p}$ = effective stress = γ_x (Chapter 10), effective overburden pressure (Chapter 10)
Q = total outflow volume of water (Chapter 3), concentrated vertical load (Chapter 7)
Q_b = axial capacity in end bearing (Chapter 11)
Q_D = load at failure of the stripe footing (Chapter 7)
Q_d = ultimate bearing capacity (Chapter 7)
Q_f = skin-friction resistance (Chapter 10), axial load capacity in skin friction (Chapter 10)
Q_g = total capacity of the group (Chapter 10)
Q_{max} = maximum value (Chapter 13)
Q_p = total end bearing (Chapter 10), axial load capacity in end bearing (Chapter 10), tip load, kN (Chapter 13)
Q_s = total skin friction capacity (Chapter 10), capacity of a reference pile that is identical to a group pile but is isolated from the group (Chapter 10), axial capacity in skin friction (Chapter 10), total load in side resistance (Chapter 11)
$Q_{s\,\text{max}}$ = maximum side resistance (Chapter 13)
Q_{ST} = load at the top of the socket (Chapter 11)
Q_t = load still in the shaft at the top of the socket (Chapter 11), given the load at the top of the socket (Chapter 13)

Q_{ult} = ultimate axial capacity of the drilled shaft (Chapter 11)
$(Q_{\text{ult}})_G$ = ultimate axial capacity of the group (Chapter 15)
$(Q_{\text{ult}})_p$ = ultimate axial capacity of an individual pile (Chapter 15)
q = total water flow rate (Chapter 3), flow rate of water from the soil (Chapter 3), bearing pressure on the foundation (Chapter 9), unit load transfer in end bearing (Chapter 10), unit end-bearing resistance (Chapter 10)
q_a = allowable soil-bearing stress (Chapter 9), net bearing pressure (Chapter 9), allowable bearing pressure (Chapter 11)
q_b = base resistance (Chapter 11), failure stress in bearing at the base of the footing (Chapter 13)
q_c = cone pressure (Chapter 4), static-cone-bearing capacity (Chapter 9)
$\overline{q}_c$ = variable values of static-cone-bearing capacity (Chapter 9)
q_d = ultimate bearing stress (Chapter 9)
$q_{\max}$ = unit end-bearing capacity (Chapter 11)
q_p = bearing capacity at the pile tip (Chapter 10)
q_u = unconfined compressive strength (Chapter 6), compressive strength (Chapter 11), average unconfined compressive strength of rock cores (Chapter 11), uniaxial compressive strength of the rock or concrete (Chapter 11)
q_{ult} = bearing stress at failure (Chapter 5)
q_{ur} = compressive strength of the rock, usually lower-bound, as a function of depth (Chapter 14)
RMR = rock mass rating (Chapter 14)
R_m = $(E_pI_p)_m$, bending stiffness of the pile at point m (Chapter 14)
RQD = rock quality designation (Chapters 4 and 14).
R^* = resistance (Chapter 8)
r = horizontal distance from point N to the line of action of the load (Chapter 7)
r_c = core recovery ratio (Chapter 11)
r_m = mean resistance or strength (Chapter 8)
S = settlement at any time (soil, piston, etc.) (Chapter 3), compression of the spring (Chapter 3), slope (Chapter 12)
S^* = loading (Chapter 8)
S_a = settlement in apparatus (Chapter 3)
S_r = degree of saturation (Chapter 3)
S_S = settlement according to the Schmertmann method (Chapter 9)
S_t = total settlement (Chapter 3), rotation of the top of the pile (Chapter 14)
S_u = ultimate settlement (Chapter 3)
s = shear strength (Chapter 4), shear resistance of soils (Chapter 6), shape factors (Chapter 7), amount of point penetration per blow, in inches (Chapter 10), parameter for cohesive intermediate geomaterial (Chapter 11)
s_m = mean value of the load (Chapter 8)

s_u = operational undrained shearing of the geomaterial beneath the base (Chapter 11)
SPT = Standard Penetration Test (Chapter 4)
T = torque (Chapter 4), relative stiffness factor (Chapter 12)
T_v = time factor (Chapter 7)
t = time (Chapter 3)
$t - z$ = load transfer curves (Chapter 13)
t_{yr} = time in years (Chapter 9)
U = average degree of consolidation (Chapter 3), pile displacement needed to develop side shear (taken as 0.1 in.) (Chapter 10)
u = water pressure (Chapter 3), porewater pressure (Chapters 3, 4, and 6), temperature (excess porewater pressure) (Chapter 7)
u' = excess porewater pressure (Chapter 3)
u_s = static porewater pressure (Chapter 3)
V = total volume (Chapter 3), vertical component (Chapter 7), shear (Chapter 11)
V_0 = volume of the measuring portion of the probe at zero reading of the pressure (Chapter 4)
V_m = corrected volume reading at the center of the straight-line portion of the pressuremeter curve (Chapter 4)
V_s = volume of solid mineral grains (Chapter 3)
V_v = volume of voids (Chapter 3)
V_w = original volume of water (Chapter 3)
v = average velocity of water flow in soil (Chapter 3), Poisson's ratio (Chapters 4, 9, and 13), inclination of the base of the foundation (Chapter 7), Poisson's ratio of the solid (Chapter 9), secant modulus (Chapter 13), Poisson's ratio of the geomaterial (Chapter 13)
v_0 = volume at the beginning of the straight-line portion of the curve (Chapter 4)
v_f = volume at the end of the straight-line portion of the curve (Chapter 4)
W = total weight (of sample) (Chapter 3), distributed load in force per unit of length along the pile (Chapter 13)
W_d = dry weight (weight after drying) (Chapter 3)
W_r = weight of ram (for double-acting hammers includes weight of casing) (Chapter 10)
W_w = weight of water (in the sample) (Chapter 3)
w = pile movement (Chapter 10), settlement of the base of the drilled shaft (Chapter 11), pile movement, *m* (Chapter 13), settlement, *m* (Chapter 13)
w_b = settlement of the footing or base of the pile (Chapter 13)
w_t = settlement at the top of the socket (Chapter 11), corresponding elastic settlement for a given load at the top of the socket (Chapter 13)

w_{t1} = settlement at the top of the socket at the end of segment 1 (Chapter 13)
x = coordinate along the pile measured from the pile measured from the top (L) (Chapter 12)
x_t = movement of a pile head (Chapter 15)
Y_t = lateral deflection at the pile head (Chapters 14 and 15)
y = deflection (Chapter 12)
y_{50} = deflection under a short-term static load at one-half of the ultimate resistance (Chapter 14)
y_c = deflection under N-cycles of load (Chapter 14)
y_s = deflection under a short-term static load (Chapter 14)
y_t = pile head (Chapter 12), deflection at the top of the pile (Chapter 14)
z = depth (Chapters 3 and 4), vertical distance between a point N within a semi-infinite mass that is elastic, homogeneous, and isotropic (Chapter 7), depth coordinate (Chapter 10), depth below ground surface (Chapter 11), ground surface (Chapter 12), pile displacement at depth x (Chapter 13), depth from the ground surface to the p-y curve (Chapter 14)
z_{bot} = depth to the bottom of the zone considered for side resistance (Chapter 11)
z_c = distance from the top of the completed column of concrete to the point in the borehole at which σ_n is desired (Chapter 11)
z_r = depth where a transition occurs (Chapter 14), depth below the rock surface (Chapter 14)
z_t = pile-head displacement (Chapter 13)
z_{tip} = pile-tip displacement (Chapter 13), pile-tip movement (Chapter 13)
z_{top} = depth to the top of the zone considered for side resistance (Chapter 11)
z_w = vertical distance below the water table (Chapter 3), depth from the top of the concrete to elevation of the water table (Chapter 11)
Σt_{seams} = summation of the thicknesses of all of the seams in the core (Chapter 11)
α = adhesion factor (Chapter 10), dimensionless factor depending on the depth–width relationship of the pile (Chapter 10), empirical adhesion coefficient (Chapter 10), empirical factor that can vary with the magnitude of undrained shear strength, which varies with depth z (Chapter 11), constant of proportionality (Chapter 11)
α_c = angle of the inclined plane with the vertical (Chapter 14)
α_p = pile-head rotation (Chapter 15)
α_r = strength reduction factor, b = diameter of the pile (Chapter 14)
α_s = rotation of the structure (Chapter 15)
α_x = coefficient that is a function of c_x (Chapter 10)

β = inclination of the ground (Chapter 7), pile composite factor (Chapter 15)
βL = nondimensional length of the pile (Chapter 14)
Δe = change in the void ratio for use in computing settlement S (Chapter 7)
$\Delta P / \Delta V$ = slope of the straight-line portion of the pressuremeter curve (Chapter 4)
ΔV_w = change in the volume of water (Chapter 3)
Δw_b = base settlement (Chapter 13)
Φ_c = friction angle at the interface of concrete and soil (Chapter 11)
σ_1 = compressive stress (Chapter 14)
σ_f = failure compressive stress in the laboratory unconfined-compression or quick triaxial test (Chapter 13)
σ_h = normal component of stress at the pile–soil interface (Chapter 10)
σ'_p = preconsolidation pressure (Chapter 11), vertical effective stress at a horizontal plane in a given depth (Chapter 11)
σ'_v = change in minor principal total stress (Chapter 10), vertical effective stress at a horizontal plane in any given depth (Chapter 11), preconsolidation pressure (Chapter 11)
σ'_{vo} = value of vertical effective stress at the elevation of the base (Chapter 11)
σ'_z = vertical effective stress in soil at depth z (Chapter 11)
$\Delta\sigma_1$ = change in the major principal total stress (Chapter 3)
$\Delta\sigma_3$ = change in a minor principal total stress (Chapter 3)
δ = friction angle between the soil and the pile (Chapter 10), thickness of individual discontinuities (Chapter 11)
ε = strain measured by the unconfined-compression or quick-triaxial test (Chapter 13), strain corresponding to compressive stress σ_1 (Chapter 14)
ε_z = strain at any depth z beneath a loaded area (Chapter 9)
γ = unit weight (Chapter 3), united weight of embankment material (Chapter 3), unit weight of soil (Chapters 4, 7, 10, and 14)
γ' = effective unit weight (Chapter 3), submerged unit weight (Chapter 7), effective unit weight of soil (Chapter 10), average effective unit weight from the ground surface to the p-y curve (Chapter 14)
γ_1 = partial load factor to ensure a safe level of loading (Chapter 8)
γ_2 = partial load factor to account for any modifications during construction, for effects of temperature, and for effects of creep (Chapter 8)
γ_c = unit weight of concrete (Chapter 11)
γ'_c = buoyant unit weight of concrete (Chapter 11)
γ_d = dry unit weight (Chapter 3)
γD_f = soil above the base of the foundation (Chapter 7)
γ_f = partial safety factor to account for deficiencies in fabrication or construction (Chapter 8)

γ_m = partial safety factor to reduce the strength of the material to a safe value (Chapter 8)
γ_p = partial safety factor to account for inadequacies in the theory or model for design (Chapter 8)
γ_s = unit weight of the solid mineral grains without any internal voids (Chapter 3)
γ_{sat} = saturated unit weight of a solid (Chapter 3)
γ_w = unit weight of water (Chapter 3)
κ = diffusivity that reflects the ability of the slab material to transmit heat (Chapter 7), reduction factor for shearing resistance along the face of the pile (Chapter 14)
λ = coefficient that is a function of pile penetration (Chapter 10)
μ = coefficient of friction (Chapter 3), modulus in the load transfer curve (Chapter 13)
μL = lateral extent influence factor for elastic settlement (Chapter 13)
ϕ = friction angle (Chapter 14)
ϕ' = friction angle from effective stress analysis (Chapter 9)
ϕ_{pl} = friction angle for plain strain conditions (Chapter 7)
ϕ_{tr} = friction angle determined from triaxial tests (Chapter 7)
φ = friction angle (Chapters 4, 6, and 10)
θ = angle of slope as measured from the horizontal (Chapter 14)
ρ = mean settlement (Chapter 9)
σ = normal stress (Chapter 3), total stress (Chapter 3)
σ' = effective stress (Chapter 3), log of effective stress (Chapter 7)
σ'_0 = depth to the water table from the ground surface (Chapter 9)
σ'_v = effective overburden pressure (Chapter 10)
σ_0 = vertical stress at the ground surface due to the applied load (Chapter 7)
σ_n = normal stress between the concrete and the borehole wall at the time of loading (Chapter 11)
σ_p = value of atmospheric pressure in units (Chapter 11)
σ'_p = preconsolidation pressure (Chapter 11)
σ_v = stress in the vertical direction (Chapter 7)
τ = shear stress (Chapter 3)
ω = angle of the pile taper (Chapter 10)
ψ = $c/\overline{p}$ for the depth of interest (Chapter 10)
ℓ = circumference of a cylindrical pile or the perimeter encompassing an H-pile (Chapter 13)

CHAPTER 1

INTRODUCTION

1.1 HISTORICAL USE OF FOUNDATIONS

Builders have realized the need for stable foundations since structures began rising above the ground. Builders in the time of the Greeks and the Romans certainly understood the need for an adequate foundation because many of their structures have remained unyielding for centuries. Portions of Roman aqueducts that carried water by gravity over large distances remain today. The Romans used stone blocks to create arched structures many meters in height that continue to stand without obvious settlement. The beautiful Pantheon, with a dome that rises 142 ft above the floor, remains steady as a tribute to builders in the time of Agrippa and Hadrian. The Colosseum in Rome, the massive buildings at Baalbek, and the Parthenon in Athens are ancient structures that would be unchanged today except for vandalism or possibly earthquakes.

Perhaps the most famous foundations of history are those of the Roman roads. The modern technique of drainage was employed. The courses below a surface course of closely fitted flat paving stones were a base of crushed stone, followed by flat slabs set in mortar, and finally rubble. The roads provided a secure means of surface transportation to far-flung provinces and accounted significantly for Roman domination for many centuries. Some portions of the Roman roads remain in use today.

1.2 KINDS OF FOUNDATIONS AND THEIR USES

1.2.1 Spread Footings and Mats

A diagram of a typical spread footing is shown in Figure 1.1. The footing is established some distance below the ground surface for several possible rea-

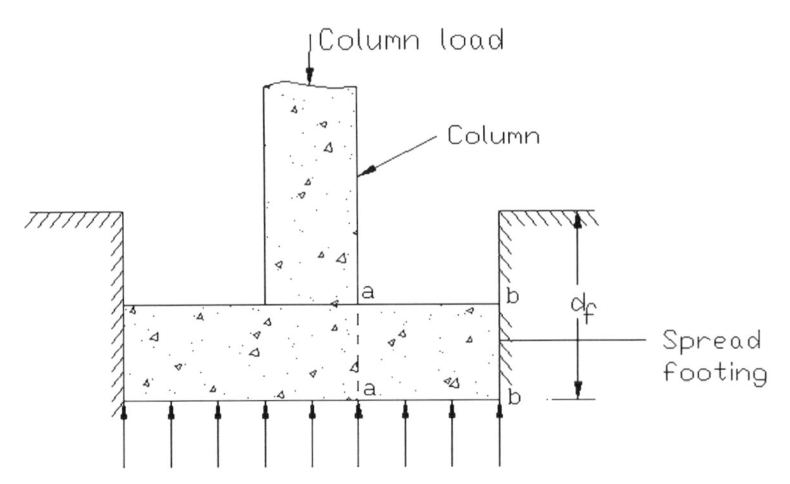

Figure 1.1 Square or continuous footing with axial loading.

sons: to get below weak near-surface soil, to get below the frost line, to eliminate expansive clay, or to eliminate the danger of scour of soil below the footing. If the lateral dimension in Figure 1.1 is large, the foundation is termed a *mat.* Mats are used instead of footings if multiple loads are supported or if the foundation is large to support a tank. The distance d_f shown in the figure is frequently small, and footings and mats are called *shallow foundations.* The geotechnical design of shallow foundations is presented in Chapter 9.

The distribution of the soil resistance shown in Figure 1.1 is usually assumed, allowing the computation of the shear and bending moment at Section a-a for design of the reinforcing steel. If the concept of *soil–structure interaction* is employed, the downward movement of the footing at Section a-a would be slightly more that the downward movement at Section b-b. If the engineer is able to compute a new distribution of soil resistance for the deformed footing, a more appropriate value for soil resistance could be computed. However, the savings due to the lesser amount of reinforcing steel needed would be far less than the cost of engineering required for such analyses. For some soils, the soil resistance would be higher at Section b-b than at Section a-a even taking the deformation of the footing into account. Unless the dimensions of the footing are very large, the engineer could assume a distribution of soil resistance to give the largest bending moment at Section a-a without substantial cost for reinforcing. The concept of soil–structure interaction is discussed in the last five chapters of this book and illustrated by application to problems where the movement of the soil under load must be considered in detail in solving a practical problem.

Rather than being square, the footing in Figure 1.1 could be continuous to support the load for a wall or from a series of columns. With a series of concentrated loads, the soil reaction would more complex than for a single column, but uniform distribution, as shown in Figure 1.1, is normally assumed.

Nominal stress distribution for the case of a square or continuous footing subjected to generalized loading is shown in Figure 1.2. Statics equations may be used to compute the distribution of vertical stresses at the base of the footing to resist the vertical load and moment. The stresses to resist the shear force may either be normal resistance at the edge of the footing or unit shear at the base of the footing. The engineer may decide to eliminate the normal resistance at the edge of the footing because of possible shrinkage or shallow penetration, allowing a straightforward computation of the shearing resistance at the base of the foundation.

A much larger shallow foundation such as a *mat,* and similar foundations are used widely. A *stress bulb* is a device sometimes used to illustrate the difference in behavior of foundations of different sizes. For equal unit loadings on a footing or mat, the stress bulb extends below the foundation to a distance about equal to the width of the foundation. Assuming the same kind of soil and a load less than the failure load, the short-term settlement of the mat would be much greater than the short-term settlement of the footing (settlement of footings and mats is discussed in Chapter 7). Thus, in the design of a mat foundation, the short-term settlement must be given appropriate consideration. Also, the structural design of the mat would plainly require more thought than that of the footing.

With regard to the design of mat foundations for residences on expansive clay, many designers use the BRAB slab recommended by the Building Research Advisory Board (1968). The mat includes a series of reinforced beams on the edges and interior of the mat, with dimensions depending on the nature

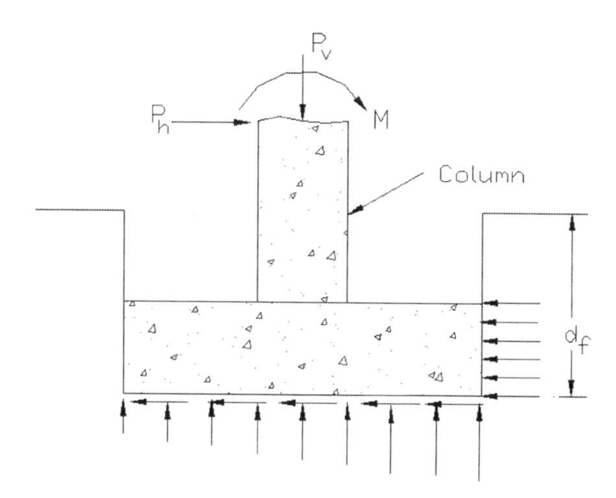

Figure 1.2 Square or continuous footing with generalized loading.

of the expansive clay. The concept is that the mat would have adequate bending stiffness to maintain a level surface even though, with time, differential movement of the clay occurs.

A view of a shallow foundation under construction is shown in Figure 1.3. The plastic sheets serve to keep the soil intact during the placement of concrete and prevent moisture from moving through the slab by capillarity. Communication cables may be placed in the slab. The deepened sections contain extra steel and serve as beams to strengthen the slab. The engineer has the responsibility to prepare specifications for construction and should be asked to inspect the construction.

1.2.2 Deep Foundations

Deep foundations are employed principally when weak or otherwise unsuitable soil exists near the ground surface and vertical loads must be carried to strong soils at depth. Deep foundations have a number of other uses, such as to resist scour; to sustain axial loading by side resistance in strata of granular soil or competent clay; to allow above-water construction when piles are driven through the legs of a template to support an offshore platform; to serve as breasting and mooring dolphins; to improve the stability of slopes; and for a number of other special purposes.

Several kinds of deep foundations are described in Chapter 5 and proposals are made regularly for other types, mostly related to unique methods of construction. The principal deep foundations are driven piles and drilled shafts.

Figure 1.3 Shallow foundations under construction.

The geotechnical design of these types of foundations under axial loading is presented in Chapters 10, 11, and 13, and the design under lateral loading is discussed in Chapters 12 and 14.

A group of piles supporting a pile cap or a mat is shown in Figure 1.4a. If the spacing between the piles is more than three or four diameters, the piles will behavior as individual piles under axial loading. For closer spacing, the concept of *pile–soil–pile interaction* must be considered. The analysis of pile groups is presented in Chapter 15. A particular problem occurs if a pile is embedded deeply in a concrete mat, as shown in Figure 1.4b. The difference in the behavior of a pile with a head free to rotate and a pile with a head fixed against rotation is dramatic. The real case for an embedded pile is that it is neither fixed nor free, but the pile-head restraint can be described with a linear or nonlinear relationship between pile-head rotation and moment.

The driving of a pile to support an offshore platform is shown in Figure 1.5. A template or jacket is set on the ocean floor with its top slightly above the water surface. The piles are driven and welded to the jacket; the deck can then be placed. In the figure, the pile is driven with a steam hammer that swings freely. The pile is marked along its length so that the engineer can prepare a driving record (number of blows required to drive the pile a given distance).

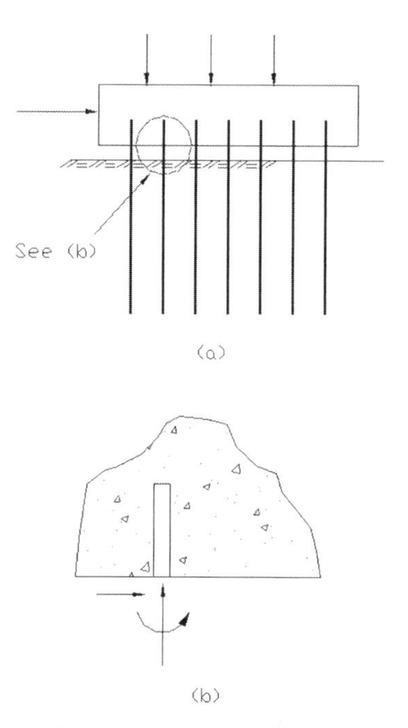

Figure 1.4 (a) Group of piles supporting a pile cap. (b) Loading of top of pile embedded in concrete for solving for the moment–rotation relationship.

Figure 1.5 Vertical pile being driven at offshore site with a swinging steam hammer.

1.2.3 Hybrid Foundations

A complex problem occurs if the mat shown in Figure 1.4a rests on the ground surface rather than with space below the mat. A solution of the problem has been presented by Zeevaert (1972) in a discussion of "compensated" foundations. The problem has received attention recently (Franke et al., 1994; Poulos, 1994; Viggiani, 2000; Cunha et al., 2001). As noted earlier, the zone of significant stresses beneath a mat will be relatively large, leading to a relatively large settlement before the working load on the mat is attained. The piles are usually friction piles, where the load is carried in side resistance, and the foundation is termed a *piled raft.*

In some cases, the piles are placed in a regular pattern; in other cases, the placement is strategic in order to lessen the differential settlement of the mat. The analysis must consider the vertical loading of the soil from the mat that affects the state of stress around the piles. Further analysis of the behavior of the mat must consider the presence of the piles that serves to reinforce the soil beneath the mat. Therefore, the modeling of the problem of the piled raft for an analytical study is done by finite elements.

The observation of the performance of piled rafts where instrumentation was employed to measure the distribution of the axial loading and where the settlement of the structure was measured carefully (Yamashita et al., 1994, 1998) provides the opportunity for a case study to validate the analytical technique. Additional data on the performance of piled rafts have been presented by Franke et al. (1994).

1.3 CONCEPTS IN DESIGN

1.3.1 Visit the Site

The value of a visit to the proposed site for the construction of a foundation from the point of view of geology is discussed in Chapter 2. Not only can information be gained about the geology of the area, but other features of the site can also be considered. The topography, evidence of erosion, possible response of the foundations of existing structures, and accessibility of construction equipment are items of interest.

The engineer will wish to consult with local building authorities to get information on criteria with regard to requirements for new construction. The locations of underground utilities should be obtained from the relevant agencies prior to the visit, and the presence of overhead lines should be observed.

1.3.2 Obtain Information on Geology at Site

Except for the very small job where foundation design is straightforward, the engineer will want to employ the appropriate *standard of care* by obtaining available information on the geology at the proposed construction site. Chapter 2 lists a number of governmental agencies that provide information on the local geology. The engineer will wish to obtain such information, which could have an impact on how the subsurface information is subsequently developed.

1.3.3 Obtain Information on Magnitude and Nature of Loads on Foundation

Obtaining information on the magnitude of the loads to be sustained by a structure would seem to be simple, but such is not the case. In many instances, when foundations are being designed, the structural design is continuing. Furthermore, a substantial difference exists between the factor of safety being employed by the structural engineer and by the foundation engineer. The foundation engineer wishes to know the *working* load to be sustained by the structure, which means the unfactored dead load and live load. Then the working load is factored upward to design the foundation, with the selected factor accounting principally for the uncertainties associated with the soil properties and the analytical procedure being employed.

In many instances, statistics play an important role in determining loads on a structure. Offshore structures are designed to resist the largest storm that may occur once in perhaps 100 years. Offshore and onshore structures may be designed to resist the maximum earthquake that could occur in the location of the proposed structure.

If uncertainties exist in determining the magnitude of the loading, detailed discussions are in order among the various engineers, probably including representatives of the owner of the structure. A detailed discussion of the safety factor, including formal recommendations by agencies that employ load-and-resistance-factor design (LRFD), is strongly advised.

Two principal problems for the foundation engineer are the adequacy of the analytical technique to be employed and the adequacy of the properties of the soil to reflect the behavior of the soil in supporting the structure. The reader will gain insight into the adequacy of the analytical procedures in studying the various chapters. With respect to soil properties, Chapters 3 and 4 will be useful. In summary, the foundation engineer frequently wishes to know as precisely as possible the load to be sustained by the foundation and, at some stage in the work, wants to employ a global factor of safety to account for uncertainties in theory and soil properties.

1.3.4 Obtain Information on Properties of Soil at Site

A number of techniques may be employed to perform a subsurface investigation, as presented in Chapter 4. If possible, a staged investigation should be employed, with the first stage aimed at identification and classification of the soils at the site. The selection of the foundation, whether shallow or deep, could be based on the results of the first-stage investigation. The second stage would focus on determining the necessary properties for the design. The work in the second stage could be minor or extensive, depending on the nature of the design and the soils at the site.

A conceptual curve can be plotted to show how the total cost of a structure can be influenced by the cost of the investigation of the soils. At some stage, the maximum amount of information could be gained to minimize the cost of the structure. Most subsurface investigations do not allow for the optimum approach due to either time or cost constraints. The knowledgeable foundation engineer will benefit the owner and all other professionals on the job by presenting lucid and convincing arguments for a subsurface investigation that is fully adequate.

1.3.5 Consider of Long-Term Effects

The foundation engineer is obligated to consider effects on the structure that could occur with time. Examples are settlement due to consolidation of clays, settlement due to compaction of sand by vibration, movement of a foundation due to swelling and shrinkage of a clay, and the adverse effects of time-related erosion.

1.3.6 Pay Attention to Analysis

The methods of analysis presented in this book contain many equations that include parameters dependent on soil properties. The available computer programs are so efficient that parametric studies can be undertaken and completed in a short time. The engineer is obligated to study the methods proposed to gain a full understanding of the purpose and appropriateness of the procedure. Full advantage should be taken of the speed of modern computers.

A simple study that is always worthwhile is to use the results of the subsurface investigation and to select lower-bound and upper-bound values of the relevant parameters. Analyses can be made with the two sets of data to allow the selection of properties to be used in the final design.

1.3.7 Provide Recommendations for Tests of Deep Foundations

Studies of the design of foundations at a particular site may indicate the need to perform a full-scale load test of the deep foundation to be employed in the construction. A justification for the test would be that no data are available on the performance of the type of foundation to be used in the soil that exists at the site. A further justification exists if the engineer can show that the cost of the deep foundations to be used without the test would be reduced by more than the cost of the test if testing is approved.

1.3.8 Observe the Behavior of the Foundation of a Completed Structure

Some very successful foundation engineers have made extensive use of formal observation. This will allow corrections to be made if the observation shows deficiencies.

Even if formal observation is not employed, many engineers make an effort, with the consent of the owner, to observe the performance of the foundation with time. On some occasions, instruments may be used to gain information on the distribution of loading to the various elements of the foundation. Such information is extremely valuable.

PROBLEMS

P1.1. Obtain a current copy of a magazine on engineering projects and write a brief exposition of the nature of one such structure, the general nature of the soil conditions, and the kind of foundation being employed for the project. If possible, list the reasons the engineer chose the particular type of foundation.

P1.2. In preparation for considering the behavior of a pile under lateral loading with respect to the pile-head restraint, (a) compute the deflection of a cantilever beam with a load at the free end and then (b) one with the end fixed against rotation but free to deflect.

P1.3. Visit a construction site in your area. Take a photograph of the foundation if allowed and describe it as well as possible.

CHAPTER 2

ENGINEERING GEOLOGY

2.1 INTRODUCTION

The geologist uses a variety of tools to study the earth, perhaps to depths of hundreds of meters, to gain information on a variety of matters, such as the occurrence of petroleum products, the prediction of earthquakes, and the science of the earth. While these matters are of general interest, the engineering geologist has a particular interest in near-surface geology, principally to guide the subsurface investigation and planning for construction.

The work of the engineering geologist is crucial in the design and construction of dams and major earthworks and is somewhat less important in the design of foundations. The design of the subsurface investigation for a project begins with a study of the geologic process that resulted in the creation of the soil and rock at the site. The investigation continues, with near-surface geology playing an important role as the final design of the foundation is made and the completed foundation is observed.

An interesting fact is that the late Dr. Karl Terzaghi, the father of modern geotechnical engineering, began his studies of soil with a study of geology and maintained an intense interest in geology throughout his career. An early paper (Terzaghi, 1913) dealt with karst geology and a later one with the limitations of subsurface investigations in revealing all information of significance (Terzaghi, 1929). In a paper about Terzaghi's method of working, Bjerrum (1960) wrote: "The intimate knowledge of the geology of the whole area is as necessary for his work as the subsoil exploration which follows." There is good reason to believe that Terzaghi's method of working remains a valid guideline.

Presented here are brief discussions of geologic processes that affect the character of near surface soils, availability of information on geology to the engineer, an example of the geology of a selected area, and geologic features that affect subsurface investigations and designs.

2.2 NATURE OF SOIL AFFECTED BY GEOLOGIC PROCESSES

The information presented here provides only a brief treatment of some aspects of geology that affect the techniques of subsurface investigation and the design of some foundations. The following sections are intended principally to promote the investment in an adequate effort to develop knowledge about the geology of a site prior to planning the subsurface studies and the subsequent design of the foundations.

2.2.1 Nature of Transported Soil

Transported by Water Almost any flowing water will transport particles of soil. The particles originate as rain hits the earth and are deposited when the velocity of flow is reduced or when particles of soil form flocs that settle. Boulders are pushed along in mountainous areas where stream beds are steep and the velocity of the water is very high. The power of flowing water can be illustrated by Figure 2.1, which shows the erosion of limestone along a

Figure 2.1 Erosion of limestone along a stream in Texas.

stream in Texas. At the other end of the scale, rivers transport particles of clay into bodies of water, lakes or the sea, where flocculation occurs and the flocs gain enough weight to fall to the lake bed or sea floor.

Thick beds of clay, with some particles of silt size, can be found on the ocean floor at many locations. An example is at the mouth of the Mississippi River. Year after year, thin layers of clay are deposited as floods along the river erode particles of soil. The weight of the soil above a particular depth causes drainage or consolidation of the clay at that depth. If deposition has ceased at a particular location, the clay will reach a state of equilibrium after a period of time with no outward flow of interstitial water, and the stratum of clay will become normally consolidated. The shear strength of the clay will be virtually zero at the surface and will increase almost linearly with depth.

If deposition continues, the underlying clay is underconsolidated, with drainage continuing. Careful sampling and testing of the clay will reveal its nature, and the engineer will consider the detailed character of the stratum of clay in designs.

Particles of sand and gravel are carried along by flowing water and are deposited with a drop in the velocity of the water. Such deposits occur near the beds of streams that exist now or did so in the past. Such deposits are usually quite variable because of the erratic nature of stream flow. The relative density of such deposits also varies significantly, depending on their historical nature. Fine sand and silt are deposited in bodies of water as the velocity of a stream in flood pushes the suspension into a lake or ocean. Deposits of sand and silt can be found in the deltas of rivers, such as near the mouth of the Mississippi, and sometimes contain decayed vegetation.

Nature creates complex patterns, and the near-surface geology rarely can be known with absolute certainty. Stream flow in the past, when the ground surface was uplifted, incised valleys that, after submergence, became filled with soft sediments. Such examples exist in the Gulf of Mexico near the coast of Texas. Offshore borings may reveal a regular pattern of soil, but such a geologic event may result in an unpleasant surprise to builders of pile-supported structures.

Transported by Wind Over geologic history, wind has played an important role in creating soil. In the deserts in some parts of the world, sand is prominent in the form of dunes, and the dunes may be in continuous motion. Rainfall is small to nonexistent, and construction is limited. If construction is necessary, the engineer faces the problem of stabilizing the wind-blown sand. Other wind-blown soil is more amenable to construction.

Dust storms in the Midwest of the United States in the first half of the 20th century transported huge quantities of fine-grained soil. Over long periods of time, deposits of great thickness were laid down. Such a soil is loess, found in the United States, in Eastern Europe, and in other places. Vegetation grew during deposition, and deposits of considerable thickness of vertically reinforced soils were developed. Loess will stand almost vertically when deep cuts are made.

If a sample of loess is placed under load in a laboratory device and water is added, the sample will undergo a significant reduction in thickness or will "collapse." Loess is termed a *collapsible* soil. Drainage is a primary concern if shallow foundations are employed. Deep foundations will normally penetrate the stratum of loess.

2.2.2 Weathering and Residual Soil

Residual soils are derived from the in-place weathering of rocks and are found on every continent, including the subcontinents of Australia and Greenland (Sowers, 1963). Residual soils exist in southeastern North America, Central America, the islands of the Caribbean, South America, southern Asia, and Africa. The soils occur in humid, warm regions that promote rapid decomposition of the parent rock.

There are many processes of weathering, depending on the climate during the geologic time at the site. The processes can be differential expansion and contraction, freezing and thawing, chemical action, rainfall and leaching, and the action of organisms.

As is to be expected, the decomposition is not uniform, with residual soil varying in thickness. At certain places in India, some areas of the parent rock resisted decomposition and were left in residual soil as boulders. Residual soil can sometimes be expansive, such as the soil shown in Figure 2.2. The site had been excavated, and drying of the clay generated the pattern of cracks shown in the foreground. The structure and physical characteristics of residual soil are related to those of the parent rock, but careful sampling and testing can obtain properties for use in design. In situ tests are frequently very useful.

Deposits of organic material can weather into peat, muck, and muskeg and can pose severe problems for the engineer.

Weathering of soils such as limestone has resulted in *karst* geology where caverns and openings in the rock are prevalent. Karst geology exists in many areas of the world and was studied by Terzaghi early in his career (1913) (see Section 2.2.5).

2.2.3 Nature of Soil Affected by Volcanic Processes

Much of the soil and rock thrown out and up by ancient volcanoes has been weathered and causes the normal problems in determining properties for design and construction of foundations. An interesting artifact from the ancient volcanoes that surround the Valley of Mexico is that lakes in the Valley were filled to a considerable depth with volcanic dust thrown up over long periods of time. The resulting soil has extremely high water contents and void ratios. Dr. Leonardo Zeevaert (1975) described to one of the authors an experience

Figure 2.2 Desiccation cracking in expansive clay.

early in his professional life when he sent data on the results of laboratory testing to his professor in the United States and got the following reply: "Son, you have misplaced the decimal point."

Zeevaert (1972) has written extensively on the design of foundations in the Valley of Mexico and elsewhere, particularly with respect to designing to accommodate the effects of earthquakes.

2.2.4 Nature of Glaciated Soil

Glaciers transport rock and other materials that are embedded in the ice or are picked up from the surfaces of rock as the glacier moves. The material carried along by a glacier is termed *drift.* Glacial deposits are of two general types: *moraines* or *till* and *outwash.*

Till is deposited directly by the glacial ice and is characterized by a wide range of particle sizes. Stones and boulders can exhibit scratches or striations due to being carried along by the ice. Moraines occur when a glacier dumps a large amount of irregular material at the edge of the ice sheet that is deposited as the glacier retreats. A large amount of material may be dumped in an area if the glacier stalls in its retreat

The outwash from melt water is mainly fine-grained material that forms a plain or apron. The material is usually stratified and may exist as long, winding ridges. Many glacial deposits create a problem for the foundation engineer because of their lack of homogeneity. A particular problem for deep foundations occurs when a glacier leaves a deposit of boulders. Not only will the deposit be difficult or impossible to map, but penetrating the boulders with piles can be extraordinarily difficult.

Varved clay occurs with seasonal deposition into lakes, but such soil is most often associated with glaciation. Each layer or varve contains a wide range of grain sizes, with gradation from coarsest at the bottom of the layer to finest at the top. Walker and Irwin (1954) write that a lake was formed when the Columbia River was dammed by ice during the last Glacial Era and that seasonal deposition occurred in the lake for hundreds of years, resulting in the Nespelem Formation found chiefly in the valley of the Columbia River and along its tributaries. While perhaps not typical of varved clay found elsewhere, the varved clay along the Columbia yielded greatly varying results in field and laboratory tests, lost strength due to even minor disturbance, was unsuitable for foundations, and was susceptible to sliding.

2.2.5 Karst Geology

The processes of solution of limestone and dolomite result in openings and cavities, some of which are extensive and important. The Edwards Limestone in Central Texas is an aquifer that provides water for many of the communities and businesses in the area. During construction of a highway, a drill exposed a cavity and alerted the highway engineers to make careful investigations for bridges in the limestone area. Several caves have been opened in Central Texas as attractions for tourists.

Some of the openings collapsed in the past and are filled with boulders, clay, or surficial soils. The weathering processes are continuing in some areas, with the result that in sections of Florida sinkholes appear suddenly, some of which may be quite large. Investigation of the subsurface conditions in karst areas is especially critical in foundation design. In addition to careful borings of the limestone or dolomite, ground-penetrating radar is sometimes appropriate.

A view of a section of limestone in an area of karst geology excavated for a roadway is shown in Figure 2.3. The cracks and joints in the limestone are evident. Rainwater can penetrate the openings, some of the limestone goes into solution, and a cavity, sometimes quite large, will develop over perhaps centuries.

2.3 AVAILABLE DATA ON REGIONS IN THE UNITED STATES

Rich sources of information have been provided by engineers and geologists working in local regions in the United States. The engineer making plans for foundations at a particular location may wish to take advantage of these studies of local areas. Some examples are given to illustrate the availability of such information. Fisk (1956) wrote about the near-surface sediments of the continental shelf off the coast of Louisiana. Included with the paper are col-

Figure 2.3 Bedding planes, cracks, and joints in sedimentary limestone.

ored drawings of various regions at present and in earlier geologic eras. The paper could be of considerable value to engineers making designs for the areas treated in the paper.

Otto (1963) wrote on the geology of the Chicago area. May and Thompson (1978) described the geology and geotechnical properties of till and related deposits in the Edmonton, Alberta, area. Trask and Rolston (1951) reported on the engineering geology of the San Francisco Bay of California.

2.4 U.S. GEOLOGICAL SURVEY AND STATE AGENCIES

The U.S. Geological Survey (USGS) maintains a vast amount of geologic information developed in the past and is adding to their collections. Geologic and topographic maps are available for many areas of the United States. The USGS released a study (1953) consisting of six maps entitled *Interpreting Geologic Maps for Engineering Purposes Hollidaysburg Quadrangle, Pennsylvania.* The maps were prepared by members of the Engineering Geology and Ground Water Branches. The first two maps in the release were a topographic map and a general-purpose geologic map. The final four maps were interpreted from the first two and showed (1) foundation and excavation conditions; (2) construction materials; (3) the water supply, both surface and underground; and (4) site selection for engineering works. The text accompanying the set stated: "This set of maps has been prepared to show the kinds of information, useful to engineers, that can be derived from ordinary geologic maps." The text and the detailed information on the right side of each map provide an excellent guide to the engineer who wishes to study an area with respect to the design and construction of foundations for a particular project.

Aerial photographs are available from the USGS and can be used to gain information on the occurrence of different types of soil (Johnson, 1951; Stevens 1951) as well as the geology of an area (Browning, 1951). The Iowa Engineering Experiment Station (1959) has published a number of aerial photographs to illustrate their use in identifying geologic conditions such as areas of sand covered by variable amounts of flood-deposited clay near the Mississippi River; erosion of shale to create a dendritic stream pattern; sinks identifying an area of limestone; almost horizontal lava flows that are cut by streams; granite or similar rock revealed by a fine fracture pattern; stable sand dunes; and alluvial fans extending out from mountains.

The current *National Atlas* is available online (http://www.nationalatlas.gov), and maps are available under various headings, including, for example, construction materials, coal fields, geologic maps, and seismic hazards. GEODE (Geo-Data Explorer) is an interactive world map (http://geode.usgs.gov) that allows the user to retrieve, display, and manipulate multiple types of information, such as satellite images, geologic maps, and other information.

Many states have an agency that provides information on the geology of the state. In Texas, the Bureau of Economic Geology has a number of publications of value to engineers—for example, a report on the engineering properties of land-resource units in the Corpus Christi area. In addition, local groups may be able to provide useful information that will assist the engineer in planning a site study.

In view of the extensive information available from governmental agencies and the importance of such information to the design of foundations, the engineer may wish to take the time to gather appropriate information for a specific project.

2.5 EXAMPLES OF THE APPLICATION OF ENGINEERING GEOLOGY

In 1997 the European Regional Technical Committee 3 "Piles," a group of the International Society for Soil Mechanics and Foundation Engineering, sponsored a conference in Brussels to present European practice in the design of axially loaded piles. Papers were submitted by engineers from a number of countries. Of special interest is that most of the articles began with a section on the geology of their country: Belgium (Holeyman et al.); the Czech Republic (Feda et al.); Denmark (Skov); Estonia (Mets); Finland (Heinonen et al.); France (Bustamante and Frank); Germany (Katzenbach and Moormann); Ireland (Lehane); Italy (Mandolini); the Netherlands (Everts and Luger); Norway (Schram Simonset and Athanasiu); Poland (Gwizdala); Romania (Manoliu); Sweden (Svensson et al.); Switzerland (Bucher); and the United Kingdom (Findlay et al.). Useful information, even if abbreviated, was presented about the geology of the various countries. The articles illustrate the importance of considering geology when designing foundations.

The late D. C. Greer, State Highway Engineer of Texas, renowned for his record of constructing highways in a big state, mailed all of the district engineers a map of the geology of Texas in August of 1946, where the presence of Taylor Marl was indicated. An accompanying document presented data on the poor behavior of asphaltic and concrete pavements on Taylor Marl and identified the material as an expansive clay. A later report in September of 1946 dealt with the performance of highways on other geologic formations and identified areas of Eagle Ford Shale and Austin Chalk. Mr. Greer urged state engineers to submit data on the performance of highways and cut slopes with respect to the local geology to allow such information to be disseminated for the benefit of engineers making highway designs.

The Nelson Mandela Bridge in Johannesburg, South Africa (Brown, 2003), was built across the city's main rail yard of 42 tracks. The design of the foundations was complicated by the existence of a graben at a depth of 50 m near the center of the bridge. The highly weathered material filling the graben was unsuitable for supporting a foundation, so the bridge was designed with two pylons with foundations at the edges of the graben.

2.6 SITE VISIT

The value of a visit to the site of a new project is difficult to overestimate. Many observations can lead to valuable insights into the problems to be encountered in the forthcoming construction. Among the things that can provide information on the site are the kinds of structures on the site and their foundations, if discernible; any observable deleterious movements of buildings in the area; the kinds of soil observed in excavations or in the sides or beds of streams; cracks in soil and desiccated soil; the quality of pavements in the area; and comments from homeowners and occupants of buildings if available.

A telling example of information to be gained occurred in 1953 when Professor Parker Trask took his graduate class in engineering geology on a tour of East Bay in California. Trask pointed to rolling hills where evidence of minor slides was apparent. Many minor scarps and slumps were observed, showing that the soil on a slope had a factor of safety close to unity. Newspaper reports during the period following the tour noted that slides had occurred when cuts were made for the construction of various projects. The resulting steep slopes reduced the safety against sliding.

PROBLEMS

P2.1. Make a field trip through an area assigned by your instructor and write a report giving all of the information that can be gleaned about the geology of the area.

P2.2. Use the USGS website and make a list of the information that could help an engineer understand the geology of a site assigned by your instructor.

P2.3. Term project. The class will be divided into groups of three to five students, and each group will be assigned an area of study where foundations are to be installed for a 10-story building with a plan view of 60 by 80 feet. Obtain relevant documents from the USGS and from local sources, and write a report presenting all pertinent information about the geology of the site.

CHAPTER 3

FUNDAMENTALS OF SOIL MECHANICS

3.1 INTRODUCTION

When designing a foundation, a geotechnical engineer is faced with three primary tasks: selection of the foundation type; computation of the load-carrying capacity; and computation of the movements of the foundation under the expected loading cases for both short-term and long-term time periods. Many factors that affect the selection of the type of foundation are discussed in Chapter 8, and the intervening chapters will present the methods recommended for design. All of these chapters assume knowledge of the fundamentals of soil mechanics. This chapter reviews the fundamentals of soil mechanics that must be mastered by any engineer who designs foundations.

3.2 DATA NEEDED FOR THE DESIGN OF FOUNDATIONS

In addition to having knowledge of soil mechanics and foundation engineering, the foundation engineer needs information on the soil profile supporting the foundation that is site specific in order to make the necessary computations of bearing capacity and settlement. The principal soil properties needed are the soil classification, the location of the water table, and the properties related to the shear strength and compressibility of the soil. In addition, the foundation engineer should anticipate events that might affect the performance of the foundation and consider them as warranted in the design process.

3.2.1 Soil and Rock Classification

The first information of interest to any designer of foundations is the soil and rock profile in which the foundations are to be constructed. This information often is the major factor that determines the type of foundation because foundations perform in different ways in different soil types and because the typical problems encountered during construction also vary in different types of soil and rock.

For purposes of foundation engineering, current design methods for determining the axial capacity of deep foundations generally classify soil and rock into five types (note that the following definitions are for foundation engineering, not for geology or other purposes). The five classifications for soil types are:

- Cohesive soils, including clays and cohesive silts.
- Cohesionless soils, including sand and gravels of moderate density.
- Cohesive intermediate geomaterials. This soil type consists of cohesive materials that are too strong to be classified as cohesive soils and too weak to be classified as rocks. It includes saprolites, partially weathered rocks, claystones, siltstones, sandstones, and other similar materials.
- Cohesionless intermediate geomaterials. This soil type consists of highly compacted granular materials, including very tight sand and gravel; naturally cemented sand and gravel; very tight mixtures of sand, gravel, and cobbles; and glacial till. Excavation in cohesionless intermediate geomaterials usually requires heavy equipment, specialized tools, and occasional blasting.
- Rock includes all cohesive materials that are too strong to be classified as cohesive materials. Rock types are further classified according to their mineralogy, fracture pattern, and uniformity.

The purpose of using these five classifications is to establish a system of soil descriptions covering the range of earth materials from the softest and most compressible to the hardest and most incompressible.

This same classification of soil types can be used for the design of foundations subjected to lateral loading. It is used to describe the soil types, but additional information on the position of the water table and the results of shear strength testing in the laboratory or field are also needed so that the proper load-transfer models can be selected by the foundation engineer.

3.2.2 Position of the Water Table

The soil and rock profile also includes information about the presence and elevation of the water table. The water table is the elevation in the soil profile at which water will exist in an open excavation or borehole, given sufficient time for steady-state conditions to be reached. The presence of the water table is of interest to the foundation designer for two principal reasons. Firstly, the position of the water table is needed to establish the profile of effective stress in the soil versus depth. Secondly, the depth of the water table determines the depths below which special procedures must be used to control the groundwater during construction. This second point is important for the engineer to consider when selecting the foundation type and for the contractor to consider when selecting the method of construction.

3.2.3 Shear Strength and Density

The shear strength of soil or rocks is the most important soil and rock property used by the foundation designer. The foundation designer will often base the selection of the foundation type on the shear strength and use the values for shear strength in computations of axial and lateral capacity of the foundation.

The manner in which shear strength of soil or rock is measured and characterized often depends on the type of soil or rock. Typically, the shear strength of cohesive and cohesionless soils is expressed using the Mohr-Coulomb shearing parameters. Advanced constitutive models that are extensions and modification of the basic Mohr-Coulomb parameters are also used in major projects where additional soil investigations and laboratory tests are made.

Any implementation of the Mohr-Coulomb shearing parameters or any of the advanced models for shearing parameters will utilize the value of effective stress. Thus, the position of the water table must be known for their application.

The density, or unit weight, of the soil or rock is a soil property that is necessary for the computation of stresses due to gravity. The density of soil is computed using the weights of the solids in the soil and the weight of water in the soil. Any changes in soil density and volume that might arise from changes in the amount of water in the soil are computed if necessary.

3.2.4 Deformability Characteristics

Soils and rocks exhibit deformability in two modes that may occur independently or in combination. The two modes are deformation in shear at constant volume and deformation in volume change in either compression or dilation. Usually, the foundation engineer considers deformation by shearing at constant volume as a short-term behavior that occurs simultaneously with the application of loading to the foundation. Conversely, the foundation engineer may consider deformation by volume change to be a long-term behavior associated with the consolidation of the soil. Both of these modes of deformation will be discussed in the appropriate chapters of this book.

3.2.5 Prediction of Changes in Conditions and the Environment

The behavior of the soils supporting the foundation is influenced strongly by the stress and groundwater conditions present in the soil and rock profile. Thus, it is important to consider what conditions existed in the soil and rock profile when the soil investigation was conducted, what changes might occur in the future as a result of construction, and what changes might occur in the future due to additional nearby development or other causes. Often, this requires guesses about a possible worst-case scenario or a reasonable-case scenario for the future.

The principal factors of concern for changes in conditions that might occur during construction are any addition or reduction of overburden to the soil profile during the project and changes in the elevation of the water table. Any change in the overburden stress will directly affect the shearing properties of the soil and may affect its deformation properties as well. The position of the water table can be changed locally by the removal of some varieties of trees and by the amount (or lack) of irrigation or landscaping at a project.

The changes in soil conditions at a site are important factors to consider. History provides many examples of successful foundation designs rendered unsuccessful by changes in soil conditions. A few of the examples are well known, but many others are known only locally because the structure did not survive and was taken down or suffered minor or repairable damage. Foundation design engineers strive to provide trouble-free foundations because repair of defective foundations is very costly and inconvenient or the damage may be hidden and the structure sold to another party.

3.3 NATURE OF SOIL

In the geotechnical investigation, a sample of soil or rock can be retrieved from a geologic stratum to be examined, classified, and tested using a variety of laboratory tests. In addition, several types of field tests are performed when necessary. It is desirable to obtain the soil and rock samples with minimal disturbance by the sampling process so that the results of testing in the laboratory are representative of the true nature of the soils in the particular stratum.

3.3.1 Grain-Size Distribution

The methods used to design foundations do not consider the size of the soil particles (also called *grains*) as a primary factor. However, the terminology used to describe grain sizes is useful when describing the type of soil.

Soil consists of three main types—clays, silts, and sands or gravels—depending on the size of the grain. The commonly used separation sizes are 2 μm to separate clay and silt, and the #200 sieve (0.074 mm) to separate silt and sand. In general, grain sizes are measured in two ways. One way is to pass the soils through a stack of wire sieves; the other way is to measure the rate of sedimentation in a hydrometer test and then infer the size of the soil grains. Measurements of grain size using sieves are used primarily for cohesionless soils such as sands and gravels. The hydrometer sedimentation test is used primarily for fine-grained soils such as silts and clays.

In a sieve analysis, the soil is passed through a stack of sieves progressing from large to small openings. The ranges in grain size are defined by the sizes of openings in testing sieves through which a soil particle can and cannot pass.

A hydrometer is a device for the measurement of fluid density. In the hydrometer sedimentation test, the density of a fluid is measured against time for a mixture of soil in water. As sedimentation occurs with the passage of time, the density of the fluid decreases because the suspended soils drop from suspension. The grain-size distribution is determined from Stokes' law[1] for the settling velocity of spherical bodies in a fluid of known density.

The types of soil with the smallest particles are *clays,* which may have particles smaller than 2 μm,[2] and the clay particles may have a variety of shapes. Usually, natural soils are composed of mixtures of several different minerals, though often one type of mineral is dominant. The behavior of clay is influenced significantly by the properties of the minerals present in the specimen, but the types of minerals are identified only in special circumstances using specialized tests.

In practice, various types of field and laboratory tests on soils from natural deposits are employed to define the soil properties used in design.

For comparison, here are the dimensions of various items:

- 0.4 to 0.66 μm—wavelength of visible light
- 1.55 μm—wavelength of light used in optical fiber
- 6 μm—anthrax spore
- 6 to 8 μm—diameter of a human red blood cell
- 7 μm—width of a strand of a spider web
- 1 to 10 μm—diameter of a typical bacterium
- $\sim$10 μm—the size of a water droplet in fog
- 0.05 μm—the maximum permissible deviation in curvature for an astronomical telescope mirror
- 17 to 181 μm—the range in diameter for human hair (flaxen to black)
- >2 μm—clay
- 2 μm to 74 μm—silt

[1]George G. Stokes (1819–1903), Fellow of the Royal Society and Lucasian Professor of Mathematics at Cambridge University.

[2]The identification of soil according to size has been proposed by a number of agencies; the sizes given here are from the American Society for Testing and Materials, D-2487-90 (1992).

3.3.2 Types of Soil and Rock

Soil and *rock* are descriptive terms that have different meanings, depending on their community of use. The following subsections present a few of the more common definitions.

Soil To a farmer, soil is the earthen material that supports life.

To a geologist, soil is an ambiguous term.

To an engineer, soil is a granular material plus the intergranular gas and fluids. This includes clay, silt, sand, and gravel. It may also contain anything found in a landfill: household waste, cinders, ashes, bedsprings, tin cans, old papers, and so on.

Rock To an engineer, rock is any individual material requiring blasting or another brute-force method to excavate. It is also classified as any material with a compressive strength greater than 725 psi (5.0 MPa). This definition is complicated by the presence of structures or defects that weaken the rock mass. The vagueness of the terms describing rock is due to the wide variety of materials encountered and the different information needs of those who use the descriptive terms.

Problems in geotechnical engineering involving the use of rock and many soils can seldom be solved by blind reliance on empirical data gathered from past projects or on sophisticated computerized analyses. Each situation is unique and requires

- Careful investigation
- Thorough analysis
- Engineering judgment based on varied experience
- Imagination to visualize the three-dimensional interplay of forces and reactions in complex systems
- Intuition to sense what cannot be deduced from scientific knowledge
- Initiative to devise new solutions for both old and new problems
- Courage to carry out the work to completion despite skeptics and the risk of the unknown

3.3.3 Mineralogy of Common Geologic Materials

Clay minerals are complex aluminum silicates formed from the weathering of feldspars, micas, and ferromagnesian minerals. The weathering reaction is

$$2KAlSi_3O_8 + 2CO_2 + 3H_2O \rightarrow 4SiO_2 + 2KHCO_3 + AlO_3 \cdot 2SiO_2 \cdot 2H_2O$$

The first product, silica, is a colloidal gel. The second product, potassium bicarbonate, is in solution. The third product, hydrous aluminum silicate, is a simplified clay mineral. Their physical form is a wide, flat sheet many times wider and longer than thick.

The three most common clay minerals, in order from largest to smallest in size, are

- Kaolinite
- Illite
- Montmorillonite or smectite—very expansive, weak clays

Other somewhat less common clay minerals are

- Chlorite
- Attapulgite
- Vermiculite
- Sepiolite
- Halloysite, a special type of kaolin in which the structure is a tube rather than a flat sheet
- Many others

Clays are arbitrarily defined as particles less than 2 μm in the long dimension. Clay particles are very small. If clay is observed under an ordinary optical microscope, the individual particles would not be seen. If a 1-in. cube of clay is sliced one million times in each direction, the resulting pieces are approximately the size of clay particles. If the surface areas of a thimbleful of clay particles are added up, the total would be 6 million square inches, the same surface area from about five truckloads of gravel.

At the microscopic level, the clay fabric may have either a dispersed (parallel) or flocculated (edge-to-side) structure. The clay fabric may have a significant effect on engineering properties. Clay soils usually contain between 10% and 50% water by weight.

Silt is commonly found in flat floodplains or around lakes. Silt is usually deposited by flowing water or dust storms. Silt is composed of ground-up rock and is usually inorganic, that is, without plant material. A dry chunk of silt is easily broken by hand and is powdery.

Clays The particles in clay soils may have two general types of structures, depending on the concentration of polyvalent ions present in the groundwater. If the particles are arranged in an edge-to-face arrangement, the clay has a *flocculated* structure, in which the individual plate-shaped particles are clustered, as shown in Figure 3.1. Particles of this type are known to have a positive charge at the edges of the particle and a negative charge on the face of the particle, as illustrated in Figure 3.1a. If the specimen of clay is remolded, thereby destroying its natural structure, the plates may become par-

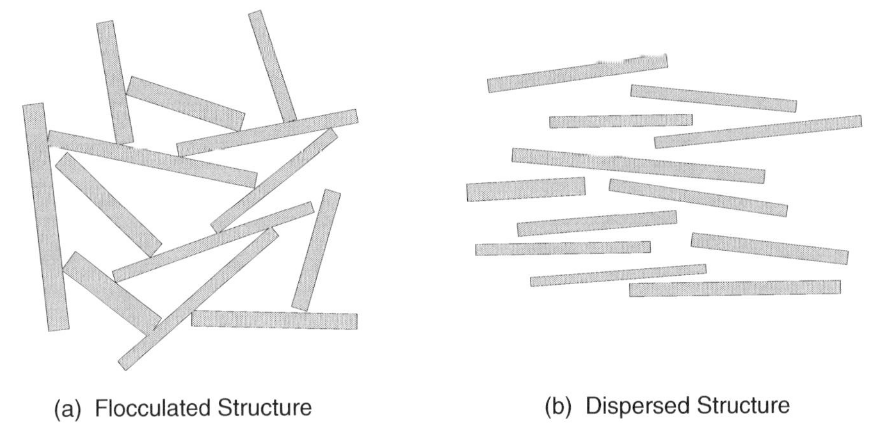

Figure 3.1 Clay structure.

allel to each other, as shown in Figure 3.1b. Frequently, remolding results in a loss of shear strength compared to the undisturbed specimen. If the particles are arranged in parallel, the clay has a *dispersed* structure.

In addition to its mineralogical composition, the strength of clay is strongly dependent on the amount of water in the specimen. The past stress history of a stratum of clay is also of great interest to the geotechnical engineer. If the stratum was deposited in water and attained considerable thickness over a long period of time, a specimen at a particular depth is *normally consolidated* if equilibrium exists, with no flow of water from the specimen due to the weight of the *overburden*. The specimen is *overconsolidated* if some of the overburden was removed by erosion or if the stratum was exposed to the atmosphere and later resubmerged. Consolidation may have occurred by *desiccation* if the stratum was uplifted and exposed to the air. Negative pore pressure develops due to *capillarity* and exposes the soil to positive compressive stress. The stratum will be overconsolidated; some formations of this sort have been resubmerged and exist in coastal areas in many parts of the world.

Silts *Silt* is a single-grained soil, in contrast to clay, where the grains are bonded due to chemical attraction. Silt grains are too small to be seen with the naked eye, ranging in size from 2 μm to 0.074 mm. On drying, a lump of silt can be easily broken with the fingers. Deep deposits of silt present problems to the engineer in the design of foundations. Chapter 2 notes problems with foundations in loess, existing principally above the water table. Deposits of silt below the water table may be organic and compressible. Because of the small grains, water will flow slowly through the soil, and drainage due to a foundation load could be lengthy. Deep foundations penetrating the deposit of silt might be recommended in lieu of shallow foundations.

Silt can hold water and is usually soft when wet. A wet silt will flatten out when shaken, and the surface will appear wet. Silt is usually found in mixtures with sand or fine sand. A mixture of sand with a smaller amount of silt is often called a *dirty sand.*

Silt is usually a poor foundation material unless it has been compressed and hardened like siltstone. Many types of silt are compressible under low foundation loads, causing settlement. As a construction material, silt is difficult to work with. It is difficult to mix with water, is fluffy when dry, and pumps under compaction equipment if too wet. Silt particles can be flat, like clay particles, or blocky, like sand particles.

Sands and Gravels *Sand* also is a single-grained soil that exists in a wide range of sizes, from fine (0.074 to 0.4 mm), to medium (0.4 to 2 mm), to coarse (2 to 4.76 mm). The grains of sand can have a variety of shapes, from subrounded to angular, and deposits of sand can have a variety of densities. Laboratory tests will reveal the effect of such variations on strength and stiffness. If the sand is loose, vibration from nearby motorized equipment can cause densification and settlement.

Sand is a granular material in which the individual particles can be seen by the naked eye. It is often classified by grain shape—for example, angular, subangular, or rounded. Sand is considered to be a favorable foundation material, but deposits are subject to erosion and scour, and need protection if they occur near a waterfront or a river.

Deposits of sand allow easy flow of water and do not hold water permanently. Water rising vertically through sand can cause quicksand. Any deposit of sand that holds water is a mixture containing finer material, either silt or clay. If a deposit of sand contains water and the deposit is surrounded by relatively impermeable deposits of clay or silt, a perched water table exists with a water table higher than exists outside the perched zone. Perched water sometimes results in construction difficulties if the perched water table is not recognized.

One problem with sand is that excavations often cave in. A slope in sand steeper than 1.5 horizontal to 1 vertical rarely occurs. The angle of the steepest natural slope is called the *angle of repose.*

The mineral composition of the grains of sand should receive careful attention. The predominant minerals in many sands are quartz and feldspar. These minerals are hard and do not crush under the range of stresses commonly encountered. However, other types of sand, such as calcareous sands, can be soft enough to crush. Calcareous sands are usually found between the Tropics of Cancer and Capricorn and present special problems for designers. One project for which the calcareous nature of a sand was of concern was an offshore oil production platform founded on piles driven into a deposit of lightly cemented calcareous soil off the coast of Western Australia. Open-ended steel pipe piles were driven to support the structure. Construction followed an intensive study to develop design parameters for axial loading. King and Lodge (1988) wrote that "it was noted with some alarm that the 1.83 m diameter piles not only drove to final penetration more easily than any of the pre installation predictions but also that the piles free-fell under their own weight with little evidence of the expected frictional resistance." The calcareous sand was composed largely of the calcium carbonate skeletal remains of marine organisms (Apthorpe et al., 1988). During pile installation, the tips of the piles apparently crushed the grains of calcareous soil and destroyed the natural cementation. The cementation present in the sand was likely a natural consequence of the deposition and prevented development of the normal axial resistance along the length of the piles. The offshore project illustrates how the strength of sand grains and natural cementation can require special attention for designs of some foundations.

Gravel, as sand, also exists in many sizes from fine (4.76 to 19 mm) to coarse (19 to 75 mm).

Driving piles into deposits of sand and gravel may present difficulties. If the deposit is loose, vibration will cause a loss of volume and allow the pile to penetrate; otherwise, driving of a pile is impossible. The problem is especially severe for offshore platforms where a specific penetration is necessary to provide tensile resistance. Various techniques may be employed, but delays in construction may be very expensive.

Cobbles range in size from 75 to 1000 mm, and the size of *boulders* is above 1000 mm. A glacier can deposit a layer of cobbles and boulders with weaker soil above and below. Often, piles cannot be driven into such materials and drilling or coring for a drilled shaft foundation is the only practical alternative (see Chapters 5 and 11).

Rock or Bedrock *Rock* or *bedrock* refers to deposits of hard, strong material that serves as an end-bearing foundation for piles or drilled shafts and, in some instances, as support for a shallow foundation. Characteristics of importance for rock are compressive strength and stress-strain characteristics of intact specimens. In some instances, the secondary structure of rock, referring to openings, joints, and cracks, is of primary importance. The secondary structure can be defined quantitatively by the *rock quality designation* (RQD), a number obtained by dividing the sum of the length of core fragments longer than 100 mm that were recovered by the overall depth that was cored (Chapter 4).

3.3.4 Water Content and Void Ratio

A soil sample is known to be saturated when it is weighed in the laboratory and its total weight is W. The sample is then placed in an oven at a temperature of 110°C and dried completely. The dry weight is measured as W_d. The loss in weight of the soil sample is the weight of water ($W_w = W - W_d$) that was in the original sample of soil. From this test, it is possible to compute the *water content* using

$$\boxed{w = \frac{W_w}{W_d} \cdot 100\%} \tag{3.1}$$

where water content is expressed as a percentage. Water contents are determined for almost every sample tested in a soils laboratory because soil properties are found to be sensitive to changes in water content.

The *void ratio* is defined as the volume of voids in a soil divided by the volume of the solid materials in a soil. The void ratio is convenient to use when calculating unit weights and strains due to consolidation. The symbol used for void ratio is e and is computed as

$$e = \frac{V_v}{V_s} \tag{3.2}$$

where V_v is the volume of voids ($V_v = V - V_s$) and V_s is the volume of soils.

3.3.5 Saturation of Soil

A soil is composed of mineral solids and voids. The voids may be filled water, other liquids, or gas (air). If all voids are filled with liquids, the soil is *saturated.*

If a stratum of clay exists above the *water table,* the position in the earth where water will collect in an open excavation, then the soil will be subject to wetting and drying, presenting a difficult problem to the engineer. If the clay is *expansive,* shrinking on drying and swelling on wetting, the problem is even greater. The design of shallow foundations in expansive clay is discussed in Chapter 9. A lump of clay on drying will be difficult to break with the fingers.

The *degree of saturation* S_r is defined as

$$S_r = \frac{V_w}{V_v} \cdot 100\% \tag{3.3}$$

3.3.6 Weight–Volume Relationships

Several relationships between weight and volume are frequently used by foundation engineers to compute the stresses in the soil due to gravity and the associated variation due to the location of the water table. This section presents the definitions of several weight–volume parameters, their interrelationships, and the equations in common use.

The *unit weight* of the soil is the weight of soil per unit volume. The most commonly used unit in the U.S. customary system is pounds per cubic foot (pcf), and the most commonly used unit in the SI system is kilonewtons per cubic meter (kN/m^3). Other combinations of weight and volume units may be used in computations where consistent units are needed.

The unit weight, γ, can be computed using one of the following relationships:

$$\gamma = \frac{W}{V} = \frac{W_w + W_d}{V_s + V_v} = \frac{V_w\gamma_w + V_s\gamma_s}{V_s + eV_s} = \frac{V_w\gamma_w + V_s\gamma_s}{V_s(1 + e)} \tag{3.4}$$

where

W = weight of the sample,
W_w = weight of water in the sample,
W_d = weight of soil after drying,
V = total volume of the soil sample,
V_s = volume of solids in the soil sample,
V_v = volume of voids in the soil sample,
γ_w = unit weight of fresh water (9.81 kN/m^3 or 62.4 pcf), and
γ_s = unit weight of the solid mineral grains without any internal voids.

This expression for unit weight makes no assumption about the amount of water present in a sample of soil or rock, and is based simply on the weight and volume of the soil or rock.

The *saturated unit weight* is the unit weight of a fully saturated soil. In this case, no air or gas is present in the voids in the soil. An expression for the saturated unit, γ_{sat}, is obtained by dividing the volume terms in the numerator and denominator of Eq. 3.4 by V_s and substituting the term for void ratio where appropriate:

$$\gamma_{sat} = \frac{e\gamma_w + \gamma_s}{1 + e} \tag{3.5}$$

The *specific gravity* of the mineral grains, G_s, is the ratio of the unit weight of solids divided by the unit weight of water:

$$G_s = \frac{\gamma_s}{\gamma_w} \tag{3.6}$$

The values for specific gravity are usually in such a narrow range (i.e., 2.60 $\leq G_s \leq$ 2.85) that an approximate value of 2.7 is usually assumed rather than performing the necessary laboratory tests for measurement.

It is convenient to rearrange the above relationship in the form

$$\gamma_s = G_s \, \gamma_w \tag{3.7}$$

and substitute Eq. 3.7 into Eq. 3.5 to obtain an expression for γ_{sat} in terms of specific gravity, void ratio, and unit weight of water:

$$\boxed{\gamma_{\text{sat}} = \frac{G_s + e}{1 + e} \gamma_w} \tag{3.8}$$

Equation 3.8 is in the form usually used in practice.

A useful expression for void ratio can be derived from the expressions for degree of saturation, water content, and specific gravity. The derivation is

$$e = \frac{V_v}{V_s}$$

$$e = \frac{V_w}{V_s} \cdot \frac{1}{S_r}$$

$$e = \frac{W_w}{\gamma_w} \cdot \frac{\gamma_s}{W_s} \cdot \frac{1}{S_r}$$

$$e = \frac{W_w}{W_s} \cdot \frac{1}{S_r} \cdot \frac{\gamma_s}{\gamma_w}$$

$$e = w \cdot \frac{1}{G_s} \cdot S_r$$

$$\boxed{e = \frac{wG_s}{S_r}} \tag{3.9}$$

The *dry unit weight*, γ_d, is the unit weight for the special case where $S_r = 0$. The dry unit weight is computed using

$$\boxed{\gamma_d = \frac{\gamma}{1 + w}} \tag{3.10}$$

where the water content w is expressed as a decimal fraction. The above expression for dry unit weight is valid for any value of w. Another useful relationship for dry unit weight in terms of specific gravity, void ratio, and unit weight of water is

$$\gamma_d = \frac{G_s}{1 + e} \gamma_w \tag{3.11}$$

Note that if S_r is not 100% then Eq. 3.8 must be changed to

$$\gamma = \frac{G_s + eS_r}{1 + e} \gamma_w \tag{3.12}$$

This is a general expression that is valid for all values of degree of saturation. When S_r is zero, this equation yields the dry unit weight, and when S_r is 100%, this equation yields the saturated unit weight.

The *effective unit weight* γ' (also called the *submerged unit weight*) is defined as

$$\gamma' = \gamma_{\text{sat}} - \gamma_w \tag{3.13}$$

where γ_{sat} is the saturated unit weight.

Some instructors ask their students to memorize the six boxed equations above. All other weight–volume relationships can be derived from these equations.

3.3.7 Atterberg Limits and the Unified Soils Classification System

Atterberg Limits The Swedish soil scientist A. Atterberg (1911) developed a method for describing quantitatively the effect of varying water content on the consistency of fine-grained soils like clays and silts. Atterberg described the stages of soil consistency and arbitrary limits shown in Table 3.1.

For many saturated clays, the relationship of volume and water content is similar to that shown in Figure 3.2. The volume of the soil sample will decrease if the water content is lowered from a high value to a lower value. At some value of water content, the solids become so tightly packed that the volume cannot decrease further and the volume remains constant. The shrinkage limit is defined as the water content below which the soil cannot shrink.

TABLE 3.1 Atterberg Limits

Stage	Description	Limit
Liquid	Slurry, pea soup to soft butter, viscous liquid	Liquid limit, *LL*
Plastic	Soft butter to stiff putty, deforms without cracking	Plastic limit, *PL*
Semisolid	Cheese, deforms permanently, but cracks	Shrinkage limit, *SL*
Solid	Hard candy, breaks on deformation	

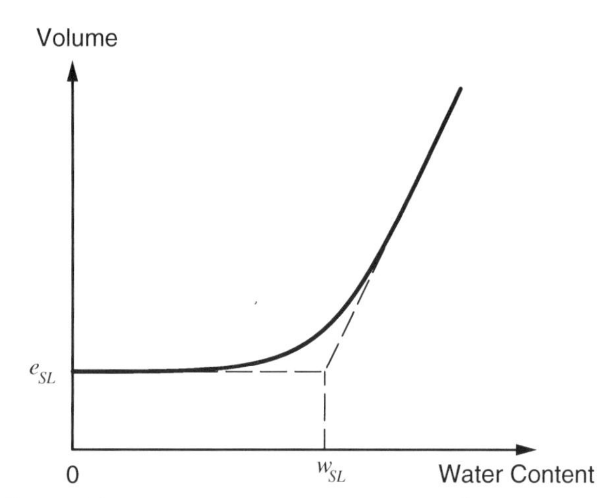

Figure 3.2 Volume versus water content of saturated clay.

Professor Arthur Casagrande of Harvard University developed a testing device to standardize the test for liquid limit. The original device was not mounted on rubber feet, as are modern standardized devices, and the results varied, depending on whether the machine was operated directly over a table leg or not. Casagrande found that consistency was improved if the machine was placed on a copy of a telephone directory. The rubber feet were added to simulate the directory, and the consistency of readings from the liquid limit machines was improved to a satisfactory point. The modern version of the liquid limit testing machine, as standardized by the American Society for Testing and Materials (ASTM), is shown in Figure 3.3.

Atterberg limits are used to classify and describe soils in several useful ways. One widely used and extremely useful soil descriptor is the plasticity index (PI), which is defined as

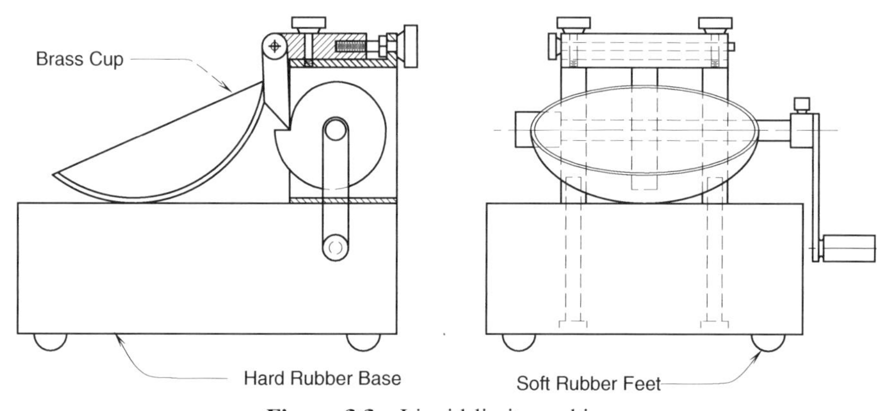

Figure 3.3 Liquid limit machine.

$$PI = LL - PL \tag{3.14}$$

In addition, the liquidity index (LI), is used to indicate the level of the water content relative to the liquid limit (LL) and plastic limit (PL).

$$LI = \frac{w - PL}{LL - PL} = \frac{w - PL}{PI} \tag{3.15}$$

If the water content of a soil is higher than the LL, then LI is greater than 1. If a soil is drier than the PL, then LI is negative. A negative value of the LI usually indicates a problem condition where the potential for expansion on wetting is severe.

Unified Soil Classification System This system was developed by the U.S. Army Corps of Engineers and the Bureau of Reclamation in 1952 and was subsequently adopted and standardized by the ASTM. Today, the Unified Soils Classification System is widely used in the English-speaking world.

The system uses a two-letter classification. The first letter is used to classify the grain size.

If more than 50% by weight is larger than the #200 sieve (0.074-mm openings), then a coarse-grained classification is used. If more than 50% is larger than a #4 sieve, then the letter G is used; otherwise, the letter S is used.

For sands and gravels, the second letter describes the gradation:

W = well graded
P = poorly graded
M = contains silt
C = contains clay

Examples of common classifications are GP, GW, GC, GM, SP, SW, SM, and SC.

If more than 50% by weight passes the #200 sieve, then the soil is fine-grained and the first letter is M for silt, C for clay, or O for organic soil.

The second letter for fine-grained soils describes their plasticity. The letter L is used for low plasticity ($LL < 50$), and the letter H for high plasticity ($LL > 50$). The Atterberg limits are tested on the fraction of soil passing the #40 sieve, not the full fraction of soil.

Examples of common classifications of fine-grained soils are ML, CL, MH, CH, OL, and OH.

If a soil is highly organic, with noticeable organic fibers, it is given a peat designation, PT.

Relative Density The degree of compaction of granular soils is measured as relative density (Table 3.2), usually expressed as a percentage:

TABLE 3.2 Classification of Cohesionless Soils by Relative Density

Designation	D_r, %	SPT N-Value
Very loose	0–15	0–4
Loose	15–30	4–10
Medium dense, firm	35–70	10–30
Dense	70–85	30–50
Very dense	85–100	50+

$$D_r = \frac{e_{\max} - e}{e_{\max} - e_{\min}} \tag{3.16}$$

Grain Size Distribution The uniformity coefficient C_u and the coefficient of curvature (called the *coefficient of gradation*) C_c were developed to help engineers evaluate and classify the gradations of granular materials. These coefficients are used to determine when different materials can be used effectively as filter materials in earth dams. Their definitions are

$$C_u = \frac{D_{60}}{D_{10}} \tag{3.17}$$

$$C_c = \frac{(D_{30})^2}{D_{60}D_{10}} \tag{3.18}$$

Well-graded soils have uniformity coefficients greater than 4 for gravels and 6 for sands, and a coefficient of curvature between 1 and 3. If the soil is not well graded, it is poorly graded.

A soil with a *gap-graded* distribution has ranges of soil particle sizes missing. In many applications, such as filter zones in earth dams used for drainage, gap-graded soils are avoided because they can be susceptible to subsurface erosion or piping.

Examples of gradation curves for well-graded and gap-graded soils are shown in Figure 3.4.

The results of the classification of the soils encountered in a borehole are of great importance to the foundation engineer because of the difference in the behavior of various soils on the imposition of loadings. The classification will guide the specification of laboratory tests, and possibly in situ tests, for acquisition of the data necessary for design.

3.4 CONCEPT OF EFFECTIVE STRESS

This chapter begins with the analysis of one-dimensional settlements, as shown in Figure 3.5. In this analysis, the stresses in the soils are uniform on

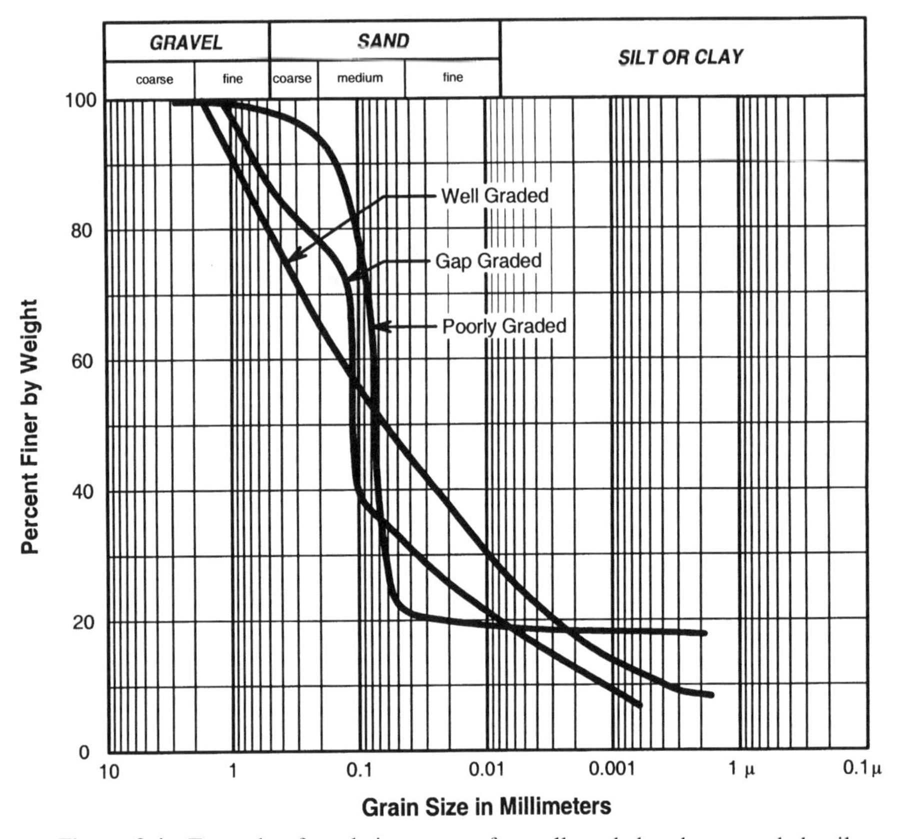

Figure 3.4 Example of gradation curves for well-graded and gap-graded soils.

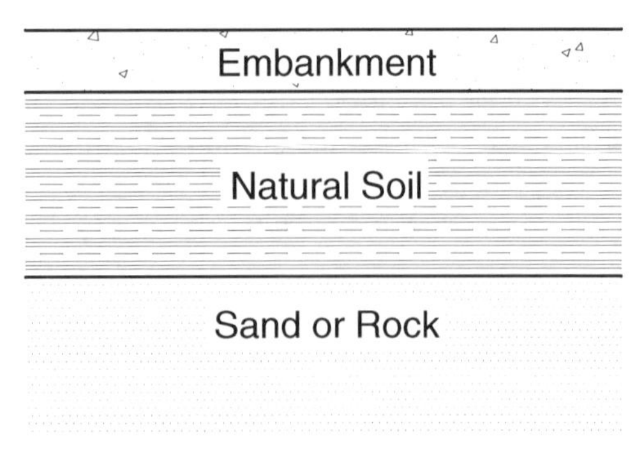

Figure 3.5 A one-dimensional settlement problem.

horizontal planes. Under these conditions, the soil cannot deform laterally. In fact, in a one-dimensional consolidation problem, neither soil deformation nor water flow can occur in the horizontal direction, only in the vertical direction.

To simplify this problem, the following conditions are assumed:

- The subsoil is homogeneous, that is, it has the same properties everywhere in a layer.
- The subsoil is *saturated,* that is, all voids between the soil particles are filled with water, and no gas bubbles are present in the spaces between the soil particles.

3.4.1 Laboratory Tests for Consolidation of Soils

To determine what would happen in the field, foundation engineers obtain soil samples from the field and test the samples in the laboratory. Because the problem involves only one-dimensional deformation and one-dimensional water flow, a one-dimensional consolidation test in the laboratory is appropriate.

The laboratory apparatus used for this purpose is called a *one-dimensional consolidation cell* or *oedometer*. An example of a consolidation cell is shown in Figure 3.6. The soil sample that is tested is a circular disk with a diameter about two to four times its thickness. Diameters of testing specimens are typically between 2 and 4.5 in. (5 to 10 cm). Such small diameters are used to reduce the cost of drilling, sampling, and laboratory testing. However, samples with larger diameters are preferred so that the properties measured in the laboratory are more representative of the soil deposit in the field.

The soil is encased in a rigid metal ring to prevent deformations due to lateral squeezing of the soil and to force the flow of water in the vertical direction. The soil sample is covered with a loading cap, and vertical loads are applied to represent applied vertical stresses in the field. The original vertical load applied on the sample in the field depends on the depth from which the sample was retrieved and the weight of the soil and fill. Since the weight and thickness of the fill may not be known during the testing phase (and for other reasons as well), a wide range of stresses will be applied in the laboratory to obtain a stress-strain curve that can be used for a range of loads in the field.

The soil is assumed to be in equilibrium under an applied load in the laboratory and that an additional load is suddenly applied. Also, this additional load is assumed to represent the weight of the load in the field. Settlement of the top cap in the consolidation cell is measured using either a dial gauge or a linear transducer, with an accuracy of 0.0001 in. (0.0025 mm). When a compressive load is applied to saturated clay, the test operator discovers that the settlement is initially very small; in fact, the initial magnitude of the settlement approximately equals the settlement that would develop if

Figure 3.6 Computer-controlled, one-dimensional consolidation testing equipment (photograph courtesy of Trautwein Soil Testing Equipment Company).

the soil sample was replaced with a rigid metal disc. This instantaneous settlement is caused by compression of the porous stones and other mechanical parts of the testing apparatus because the soil is nearly incompressible during the short interval after the application of stress.

If the applied load is sustained, the test specimen compresses and the amount of compression change with time in a manner similar to that shown in Figure 3.7. After a period of time, the compression essentially ceases and the soil is again in equilibrium. Similar settlement curves versus time are obtained for each level of loading, and the load is increased again. After the maximum consolidation stress of interest has been reached, the level of consolidation stress is lowered in stages, and the rebound of the specimen under each stage is measured, until the test specimen is unloaded completely.

For samples of usual dimensions in the laboratory, the time needed to achieve equilibrium varies from a few minutes for sands to several days for some clays. The delay is caused by the time needed for water to squeeze out of the specimen of soil. The more pervious soils come to equilibrium quickly, whereas relatively impervious clays are comparatively slow to equilibrate.

As in pipe-flow hydraulics, where the rate of water flow in a pipe depends on the diameter of the pipe, the rate at which water can be squeezed from a saturated specimen of soil depends on the sizes of the openings between the soil particles. Also, as in hydraulics, the rate of flow for a given pore (pipe) size decreases as the total flow distance increases, so the time required for equilibrium for a thick layer of soil in the field is much longer than for a

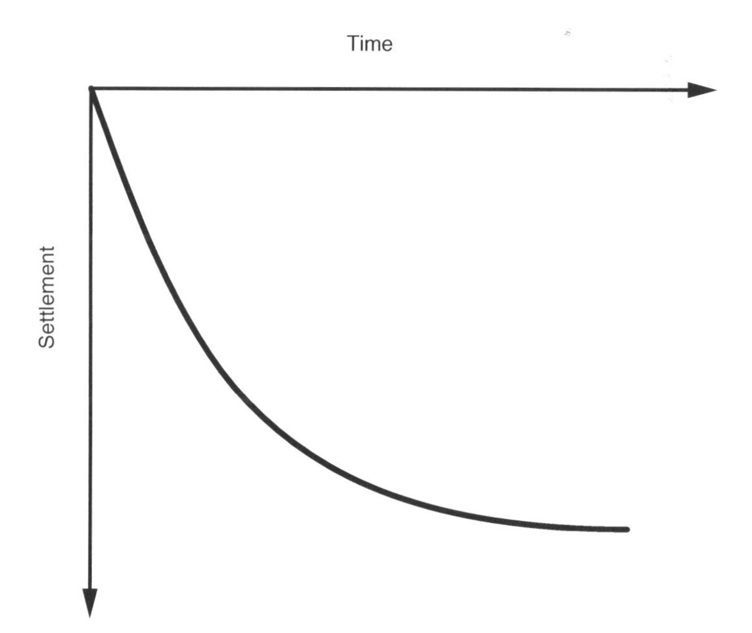

Figure 3.7 Curve of settlement versus time for a saturated soil.

small specimen of the same soil in the laboratory. Delayed settlement can be a serious problem for the design engineer if the settlement is still continuing after a structure has been completed on a stratum of consolidating soil.

3.4.2 Spring and Piston Model of Consolidation

The mechanics of consolidation will be presented using a simple model of a spring and piston, shown in Figure 3.8, where a saturated soil is represented by a piston that moves up and down in a cylinder. The spring supporting the piston inside the cylinder represents the compression of the soil particles under an external applied load. To represent a saturated soil, the cylinder is also saturated; that is, it is filled with water without gas bubbles. Water can escape from the cylinder only through the valve, as shown. Included in this model is a pressure gage that can measure the pressure of the water at any time in the cylinder.

If an external load P is applied to the piston with the valve closed, one would discover (just as in a laboratory consolidation test on a saturated soil sample) that immediately after loading, no settlement results. Instead, the force applied to the piston P must be balanced by the force supporting the piston. The force supporting the piston is the combination of the force in the spring P_s and the force in the water P_w. If the total area inside the cylinder is A, the area of the wire in the spring is A_s, and the area of the piston covered by water is A_w, then equilibrium of forces requires that

$$A = A_s + A_w \tag{3.19}$$

The spring constant k is defined as

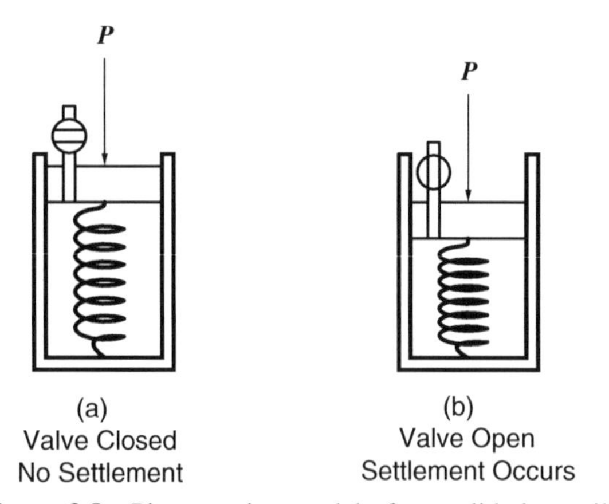

Figure 3.8 Piston-spring model of consolidating soil.

$$k = \frac{P_s}{S} \tag{3.20}$$

where S is the compression of the spring (equal to the settlement of the piston).

The bulk modulus of the water is defined as the change in volume of the water that results from the water pressure u resulting from the application of the external force P

$$C_w = \frac{\Delta V_w / V_w}{u} \tag{3.21}$$

where ΔV_w is the change in volume of water, V_w is the original volume of water, and u is the pressure developed in the water. If the area of the cylinder is uniform, then the change in the volume of water is equal to the settlement times the area of the water:

$$\Delta V_w = S\,A_w \tag{3.22}$$

An expression for the water pressure due to the change in volume of the water under the applied external force is obtained by inserting Eq. 3.22 into Eq. 3.21 and rearranging:

$$u = \frac{S\,A_w}{C_w V_w} \tag{3.23}$$

The force carried by the water is equal to the water pressure acting over the area of the piston:

$$P_w = uA_w \tag{3.24}$$

Equilibrium of forces requires the externally applied force to equal the sum of forces carried in the spring and water. This is expressed as

$$P = P_s + P_w \tag{3.25a}$$

$$P = P_s + uA_w \tag{3.25b}$$

Stresses are found by dividing the forces by the area of the cylinder. Thus, total stress σ applied to the cylinder is

$$\sigma = \frac{P}{A} \tag{3.26}$$

The stress carried by the spring is

$$\sigma' = \frac{P_s}{A} \tag{3.27}$$

The porewater pressure is

$$u_w = \frac{P_w}{A} = \frac{uA_w}{A} \approx u \tag{3.28}$$

because A_s is much less than A_w. In this model, A_s is much less than A_w by choice, but studies of real soils have found that this is an excellent approximation. Thus, after dividing both sides by the area of the piston, Eq. 3.25b reduces to

$$\sigma = \sigma' + u \tag{3.29}$$

Usually this equation is written as

$$\boxed{\sigma' = \sigma - u} \tag{3.30}$$

Equation 3.30 is called *Terzaghi's effective stress equation,* and σ' is called the *effective stress*. Other notations for effective stress in common use include the symbols $\overline{\sigma}$ and $\overline{p}$. This text will use σ' to denote effective stress in most situations and use the prime (′) notation applied to other symbols to indicate conditions under which stresses are effective stresses.

When the spring and piston model is applied to real soils, the spring represents the compressible framework of soil particles and P_s is the force carried by this framework. The effective stress is then the ratio of the force carried in the particle framework to the total area of the soil.

The settlement of the piston S appears in both Eq. 3.20 and Eq. 3.23. Solving both equations for S, one obtains

$$S = \frac{P_s}{k} = \frac{C_w V_w u}{A_w} \tag{3.31}$$

Again, dividing both sides by the piston area A and multiplying both sides by the spring constant k, one obtains an expression for effective stress:

$$\sigma' = \frac{P_s}{A} = \frac{k\, C_w\, V_w\, u}{A\, A_w} \tag{3.32}$$

If one inserts values of k obtained by testing real soils and the appropriate values for the bulk modulus of water C_w, the volume of water V_w, the piston area A, and the water area A_w, one finds that

$$\sigma' \approx \left(\frac{1}{1000}\right) u \tag{3.33}$$

In other words, when the piston is loaded and the valve is closed, approximately 99.9% of the load goes into the water and none into the spring.

Laboratory measurement of the consolidation properties of saturated soils has found that if a pressure σ is applied to the soil and drainage is not allowed, then the resulting porewater pressure u equals the applied pressure σ. Insertion of typical values into Eq. 3.31 demonstrates why the settlement is essentially zero when a saturated soil is loaded without drainage.

If the valve is opened, the water in the cylinder escapes, settlement occurs, and the spring compresses under the external force. Settlement causes the spring (representing the soil particle framework) to compress and to take on an increasing fraction of the externally applied load. Eventually, enough water will escape so that the spring carries the full externally applied load and the porewater pressure is zero. The spring and piston model of consolidating soil is then in equilibrium. One can now calculate the spring constant of the model by dividing P by the measured settlement S.

With the known value of the spring constant, piston settlements can be computed for any given applied load. In addition, measurement of the time rate of settlement in the laboratory and computing the soil property that characterizes the time rate of settlement (represented by the opening in the valve in the model) would allow the computation of the time rate of settlement in the field using an appropriate theory.

3.4.3 Determination of Initial Total Stresses

A homogeneous soil deposit with a horizontal ground surface is shown in Figure 3.9, and the vertical stress at depth z is desired. A column of soil of area A as shown may be considered. Static equilibrium in a vertical direction requires that the weight of the column W plus any shear forces on the sides to equal the force on the bottom. Since the soil deposit is homogeneous and extends indefinitely in the horizontal direction, the stresses on all vertical planes are the same. Thus, if any shear exists on vertical planes, the shear on the outside of one face must be identical to the shear on the outside of the opposite face, so the shear forces are in equilibrium. Actually, if the soil

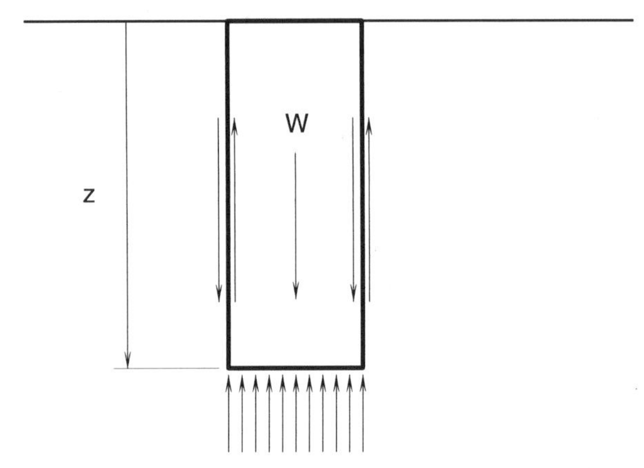

Figure 3.9 Equilibrium of vertical forces for a vertical prism of soil.

deposit is formed by uniform deposition over the surface, no shearing stresses on vertical planes will occur. Thus, for force equilibrium in the vertical direction

$$W = \sigma A \tag{3.34}$$

where σ is the total stress on the base. The stress is then

$$\sigma = \frac{W}{A} \tag{3.35}$$

As a matter of convenience, substitute

$$W = \gamma v = \gamma A z \tag{3.36}$$

where v is the volume of the column of soils and γ is the *total unit weight* (total weight per unit volume). After inserting Eq. 3.36 into Eq. 3.35, the following equation results

$$\sigma = \gamma z \tag{3.37}$$

Equation 3.37 is valid only if γ is constant with depth. For real (nonhomogeneous) soils, the soil may be considered homogeneous on a microscopic scale, that is, in differential elements, so

$$d\sigma = \gamma \, dz \tag{3.38}$$

and

$$\sigma = \int_0^z \gamma dz \tag{3.39}$$

Equation 3.39 is valid for any soils with a horizontal ground surface.

The total unit weight γ can be found for soil samples by measuring the weight and volume of the soil samples, but γ is computed more conveniently from other parameters that are measured for other purposes.

3.4.4 Calculation of Total and Effective Stresses

The total vertical stress at a depth of 10 ft in a saturated clay that has an average water content of 22% will be calculated below as an example calculation of total and effective stresses:

$$e = \frac{wG_s}{S_r} = \frac{(0.22)(2.70)}{1.00} = 0.594$$

$$\gamma = \frac{G_s + eS_r}{1 + e}\gamma_w = \frac{2.70 + (0.594)\,(1.00)}{1 + 0.594}\,62.4 \text{ pcf} = 129.0 \text{ pcf}$$

$$\sigma = \gamma z = (129.0 \text{ pcf})\,(10 \text{ ft}) = 1290 \text{ psf}$$

Previously, it was noted that compression of clay is controlled by changes in effective stress, not total stress. The effective stress is calculated using Eq. 3.30. When the water table is static (i.e., no flow of groundwater in the vertical direction), the water pressure increases linearly with depth, just as if no soil were present. The depth in the soil where the top of the equivalent free water surface exists is the *water table*. At the water table, the water pressure is zero gage pressure (1 atmosphere absolute). Below the water table the water pressure u is equal to

$$u = z_w\,\gamma_w \tag{3.40}$$

where z_w is the vertical distance below the water table. In a soil in the field, the water table is at a given depth and there is no sudden change in water content or any other property at the water table. In highly pervious soils, like sands and gravels, the water table is located by drilling an open hole in the soil and letting water flow into it from the soil until equilibrium occurs. The water level in the hole is then easily measured. For clays, the time needed to obtain equilibrium may be many days. To speed up equalization in clays, a *piezometer* may be used. A simple piezometer is formed in the field by drilling a hole, packing the bottom of the hole with sand, embedding a small-diameter tube with its bottom end in the sand and its top end at the surface, and then backfilling the rest of the hole with clay. The water level will equilibrate in the small-diameter tube much faster than in a large borehole because less water is needed to fill the tube. Once the water table is located, the pore water pressures below the water table are calculated using Eq. 3.40.

In many practical problems the water table moves up and down with the seasons, so the level of the water table at the time of the soil exploration may indicate only one point within its range of fluctuation.

To simplify the initial calculation of water pressure, the water table in the sample problem was assumed to be at the surface. Then at a depth of 10 ft the water pressure would be

$$u = (10\text{ ft})(62.4\text{ pcf}) = 624\text{ psf}$$

The effective stress is then computed using Eq. 3.30:

$$\sigma' = (1290\text{ psf} - 624\text{ psf} = 666\text{ psf}$$

More generally, Eqs. 3.30 and 3.40 may be combined as follows:

$$\begin{aligned}\sigma' &= \sigma - u \\ &= \gamma z - \gamma_w z_w \\ &= \gamma(z - z_w) + \gamma z_w - \gamma_w z_w\end{aligned}$$

The term $\gamma\ (z - z_w)$ is clearly equal to the total stress and to the effective stress at the position of the water table (where $u_w = 0$). The terms $(\gamma z_w - \gamma_w z_w)$ may be combined to yield $\gamma'\ z_w$, where γ' is the effective unit weight, as found according to Archimedes' principle.

The effective stress below the water table is then the sum of the effective stresses above and below the water table:

$$\sigma' = \gamma(z - z_w) + \gamma' z_w \tag{3.41}$$

The expression for submerged unit weight is obtained by subtracting γ_w from γ_{sat} using Eq. 3.8 (or Eq. 3.12 if S_r is less than 100%) to derive an expression that does not use w or S_r:

$$\begin{aligned}\gamma' &= \gamma - \gamma_w \\ &= \frac{G_s + e}{1 + e}\gamma_w - \gamma_w \\ &= \frac{G_s + e}{1 + e}\gamma_w - \frac{1 + e}{1 + e}\gamma_w\end{aligned}$$

$$\gamma' = \frac{G_s - 1}{1 + e} \gamma_w \tag{3.42}$$

The effective stresses in the soil prior to construction of the embankment are now known. Next, to determine the strains caused by the placement of the embankment fill, the effective stresses must be determined after settlement has been completed. This calculation is the same as the one used previously, except that now there is another layer of soil on the surface, so the depths are measured from the new surface.

The only question relates to the change in density that occurs during settlement. Since no solids are lost, the submerged weight of all the original soil above any depth remains the same (the loss of porewater during consolidation causes a decrease in void ratio and thus an increase in submerged unit weight, but this effect is exactly counterbalanced by the decreased thickness of the submerged soil above the depth in question).

However, if the elevation of the water table remains constant as the embankment settles, then some of the soil (including embankment soil) that was above the water table when the embankment was first placed will submerge below the water table after settlement has been completed. The amount of soil that is converted from total unit weight to submerged unit weight (Eq. 3.41) depends on the settlement that occurs but calculation of settlement is needed. Thus, a trial solution is sometimes needed where the effect of submergence is ignored in the first calculation of settlement. The first computed settlement is used to estimate a submergence correction, and another (better) settlement is calculated. The process may be repeated until a settlement of satisfactory precision has been calculated from a plot of computed settlement versus assumed settlement (usually only two trials).

3.4.5 The Role of Effective Stress in Soil Mechanics

The influence of effective stress is found throughout soil mechanics. Effective stress is an important factor in the consolidation of soil under the application of external loadings, as illustrated in the preceding discussion of the spring piston model, and it is an important factor in the development of the shearing strength of soils and rocks.

3.5 ANALYSIS OF CONSOLIDATION AND SETTLEMENT

3.5.1 Time Rates of Settlement

When a soil is subjected to an increase in effective stress, the porewater is squeezed out in a manner similar to water being squeezed from a sponge, and the basic concepts presented above govern the time rate of settlement of the surface of a clay layer. The following sections describe the theory of one-dimensional consolidation for the case of instantaneous loading. In cases where the loading of the surface of the consolidating layer of soil varies over time, more advanced analytical techniques have been developed. These advanced theories of consolidation also may include more nonlinear soil properties than the simple theory that follows.

The theory to be presented herein is *Terzaghi's theory of one-dimensional consolidation,* first presented in 1923. It remains the most commonly used theory for computing the time rate of settlement, even though it contains simplifying assumptions that are not satisfied in reality. The theory yields results of satisfactory accuracy when applied to predicting time rates of settlement of embankments on soft, saturated, homogeneous clays.

Primary Consolidation Water is forced out of a porous body like a sponge by the water pressure developed inside the sponge. However, in the case of thick layers of soil, the water pressure increases with depth, as in a quiet body of water, even though no flow is present in the water. Instead, water is forced to flow by applying an external pressure to the soil, but it is not the total water pressure that governs. In a soil where no water flow is occurring, the porewater pressure is termed the *static porewater pressure* u_s and is equal to the product of the unit weight of water and the depth below the water table:

$$u_s = \gamma_w z_w \tag{3.43}$$

As discussed previously, the application of the load from an embankment increases the total porewater pressure and flow begins. The flow of water is caused by the part of the total pore pressure that is in excess of the static value. For convenience, the excess porewater pressure u' is defined as

$$u' = u - u_s \tag{3.44}$$

where u is the total porewater pressure and u_s is the static water pressure. The rate of outflow of water is also controlled by how far the water must flow to exit and by the size of the openings in the soil. The average velocity of water flow in the soil v varies directly with u' and inversely with flow distance L:

$$v \propto \left(\frac{du'}{dL}\right) \tag{3.45}$$

The total rate of water flow (flow volume per unit of time), q, is

$$q = v\,A \tag{3.46}$$

where v is the average flow velocity and A is the cross-sectional area.

Actual experiments with soils show that the flow rate varies with the permeability of the soil and that the total rate of water flow is

$$q = \frac{k\,du'}{\gamma_w\,ds}A \tag{3.47}$$

where k is a constant of proportionality termed the *coefficient of permeability* or the *coefficient of hydraulic conductivity,* and s is now used to indicate distance in the direction of flow. Equation 3.47 is one form of *Darcy's law,* which states that water flows in response to an energy gradient in the soil. Those familiar with fluid mechanics may also know this, as flow will occur in response to a differential in hydraulic head.

Darcy's law is applied to the one-dimensional consolidation problem first in a qualitative sense. A clay layer is underlain by a freely draining sand (Figure 3.10). The water table is at the surface. An embankment is put into place in a very short period of time, and the total pore pressure increases by an amount γH, where γ is the unit weight of the embankment material and H is the thickness of the embankment. Excess pore pressures are developed in the sand layers too, but the sand consolidates so fast that the excess porewater pressure in the sand is dissipated by the time the construction of the embankment is complete, even though very little consolidation has occurred in the clay at this stage. Thus, the sand is *freely draining* compared with the clay, and u' is set equal to zero in the sand. The initial excess porewater pressure in the clay, u'_0, equals γH and is independent of depth. Because only the excess porewater pressure causes water flow, the value of u' is plotted

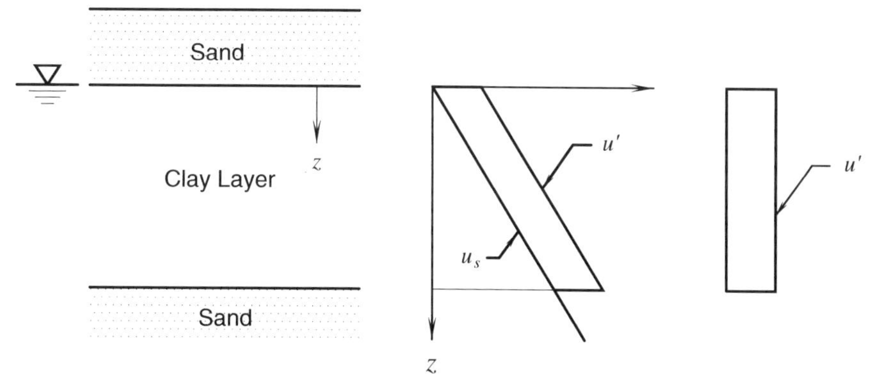

Figure 3.10 Pore pressures in a clay layer loaded instantaneously.

versus depth, as in Figure 3.10, and the total porewater pressure, u, and the static porewater pressure, u_s are ignored.

At the upper and lower boundaries of the clay layer (Figure 3.10) where drainage occurs, the original excess porewater pressure transitions between u'_0 in the clay layer and zero at the boundary with the sand over a very small distance. Thus, at time zero, just after the embankment has been put in place, the excess porewater pressure gradient, du'/dz, at both drainage boundaries approaches infinity, and thus the rate of outflow of porewater is also nearly infinite (Eq. 3.47). The total outflow volume of water Q is

$$Q = \int_0^t q \, dt \tag{3.48}$$

where t is time. Clearly, an infinite value of q can exist only instantaneously; then the flow rate drops and finally becomes zero when equilibrium is again established. The flow rate of water from the soil q decreases because the excess porewater pressure dissipates, thereby decreasing the gradient, du'/dz, and the resulting flow (Eq. 3.47). The shapes of the curves of q and Q versus time must be as shown in Figure 3.11. The settlement of the surface is found by dividing Q by A, the horizontal area of the soil deposit from which the flow Q emanates. The excess porewater pressures must vary as indicated in Figure 3.12, where time increases in the order $t_0, t_1, t_2, \ldots, t_\infty$.

Development of a mathematical theory to yield numerical values for these curves requires mathematics for solution of differential equations using Fourier series. Solutions may be developed for a variety of different initial excess porewater pressure distributions. The following solution is for the case of a uniform distribution of excess porewater pressure versus depth with drainage from the upper and lower boundaries.

The settlement at any time S is equal to the average degree of consolidation U times the ultimate settlement S_u:

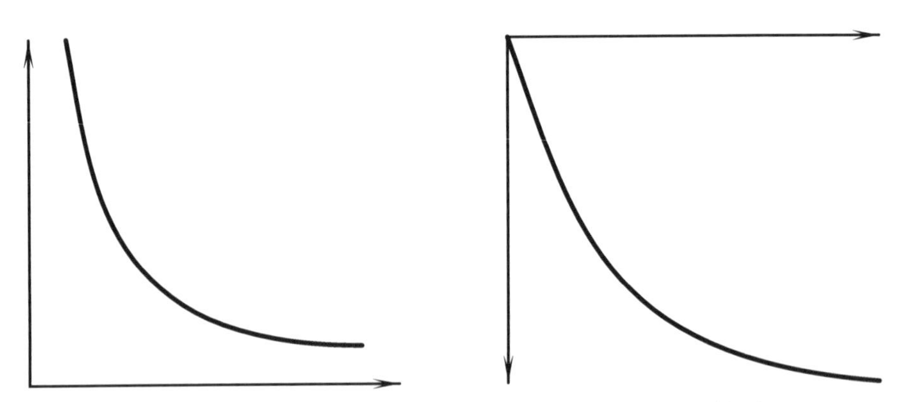

Figure 3.11 Shapes of q-t and Q-t diagrams derived intuitively.

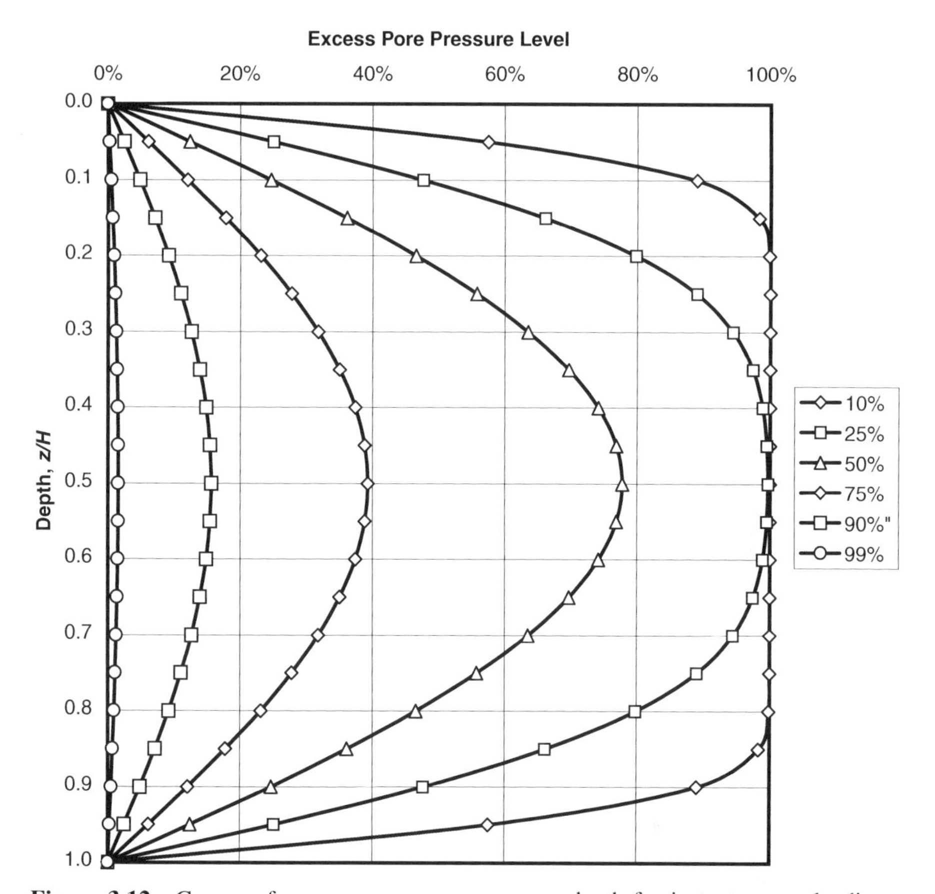

Figure 3.12 Curves of excess pore pressure versus depth for instantaneous loading.

$$S = US_u \tag{3.49}$$

In this equation, S_u is the settlement calculated previously and U has a value between zero and unity. U is given by

$$U = 1 - \sum_{m=0,1,2\ldots}^{\infty} \frac{2}{M^2} e^{-M^2T} \tag{3.50}$$

where M is

$$M = \frac{\pi}{2}(2m + 1) \tag{3.51}$$

e is the base of Napierian (natural) logarithms, and T is a dimensionless *time factor* given by

$$T = \frac{c_v\, t}{(H/n)^2} \tag{3.52}$$

where

c_v = *coefficient of consolidation,*
t = time (equal to zero at the instant the embankment is put in place),
H = total thickness of the clay layer, and
n = number of drainage boundaries.

The coefficient of consolidation has units of length^2/time and is given by the equation

$$c_v = \frac{k(1 + e)}{a_v \gamma_w} \tag{3.53}$$

where

k = coefficient of permeability,
e = void ratio,
a_v = *coefficient of compressibility,* and
γ_w = unit weight of water.

The coefficient of compressibility is the slope of the curve of void ratio versus effective stress and is given by

$$a_v = -\frac{de}{d\sigma'} \tag{3.54}$$

The profile of excess porewater pressure versus depth is calculated using

$$u' = \sum_{m=0,1,2\ldots}^{\infty} \frac{2u_0'}{M} \sin\left(\frac{Mz}{H/n}\right) e^{-M^2T} \tag{3.55}$$

The following is a sample calculation of the settlement–time curve. The soil profile in this problem is a homogeneous layer of saturated clay with a thickness of 10 ft. The compressible stratum is overlain and underlain by freely draining sand layers, and the water table is at the upper surface of the clay and is assumed to remain at the interface of the compressible layer and the upper sand layer. The upper sand layer has a submerged unit weight of 70 pcf. The clay is normally consolidated and has an average water content of 40%, a compression index of 0.35, and a coefficient of consolidation of

0.05 ft²/day. A 15-ft-thick wide embankment is rapidly put in place. Its unit weight is 125 pcf. In this problem, the clay layer will not be subdivided into sublayers.

The solution begins with the computation of the initial void ratio, effective unit weight, and initial effective stress at the middle of the clay layer:

$$e_0 = \frac{w_0 G_s}{S_r} = \frac{(0.40)(2.70)}{1.00} = 1.08$$

$$\gamma' = \frac{G_s - 1}{1 + e_0}\gamma_w = \frac{2.70 - 1}{1 + 1.08} 62.4 \text{ pcf} = 51 \text{ pcf}$$

$$\sigma'_0 = \sum_{j=1}^{\text{layers}} H_j \gamma'_j = (10 \text{ ft})(70 \text{ pcf}) + (5 \text{ ft})(51 \text{ pcf}) = 955 \text{ psf}$$

The change in vertical effective stress due to the construction of the wide embankment is

$$\Delta\sigma' = (15 \text{ ft})(125 \text{ pcf}) = 1875 \text{ psf}$$

The final vertical settlement is

$$S_u = C_c \frac{H}{1 + e_0} \log \left| \frac{\sigma'_f}{\sigma'_0} \right|$$

$$= \frac{(0.35)(120 \text{ in.})}{1 + 1.08} \log \left| \frac{955 \text{ psf} + 1{,}875 \text{ psf}}{955} \right| = 9.5 \text{ in.}$$

Values of T as a function of U are given in Table 3.3.

TABLE 3.3 Dimensionless Time Factors

U, %	T	U, %	T
0%	0	55%	0.238909
1%	7.85E-05	60%	0.286399
5%	0.001964	65%	0.340414
10%	0.007854	70%	0.402851
15%	0.017672	75%	0.47673
20%	0.031416	80%	0.567164
25%	0.049087	85%	0.683757
30%	0.070686	90%	0.848085
35%	0.096212	95%	1.129007
40%	0.125673	99%	1.781288
45%	0.159121	99.9%	2.714491
50%	0.196731	99.99%	3.647693

In lieu of using the values of T versus U in Table 3.3, values for T can be approximated for selected values of U by the following equations:

For values of U less than 60%:
$$T = \frac{\pi}{4} U^2 \tag{3.56}$$

For values of U between 60% and 99%:
$$T = 1.781 - 0.933 \log(100 - U\%) \tag{3.57}$$

The easiest way to calculate a complete S-t curve is to use Table 3.3 to calculate values of S and t for given values of U. For example, if $U = 50\%$, then $T = 0.197$ and the settlement is (from Eq. 3.49)

$$S = US_u = (0.50)(9.5 \text{ in.}) = 4.75 \text{ in.}$$

and the corresponding time is (from Eq. 3.52)

$$t = \frac{TH^2}{c_v n} = \frac{(0.197)(10 \text{ ft})^2}{(0.05 \text{ ft}^2/\text{day})(2)^2} = 98.5 \text{ days}$$

Other points on the S-t curve corresponding to different values of U are calculated in a similar way.

Calculations like those shown above are often made in engineering practice, but many settlement problems are more complex than the case with instantaneous loading that was considered. Some of the more common factors that require the application of computer-based solutions for problems of time rate of settlement occur when fill is placed and/or removed versus time, consolidation properties exhibit significant nonlinear properties, artesian conditions are affected by construction activities, and consolidation in the field is accelerated using wick drains.

Secondary Compression After sufficient time has elapsed, the curve of settlement versus logarithm of time flattens. Theoretically, the curve should become asymptotically flat, but observations both in laboratory tests and in the field find a curve with a definite downward slope. This range of settlement with time is called *secondary compression* and includes all settlements beyond primary consolidation.

The slope of the secondary compression line is expressed as the coefficient of secondary compression C_α:

$$C_\alpha = \frac{\Delta e}{\log\left|\frac{t_2}{t_1}\right|} \tag{3.58}$$

The coefficient of secondary compression is used in settlement computations in a manner similar to that of the compression, reloading, and rebound indices, except that time is used in place of effective stress when computing the settlement:

$$S = C_\alpha \frac{H}{1 + e} \log\left|\frac{t_2}{t_1}\right| \tag{3.59}$$

The terms *consolidation* and *compression* are used carefully in this book. *Compression* includes any type of settlement due to a decrease in the volume of soil. *Consolidation* refers to settlement due to the squeezing of water from the soil and the associated dissipation of excess pore water pressures.

3.5.2 One-Dimensional Consolidation Testing

It is not realistic to measure the consolidation characteristics of many soils in the field because the time necessary to do so is usually many years. Instead, soil samples are obtained as part of the field investigation program and representative samples are tested in the laboratory. The time required to measure soil properties in the laboratory is usually a few days or a few weeks. The testing procedures used in the laboratory are usually varied, depending on the type of soil, the nature of the problem in the field, the type of equipment available in the laboratory, and other factors.

In the following discussion, a soil sample to be tested is assumed to be a saturated clay contained in a thin-walled steel tube used for sampling. The sampling tube is typically 3 in. (75 mm) in diameter. The consolidation cell to be used is shown in Figure 3.6. The usual steps in preparing a consolidation test are as follows:

- The cell is dismantled and cleaned prior to testing.
- The inside surface of the confining ring is given a thin coating of silicone grease to minimize friction between the ring and soil and to seal the interface from seepage in order to force the fluid flow to be in the vertical direction.
- The ring is weighed and has a weight of 77.81 g.
- The soil is trimmed into the ring so that it fills the ring as exactly as possible (volume = 60.3 cc), and the ring and soil are weighed together (176.50 g).

- The soil is covered, at the top and bottom, with a single layer of filter paper to keep the soil from migrating into the porous stones, and the consolidation cell is assembled.
- The cell is placed in a loading frame, and a small seating pressure (60 psf or 3 kPa) is applied. A dial indicator or displacement transducer, capable of reading to 0.0001 in., is mounted to record the settlement of the loading cap.
- The tank around the ring is filled with water and the test begins.

The usual steps taken during the consolidation test are as follows:

- If the soil swells under the seating load, then the seating load is increased until equilibrium is established.
- The applied pressure is then increased in increments, usually with the stress doubled each time.
- Settlement observations are recorded at various times after the application of each load level. A set of settlement readings obtained for a pressure of 16,000 psf (766 kPa) are shown in Figure 3.13 as an example. When testing most clay soils are tested, each load is left in place for 24 hours.
- After the maximum pressure has been applied, the load acting on the soil specimen is unloaded in four decrements (e.g., down from 64,000 psf to 16,000 psf to 4000 psf).
- When a suitably low pressure is reached, the rest of the remaining load is removed in a single step and the apparatus is dismantled rapidly to minimize any additional moisture taken in by the soil.

The steps taken after the consolidation test is completed are as follows:

- The confining ring is dried on the outside, and then the ring and the soil are weighed together (156.66 g in Figure 3.14).
- The ring and soil are then oven-dried for 24 hours, and their combined dry weight is measured (137.44 g in Figure 3.14). In the example shown in Figure 3.14, the dry weight of the solids (137.44 − 77.81 = 59.63 g) is calculated, as are the volume of solids (59.63/2.80 = 21.3 cc), volume of voids at the beginning (60.3 − 21.3 = 39.0 cc), and original void ratio (39.0/21.3 = 1.83).

Because the settlements under each pressure were measured, the change in void ratio from the beginning of the test to that pressure is calculated using Eq. 3.60:

ONE-DIMENSIONAL CONSOLIDATION TEST

Loading Test Data

Load Number	10
Total Load on Sample	16,000 psf
Date Load Applied	3/26/04
Load Applied by	JDC

Time		Dial Readings	
Clock	Elapsed, min.	Original	Adjusted
10:08 a.m.	0	2199	2301
	0.1	2163	2337
	0.25	2150	2350
	0.5	2135	2365
10:09	1	2113	2387
10:10	2	2083	2417
10:12	4	2050	2450
10:16	8	2019	2481
10:23	15	2000	2500
10:38	30	1985	2515
11:08	60	1972	2528
12:08 p.m.	120	1961	2539
2:08	240	1951	2549
6:08	480	1943	2557
Tues. 7:57 a.m.	1309	1933	2567

Figure 3.13 Example data form for recording the time-settlement data during a one-dimensional consolidation test.

$$S_j = \left(\frac{\Delta e}{1 + e_0} H\right)_j \tag{3.60}$$

$$\Delta e = \frac{1 + e}{H} S \tag{3.61}$$

The total settlement S_t measured in the laboratory is the sum of the settlement of the soil S and the settlement in the apparatus S_a, so S is calculated as

ONE DIMENSIONAL CONSOLIDATION TEST

Project ______________ Boring No. ______________ Sample No. ______________

Sample Description: *Norwegian Quick Clay from Aserum*			
Consolidation Frame No.	1	Consolidation Ring No.	1
Diameter of Ring	2.500 in.	Operator:	class
Height of Ring	0.748 in.	Specific Gravity	2.80

	Before	After	Trimmings
Weight, wet soil plus tare, gm	176.50	156.66	73.42
Weight, dry soil plus tare, gm	137.44	137.44	67.97
Weight, moisture, gm	39.06	19.22	5.45
Weight, tare, gm	77.81	77.81	59.62
Weight, dry soil, gm	59.63	59.63	8.35
Water Content, *w*, %	65.8%	32.3%	65.2%

Initial Conditions	
Volume of Ring	60.3 cc
Volume of Solids = W_d/γ_s	21.3 cc
Volume of Voids	39.0 cc
Void Ratio	1.830
Volume of Water	39.0 cc
Initial S_r	100%

Figure 3.14 Sample data form for a one-dimensional consolidation test.

$$S = S_t - S_a \tag{3.62}$$

The apparatus settlement is caused by compression of the porous stones above and below the specimen (Figure 3.6) and the filter paper, and to a trivial extent by compression of other parts of the apparatus. Numerical values for S_a are usually obtained by literally performing a full consolidation test with no soil in the ring. Either a metal block is used in place of the sample or the loading cap bears directly on the lower drainage stone. The void ratios are then calculated for each consolidation pressure by subtracting Δe from e_0. A laboratory report form and data for the test specified in Figure 3.14, are shown in Figure 3.15.

In principle, calculation of the coefficient of consolidation should be simple. For any given pressure in the laboratory, say the nth pressure, the settlement at the end of the $n - 1$ load is $S_{u,n-1} = S_{0,n}$, where u again means ultimate and 0 now means time zero under the nth load. The settlement at the end of consolidation under the nth load is $S_{u,n}$. At $U = 50\%$,

ONE-DIMENSIONAL CONSOLIDATION TEST - Calculation Sheet
Height of Solids = 0.2655 inch
Initial Void Ratio = 1.830

Load No. (1)	Beam Load, lb (2)	Stress psf (3)	Decrease in Height of Voids (0.0001's) (4)	Machine Deflection (0.0001's) (5)	Net Dec. in Height of Voids $\Sigma\Delta H$ (6)	$\frac{\Sigma\Delta H}{H_s}$ (7)	Void Ratio, e (8)
1		31	8	0	8	0.003	1.827
2		62	12	3	9	0.003	1.827
3		125	42	5	37	0.014	1.816
4		250	88	11	77	0.029	1.801
5		500	286	18	268	0.101	1.729
6		1,000	1079	29	1050	0.396	1.424
7		2,000	1571	44	1527	0.576	1.254
8		4,000	1911	60	1851	0.697	1.133
9		8,000	2214	78	2136	0.805	1.025
10		16,000	2497	100	2397	0.903	0.927
11		32,000	2764	127	2637	0.994	0.836
12		64,000	3033	159	2874	1.082	0.748
12a		64,000	3098	159	2939	1.107	7.723
13		16,000	3011	118	2893	1.089	0.741
14		4,000	2950	92	2858	1.077	0.753
15		1,000	2849	69	2780	1.047	0.783
16		250	2723	56	2667	1.004	0.826
17		31	2548	48	2500	0.942	0.888

Figure 3.15 Sample data form for calculation of the void ratios at various consolidation pressures.

$$S = S_{0,n} + 0.50\,(S_{u,n} - S_{0,n}) = 1/2(S_{0,n} + S_{u,n}).$$

The engineer reads the time needed to achieve 50% consolidation, $t_{50,n}$, from the laboratory *S-t* curve. At $U = 50\%$, $T = 0.197$ (from Table 3.3), so from Eq. 3.52

$$t = \frac{T(H/n)^2}{c_v} = \frac{0.197(H/n)^2}{c_v} \tag{3.63}$$

Once c_v has been calculated, the entire theoretical *S-t* curve can be calculated by successively entering the appropriate values of T into Eq. 3.63 and plotting the theoretical curve on the same graph for comparison with the measured curve. If the theory is correct, then the theoretical and experimental curves should be identical. The theoretical curve is calculated as follows using the symbols defined previously. A table is prepared with values of U and T (from Table 3.3) in the first two columns. For each value of U, the settlement S is calculated at time t using

$$S = S_0 + U\,(S_u - S_0)$$

The corresponding values of t are computed using Eq. 3.63, where H is the average thickness of the specimen for consolidation under the given load. The *S-t* data from Figure 3.13 are plotted in 3.16. For this load, $S_0 = 2301$ dial divisions (each equal to 0.0001 in.) and $S_u = 2567$ divisions. Thus, $S_{50} = 2434$ and $H = 0.750$ in. $- (2434 - 100) \times 10^{-4} = 0.517$ in. (100 is the machine deflection from Figure 3.13). From Figure 3.16, $t_{50} = 2.9$ minutes. Thus,

$$c_v = \frac{T(H/n)^2}{t} = \frac{(0.197)(0.517\text{ in.}/2)^2}{2.9\text{ min}} = 0.00455\text{ in.}^2/\text{min}$$

The dimensionless time factors needed to calculate a theoretical curve of settlement versus time are presented in Table 3.3

The theoretical curve is shown in Figure 3.16, and it clearly does not fit the measured curve very well. The only points in common are at time zero, where both curves must start at $S = 2301$ divisions; at $U = 50\%$ where the calculations force the two curves to be identical; and at times indefinitely.

To examine the discrepancies in more detail, the data are redrawn to a semi-log scale in Figure 3.17. The most obvious discrepancy is found over long time period where the experimental curve continues sloping downward, whereas the theoretical curve should level off. Apparently, the Terzaghi model is not quite correct over long times. Apparently, the soil compresses, perhaps like a spring, but then it *creeps,* causing continued settlement at large times.

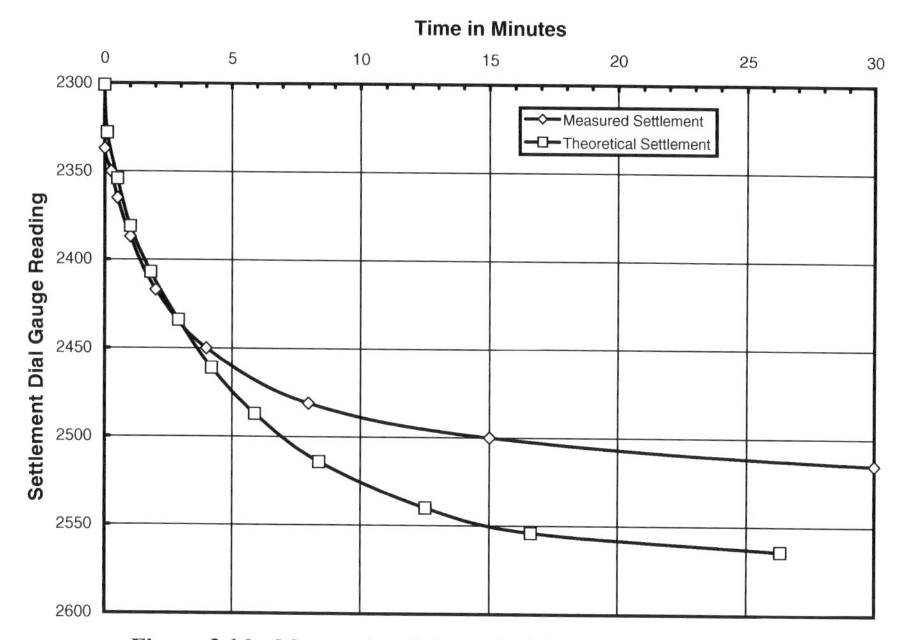

Figure 3.16 Measured and theoretical time settlement curves.

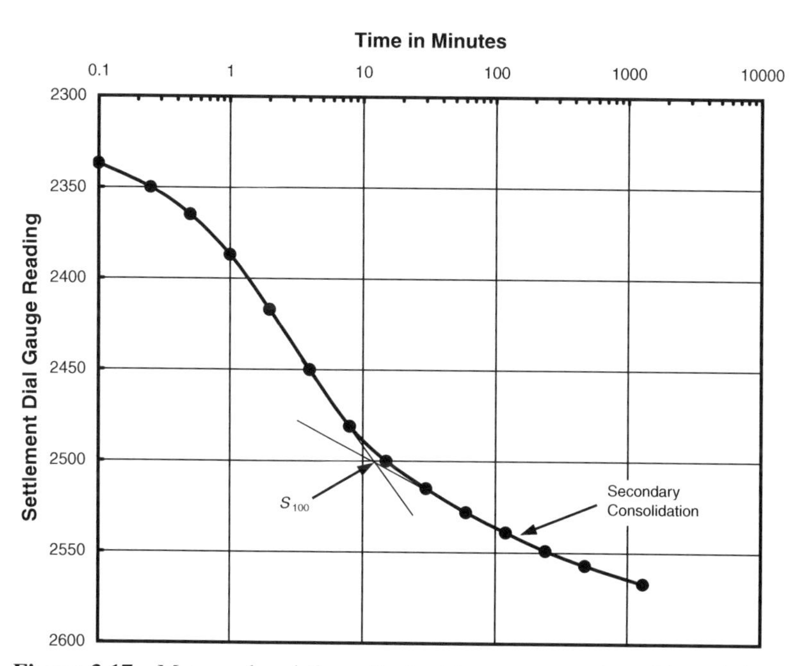

Figure 3.17 Measured and theoretical settlement versus logarithm of time.

Terzaghi recognized this effect. He termed the part of consolidation (settlement) that obeys his theory the *primary consolidation* and the postprimary creep the *secondary compression*. Casagrande and Fadum (1940) proposed a simple construction to separate primary from secondary settlements. They drew one tangent line to the secondary settlement curve and another to the primary curve at the point of inflection (where the direction of curvature reverses), as shown in Figure 3.17, and found that the intersection of these two lines defines approximately the settlement at the end of primary consolidation. This settlement is defined as $S_{100,n}$ where 100 indicated the point of 100% primary consolidation.

Further studies indicate some non-Terzaghian consolidation right at the beginning too. The correction is based on Eq. 3.56. The correction, also originated by Casagrande and Fadum (1940), is performed as follows: a point is selected on the laboratory S-log(t) curve near $U = 50\%$ (Figure 3.17), and the time is recorded as $4t$ and the settlement as S_{4t}. The point on the curve is now located corresponding to t and settlement S_t (this symbol was used previously for a different variable) is recorded. The corrected settlement at the beginning of primary settlement S_0 is

$$S_0 = S_t - (S_{4t} - S_t) \tag{3.64}$$

where the subscript n's have been left off for simplicity.

The coefficient of consolidation c_v is now calculated as before, but the $U = 50\%$ point is defined at S_{50} where

$$S_{50} = \frac{S_0 + S_{100}}{2} \tag{3.65}$$

and S_0 and S_{100} are the settlements at the beginning and end of primary consolidation. The theoretical curve calculated using this value of c_v is compared with the measured curve in Figure 3.17, and the curves compare very well for the range $20 \leq U \leq 90\%$. More complicated theories exist that fit the experimental curves better, but the mathematics associated with their use is formidable and the theories have not been used successfully in engineering practice.

3.5.3 The Consolidation Curve

Several parts of the analysis must be altered to conform to the methods normally used in engineering practice. First, the σ'-ε curve is not plotted in the conventional manner but rather is rotated, as shown in Figure 3.18, because then the strain scale reads in the same direction in which settlement occurs. Furthermore, it is common practice to plot σ' on a logarithmic scale as shown in Figure 3.19.

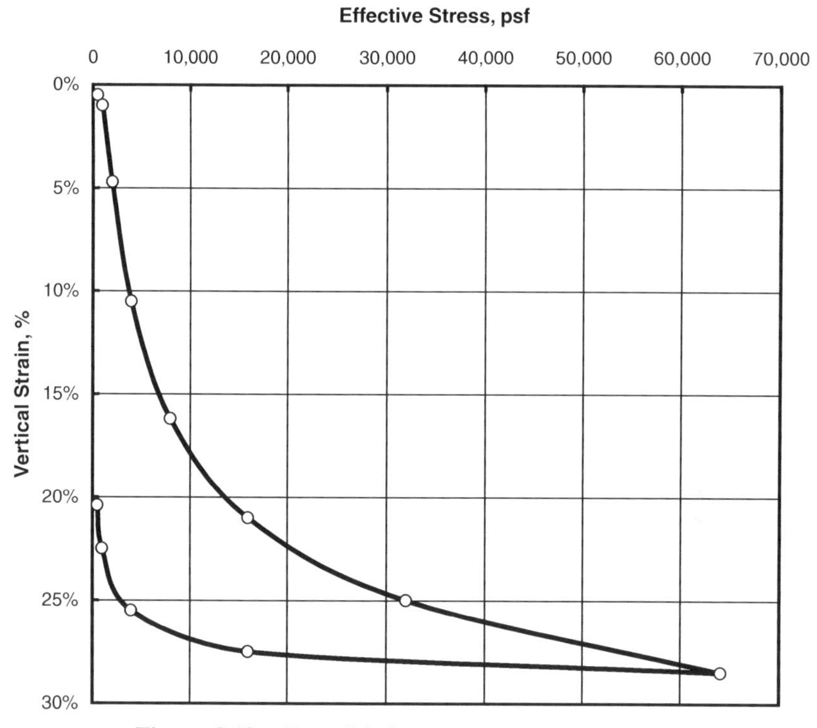

Figure 3.18 Consolidation curve on a natural scale.

Determining the vertical strain analytically rather than by reading it from a curve is sometimes convenient. If a one-dimensional consolidation test is performed on a specimen of very soft clay, one will obtain a curve such as that shown in Figure 3.20. If the soil is in equilibrium in nature at a stress such as σ'_0 and the stress is increased, by placement of an embankment to σ'_f, the curve between σ'_0 and σ'_f can be approximated by a straight line. The slope of the line is C_ε, termed the *strain index*. It is defined as

$$C_\varepsilon = \frac{\varepsilon_f - \varepsilon_0}{\log \sigma'_f - \log \sigma'_0} = \frac{\varepsilon_f - \varepsilon_0}{\log \dfrac{\sigma'_f}{\sigma'_0}} \qquad (3.66)$$

The strain caused by the loading is $\varepsilon = \varepsilon_f - \varepsilon_0$, so equations from Eqs. 3.68 and 3.66 on the settlement of a layer j with thickness H can be computed using

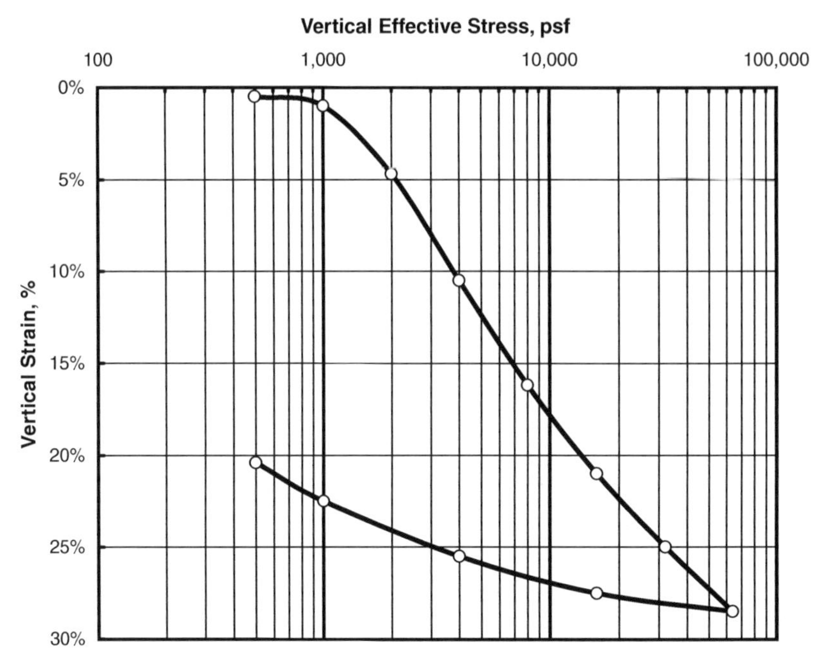

Figure 3.19 Consolidation curve on a semilog scale.

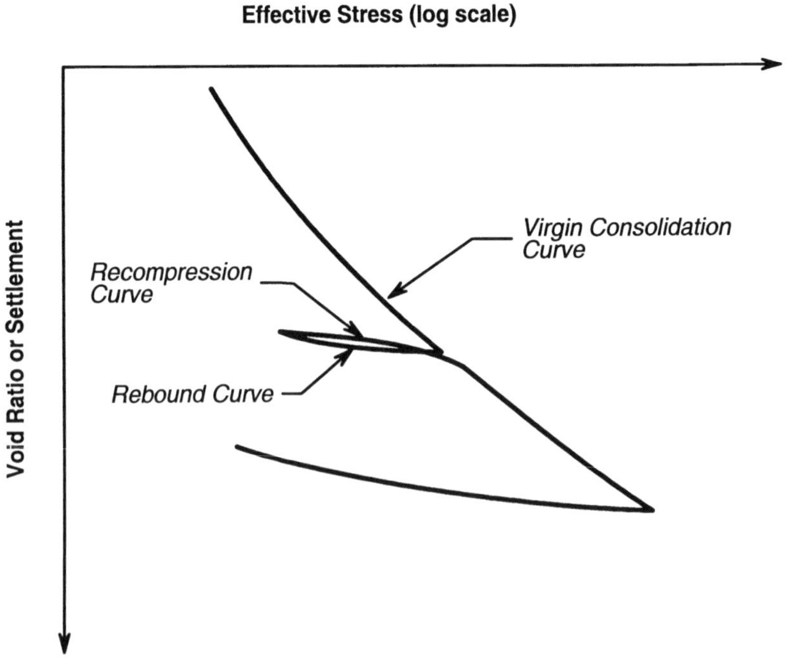

Effective Stress (log scale)

Void Ratio or Settlement

Virgin Consolidation Curve

Recompression Curve

Rebound Curve

Figure 3.20 Virgin consolidation curve for vertical strain versus logarithm of vertical effective stress.

$$S_j = \left(C_\varepsilon H \log \frac{\sigma'_f}{\sigma'_0} \right)_j \tag{3.67}$$

A soil that has been deposited from suspension and consolidated to some stress, clearly the maximum stress to which the soil has ever been subjected, is said to be *normally consolidated,* and the associated σ–ε relationship is termed the *virgin consolidation curve.*

Compression of a soil causes largely irreversible movements between particles, so if the soil is unloaded, the resulting *rebound curve,* or *swelling curve,* has a flatter slope than the virgin curve and the *recompression curve* (also called the *reloading curve*) forms a hysteresis loop with the rebound curve. Soils now existing under effective stresses smaller than the maximum effective stress at some time in the past are said to be *overconsolidated.*

If the soil is undergoing consolidation down the virgin curve, it is *underconsolidated.* Obviously, this is a transient condition unless deposition of sediment is occurring at a fast enough rate that the excess porewater pressures cannot be dissipated.

The samples tested for Figure 3.20 started as a slurry, and the strain is calculated using the initial height of the unconsolidated slurry.

3.5.4 Calculation of Total Settlement

Among the first problems is to investigate the *spring constant* of the soil. A soil sample of height H is placed in an apparatus such as the one shown in Figure 3.6. A series of loads are applied, and for each load the settlement and the elapsed time are measured when equilibrium is established. For convenience, S is divided by H to obtain strain ε and a stress-strain curve is plotted. The curve is likely to have the shape shown in Figure 3.21. The spring representing the model must be nonlinear to fit measured soil properties.

If the stress-strain curve is really representative of the soil in the field, the curve should be useful for analysis. A thin stratum of soil at depth z_j in the soil deposit is considered. The stratum has a thickness H_j, and before placement of the embankment, it is in equilibrium under a vertical stress, σ'_j. The condition of the soils is indicated by point j in Figure 3.21. If a stress $\Delta\sigma'$ from the embankment is added, the strain $\Delta\varepsilon$ results. The settlement of the surface caused by the compression of the jth layer is then

$$S_j = (\Delta\varepsilon)_j H_j \tag{3.68}$$

The settlement of the surface caused by compression of the entire subsoil is then

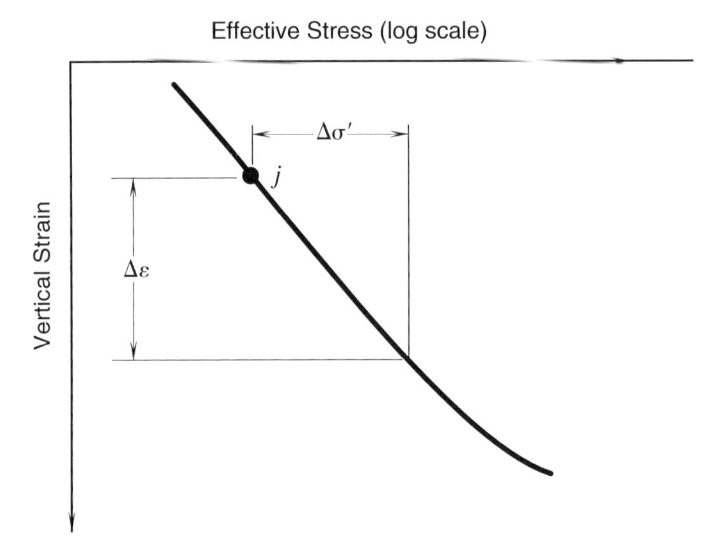

Figure 3.21 One-dimensional stress-strain curve for soil.

$$S = \sum_{j=1}^{\text{layers}} S_j \tag{3.69}$$

where

$$H = \sum_{j=1}^{\text{layers}} H_j \tag{3.70}$$

Although the foregoing analysis is simple, many necessary quantities remain undefined. For example, how is the added effective stress $\Delta\sigma'$ calculated with depth below a foundation? How is the initial effective stress calculated? How are the proper σ'-ε curves obtained? These quantities and many others must be determined before any realistic field problems can be analyzed.

3.5.5 Calculation of Settlement Due to Consolidation

Sufficient data have now been presented so that the total settlement under a wide embankment can be calculated. To review, the settlement is computed as follows:

1. The subsoil is divided into a series of layers.
2. The quantities γ, γ' value and the σ'-ε curve are measured for each layer.

3. The initial effective stresses are calculated at the center of each layer (the stress at the center may be considered an average stress for the layer).
4. The stress added by the embankment is determined.
5. The strain resulting in each layer from the applied stress is read from the appropriate σ'-ε curve.
6. The settlement of each layer is then calculated using Eq. 3.68.
7. The settlement of the original ground surface is calculated from Eq. 3.69.
8. If the settlement is large enough so that submergence effects are important, then successive trials of Steps 4 through 7 are used until a settlement of desired accuracy has been determined.

3.5.6 Reconstruction of the Field Consolidation Curve

The consolidation curve measured in the testing laboratory is not used directly for computation of total settlement because it represents the behavior of soil in the laboratory, not in the field. Instead, the laboratory curve must be corrected to obtain the *field consolidation curve*. This procedure is called *reconstruction of the field consolidation curve.* The reasons for this correction are as follows.

A soil in the field is assumed to have been consolidated to σ'_{max} and then rebounded to point A, and an embankment is assumed to be placed so that the effective stress will increase again. In calculating the settlement (Eq. 3.68), the strains must be calculated in terms of the height of the sample (or stratum j in the field) at point A, not the original height, so the slopes of the reloading are virgin curves that must be suitably corrected. Similar corrections are required if a soil sample is to be used as representative of a stratum, but does not come from the center of the stratum, and for various other applications. The occasional inconveniences due to changed total heights are minor, and some engineers use the types of graphs shown in Figure 3.19, but most geotechnical engineers prefer to avoid confusion by calculating strains in terms of a constant height. The height selected is the *height of solids* H_s, a fictitious height defined as

$$H_s = \frac{V_s}{A} \tag{3.71}$$

where V_s is the volume of solid mineral grains and A is the cross-sectional area of the test specimen. The vertical strains are calculated in terms of H_s and are

$$\begin{aligned}
\frac{\Delta H}{H_s} &= \frac{(\Delta H)A}{H_s A} \\
&= \frac{\Delta V}{V_s} \\
&= \frac{\Delta(V_v + V_s)}{V_s} \\
&= \frac{\Delta V_v + \Delta V_s}{V_s} \\
&= \frac{\Delta V_v}{V_s} = \Delta e
\end{aligned} \tag{3.72}$$

because $\Delta V_s = 0$. Thus, void ratios are plotted instead of strains, and the change in void ratio is a strain expressed in terms of the constant height of solids. The settlement equation (Eq. 3.68) becomes

$$\begin{aligned}
S &= \frac{\Delta V}{A} \\
&= \frac{\Delta V_v}{V_s} \cdot \frac{V_s}{A} \\
&= \Delta e \frac{V_s}{A} \cdot \frac{H}{H} \\
&= \Delta e \frac{HV_s}{V} \\
&= \Delta e \frac{HV_s}{V_s + V_v} \\
&= \Delta e H \frac{1}{1 + \dfrac{V_v}{V_s}}
\end{aligned}$$

Thus

$$S_j = \left(\frac{\Delta e}{1 + e} H\right)_j \tag{3.73}$$

The e-log σ' curve for an overconsolidated clay may have the appearance of the curve shown in Figure 3.22. If a structure is placed on the soil deposit so that σ'_0 increases to σ'_f, the settlement is then

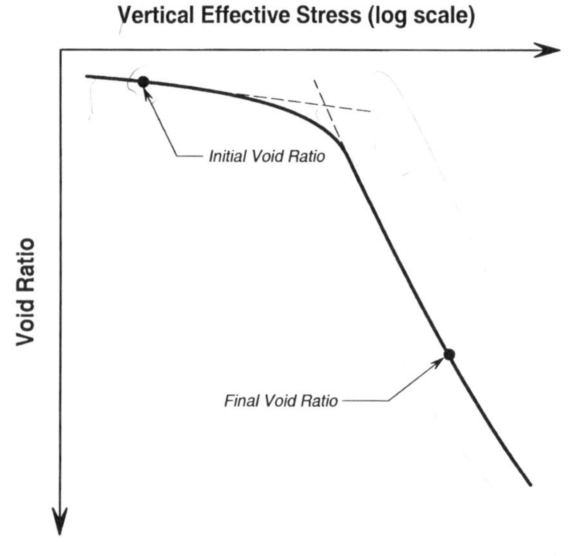

Figure 3.22 Consolidation curve for overconsolidated clay.

$$S_j = \left(\frac{e_0 - e_f}{1 + e_0} H_0 \right)_j \tag{3.74}$$

If the soil is normally consolidated and the virgin curve is approximately linear on the log scale, then the slope of the virgin curve on the e-log σ' plot C_c is defined as follows:

$$\boxed{C_c = \frac{e_1 - e_2}{\log \sigma'_2 - \log \sigma'_1} = \frac{e_1 - e_2}{\log \left| \frac{\sigma'_2}{\sigma'_1} \right|}} \tag{3.75}$$

where points 1 and 2 are both on the virgin curve. The slope, C_c, is termed the *compression index* and is defined in such a way that it is positive even though the curve slopes downward.

The change in void ratio Δe can be computed from C_c by rearranging Eq. 3.75:

$$\Delta e = e_1 - e_2 = C_c \log \left| \frac{\sigma'_2}{\sigma'_1} \right| \tag{3.76}$$

The settlement S that results from increasing the loading from σ'_1 to σ'_2 is computed by combining Eqs. 3.74 and 3.75 using

$$S = \frac{e_1 - e_2}{1 + e_1} H_1 = C_c \frac{H_1}{1 + e_1} \log \left| \frac{\sigma_2'}{\sigma_1'} \right| \tag{3.77}$$

In the usual notation, the initial condition of the normally consolidated soil is denoted by the subscript 0 and the final condition by the subscript f. The settlement for layer j is then

$$S_j = \left(C_c \frac{H_0}{1 + e_0} \log \left| \frac{\sigma_f'}{\sigma_0'} \right| \right)_j \tag{3.78}$$

If the soil is overconsolidated in the field and the applied loads will make $\sigma_f' < \sigma_{max}'$, then the settlement can be calculated analytically using an equation of the same form as Eq. 3.78, but with C_c replaced by the slope of a reloading curve C_r so that

$$S_j = \left(C_r \frac{H_0}{1 + e_0} \log \left| \frac{\sigma_f'}{\sigma_0'} \right| \right)_j \tag{3.79}$$

If $\sigma_f' > \sigma_{\max}'$, then the settlement can be calculated from point 0 to point i (Figure 3.22) and then from i to f:

$$\boxed{S_j = \left(C_r \frac{H_0}{1 + e_0} \log \left| \frac{\sigma_i'}{\sigma_0'} \right| \right)_j + \left(C_c \frac{H_0}{1 + e_0} \log \left| \frac{\sigma_f'}{\sigma_i'} \right| \right)_j} \tag{3.80}$$

The constant height of solids H_s is

$$H_s = \frac{H_0}{1 + e_0} \tag{3.81}$$

and

$$\frac{H_0}{1 + e_0} = \frac{H_i}{1 + e_i} = \frac{H_f}{1 + e_f} = H_s \tag{3.82}$$

Equation 3.79 is often used for inexpensive projects where no laboratory measurements can be afforded and a value for C_r is assumed based on previous experience from other projects in the area for which laboratory testing has been performed. Equation 3.80 is used where experience (or laboratory tests) shows that a soil has been overconsolidated by a fixed amount independent of depth—for example, by erosion of a known thickness of surface soil—so the stress, $\sigma_i' - \sigma_0'$, is known.

3.5.7 Effects of Sample Disturbance on Consolidation Properties

One problem remains: whether or not the laboratory tests give properties that can be used directly in the field. More explicitly, does taking a soil sample from the ground and putting it into the consolidation cell in the laboratory cause a change in properties? Field sampling of soils will be discussed with respect to the effects of disturbance on the results of tests in the laboratory.

The most extreme disturbance would be obtained by physically remolding a soil. An e-log σ curve for a hand-carved sample (minimum sampling disturbance) is shown in Figure 3.23 for a sample of Leda clay from Toronto, Canada, and also is a curve is shown for a sample that was completely remolded prior to testing. Clearly, using a seriously disturbed sample of this soil for a field design would lead to large errors in predicting total settlement. As a simple example, a clay is considered that is 40 ft thick in the field. To simplify the calculations (with some loss of accuracy), the layer will not be subdivided. An average initial effective stress of 1000 psf is assumed along

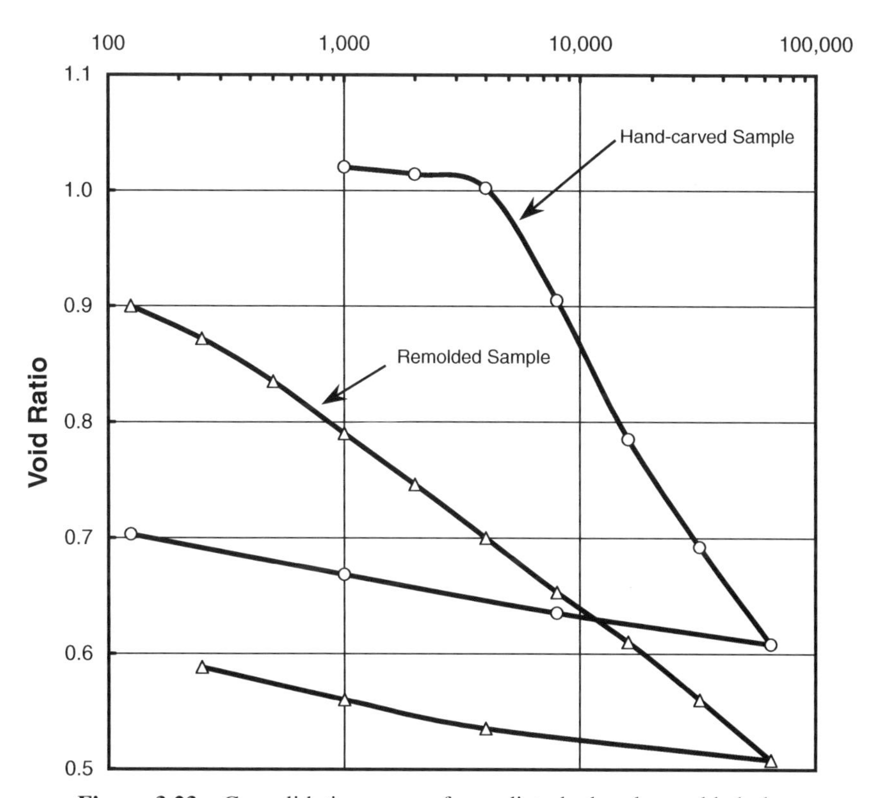

Figure 3.23 Consolidation curves for undisturbed and remolded clay.

with an increase in effective stress of 3000 psf. Eq. 3.60 will be used to calculate settlements. The calculated settlement for the hand carved specimen is

$$S = \frac{1.020 - 1.002}{1 + 1.020} 480 \text{ in.} = 4.3 \text{ in.}$$

If the curve for the remolded sample is used (which started at the same e_0 but settled under the first applied load), the settlement is

$$S = \frac{1.020 - 0.707}{1 + 1.020} 480 \text{ in.} = 74 \text{ in.}$$

Gross errors in settlement predictions are clearly possible with some soils if disturbance is severe. Most soils are less sensitive to disturbance, and careful sampling and handling will reduce the effects of disturbance further.

It would be useful if the position of the "undisturbed" consolidation curve (e-log σ') could be estimated from a test on a disturbed sample, but it seems obvious that there is no way of reconstructing the curve of the hand-carved sample, shown in Figure 3.23, from the curve of the remolded sample. A less severe example of disturbance is shown in Figure 3.24. The soil here was a red-colored plastic clay from Fond du Lac, Wisconsin. Samples were taken alternately with 3-in.- and 2-in.-diameter thin-walled samplers. The 3-in. sampler apparently caused less disturbance than the 2-in. sampler. If the sample is only slightly disturbed, it might be possible to reconstruct an approximate undisturbed (field) curve. Figure 3.25 shows the results of a one-dimensional consolidation test on clay that started as a suspension. After consolidating at 1000 psf (point 1), the sample was allowed to swell in steps back to 20 psf and was then reloaded. Note that the reloading curve makes its sharpest curvature near the maximum previous consolidation pressure ($\sigma'_{max} = 1000$ psf). Based on this fact, Casagrande (1936) recommended a construction for finding σ'_{max} (a first step in reconstructing a field curve). His construction is shown in Figure 3.26. Casagrande located the point of sharpest curvature on the e-log σ' laboratory curve and drew three lines at that point: a horizontal line, a tangent line, and a line bisecting the angle between the first two lines. Another line is drawn tangent to the virgin consolidation curve, and the intersection of this line and the bisecting line is taken as the best estimate of σ'_{max}. The method clearly does not work for badly disturbed samples, such as, the remolded samples in Figure 3.23, and the application of the Casagrande construction to undisturbed samples where the curve bends sharply down at σ'_{max} (Figure 3.23) is uncertain, so many engineers prefer to estimate the value of σ'_{max} by eye.

The next step is to calculate σ'_0 in the field. The void ratio in the field e_0 is the same as the initial void ratio in the laboratory because no change in water content is allowed during sampling or storage, and none is allowed

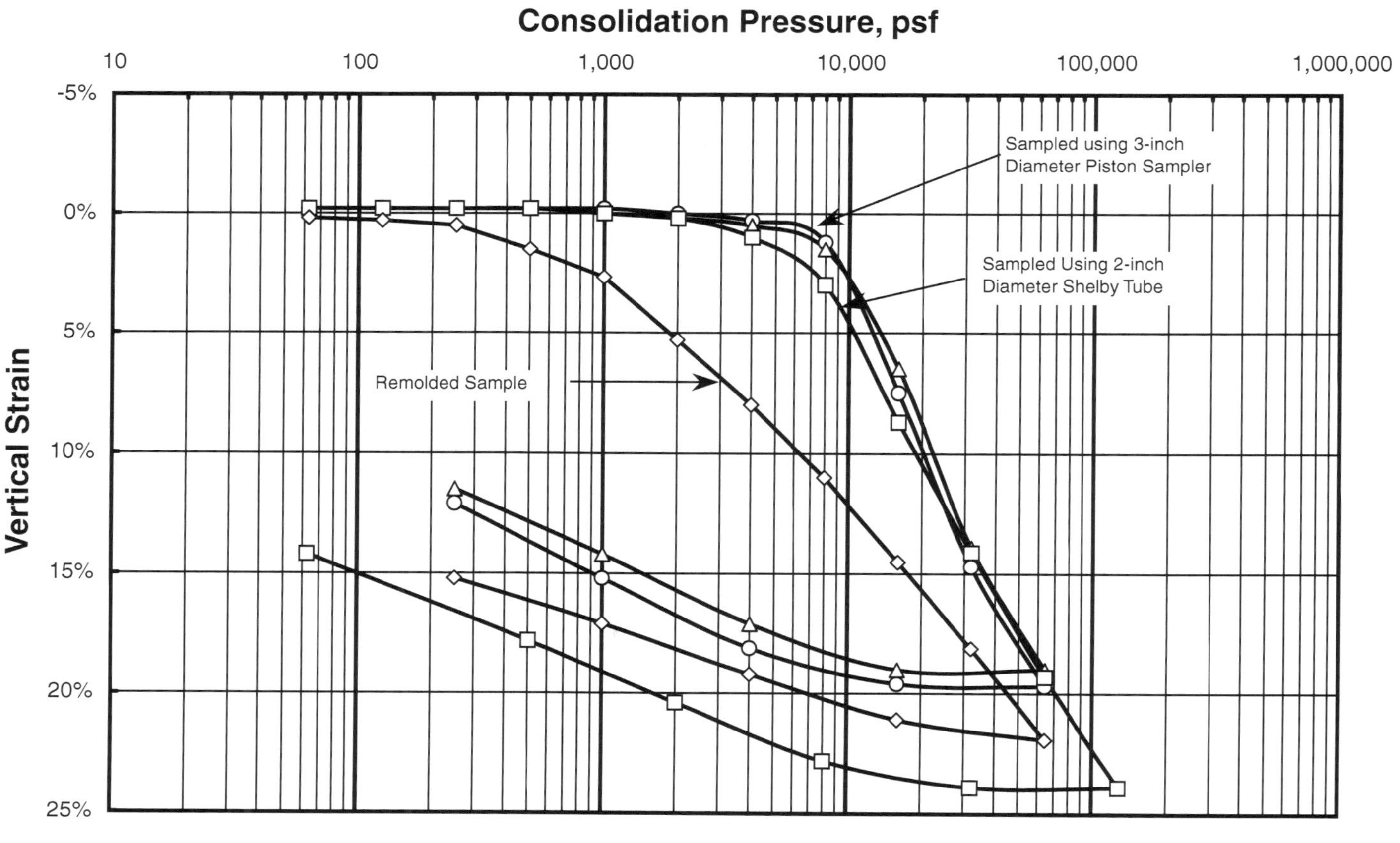

Figure 3.24 One-dimensional consolidation curves on tube samples and a remolded sample of clay.

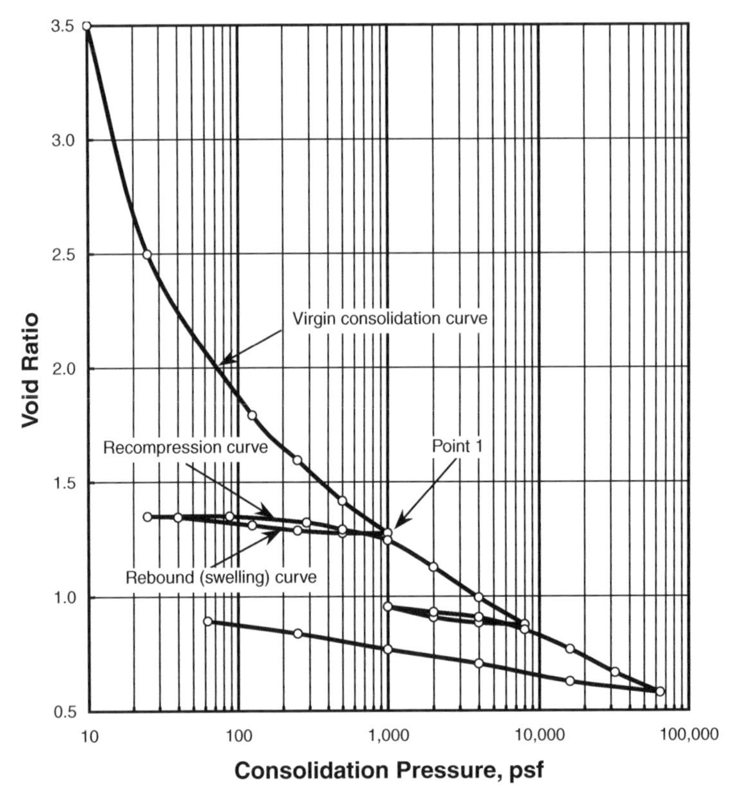

Figure 3.25 One-dimensional consolidation curve for a specimen of clay that was sedimented from suspension.

when the soil is equilibrating under the first load in the laboratory. If σ'_{max}, determined from the laboratory curve, is equal to σ'_0, the soil is normally consolidated. The field curve must then pass through the point (e_0, σ'_0) and will merge smoothly with the laboratory virgin curve.

A question arises as to how close σ'_0 and σ'_{max} are likely to be, considering the difficulty of estimating σ'_{max} from a laboratory curve and considering that the value of σ'_0 depends on the elevation of the water table, which may be difficult or (or uneconomical) to locate in the field. As one example of a relevant study, Skempton (1948) found a soil deposit known to be normally consolidated by geological evidence. He took samples, performed laboratory one-dimensional consolidation tests, and compared σ'_0 with σ'_{max}. His data are summarized in Table 3.4. Samples were taken with a 4 1/8-in.-diameter sampler with an area ratio of 23%.

If σ_0 is substantially less than σ'_{max}, then the soil is clearly overconsolidated. For example, Figure 3.27 shows the consolidation curve for stiff clay where σ_0 is about 230 psf and σ'_{max} is about 5300 psf. In this figure, the laboratory

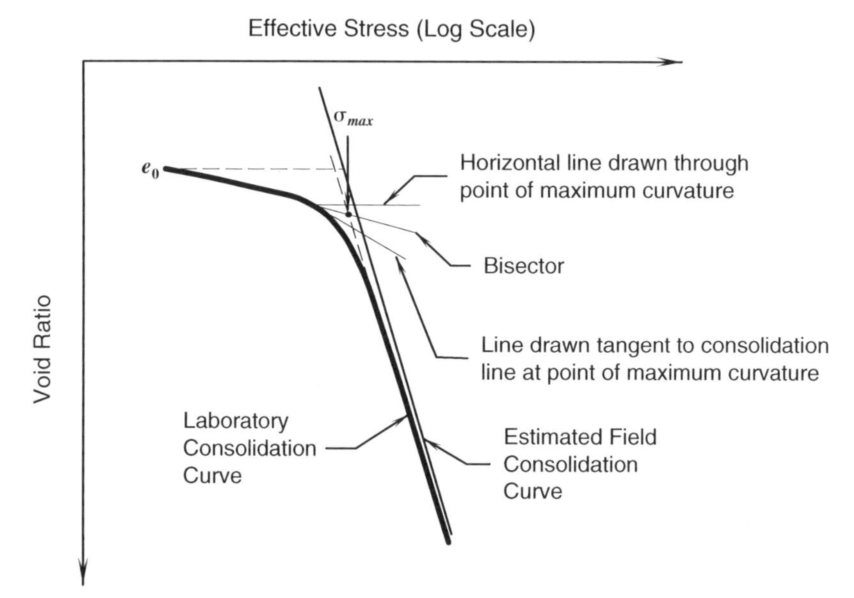

Figure 3.26 Casagrande's construction to find the maximum previous consolidation pressure.

curve passes below the (e_0, σ'_0) point, so the sample is disturbed. The procedure to use in estimating the location of the field curves is the following: the soil in the field was assumed to be consolidated to σ'_{max} and then rebounded directly to σ'_0. The slopes of the swelling curves in the field and in the laboratory are also assumed to be the same. The location of the field swelling curve is then estimated by passing a curved line, parallel to the laboratory swelling (rebound) curve, from a point on the σ'_{max} line (Figure 3.27) back to (e_0, σ'_0). The field reloading curve then starts at (e_0, σ'_0), forms a hysteresis loop with the swelling curve, passes below the point on the virgin curve corresponding to the original consolidation at σ'_{max}, and merges with the laboratory virgin curve or its extension. This construction, originated by

TABLE 3.4 Comparison of σ'_0 and σ'_{max} for a Normally Consolidated Estuarine Clay from Grosport, England

Boring	Sample	Depth, ft	σ'_0, psf	σ'_{max}, psf	$\frac{\sigma'_0 - \sigma'_{max}}{\sigma'_0}$
1	2	21	740	560	+18
1	6	61	2480	2440	+2
3	2	19	680	600	+12
3	7	56	2100	1680	+20
4	2	17	640	880	−38

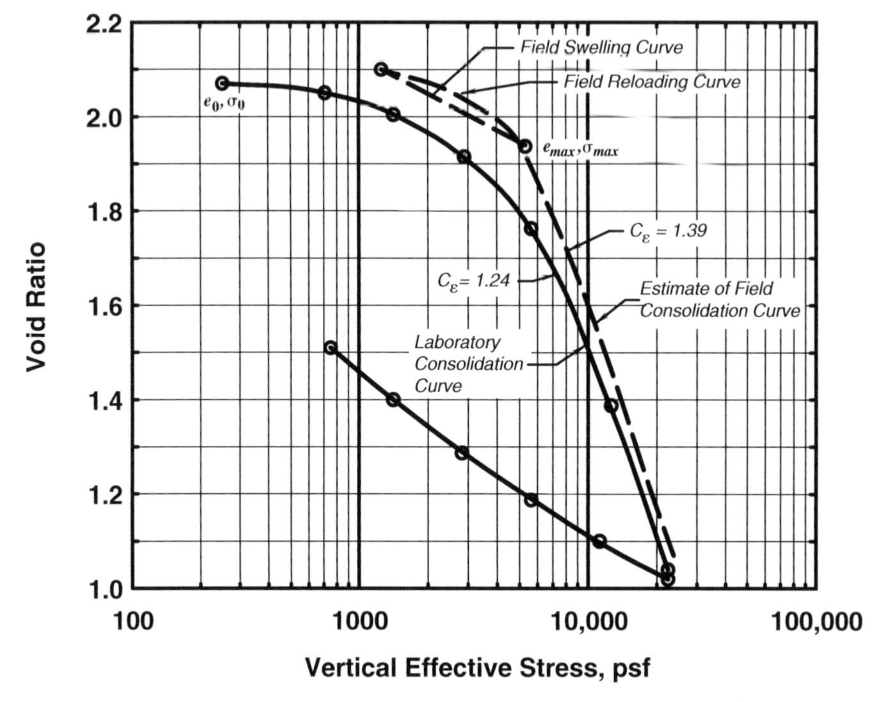

Figure 3.27 One-dimensional consolidation curve for an overconsolidated clay.

Schmertmann (1955), was used to estimate the location of the field curves in Figures 3.24 and 3.27.

The time rate of settlement is controlled by the estimated value of the coefficient of consolidation in the same way that the total settlement is controlled by the consolidation curve. The effect of complete remolding on the c_v of a sensitive clay is shown in Figure 3.28 (the same test as for Figure 3.23). At low confining pressure, remolding caused a severe reduction in c_v. If time rate or settlement calculations are important on a project, then high-quality undisturbed samples should be used for testing.

3.5.8 Correlation of Consolidation Indices with Index Tests

On small projects, it may not be economically possible to perform laboratory tests. Consequently, geotechnical design computations are carried out using estimated values for soil properties. On other projects, the cost of the structure is such that a few laboratory tests can be performed but the soils under the structure are so highly variable that no reasonable number of consolidation tests would provide the needed information. In such cases, it would be desirable if the slope of the consolidation curve could be correlated with some soil property that is simpler to measure (and less expensive). These correla-

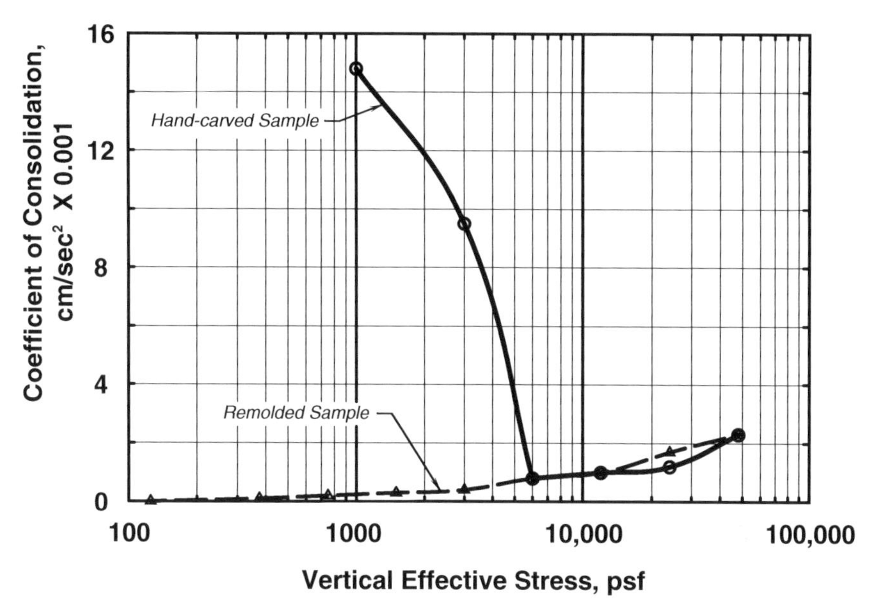

Figure 3.28 Coefficients of consolidation of leda clay.

tions may be made by an individual firm for soil deposits encountered in their locale, or correlations may be developed that apply to soils with a common geologic origin. Several examples will be cited.

Peck and Reed (1954) reported that

$$C_c = 0.18\ e_0 \quad \text{for soft clays in Chicago} \tag{3.83}$$

Moran et al. (1958) showed that

$$C_c = 0.01\ w_0 \quad \text{for highly organic clays} \tag{3.84}$$

where w_0 is the field water content in percent. Terzaghi and Peck (1967), based on a paper by Skempton (1944), showed that

$$C_c = 0.009\ (LL - 10) \tag{3.85}$$

where LL is the liquid limit.

None of these correlations are exact, but when used with the soils for which the correlation was developed, they are often accurate within ±30%. Considering that soil properties sometimes change by more than 30% within a few feet or meters or less, such correlations may be sufficiently accurate and may yield far more accurate estimates than would be obtained by investing the same money in a few consolidation tests.

A major source of uncertainty in using such correlations is the issue of whether or not the soil is normally consolidated in the field. Since the reloading index is usually four to six times less than the compression index, it is important to know whether to use the reloading index or the compression index. If no laboratory consolidation test results are available, the best estimate of whether a soil is normally consolidated or overconsolidated is obtained by comparing the strength of the soil in the field with the strength expected for a normally consolidated soil.

3.5.9 Comments on Accuracy of Settlement Computations

Accuracy of Predictions of Total Settlement One-dimensional settlement analyses are used in estimating total settlements, and the rates at which they occur, under the central parts of loads that are wide compared to the depth of the compressible layer. For soft, normally consolidated clays, the accuracy of the prediction of total settlement is mainly limited by the natural variability in soil properties coupled with a limited number of tests, as dictated by economic constraints. The accuracy of prediction is adequate for real design work.

For stiff, overconsolidated clays, the accuracy of the prediction of total settlement seems to be affected by sampling disturbance, which increases the slopes of reload curves in the laboratory. Although the predicted settlements are likely to be too large, the total settlements are small enough that they are often not a serious problem. The only serious problems are with lightly overconsolidated clays, where the engineer may be uncertain whether the loading will be entirely on a reloading curve or partially on a virgin curve.

Prediction of Time Rate of Settlement During Secondary Compression Predictions of time rates of settlement are relatively inaccurate in most cases. In the special case of saturated, homogeneous, normally consolidated clays, the prediction of the rate of primary consolidation is based on Terzaghi's theory and appears to be reasonably accurate. Further, field studies suggest that the slope of the secondary settlement curve (S-log t) in the field is about the same as in the laboratory, so the secondary settlement in the field is estimated just by plotting the primary field S-log t curve and then drawing in the secondary curve tangent to the laboratory curve, as suggested in Figure 3.29.

In most field problems, the simplifying assumptions of Terzaghi's theory are not satisfied. Estimates of the times required for consolidation in such cases are usually based on previous experience in the area, although more advanced analytical techniques than those considered here have shown much promise.

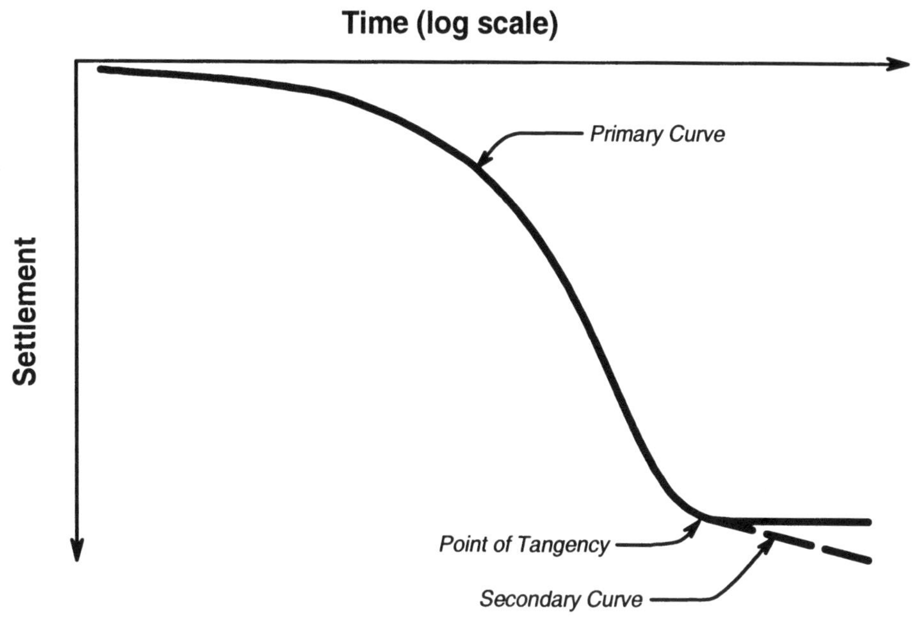

Figure 3.29 Method for predicting secondary settlements.

3.6 SHEAR STRENGTH OF SOILS

3.6.1 Introduction

The following discussion is an introduction to the shear strength of soils, a topic requiring a book to discuss thoroughly. Throughout this discussion, shear strength will be discussed in terms of the Mohr-Coulomb failure criteria: cohesion and friction. Different methods of shear testing are introduced in this discussion. The reader should learn when various shear tests may be unsuitable for characterization of shear strength for an engineering design because the particular design methodology was developed originally using a single, specific form of shear testing.

3.6.2 Friction Between Two Surfaces in Contact

This discussion of the mechanics involved in the shear strength of soils begins with fundamental concepts based on Amonton's law of friction. A block resting on a horizontal plane is shown in Figure 3.30. A vertical force N and a shearing force F are applied to the block. If the shearing force F is gradually increased until sliding occurs, then F exceeds the friction force holding the block in place. The magnitude of F is

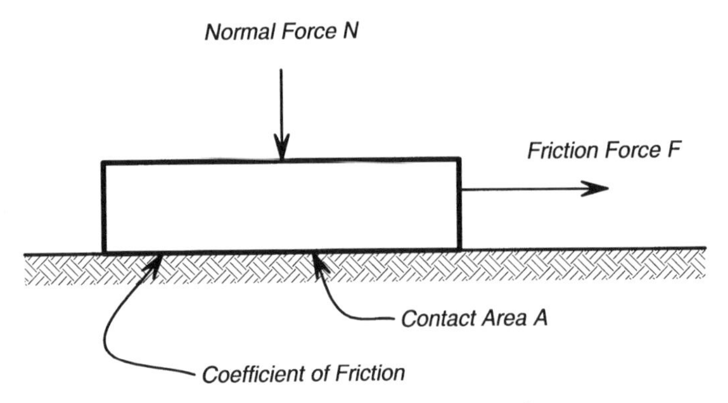

Figure 3.30 Block sliding on a plane.

$$F = \mu N \tag{3.86}$$

where μ is the coefficient of friction between the block and the plane. Dividing both sides of Eq. 3.86 by the area of the block, A, we obtain

$$\frac{F}{A} = \frac{N}{A}\,\mu \tag{3.87}$$

or, in terms of the shear and normal stresses, τ and σ,

$$\tau = \sigma\,\mu \tag{3.88}$$

The angle of the resultant force or stress acting on the plane is denoted by the Greek letter ϕ, shown in Figure 3.31. Eq. 3.88 can be written in the form

$$\tau = \sigma_n \tan(\phi) \tag{3.89}$$

The above equation shows that the shear stress at failure is proportional to the applied normal stress and that the relationship between these stresses is a straight line inclined at the angle ϕ, as shown in Figure 3.32.

Relationship Between Principal Stresses at Failure. The relationship between the principal stresses at failure when the failure envelope has a cohesion intercept is obtained by solving Eq. 3.94 for the desired principal stress and then using the trigonometric identity

$$\frac{1 + \sin\phi}{1 - \sin\phi} = \tan^2\left(45^\circ + \frac{\phi}{2}\right) \tag{3.90}$$

The resulting equations are

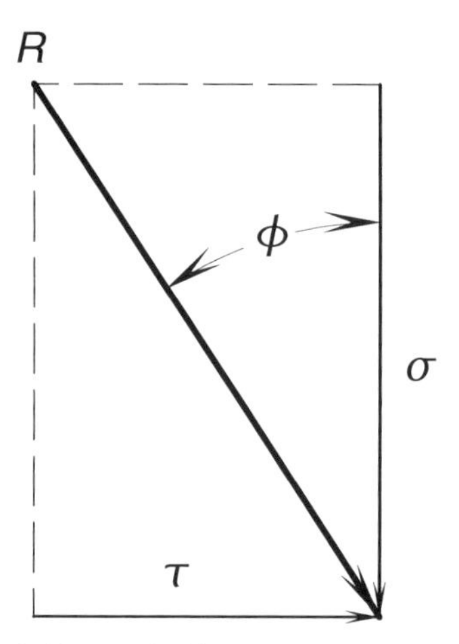

Figure 3.31 Definition of the friction angle.

$$\sigma_1 = \sigma_3 \tan^2\left(45° + \frac{\phi}{2}\right) + 2c \tan\left(45° + \frac{\phi}{2}\right) \tag{3.91}$$

$$\sigma_3 = \sigma_1 \tan^2\left(45° - \frac{\phi}{2}\right) + 2c \tan\left(45° - \frac{\phi}{2}\right) \tag{3.92}$$

Equations 3.99 and 3.100 apply to circles tangent to linear failure envelopes whether the envelopes are expressed in terms of total stresses or effective stresses, provided, of course, that the symbols are adjusted for the type of stress used and for the type of test.

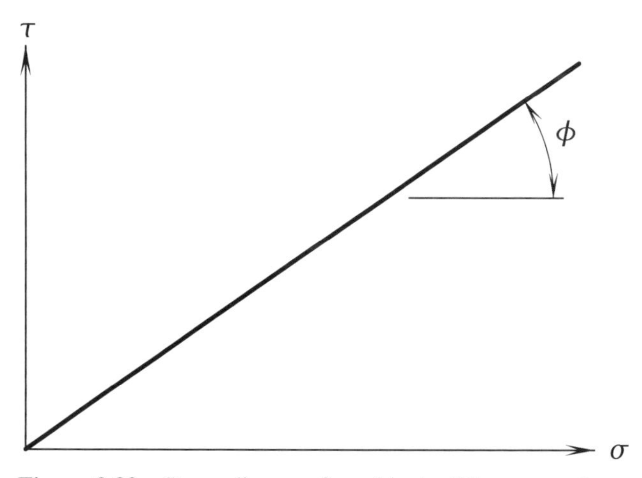

Figure 3.32 Stress diagram for a block sliding on a plane.

3.6.3 Direct Shear Testing

A method for testing soils that is similar to the sliding block test is the direct shear test. A cross-sectional view through a direct shear apparatus is shown in Figure 3.33. In this test, the soil is placed in a box that is split horizontally in a manner that allows the upper and lower halves to be displaced relative to one another while a vertical stress is applied to the upper surface of the test specimen. The shape of the shear box may be square or circular in plan. The soil may be consolidated prior to shearing if the soil being tested is clay. During the test, the horizontal displacement of the shear box is increased and the shear force is measured until failure occurs or until maximum displacement is reached.

If a series of direct shear tests are performed on a dry sand using various vertical pressures, the shear stress at failure can be plotted versus the vertical normal stress in a diagram like that shown in Figure 3.34. By analogy with the block sliding on the plane, the slope of the line in this diagram is designated by the Greek letter ϕ, and the angle is called the *angle of internal friction.*

The typical results for a direct shear test series on stiff clay are shown in Figure 3.35. In this diagram, the shear strength of the soil consists of two parts. The first part is indicated by the intercept of the vertical axis, labeled c, called the *cohesion.* The second part is indicated by the slope of the line

Figure 3.33 Direct shear test.

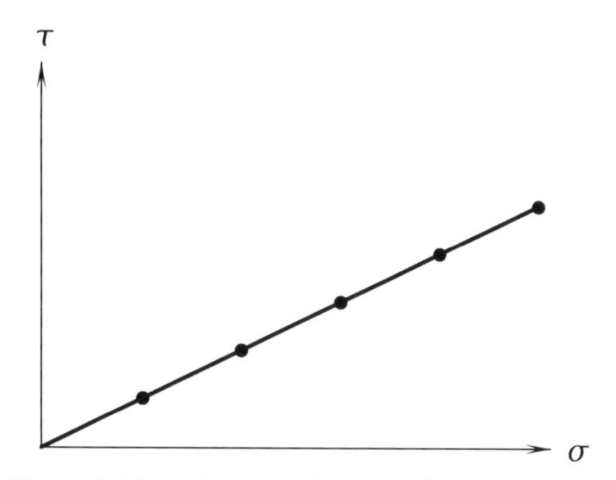

Figure 3.34 Failure envelope for direct shear tests.

and is called the *internal friction.* The shear strength of the soil is then given by the equation

$$\tau = c + \sigma_n \tan(\phi) \tag{3.93}$$

The lines in Figures 3.34 and 3.35 represent the relationship between shearing stress and normal stress at failure. It is not possible to have a stable state of stress if values plot above these lines. Because these lines envelop all stable states of stress, they are called the *failure envelope.*

3.6.4 Triaxial Shear Testing

Triaxial shear tests are performed on solid cylindrical specimens of soil. The height of the test specimen is usually about twice its diameter. The diameter varies from about 1.3 in. or (33 mm) to 4 in. (100 mm) for more common specimens. In a typical *triaxial cell,* the soil specimen is held between the

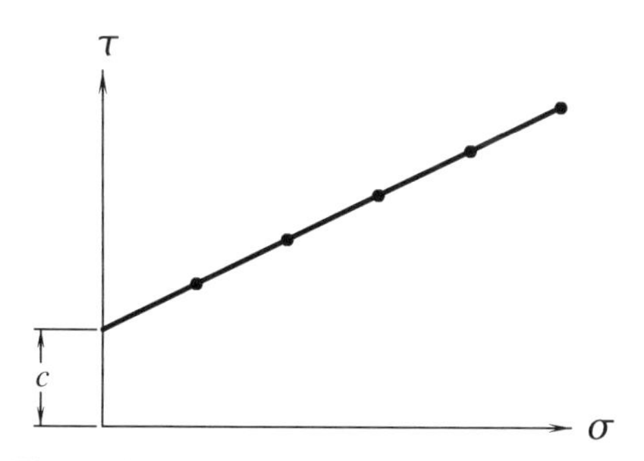

Figure 3.35 Direct shear envelope on clay.

base pedestal and the *top cap* of the triaxial cell and is laterally confined in a thin, impervious rubber membrane. The membrane is sealed to the top cap and base pedestal using silicone grease and rubber O-rings. A photograph of a modern triaxial cell is shown in Figure 3.36. A more detailed discussion of the design and construction of triaxial cells and associated equipment is presented by Bishop and Henkel (1957) and by Andersen and Symons (1960).

The pressure in the triaxial cell confines the specimen under a hydrostatic stress. It is not possible to develop a shear stress on the rubber membrane

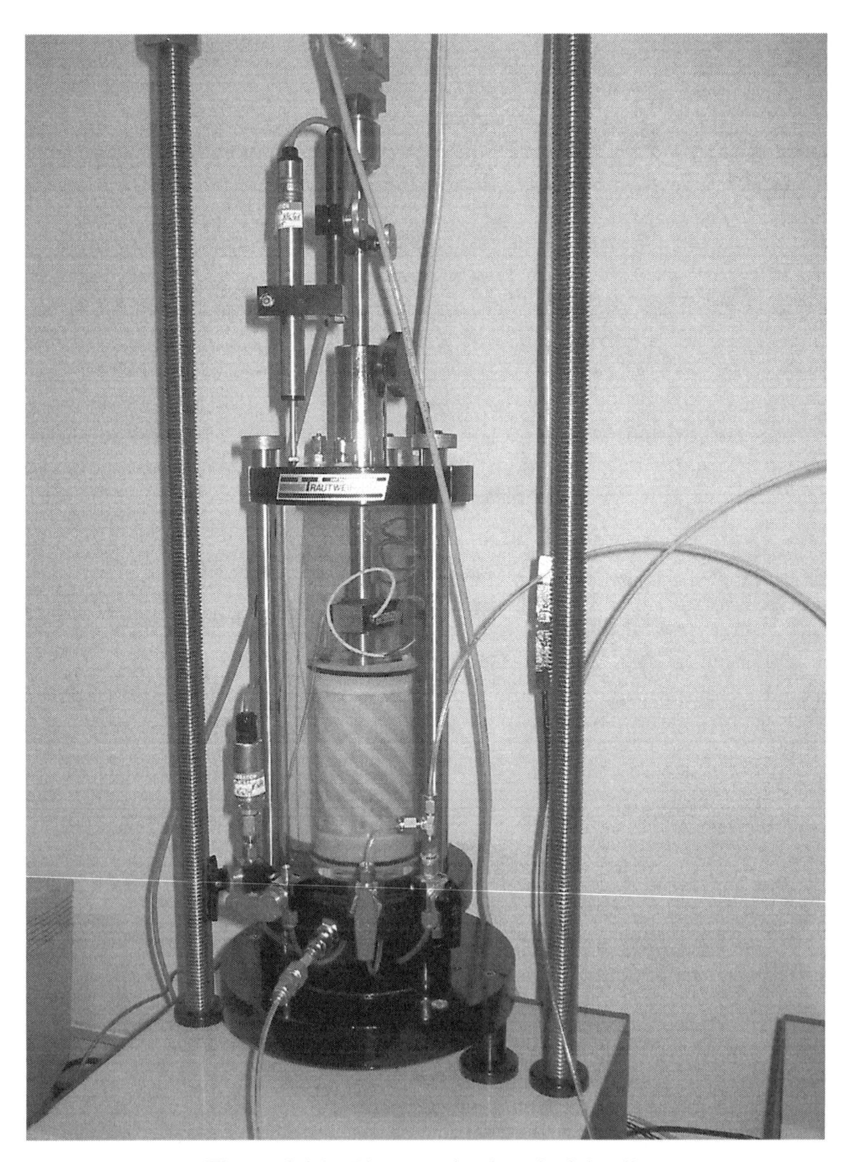

Figure 3.36 Photograph of a triaxial cell.

covering the sides of the specimen because it is flexible; thus, the exterior vertical surface of the specimen is a principal surface. If the vertical surfaces of the specimen are principal surfaces, then any horizontal plane through the specimen is also a principal surface. When a compressive load is applied through the loading piston, the vertical stress acting on horizontal planes is the maximum principal stress (σ_1) and the horizontal stress acting on vertical planes is the minimum principal stress (σ_3). In the triaxial test, the intermediate principal stress (σ_2) is equal to the minor principal stress. The axial stress applied to the soil specimen by the loading piston is ($\sigma_1 - \sigma_3$), and this quantity is called the *principal stress difference*.

If the loading piston is attached to the top cap, it is possible to apply a tensile load to the specimen and make the horizontal stress the major principal stress. This makes all vertical planes the major principal planes, with the intermediate principal stress equal to the major principal stress. Tests performed in this manner are called *extension tests*. In extension tests, the stress on horizontal planes is still compressive; an extension test is not a tensile test.

Triaxial tests are performed in two or more stages. In the first stage, the specimen is subjected to an initial state of stress. This stress may be hydrostatic (also called *isotropic*) or may be made to simulate the in situ state of stress by using different values for the vertical and radial stresses (this state is called *anisotropic*). For simplicity in the following discussion, the assumption is made that the initial state of stress is hydrostatic.

The triaxial test specimen may be allowed to consolidate after the confining stress is applied. If consolidation is permitted, multiple stages of consolidation pressures may be used. The specimen is sheared in the final stage and may or may not be allowed to drain during shearing. If the specimen is allowed to drain, it is possible to measure any volume change in the test specimen during the shearing stage.

There are three possible types of triaxial tests, depending on the combination of drainage conditions during the application of confining stresses and during shear. The three types of tests are consolidated-drained, consolidated-undrained, and unconsolidated-undrained. Short descriptions of these types of triaxial tests follow.

Consolidated-Drained Triaxial Tests In this type of test, the soil specimen is allowed to consolidate completely under the initial state of stress prior to initiating shear. During shearing, either the axial deformation is applied so slowly that the porewater pressures have time to dissipate (the drains are left open) or else the axial stress is increased in small increments and each increment of pressure is maintained constant until the porewater pressure has dissipated. The amount of time required to shear the soil while allowing for dissipation of porewater pressures is determined from the consolidation properties of the test specimen. Common names for this type of test include *consolidated-drained* (*CD-test*), *drained* (*D-test*), and *slow* (*S-test*).

The amount of time required to complete a consolidated-drained test is long compared to that of the other two types of triaxial tests.

Consolidated-Undrained Triaxial Tests In this type of test, the soil specimen is again allowed to consolidate fully under the initial state of stress in manner similar to that used in the consolidated-drained test. During shear, however, the drainage lines are closed and the specimen is sheared to failure under undrained conditions. This test is commonly referred to as a *consolidated-undrained test* (*CU-test*), a *consolidated-quick test* (*CQ-test*), or an *R-test.*

There are two varieties of consolidated-undrained tests. In one type, the specimen is sheared quickly in about 10 to 15 minutes. In the second type, the porewater pressures developed in the soil specimen are measured using porewater pressure transducers connected to drainage lines from the specimen. When the porewater pressures are measured, it is necessary to shear the specimen slowly enough for the porewater pressures to equilibrate throughout the test specimen. The additional time required for shearing is determined from the time needed to complete consolidation of the test specimen prior to shearing. While the time required for porewater pressure equalization in the specimen is often on the order of several hours, the testing time is usually much shorter than the time required to complete shearing in the consolidated-drained test.

Unconsolidated-Undrained Triaxial Tests In this type of test, the soil specimen is not allowed to consolidate under the initial state of stress or to drain during shear. This test is commonly called an *unconsolidated-undrained test* (*UU-test*) or a *quick test* (*Q-test*).

Unconsolidated-drained tests cannot be performed because consolidation would occur whenever the drained specimens were opened during the shearing stage.

Test Nomenclature. The above notations were developed at various institutions at various times. The *quick* versus *slow* type designation was developed at Harvard University during the 1930s, when quick tests meant that the specimen was sheared too quickly for the porewater to get out even though the drainage lines were open throughout the test. Drained test were performed slowly, so they were called *slow tests.* Actually, drained tests can be performed quite rapidly if the soil has high permeability (hydraulic conductivity), and undrained tests can be performed very slowly—for example, to study the creep-deformation properties at constant water content—so the quick versus slow designation soon became meaningless, though still widely used. Between World War II and the early 1960s, much of the important research on the shearing properties of soils was performed at the Imperial College of Science and Technology and at the Norwegian Geotechnical Institute, where the more descriptive terms—*drained* versus *undrained* and *consolidated* versus *unconsolidated*—were developed. These latter terms have a disadvantage in conversation since the prefix *un-* may be either slurred, and thus missed, or else emphasized so much that the sentence is disrupted. Confusion in spoken discussion can be avoided by using the terms proposed by Casagrande (1960):

- Q-test = unconsolidated-undrained (does not mean quick)
- R-test = consolidated-undrained.
- S-test = consolidated-drained (does not mean slow)

The letter R was selected by Professor Arthur Casagrande of Harvard University for consolidated-undrained tests because it is between the letters Q and S in the alphabet. Casagrande's terminology has not been accepted by all geotechnical engineers for a variety of reasons. The reader should be aware of the terms that are commonly used to refer to the different types of tests and to understand the procedures employed in the testing.

In this text, Q, R, and S will be used with symbols representing various soil parameters to designate the type of test used to define the parameter. A bar over the symbol or an apostrophe will designate that effective stresses were used. Thus, ϕ'_R and c'_R are the effective stress angle of internal friction and cohesion obtained using R-type triaxial tests.

3.6.5 Drained Triaxial Tests on Sand

The changes in stress conditions in a sand specimen during shearing may be understood by plotting the states of stress on a Mohr-Coulomb diagram. During Stage 1, a hydrostatic pressure is applied and the specimen is allowed to consolidate completely. The three principal stresses are all equal to σ_3, and Mohr's circle plots as a single point in Figure 3.37a.

In Stage 2a (Figure 3.37b) a stress difference is applied representing an intermediate state in loading where the applied stress difference is not large enough to cause failure. Shear stresses exist on all inclined planes through the specimen. In Stage 2b (Figure 3.37b), the axial stress has been increased to its maximum value and the soil specimen fails.

If such tests are performed on a series of identical specimens under various cell pressures and all failure circles are plotted on a single diagram, a single

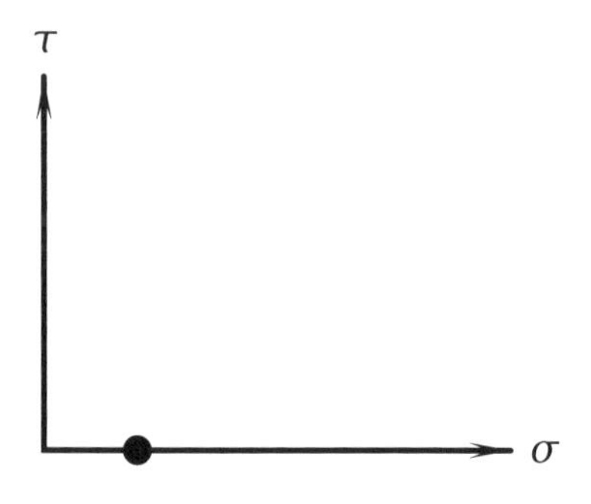

(a) Initial Isotropic State of Stress

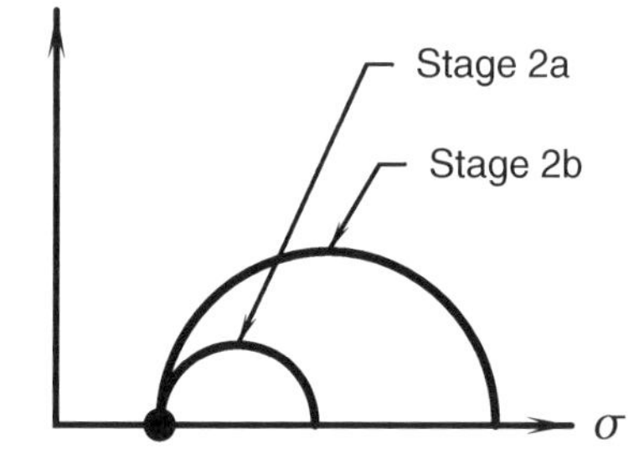

(b) Application of Stress Difference

Figure 3.37 Mohr diagram for isotropic state of stress.

line can be drawn from the origin that is tangent to each circle. Such a series of circles is shown in Figure 3.38. Specimens of sand do not fail until some part of the stress circle touches this line. Therefore, this line is called a *failure envelope.* By analogy with the direct shear tests and, more exactly, by consideration of the stresses on the planes of failure, the slope of this line represents the obliquity of the resultant stress on the failure plane. As before, the slope of this line is called the *angle of internal friction* of the sand.

The relationships between stresses, strains, and the angle of internal friction vary from one sand to another and with the density of an individual sand. At the levels of confining stress usually encountered in foundation engineering, angular sands tend to have higher angles of internal friction than sands with rounded grains. Dense sands have higher angles of internal friction than loose sands.

A classic study of the shearing behavior of fine sands was reported by Bjerrum et al. (1961). The study concerned both the drained and undrained behavior of a fine sand that is widely found in Norwegian fjords. It was found that when a dense sand is sheared under drained conditions, it usually undergoes a small decrease in volume at small strains, but the denseness of the packing prevents significant volume decreases. As the sand is strained further, the grains in the zone of failure must roll up and over one another. As a result, the sand expands with further strain. The relationship between volumetric and axial strain for different densities of sand is shown in the lower half of Figure 3.39. Loose sands undergo a decrease in volume throughout the shear test.

The expansion of dense sand with increasing strain has a weakening effect and the specimens fail at relatively low strains, as seen in Figure 3.40. The densification of initially loose specimens with increasing strain makes such sands *strain-hardening* materials, and failure occurs at larger strains.

The relationship between angle of internal friction and initial porosity for fine quartz sand is shown in Figure 3.41. For this sand, the angle of internal

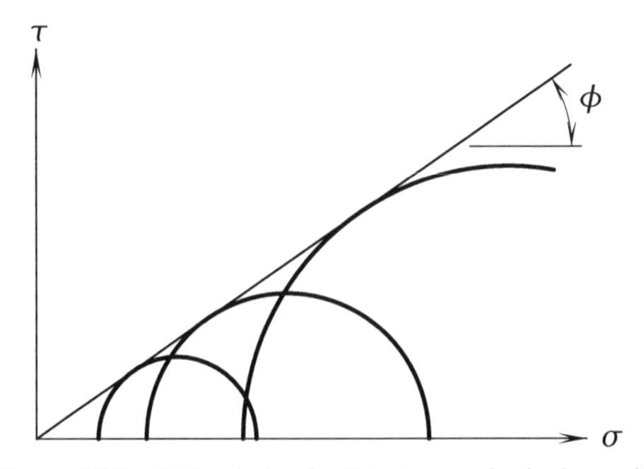

Figure 3.38 Mohr circles for *S*-tests on cohesionless soil.

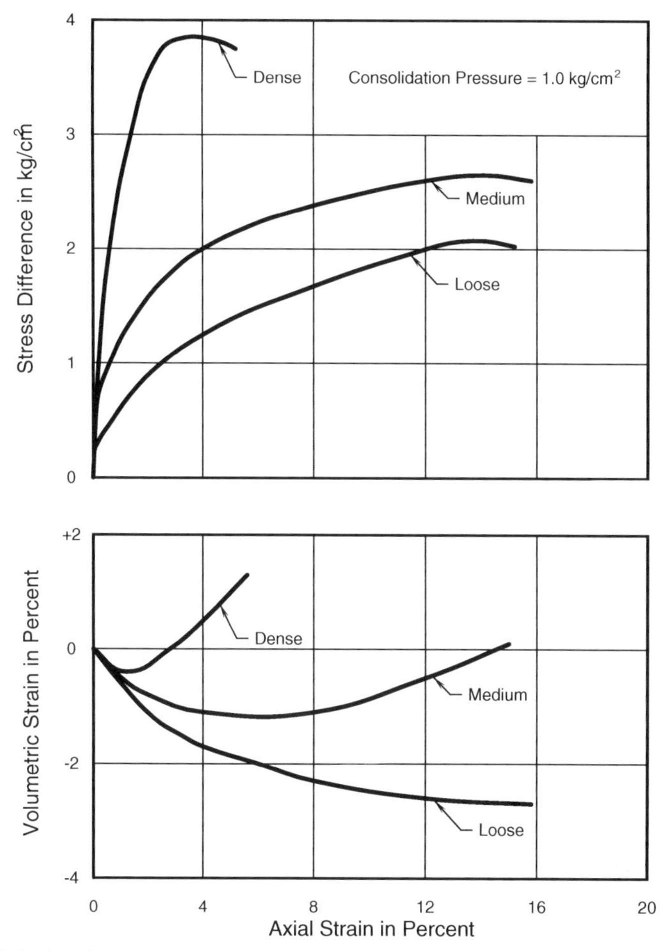

Figure 3.39 Typical volume change versus axial strain curves for fine sand (from Bjerrum et al., 1961).

friction decreases gradually until the initial porosity is about 46 or 47% and then decreases rapidly. Sand with such high initial porosities is not common, but the occasional catastrophic landslides in deposits of loose sand suggest that low angles of internal friction can occur in nature.

The data for porosity shown in Figure 3.41 were converted to void ratios and redrawn in Figure 3.42. Also drawn in this figure are vertical lines denoting the ranges of relative density for the sand that was tested. The maximum and minimum void ratios for relative density were evaluated on dry samples of sand and the shear tests were performed on saturated sand. For these data, the angle of friction is found to drop when the void ratio is larger than the maximum value measured on dry sand for relative density evaluations. These extremely low friction angles were possible because the test

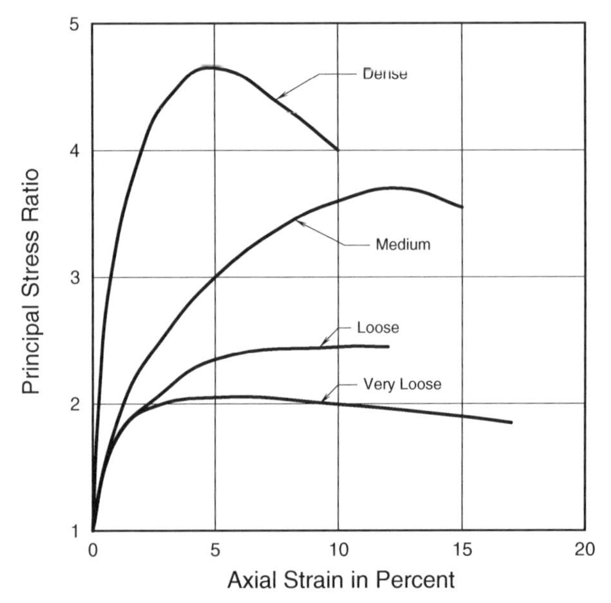

Figure 3.40 Typical axial stress difference versus axial strain curves for fine sand (from Bjerrum et al., 1961).

specimens were formed by deposition in water. Thus, the loosest sand specimens were normally consolidated and never subjected to the higher stresses than can exist in dry sand. The lesson to be learned from this study is that the range of friction angles that may exist for saturated sands in nature can be both larger and smaller than the values measured in the laboratory if the sand is permitted to dry.

The application of shearing deformation results in densification of loose sands and expansion of dense sands. Available data suggest that, for a given sand and given confining pressure, the density at high strains is the same for all specimens. Analysis of the stress-strain curves suggests that loose sands get stronger with strain (because of an increase in density) and approach a given strength. The data suggest that the strength of a dense specimen of the same sand peaks at low strains where interlocking of grains is maximum and then decreases and approaches the same limiting level of strength as for the loose sands. The strength of dense sands decreases toward the limiting level and the strength of loose sands increases toward the limiting level. The strength of a soil at large strains, where neither volume nor strength is changing with strain, is termed the *ultimate strength*.

3.6.6 Triaxial Shear Testing of Saturated Clays

Triaxial Compression Tests on Normally Consolidated Specimens

Drained Tests. Drained (S-type) triaxial compression tests on saturated specimens are performed in the same manner in which they are performed for

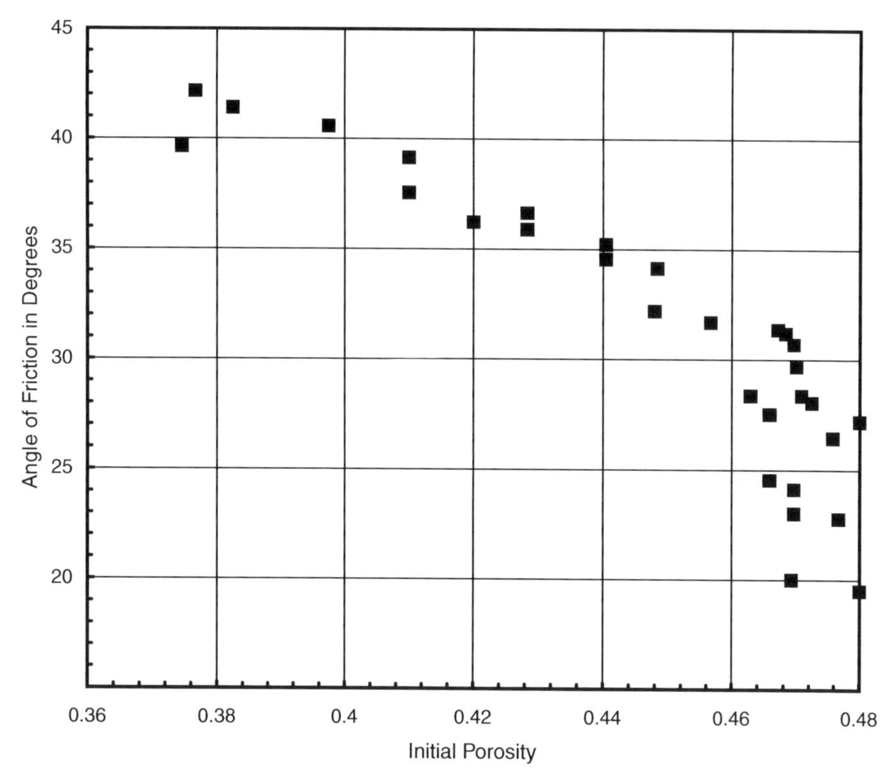

Figure 3.41 Angle of friction versus initial porosity for fine sand (from Bjerrum et al., 1961).

sand, with the allowance of greater durations of time that allow adequate dissipation of the excess porewater pressures generated during shear. For tests performed using a constant rate of deformation, the time to failure may vary from a minimum of about 6 hours to over 6 months, depending on the coefficient of permeability of the clay.

The type of Mohr-Coulomb diagram obtained from such tests is shown in Figure 3.43. Normally consolidated clays are consolidated from dilute suspensions that have essentially zero shear strength. Hence, the application of a very small confining pressure results in the development of a very small strength, and the failure envelope passes through the origin. The slope of the failure envelope is denoted by ϕ'_S—the effective stress angle of internal friction determined using S-type tests.

The curves for compressive stress–axial strain and for volumetric strain–axial strain for drained tests on specimens of saturated, normally consolidated clay are similar to the curves for sand shown in Figure 3.39, except that the specimens of clay may undergo volume decreases of more than 10% during shear. The decrease in volume during shear means that the relatively loose soil structure is breaking down under the action of shearing deformation.

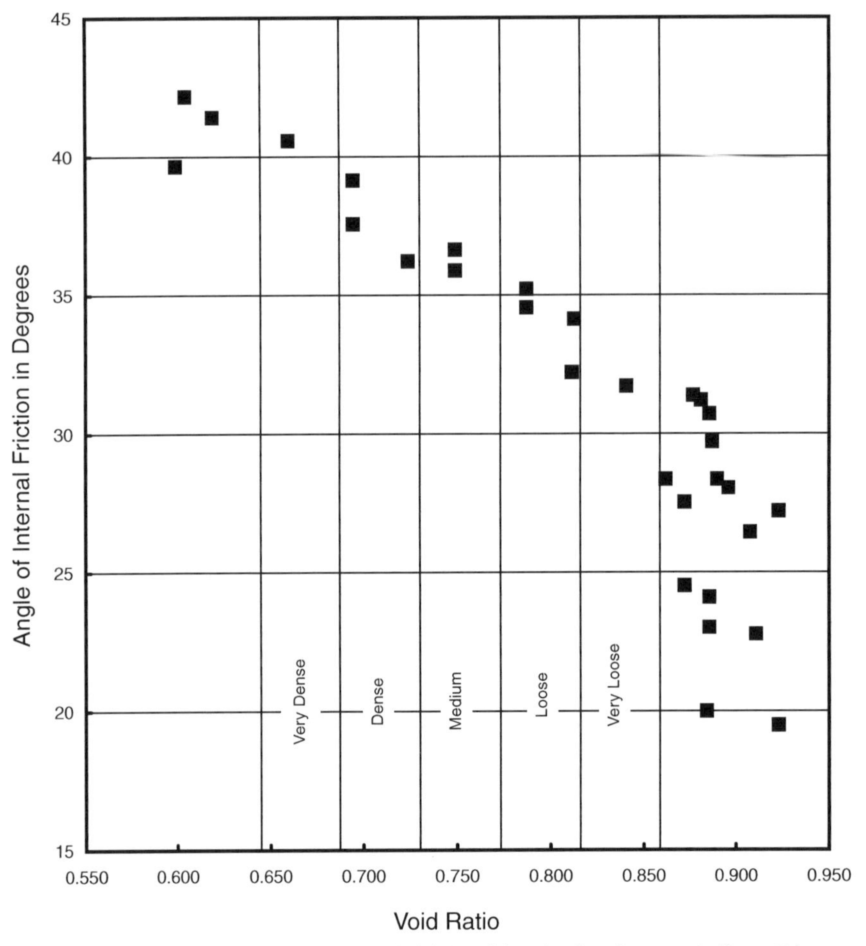

Figure 3.42 Angle of friction versus initial void ratio for fine sand (from Bjerrum et al., 1961).

Unconsolidated-Undrained Tests, $\phi = 0$ Case. If a saturated specimen of soil is subjected to a change in hydrostatic total stress under undrained conditions, such as by increasing the pressure in a triaxial cell without allowing the specimen to drain, the porewater pressure changes by an amount equal to the change in hydrostatic pressure. Direct observations of this behavior are possible. A soil specimen is assumed to be taken from the field and placed in a triaxial cell. An initial cell pressure in the specimen is measured. The difference between the cell pressure and the porewater pressure is the effective stress. If the cell pressure is increased, the porewater pressure will increase by an equal amount, and no change in effective stress will occur if both the specimen and the measuring system are saturated. Thus, according to the

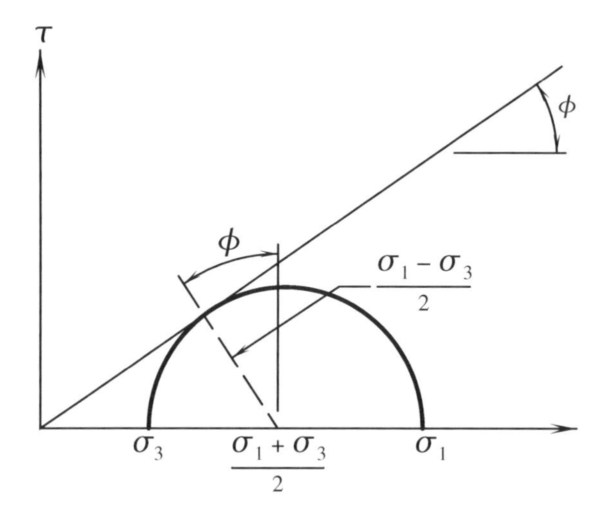

$$\sin\phi = \frac{\sigma_1 - \sigma_3}{\sigma_1 + \sigma_3}$$

$$\frac{\sigma_1}{\sigma_3} = \frac{1 + \sin\phi}{1 - \sin\phi} = \tan^2\left(45° + \frac{\phi}{2}\right)$$

Figure 3.43 Mohr's circles for *S*-tests.

principal of effective stress, no change in shearing strength is possible. The confining pressure in the triaxial cell, therefore, has no effect on the shearing strength of the soil. A total stress failure diagram for a saturated soil is shown in Figure 3.44. The dashed circle is the Mohr's circle for effective stress at failure. The circles shown with solid lines are the total stress circles of stress. The shear strength is independent of the cell pressure, all these circles of stress have the same diameter, and the tangent to the failure envelope is horizontal; thus, $\phi = 0$. Actual experimental data have been published for both sands and clays showing that the angle of internal friction is zero experimentally as well as theoretically (Bishop and Eldin, 1950).

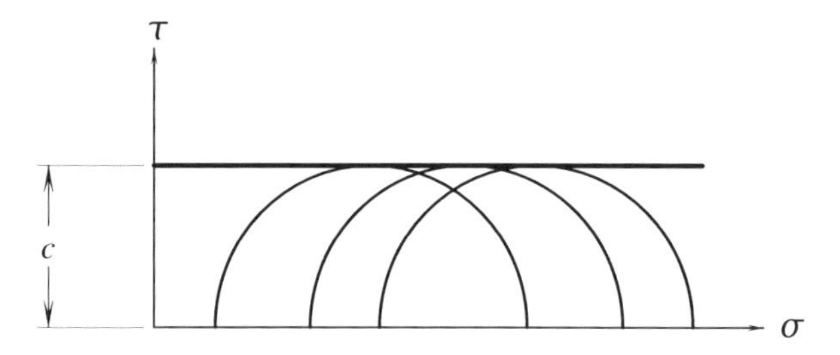

Figure 3.44 Failure diagram for *UU*-tests on saturated soil.

If an undisturbed sample of a saturated soil can be obtained, the strength of the soil in situ can be obtained by performing a Q-type triaxial test. If the test specimen remains saturated under zero gage pressure, the shear strength of the soil may be measured using an unconfined compression test because the diameter of the circle of stress passing through the origin is the same as the diameter of all other circles of stress. Thus, the $\phi = 0$ concept is the justification for using unconfined compression tests to measure the shear strength of a sample of saturated clay. The unconfined compression test is justified only if the samples remain saturated.

In practice, samples of soil often become slightly unsaturated under conditions of zero total stress and are subject to some sampling disturbance. Both factors cause a loss of strength that cannot be estimated without more detailed investigations.

Consolidated-Undrained (R-*Type*) *Tests.* For consolidated-undrained (R-type) triaxial compression tests on samples of saturated clay, the testing procedure differs, depending on whether or not porewater pressures are to be measured. If porewater pressures are not measured, the specimen is consolidated under the desired initial state of stress, the drainage connections are closed, and the specimen is sheared to failure over a time period that varies from a few minutes to perhaps half an hour. The time to failure varies, depending on whether or not complete stress-strain data are recorded.

If porewater pressures are to be measured, it is absolutely necessary that the porewater pressure connections be saturated and free of air bubbles. The presence of a small amount of free air in the drainage lines or fittings will usually have a negligible effect on shearing strength but will cause the measured porewater pressure to be considerably smaller than the actual porewater pressure in the specimen. Two approaches are used in practice to saturate the drainage system. The preferred method is to consolidate the specimen first, then simultaneously increase the confining pressure and porewater back pressure by the same amount. No change in effective confining stress and no change in strength ($\phi = 0$ condition) occur because the porewater pressure and cell pressure are increased by the same amount.

The second method used is to apply back pressure to the specimen first under a low effective confining stress, usually around 1 psi, then consolidate the specimen to the final effective confining pressure under the back pressure. The two methods work equally well for saturated clays of low plasticity. The first method usually requires back pressures in the range of 20 to 40 psi and works best on specimens that are initially partly saturated and on clays of high plasticity that are expansive. The second method may require back pressures of up to 100 psi to be effective in saturating the test specimen. In no case does the second method work better than the first method for common geotechnical applications. The increased pressure in the porewater pressure system dissolves any air bubbles that might be present and ensures saturation. When the system is properly saturated, the application of an increment of hydrostatic cell pressure results in the immediate development of an equal porewater pressure. Thus, the measurement of the excess porewater pressure resulting from an increment of cell pressure is a sensitive check on whether the specimen and the measuring system are properly saturated.

After ensuring that the specimen and the system are saturated, the specimen is loaded to failure with simultaneous measurements of the porewater pressure. The time to failure varies from about an hour for pervious clays to over a month for relatively impervious clays. The steps in the three-stage *R*-test are depicted in Figure 3.45, where additional symbols have been used to clarify the stresses. In Stage 1, the specimen is consolidated under a hydrostatic stress (σ_{3c}), the excess porewater pressure is zero, and the initial effective stress is σ'_{3i}. In Stage 2, an increment of hydrostatic pressure ($\Delta\sigma_3$) is applied, and if the specimen is saturated, the porewater pressure increases by the same amount. In Stage 3, the specimen is subjected to a compressive stress ($\sigma_1 - \sigma_3$) and an additional increment of porewater pressure (u_2) is generated. The resulting state of stress is found by adding the stresses from Stages 1, 2, and 3.

The three-stage *R*-test depicted in Figure 3.45 is the most common type of *R*-test, but several modifications are possible. For example, initial consolidation may be performed under conditions of no lateral deformation to approximate in situ conditions. During the consolidation stage of such a test, axial and volumetric strains are measured. Lateral strain is zero when the volumetric strain and axial strain are equal, so a stress difference is applied during consolidation that is just sufficient to produce zero lateral strain. The common assumption is that such specimens are consolidated under K_0 conditions.

As an example, a specimen is consolidated to 500 psf (σ_{3c}) during Stage 1, subjected to an increase of cell pressure and porewater pressure of $\sigma_3 = \Delta u = u_1 = 600$ psf (Stage 2), and then subjected to a stress difference of

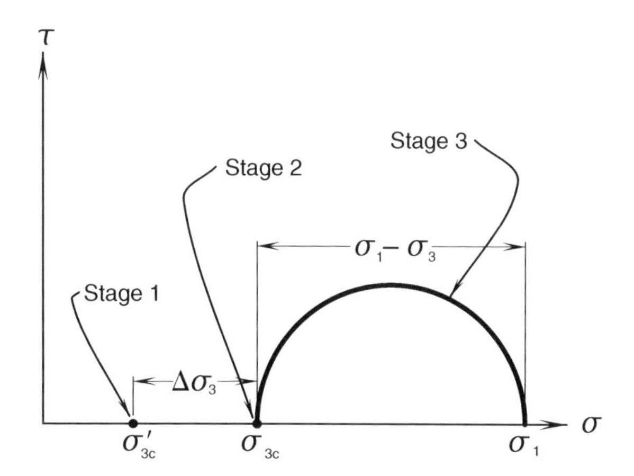

Figure 3.45 States of stress during the *R*-test.

$(\sigma_1 - \sigma_3)$ of 200 psf that generates a porewater pressure (u_2) of 150 psf (Stage 3). The resulting stresses are

$$\sigma_1 - \sigma_3 = 200 \text{ psf}$$
$$u = 600 + 150 = 750 \text{ psf}$$
$$\sigma_3 = 500 + 600 = 1100 \text{ psf}$$
$$\sigma_3' = 1100 - 750 = 350 \text{ psf}$$
$$\sigma_1' = 350 + 200 = 550 \text{ psf}$$

In another modification, either the lateral stress is increased with a constant axial stress until failure occurs or the lateral stress is maintained constant and the axial stress is decreased until failure occurs. Such tests are called *extension tests.* All stresses must be compressive in an extension test. An extension test is not a tensile test, it is a compressive test, with the lateral stress exceeding the axial stress.

For normally consolidated clay, the tendency of the soil structure to break down during shear leads to a transfer of part of the effective stress to the porewater; thus, positive porewater pressures are developed. A typical stress-strain curve for an *R*-type test is shown in Figure 3.46. The strain at the peak stress difference (often defined as the point of failure) may vary from less than 1% to 20%.

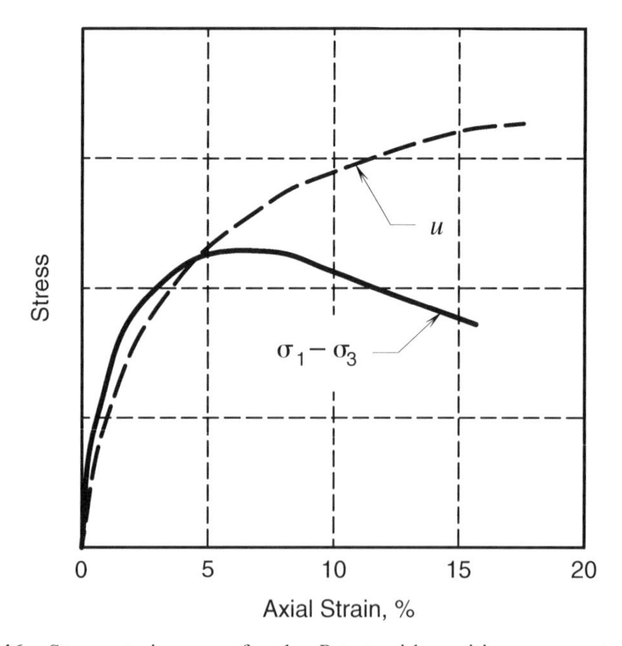

Figure 3.46 Stress-strain curve for the *R*-test with positive pore water pressure.

A comparison of the stress-strain properties of clays under drained and undrained conditions is of interest. Under drained conditions, the soil structure is broken down during shear and the volume decreases, that is, the density increases during shear. The continuously decreasing volume (increasing density) causes the soil to gain strength as it is deformed. Thus, the soil may be considered a strain-hardening material under these conditions of drainage. Eventually, the applied stress exceeds the strength and the specimen fails. The large strains at failure result from the continuous increase of strain.

In an undrained test on a normally consolidated specimen, the total confining pressure is constant. Thus, the continuously increasing porewater pressure (Figure 3.46) causes a reduction of the effective confining pressure (σ_3'). The strength of the soil is controlled by the effective stress, not the total stress; thus, continuous reduction in effective confining pressure causes the soil to lose strength as shearing deformations are applied. Hence, the soil in an undrained test is strain weakening, and the strains at failure (at peak stress difference) are considerably smaller than those obtained from drained tests on initially identical specimens (this argument is restricted to specimens that decrease in volume during drained shear and have positive porewater pressures during undrained shear, which is generally true for normally or lightly overconsolidated clays).

Mohr-Coulomb failure envelopes cannot be plotted in terms of total stresses because the strength is not a function of the state of total stress (see the discussion of Q-type tests). Failure envelopes may be plotted in terms of effective stresses at failure or, for tests in which porewater pressures are not measured, in terms of the consolidation pressure. In the first case, σ_1' and σ_3' are calculated at failure, and the failure circle is plotted in a Mohr-Coulomb diagram like that shown in Figure 3.47. The effective stress failure envelope usually passes through the origin for normally consolidated clays.

For many field problems, it is not possible to determine the change in porewater pressure in situ during loading. Thus, it may be expedient to assume that the porewater pressures generated in situ are equal to those generated in an undisturbed specimen of soil during laboratory shear testing. Applying this assumption, it is unnecessary to measure porewater pressures in the laboratory, thus substantially reducing the cost of laboratory tests. The Mohr-Coulomb diagram is then plotted using $\sigma_3 = \sigma_{3i}'$ and $\sigma_1 = \sigma_{3i}' + (\sigma_1 - \sigma_3)_{\max}$. This circle is a total-stress circle if the cell pressure is not increased after final consolidation (to check saturation). If the cell pressure is increased after consolidation, then the circle is not a total-stress circle. No problem will develop if σ_3 is always taken as σ_{3i}', where σ_{3i}' is the minor principal effective stress at the moment that shearing begins. The slope of the failure envelope expressed in terms of σ_{3i}' is designated as ϕ_R.

As typical calculations, a specimen is assumed to be consolidated under a stress σ_{3c} = 1000 psf, the cell pressure is increased by 750 psf with a simultaneous increase of porewater pressure of 750 psf, and, at failure, $(\sigma_1 -$

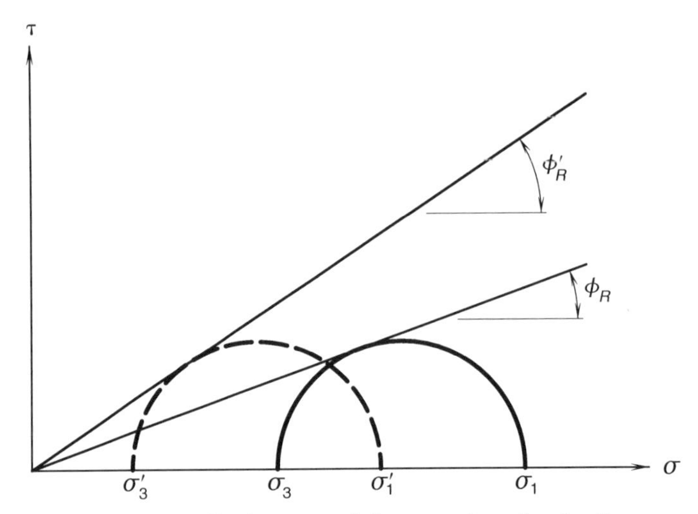

Figure 3.47 Effective stress failure envelope for the *R*-test.

σ_3) = 600 psf and u_2 = 700 psf. The computation of ϕ'_R and ϕ_R is shown below.

$$\sigma_1 = 1000 + 600 = 1600 \text{ psf}$$
$$\sigma'_3 = 1000 - 700 = 300 \text{ psf}$$
$$\sigma'_1 = 300 + 600 = 900 \text{ psf}$$

$$\text{Applying } \sigma_1 = \sigma_3 \tan^2(45° + \phi'/2) = \sigma_3 = 1 + \sin\phi'/1 - \sin\phi'$$

$$\tan^2 (45 + \phi_R/2) = 1600/1000 = 1.60$$
$$\phi_R = 13.3°$$
$$\tan^2(45 + \phi'_R/2) = 900/300 = 3.00$$
$$\phi'_R = 30.0°$$

Whether the failure envelopes are defined in terms of the initial effective stress or the effective stress at failure, the diameter of Mohr's circle is the same. The horizontal distance between the two circles for a single test is equal to the porewater pressure (u_2) generated during shear.

The ratio between the slopes of the envelopes defined in terms of initial effective stress and effective stress at failure can be obtained by solving for sin ϕ and sin ϕ', dividing one equation by the other, and substituting $A(\sigma_1 - \sigma_3)$ for u_2, where A is an experimentally determined parameter. Solving for sin ϕ',

$$\sin \phi' = \frac{\sin \phi}{1 - 2A \sin \phi} \tag{3.94}$$

If the A-coefficient has a value of 1, then the R'-envelope has a slope about two times the slope of the R-envelope for the range in ϕ usually encountered.

Attempts to correlate the effective-stress angle of internal friction with one of the index parameters have not been successful. The relationship between ϕ' and the plasticity index is suggested by the points plotted in Figure 3.48. Most soils represented in this diagram are from Norway because much of the available knowledge about the shearing properties of undisturbed clay has come from the Norwegian Geotechnical Institute in Oslo. Apparently most of these soils have effective-stress angles of internal friction between about 23° and 36°. Correlations between the effective-stress angle of internal friction of undisturbed clay and any of the common index properties are not likely to be very satisfactory because the index properties are determined using completely remolded soil. Any structural effects in the soil or effects related to natural cementation would influence the undisturbed properties but not the index properties. Further, there are serious difficulties associated with the measurement of porewater pressures in clays of low permeability, and these difficulties influence the evaluation of ϕ'. Errors associated with sampling

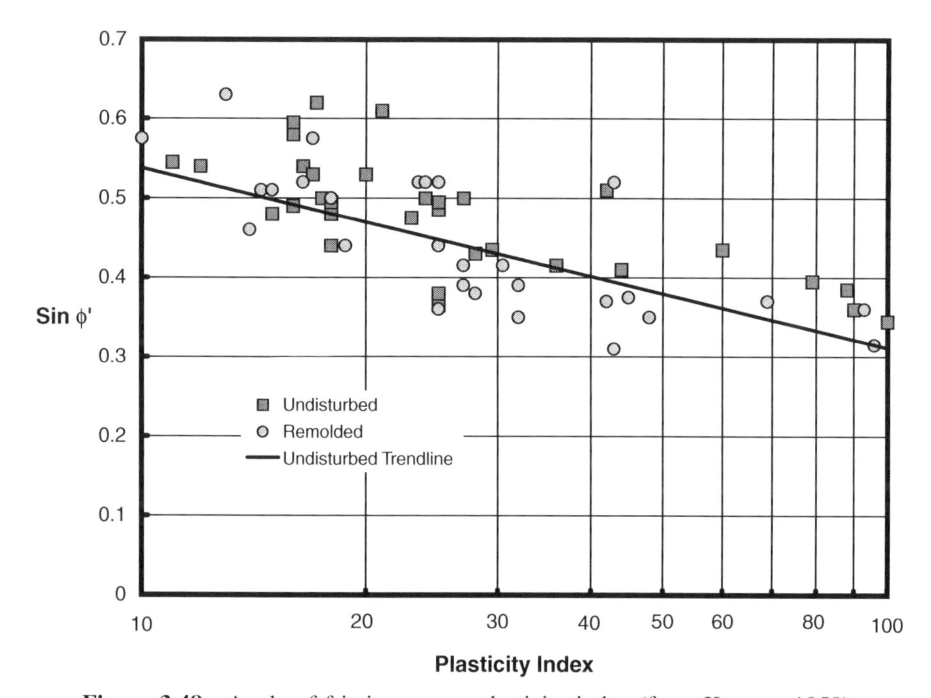

Figure 3.48 Angle of friction versus plasticity index (from Kenney, 1959).

disturbance, nonuniform stress and strain conditions in the triaxial cell, and other experimental problems also influence the results of these shear tests. Investigations into the shearing characteristics of undisturbed cohesive soils are in progress, but the problems are so complex that a simple solution of the problem is unlikely.

Drained versus Undrained Values of ϕ'. In performing drained triaxial compression tests, there is little ambiguity regarding the method of defining failure; the several failure criteria maximize at the same point. For undrained tests, however, the two common failure criteria of peak $(\sigma_1 - \sigma_3)$ and peak (σ_1'/σ_3') result in a higher value of ϕ'. Available data suggest that the effective stress angles of internal friction of drained (ϕ_S') and undrained (ϕ_R') soil samples correlate better when the peak (σ_1'/σ_3') failure criterion is used with the undrained tests. Field data are insufficient to judge which failure criterion correlates best with field conditions.

c/p Ratios. The previous discussion of the shearing properties of normally consolidated, saturated clay has described how the shear strength of the soil increases linearly with the logarithm of consolidation pressure. In a normally consolidated soil deposit, the consolidation pressure increases almost linearly with depth. Thus, the shear strength of the soil will also increase almost linearly with depth.

Early investigations of the increase of shear strength with depth were based on unconfined compression tests or Q-type triaxial tests. The soils studied were saturated in all cases, causing $\phi = 0$. If the strength of the soil increased linearly with overburden pressure in the field, then the shear strength profile versus depth could be defined with a single parameter, c/p. The strength c is actually the undrained shear strength, and p is the vertical effective overburden pressure. Occasionally, the notation s_u/σ_v' is used in place of c/p.

Early studies of the variation of strength with depth, using the unconfined compression test, were complicated by the problem of sampling disturbance. Disturbance of the soil samples during sampling operations and transport to the laboratory resulted in a decrease in the strength of the samples, and laboratory strength values were significantly less than field strength values. To avoid the problem of sampling disturbance, attempts were made to develop techniques for measuring the strength of the soil in situ under undrained conditions. Probably the most common way of measuring in situ strength is to use the field vane device, of which is shown in Figure 3.49. The vane is made of two thin steel fins set at right angles to each other and attached to a thin rod. The vane is pressed into the soil to the desired depth, and then a torque is applied to the rod to rotate the vane about its vertical axis. The soil fails on a vertical cylindrical surface. The area of this surface is calculated from the dimensions of the vane. Knowledge of the torque required to cause failure, the area of the failure surface, and the radius of the vane makes it possible to calculate the undrained shear strength. The variation of undrained

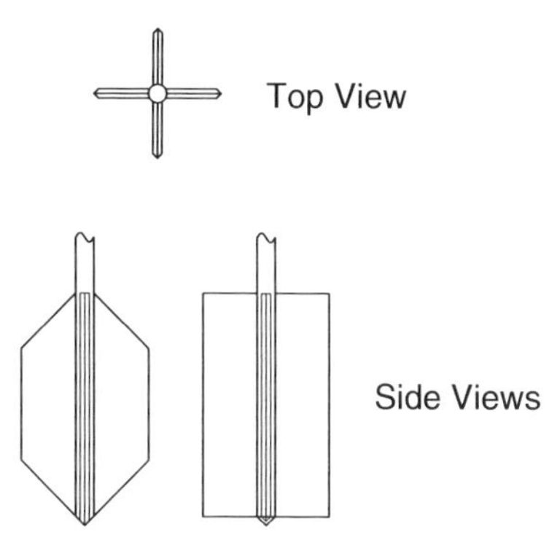

Figure 3.49 Field vane apparatus.

shear strength with depth for a uniform deposit of clay in Drammen, Norway, is shown in Figure 3.50, where the undrained shear strength is shown to increase linearly with depth for depths greater than about 7 m. For shallower depths, the soil has been desiccated and has thus been strengthened. The remolded strengths plotted in Figure 3.50 were obtained by rotating the vane several times and then measuring the strength again.

In the early investigations, the vane strength was always greater than the strength determined using unconfined compression tests because all samples taken into the laboratory were disturbed by the sampling operation. As specialized sampling techniques were developed, the laboratory strengths became higher (due to less disturbance); recent studies have shown that the laboratory strengths exceed the vane strengths. As a result of variations of soil properties with depth, it is difficult to make accurate comparisons of vane and unconfined compression strengths, but one recent study suggested that the unconfined compression tests yielded strengths about 30% higher than the vane strengths. The cause of this discrepancy is not fully understood, but part of it may result from the vane device measuring shear strength on vertical planes (the area of the end of the vane is small), the high rate of shearing in the zone of failure, and the degree of plasticity of the soil. The normal effective stress on a vertical plane is K_0 times the normal effective stress on the horizontal plane, where K_0 is termed the *coefficient of earth pressure at rest* and typically has a value of about 0.5 for a normally consolidated clay. If the strength is a function of the normal effective stress, then the vane strength should be lower than the strength on diagonal planes as measured using the unconfined compression apparatus.

In situ shear strengths may also be determined by pushing a cone down into the soil and measuring the resistance. Most of the research had been done for the Dutch cone penetrometer. Beginning in the early 1980s, pore pressure measuring equipment was added to Dutch cone equipment and was used to identify changes in stratigraphy in the soil profile. The cone penetrometer test

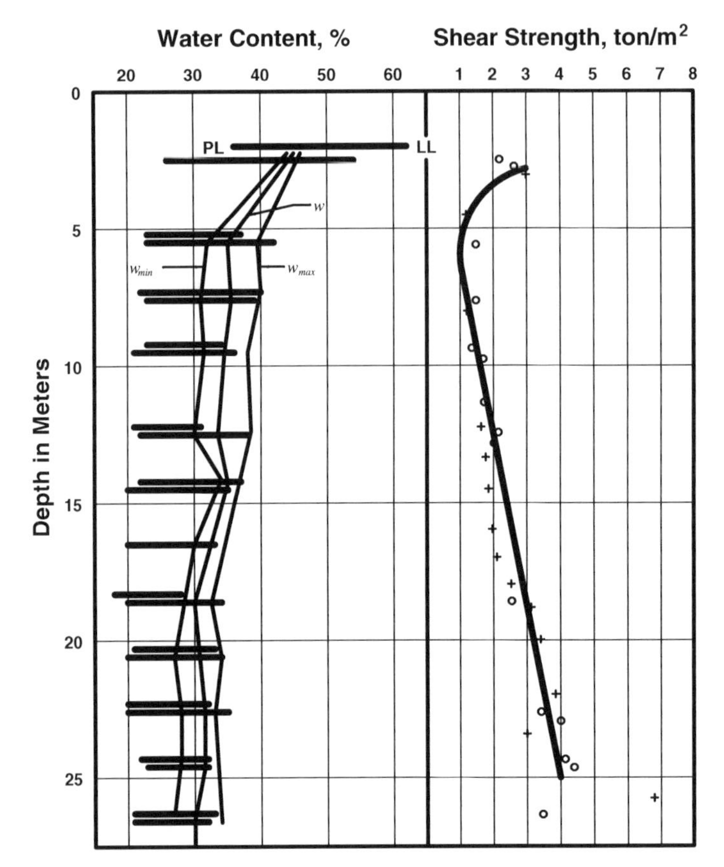

Figure 3.50 Strength profile from Drammen, Norway.

is favorable in that each cone sounding is a model pile load test. However, the use of shear strengths determined from cone soundings for other applications, such as slope stability, should be done with caution.

The c/p ratio has proved to be a useful parameter in foundation engineering in areas where deep deposits of soft clays are encountered. Correlations between the c/p ratio of normally consolidated soils and the plasticity index have been developed (Figure 3.51). Additional research has established the relationship between c/p ratio and overconsolidation ratio for a variety of soils (Figure 3.52).

Use of the c/p ratio for design purposes was formalized by Ladd and Foott (1974) in the SHANSEP (Stress History And Normalized Soil Engineering Properties) method. This method is a synthesis of the observations and methods of several researchers into a relatively straightforward approach that can be used if one has the benefit of a complete field investigation and high-

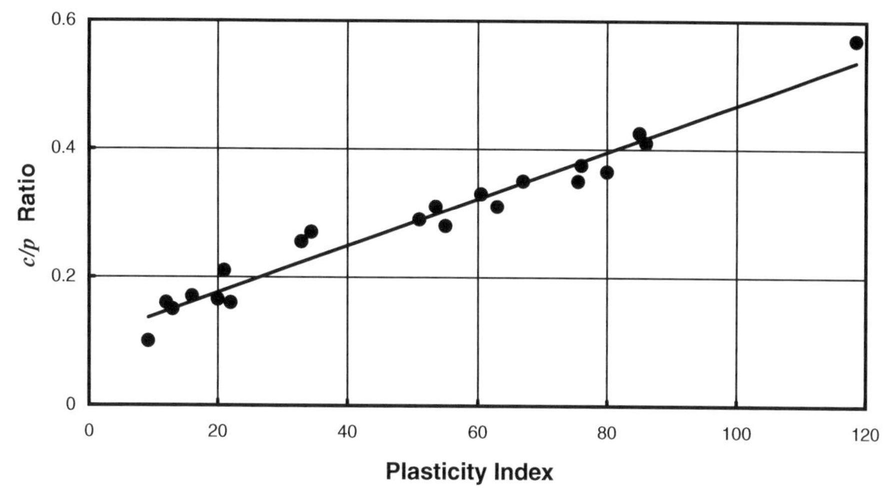

Figure 3.51 The c/p ratio versus the plasticity index.

quality laboratory tests. The difficulty in using the SHANSEP method is in establishing the stress history, consolidation characteristics, and overconsolidation ratio of the soil. The SHANSEP method is expensive, but its use can be justified for large projects where the soil profile is relatively uniform over the site.

Sensitivity. The remolded strength is shown to be less than the undisturbed strength in Figure 3.51. A quantitative measure of the strength loss on remolding is the sensitivity, which is the ratio of the undisturbed strength to the remolded strength at the same void ratio. The symbol for this parameter is S_t. The classification of the sensitivity of clays, as proposed by Skempton and Northey (1952) and modified by Rosenqvist (1953), is shown in Table 3.5.

Bjerrum (1954) has reported sensitivities as high as 1000 in some of the Norwegian clays. Because the undisturbed strengths of the Norwegian clays are low, such high sensitivities mean that the remolded clays turn into viscous fluids. If such clays are disturbed in situ, landslides may result in which the clay flows away like a viscous fluid. Many landslides of this type have taken place in Norway, in Sweden, and along the St. Lawrence River Valley in Canada. Bjerrum (1954) has shown that the sensitivity of Norwegian clays can be correlated with the liquidity index (Figure 3.53). This correlation is useful because it warns the engineer to use special precautions with deposits where the natural water content is equal to the liquid limit or higher. Apparently, special sampling techniques are required when such soils are encountered. Otherwise, the shear strengths measured in the laboratory may be only a small fraction of the strengths in the field.

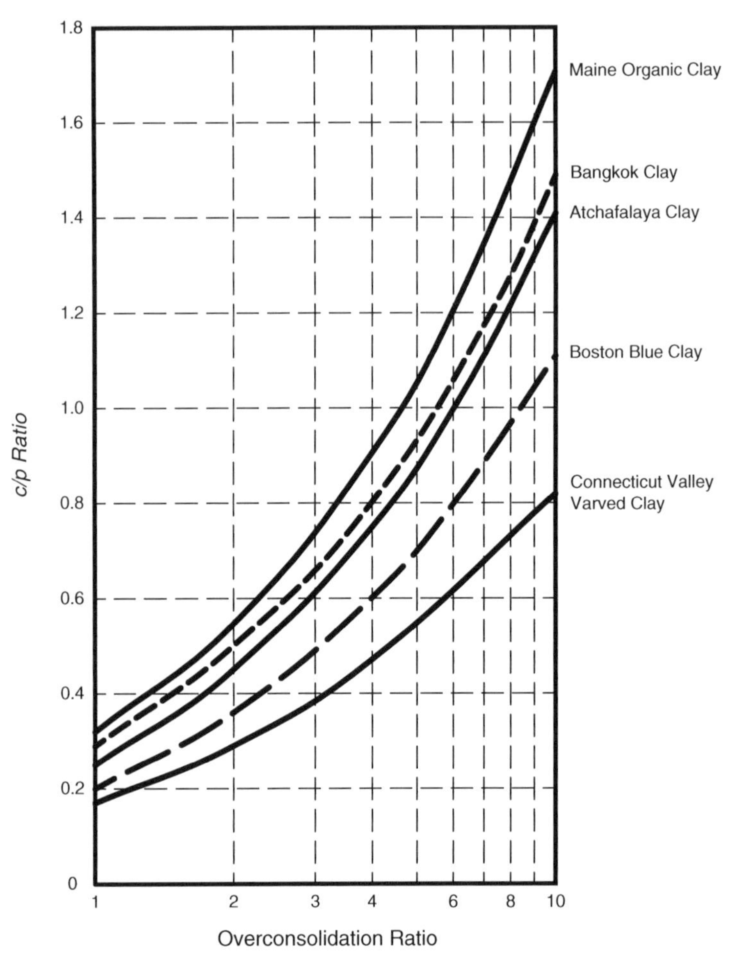

Figure 3.52 The *c/p* ratio versus the logarithm of the overconsolidation ratio (from Ladd and Foott, 1974).

TABLE 3.5 Classification of Sensitivity

Sensitivity	Classification
1	Insensitive
1–2	Slightly sensitive
2–4	Medium sensitive
4–8	Very sensitive
8–16	Slightly quick
16–32	Medium quick
32–64	Very quick
64–∞	Extra quick

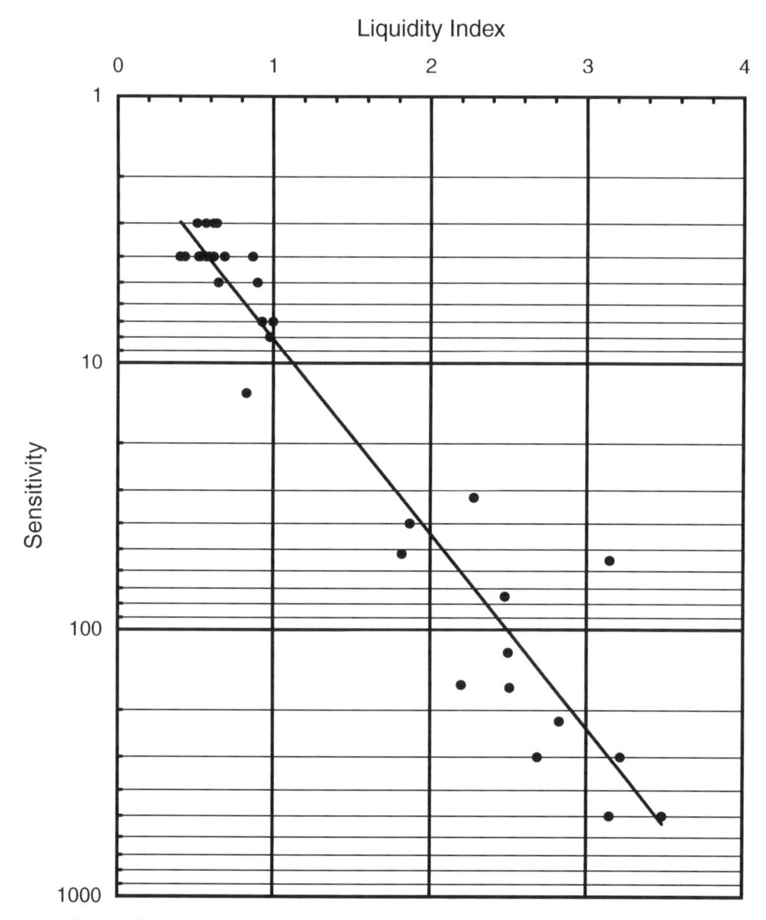

Figure 3.53 Sensitivity of marine clay from Norway (from Bjerrum, 1954).

Thixotropy. If a series of identical specimens of clay are remolded and then allowed to rest at a constant void ratio for varying periods of time before shearing, the strength of the remolded clay will be found to increase with time. The strength increase with time is reversible; that is, a specimen can be allowed to "set up" for some period of time and then remolded again, and the remolded strength will be same as the strength when the soil was originally remolded. Thus, cyclically allowing the specimen to set up, remolding it, and then letting it set up again, will result in strength changes such as those shown in Figure 3.54.

Some of the experimental results obtained by Skempton and Northey (1952) in an early investigation of thixotropy are shown in Figure 3.55. These results are typical of those obtained when specimens are remolded at water contents near the liquid limit.

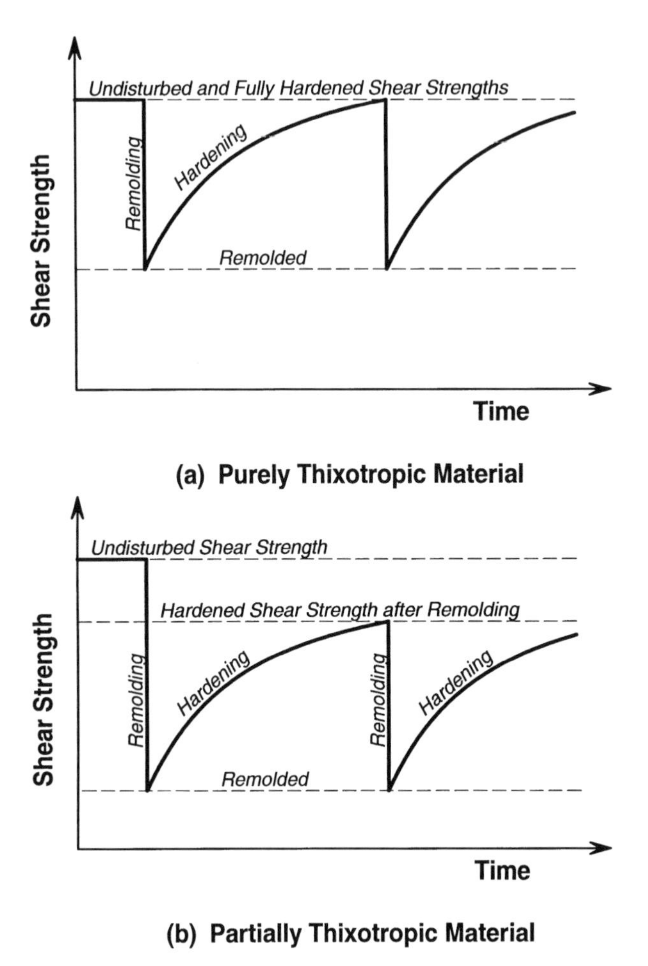

Figure 3.54 Full and partial thixotropic regain (from Skempton and Northey, 1952).

Specimens remolded at low water contents usually have only a small thixotropic increase in strength. Clay suspensions often undergo thixotropic effects that are many times greater than the effects shown here.

Apparently, clay can develop sensitivity as a result of thixotropic strength increases. However, for quick clays, like those found in Norway, thixotropy accounts for only a small part of the total sensitivity.

Relationship Between Shear Strength and Water Content for Normally Consolidated Clays. As discussed previously, if a series of R-type triaxial compression tests are performed on identical specimens of clay that have been sedimented from suspensions and normally consolidated, the R-envelope is often nearly a straight line, which passes through the origin. In such a case, the shear strength of the soil can be shown to be a constant percentage of the consolidation pressure given by

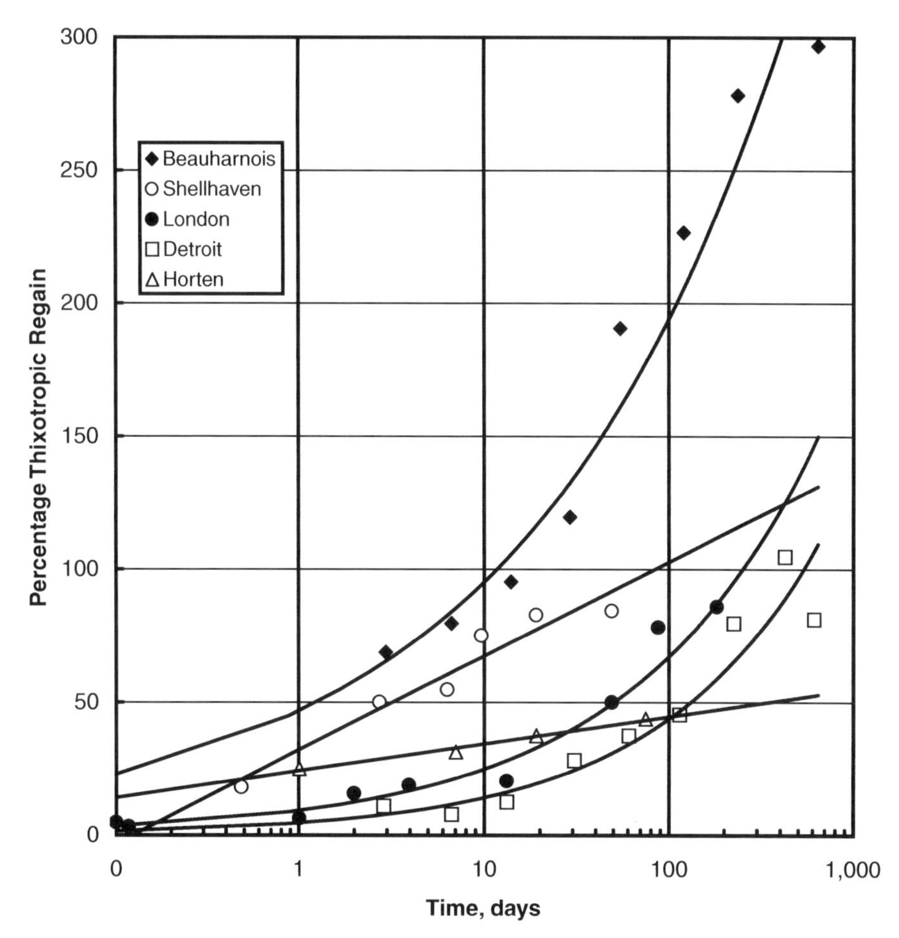

Figure 3.55 Thixotropic regain in some typical clays (from Skempton and Northey, 1952).

$$\frac{\tau}{\sigma_{3c}} = \frac{\sin \phi_R \cos \phi_R}{1 - \sin \phi_R} \tag{3.95}$$

If the undrained shear strength and consolidation pressure are both plotted against water content (for saturated clay only; if clay is not saturated, use the void ratio), the two curves will be parallel. One such relationship is shown in Figure 3.55. If a specimen is consolidated to Point *a* and subjected to undrained shearing, the strength is given by Point *b*.

The type of relationship shown in Figure 3.55 is useful for many practical problems. The slope of the virgin consolidation curve can be obtained from one-dimensional consolidation tests. If the shear strength of the samples is determined using either unconfined compression tests or *Q*-type triaxial tests, the data can often be conveniently plotted on a diagram such as the one shown in Figure 3.55. The assumption is made, for example, that 50 unconfined compression tests are performed on samples from a particular stratum. Plotting 50 Mohr's circles for the 50 unconfined compression tests on a Mohr-Coulomb diagram would produce an indecipherable tangle of lines, all passing though the origin, and meaningful interpretation of the data would probably not be possible. Alternatively, 50 values of unconfined compressive strength could be plotted versus water content in a diagram similar to Figure 3.55, and meaningful results might be obtained. The strengths could also be plotted versus depth (see Figure 3.50). The geotechnical engineer usually tries several methods of plotting the data and chooses the method that yields the most meaningful results for the particular problem under investigation.

If the sensitivity of the clay is independent of the consolidation pressure within the range of consolidation pressures encountered at the site, then the curve of remolded strength versus water content will have the same slope as the virgin curve and undisturbed strength curve. However, the sensitivity usually decreases as the consolidation pressure increases, so the remolded-strength curve is expected to be slightly flatter than the other two curves. If the undisturbed and remolded strength curves can be defined accurately, then the degree of disturbance of partially disturbed samples can be defined in terms of the relative position of the strength of the partially disturbed sample with respect to the curves of undisturbed and remolded strength.

Overconsolidated Saturated Clays The previous discussion of clays applies to normally consolidated clays. In nature, most soil deposits vary from slightly to heavily overconsolidated. Lightly overconsolidated sediments are usually treated as if they were normally consolidated. However, the properties of heavily overconsolidated clays are quite different from those of normally consolidated clays, and engineering problems of a different nature are encountered.

Stress-Strain Properties and Pore Water Pressures. In the same way that the properties of normally consolidated clays are comparable to those of loose sands, the properties of overconsolidated clays are comparable to those of dense sands. If a specimen of clay is consolidated under a high pressure and then allowed to swell under a much lower pressure, the clay will be considerably more dense than a normally consolidated specimen at the same final consolidation pressure; that is, a specimen consolidated to 1000 psi and then allowed to swell under 10 psi is far denser than a normally consolidated specimen at 10 psi. As a result of its more dense structure, heavily overconsolidated clay tends to expand in volume (i.e., dilate) during shear in a manner similar to that of dense sand during shear. Figure 3.56a shows the results of a drained triaxial compression test on a specimen of kaolinite that had been sedimented from a dilute suspension, consolidated to 120 psi, and then rebounded to 10 psi before shear. Just like the dense sands, the soil appears to decrease in volume at small strains and then to expand throughout the re-

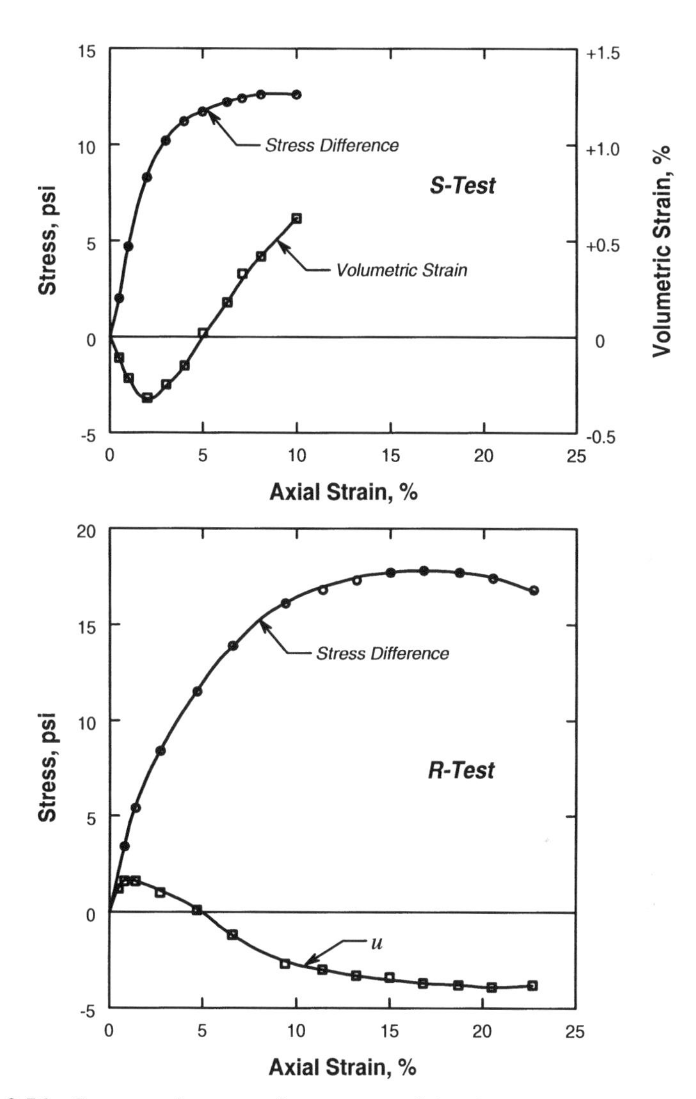

Figure 3.56 Stress-strain curves for overconsolidated kaolinite (from Olson, 1974).

mainder of the test. As for dense sands, the first application of compressive stress results in a small densification, and larger shearing deformations cause the particles to ride up over each other and the specimen to dilate.

The stress-strain curves for an identical specimen of kaolinite are shown in Figure 3.56b. This specimen was consolidated to 120 psi, rebounded to 10 psi, and then sheared under undrained conditions with pore pressure measurements. In the early part of the stress-strain curve where the soil tends to compress, the porewater pressures are positive but they become negative when the soil tends to expand. Comparison of the failure strains for the two specimens is of interest. In the drained test, the expansion of the soil has a weakening effect because water is drawn into the specimen. Thus, the greater the strain, the weaker the specimen. Failure occurs at about 10% strain. In the undrained test, the development of porewater pressures has a strengthening effect due to an increase in the effective confining pressure, and the specimen became stronger as strain increases. Thus, failure occurs at a strain of 18%, 8% above the strain at which the specimen fails during drained shear.

In considering the effect of overconsolidation on the properties of clays, having a parameter that expresses the degree to which a specimen has been overconsolidated is convenient. For this purpose, the *overconsolidation ratio* (OCR) is defined as the ratio of the maximum consolidation pressure to which a specimen has been subjected to the consolidation pressure just before shear. Thus, the overconsolidation ratio of the specimens is 120/10 = 12.

The assignment of parameters to the porewater pressures is also convenient. For this purpose, it is convenient to use Skempton's (1954) equation, which describes the changes in porewater pressure due to changes in the state of stress:

$$u = B[\Delta\sigma_3 + A(\Delta\sigma_1 - \Delta\sigma_3)] \qquad (3.96)$$

where

u = porewater pressure resulting from the application of stresses,
$\Delta\sigma_1$ = change in major principal total stress,
$\Delta\sigma_3$ = change in minor principal total stress,
A = A-coefficient, and
B = B-coefficient.

If a saturated specimen of clay is subjected to a change in hydrostatic stress, then the porewater pressure changes by B times the change in hydrostatic stress. However, it has already been concluded that the change in porewater pressure is equal to the change in hydrostatic pressure if the soil is saturated, and that the compressibility of the water is much less than that of the soil structure; thus, $B = 1$ for a saturated clay.

The A-coefficient expresses the influence of shearing stress on the porewater pressure. The A-coefficient at the point of failure is defined as A_f. The range in values of A_f for typical clays as a function of the OCR is shown in Figure 3.57. For normally consolidated clays, A_f is usually between 0.7 and 1.5, with the higher values found for the more sensitive clays.

In Figure 3.58, σ' indicates that Mohr's circle is plotted in terms of effective stress at failure, and σ indicates that the circle is plotted with $\sigma_3 = \sigma_{3i}$; the first case is used for R'-envelopes and the second for R-envelopes. The positions of the R'-circles for various values of A_f are shown. Because both the R- and R'-envelopes have positive slopes, the R'-envelope will be above

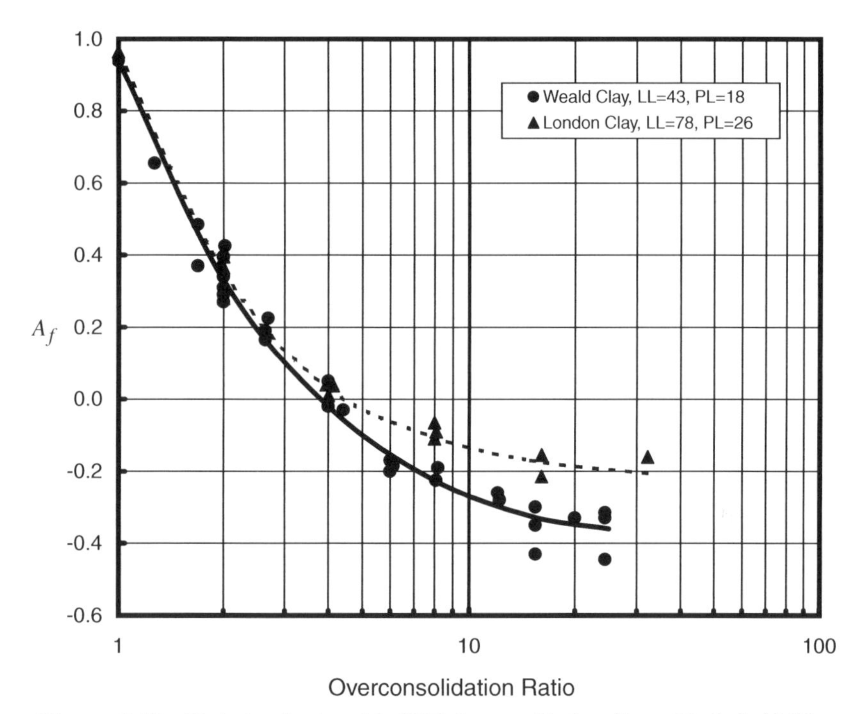

Figure 3.57 Variation in A_f with OCR for weald clay (from Henkel, 1956).

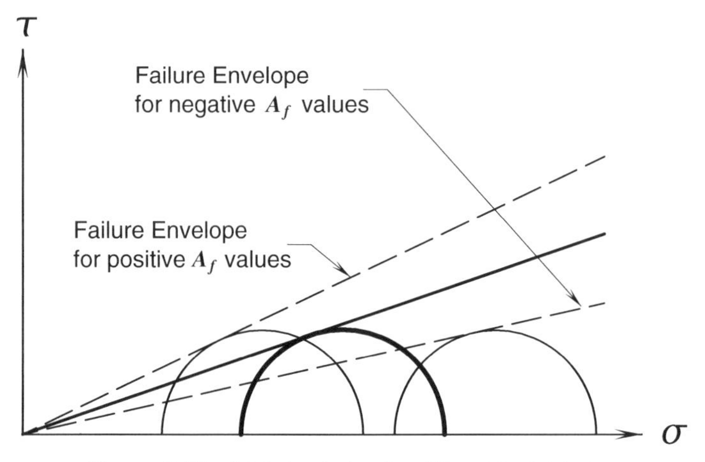

Figure 3.58 Mohr's circles for R-bar triaxial tests.

the R-envelope when A_f is significantly greater than zero and will be below the R-envelope when the A_f-coefficient is zero or negative.

At this stage of the discussion, a rather bothersome problem arises in plotting the Mohr-Coulomb diagrams. When only the results of normally consolidated tests or overconsolidated tests are plotted on a single failure diagram, there is not much difficulty in distinguishing the circles, though even in these cases, it is difficult to determine visually the scatter of the circles from the failure envelope if more than about 10 circles are plotted on a single diagram. However, when circles representing tests on both normally consolidated and overconsolidated specimens are plotted on the same diagram, or when R'- and S-circles are plotted together, or when many circles are plotted, the diagram becomes a maze of lines and rational interpretation is very difficult. Hence, it is now convenient to define a *modified Mohr-Coulomb diagram* where this problem will be eliminated.

A Mohr's circle tangent to a failure envelope is shown in Figure 3.59. The relationship between the principal stresses derived from the geometry of the circle and the failure envelope are

$$\cos \phi = \frac{\frac{1}{2}(\sigma_1 - \sigma_3)}{\frac{1}{2}(\sigma_1 + \sigma_3)\tan \phi + c}$$

Rearranging,

$$\tfrac{1}{2}(\sigma_1 - \sigma_3) = c \cos \phi + \tfrac{1}{2}(\sigma_1 + \sigma_3)\sin \phi \tag{3.97}$$

If, instead of plotting Mohr's circles on a diagram of τ versus σ, points are plotted on a plot of $\frac{1}{2}(\sigma_1 - \sigma_3)$ versus $\frac{1}{2}(\sigma_1 + \sigma_3)$, then a failure envelope

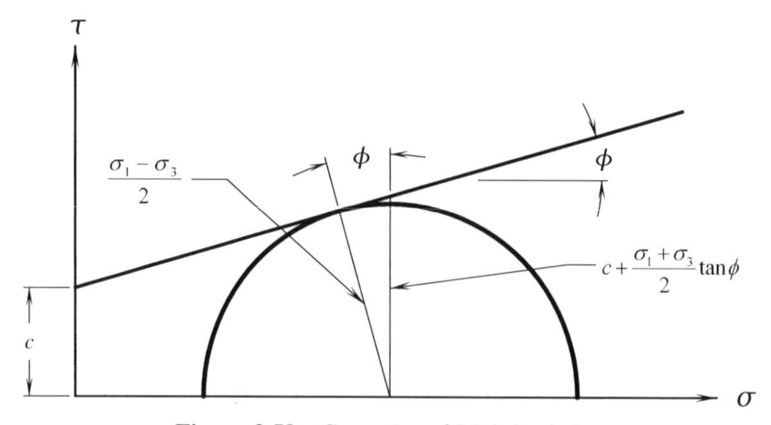

Figure 3.59 Geometry of Mohr's circle.

is obtained with an intercept equal to $c \cos(\phi)$ and a slope equal to $\sin(\phi)$. If the intercept is designated as d and the slope as ψ, Eq. 3.97 would be written in the form

$$\tfrac{1}{2}(\sigma_1 - \sigma_3) = d + \tfrac{1}{2}(\sigma_1 + \sigma_3)\tan \psi \tag{3.98}$$

The standard Mohr-Coulomb shearing parameters are obtained from

$$\sin \phi = \tan \psi \tag{3.99}$$

$$c = \frac{d}{\cos \phi} \tag{3.100}$$

Probably the largest advantage of using a modified Mohr-Coulomb diagram is that each test is represented by a single point rather that a circle. Thus, different types of tests can be plotted on the same diagram and clearly differentiated from each other by using different symbols. Furthermore, when the envelopes are straight lines, linear regression analysis can easily be used to determine the parameters of the failure envelope best fitting the experimental points.

Overconsolidation by Desiccation and Weathering. The strength profile of a soil deposit in Norway reported by Moum and Rosenqvist (1957) is shown in Figure 3.60. The increased strength of the upper part of the deposit could not have resulted from consolidation under the existing overburden pressure or from any previous overburden pressure that was eroded away. The increased strength of the upper part of this deposit and of many deposits around the world is the result of desiccation (drying) and chemical alteration (weathering).

Desiccation causes a reduction in the void ratio of the soil. According to the principle of effective stress, the void ratio can be reduced only if the effective stress is increased. Because the total stress at any given depth near the surface is small, the reduction in void ratio and the increase in effective stress must result from the development of negative porewater pressures. In highly plastic clays, this negative porewater pressure would have to be more than 100,000 psf to explain the strength of desiccated crusts.

Strengthened crusts may also result from chemical changes brought on by weathering. These changes may involve simple cation exchange reactions, such as weathering of feldspar to release potassium, which then replaces the adsorbed sodium cations to convert the sodium clay to potassium clay. In other cases, weathering causes the precipitation of cement at the point of contact between the particles. Such cements are commonly calcium carbonate or ferric oxide.

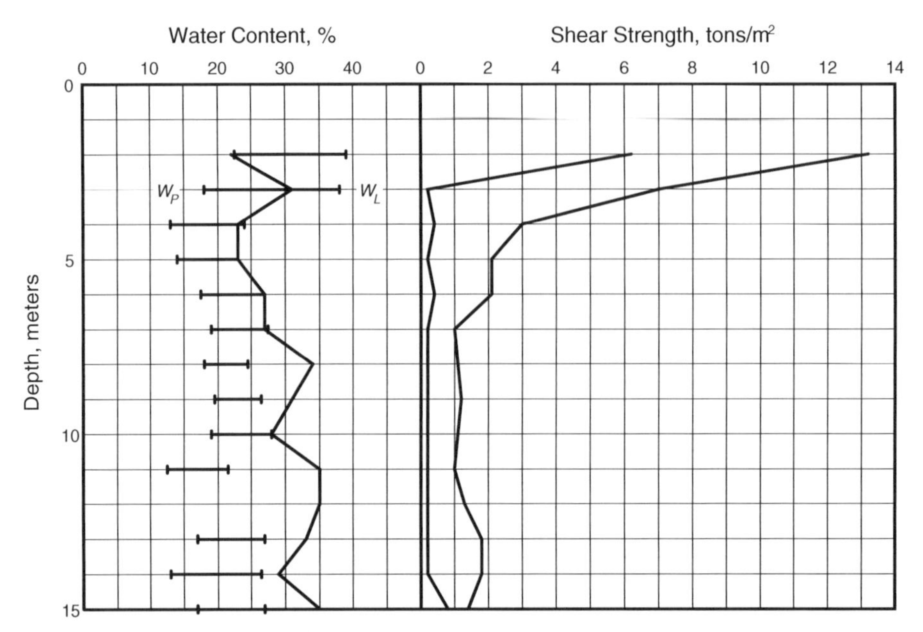

Figure 3.60 Strength profile showing gain in shear strength due to desiccation (from Moum and Rosenqvist, 1957).

Fissured Clays. In a natural soil deposit, the soil may be consolidated under conditions of approximately zero lateral strain. In such a case, the lateral stress is equal to the vertical stress times K_0, the coefficient of earth pressure at rest, which may be about 0.5 for a normally consolidated soil. If an element of soil is buried so deeply that the overburden pressure is 1000 psi, then the lateral stress is about 500 psi. At this stage, the soil is normally consolidated. If erosion begins to remove the overburden, the vertical pressure is reduced and approaches zero as a limit. The lateral stress is also released, but does not reduce by as much as the overburden pressure. Thus, a stage is eventually reached where the lateral and vertical stresses are nearly equal. Further erosion may reduce the vertical stress to such a low level that the large lateral stresses can cause shearing deformations to occur. For example, if $\phi' = 30°$ and there is a zero cohesion intercept, the soil will undergo a shearing failure when the lateral stress (σ_1') is three times the vertical stress (σ_3'). The shearing deformations result in the formation of a series of shearing planes. If significant deformations occur along these planes during unloading, planes of weakness will occur in the resultant soil deposit. The formation of such fissures is more likely to occur if the lateral stress is relieved by erosion of river valleys at some distance horizontally from the element of soil under consideration.

Fissures in the soil can also form as a result of tensile stresses developed during desiccation. Such fissures are approximately vertical and tend to have a hexagonal configuration when viewed from above.

The *intact* part of the fissured soil, that is, the part between fissures, may be relatively strong. Some deposits are subjected to cementation during the geologic times involved in the formation of the soil deposit, and the intact part of the soil becomes very strong indeed. The strength of the deposit as a whole is then determined by the strength along the fissures. Often an unconfined specimen of soil in the laboratory will fall apart within a few minutes after the stresses are relieved. Thus, the unconfined compression strength is zero. The field strength is well above zero, as proved by stable slopes in the field. Thus, the unconfined compression strength is too conservative as a measure of the shear strength in the field.

If a confining pressure is applied, the fissures are held closed and the strength is increased significantly over that of unconfined specimens. Triaxial shear testing of some fissured clays reveals an apparently anomalous behavior where the failure envelope has a finite slope, a condition in direct violation of the $\phi = 0$ condition. Examination of such test specimens after shear usually finds that fissures had been partially open at the confining pressures used in testing. This caused soil-to-soil contact to occur over only part of the total area of the shear planes. As the level of confining pressure is increased in the testing series, the area of soil-to-soil contact also increases with increasing confining pressure, thereby causing the measured strength of the soil to increase. If confining pressures are used that are large enough to keep any fissures closed throughout the test, then the $\phi = 0$ condition will be attained.

Fissured clays pose a serious problem to the engineer who relies on laboratory data for field design. Laboratory technicians may discard samples that break apart during attempts to prepare specimens for shear and test only intact specimens. If the fissures are far enough apart so that intact specimens can be obtained, the laboratory strengths will be those of the intact material, while failure in the field will take place along the fissures. Thus, the laboratory strengths are in error in the unconservative direction. If fissured specimens are tested, their strength is found to depend greatly on the orientation and position of the fissures in the specimens, and erratic test results are obtained. Since the strength in the field also depends on the orientation of the fissures, a rational field design is also made very difficult, if possible at all. Meaningful laboratory tests can be obtained only if the specimens are large compared with the spacing of the fissures. Samples of this size are seldom available, and the required equipment is generally not available.

Fissured clays have been selected as one example of the types of problems encountered in testing real soils as opposed to the artificially prepared specimens usually used for research. Each type of natural soil deposit has its own particular idiosyncrasies, and both the testing method (if any useful tests can be performed) and the interpretation of the test data must be adapted to fit the deposit and the requirements of the particular job. Useful laboratory shear strength tests on soil cannot consist of the standard performance and routine interpretation of tests in the same way many laboratories perform tests on asphalt, concrete, and steel.

Orientation of Failure Planes. According to Mohr's theory of stresses, the angle θ between the plane on which the maximum principal stress acts and the failure plane is half of the central angle in the Mohr diagram shown in Figure 3.44. Hence, the failure planes should be oriented as shown in Figure 3.61. It would appear, therefore, that the position of the failure planes could be used as a measure of ϕ'.

However, the previous discussion has shown that the failure envelope can have a wide range of slopes, depending on whether total or effective stresses are used and on whether the soil is normally consolidated or overconsolidated. The question arises as to which ϕ' correlates with the position of the failure planes and, therefore, which ϕ' can be called the true angle of internal friction.

The orientation of the failure planes in triaxial specimens has been measured for hundreds of specimens by several qualified researchers, but no simple, definite conclusion can be drawn from their data. The orientation of the planes may be altered considerably by anisotropy resulting from either the method of preparation of the soil specimens or the type of deformation allowed during consolidation. Measurements also show that the orientation of the shear planes changes with increased deformation. Thus, the orientation must be measured at the moment of failure with all the stresses in place, a very difficult procedure when the specimen is encased in a nearly opaque rubber membrane inside a triaxial cell. In addition, some specimens fail by uniform bulging, and either shear planes do not form or else they form at strains much greater than the failure strain. Other specimens fail by splitting down the middle. For these and various other reasons, the orientation of the failure planes cannot be used as a measure of soil behavior. As a necessary

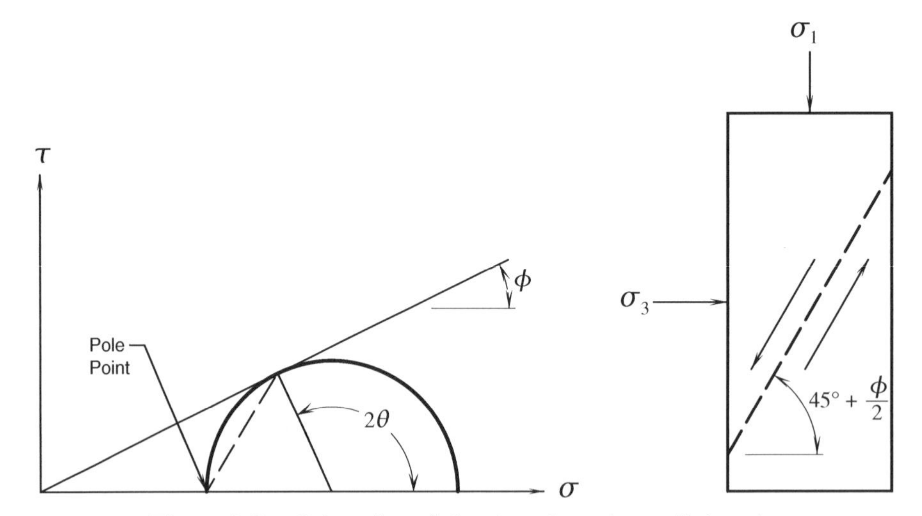

Figure 3.61 Orientation of shearing planes in a soil deposit.

corollary of this observation, the determination of the position of failure planes in the field is also not possible.

Influence of Intermediate Principal Stress on Strength In triaxial compression tests the intermediate principal stress is equal to the minor principal stress, while in triaxial extension tests the intermediate principal stress is equal to the major principal stress. In slope stability problems in the field, the intermediate principal stress actually has an intermediate value. In such problems, it may be better to define the intermediate principal stress as the stress required to maintain a condition of zero strain parallel to the slope. Naturally, the question arises about the influence of the intermediate principal stress on the shear strength of the soil.

The difficulty of obtaining precise answers to this question is an experimental one: because the construction of an apparatus that allows independent control of the three principal stresses and strains for a material as compressible as soil is very difficult. In fact, a generally satisfactory device has never been constructed, though some notable attempts have been made. (As the versatility of the apparatus approaches the desired value, both the size of the apparatus and its cost approach infinity.)

Available information suggests that the angle of internal friction in plane strain may be several degrees higher than the angle measured in triaxial tests. Because this error is on the safe side, and because experimental data from natural deposits usually scatter considerably, the influence of the intermediate principal stress is usually ignored.

Unsaturated Soils. The shear strength of unsaturated soil is a function of the type of soil and its degree of saturation. Soils with low degrees of saturation often behave as frictional materials under undrained shearing conditions. In contrast, soil with a high degree of saturation often behaves as if the degree of saturation is 100%.

The type of laboratory testing necessary to define the shearing characteristics of unsaturated soils is often beyond the capabilities of many soil testing facilities. Readers are referred to the work of Fredlund and Rahardjo (1993) for further information about the mechanics and shearing properties of unsaturated soils.

3.6.7 The SHANSEP Method

Between the 1950s and 1970s, geotechnical engineers recognized that the undrained shear strength of many soils followed a characteristic pattern. Initially, many engineers utilized this knowledge through their local experience but did not formalize their thinking in any particular manner. Eventually, several well-known geotechnical engineers began to present graphs of the shear strength that became popular. Foremost among these engineers were Professor Alan W. Bishop of the Imperial College of Science and Technology in London and Professor Charles C. Ladd of the Massachusetts Institute of Technology. In 1974, Professor Ladd formalized a system to present and characterize the undrained shear strength of soils. This system is known as the SHANSEP (Stress History And Normalized Soil Engineering Properties) system. This system is probably the most widely used system for characterizing the shear strength of soils for engineering in practice in the English-speaking world. The following is a brief presentation of the SHANSEP system.

The SHANSEP system is based on the observation that the shear strength of many soils can be normalized with respect to the vertical consolidation pressure. One example of normalized behavior is illustrated in Figure 3.62. Here, two stress-strain curves measured in consolidated-undrained (CU) tri-

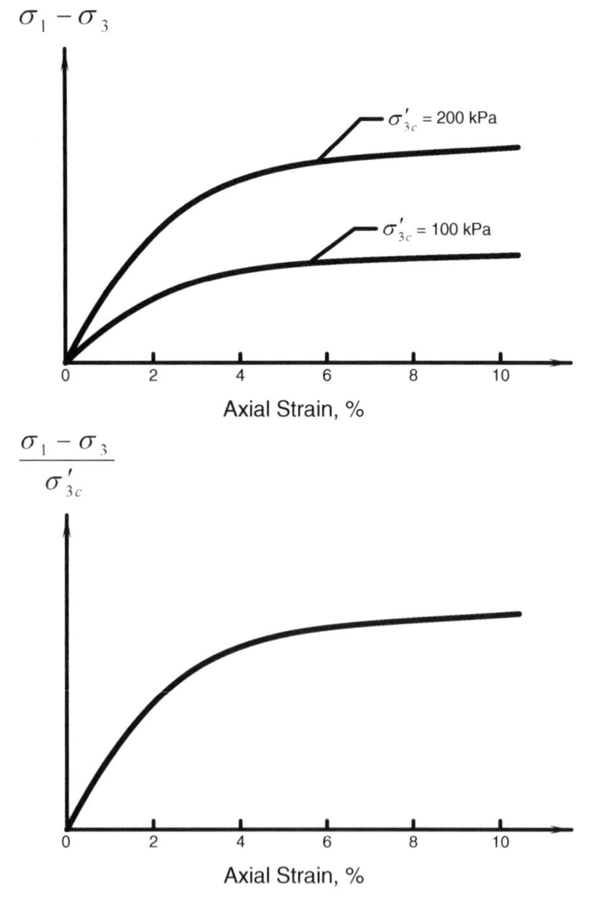

Figure 3.62 Example of normalized stress-strain behavior in idealized triaxial tests on homogeneous clays (from Ladd and Foott, 1974).

axial tests are plotted. In the upper half of the figure, the unnormalized stress-strain curves are plotted. In the lower half of the figure, the normalized stress-strain curve is plotted. In the normalized plot, the strain axis is unchanged and the stress axis is normalized by dividing the axial stress difference by the vertical consolidation pressure.

When normalized plots are made using CU triaxial data, unique curves are obtained for each value of the OCR. As the overconsolidation ratio increases from 1 to higher values, the strain at peak stress decreases and the normalized stress-strain curves plots higher on the figure.

If a soil exhibits normalized behavior, a figure of strength ratio versus log OCR can be plotted. The strength ratio is defined as the undrained shear strength c divided by the vertical effective stress. Traditionally, the most commonly used symbol for strength ratio is c/p'. Other symbols are also used.

The SHANSEP procedure is as follows:

1. A soil investigation is conducted, and a sufficient number of samples are obtained from the soil strata of interest.
2. A series of one-dimensional consolidation tests are run on samples from various depths to define the overconsolidation ratio versus depth in the soil profile.
3. A series of CU test series are performed at confining pressures corresponding to OCR values of 1, 1.5, 2, 4, and 6. Often, this testing program requires that the test specimens be consolidated to stresses well above those found in the field and then rebounded to lower levels of effective stress to obtain the desired OCRs.
4. Figures similar to those shown above are plotted using the test data.
5. The SHANSEP figures can be used to estimate undrained shear strengths in many situations. The general procedure used to determine undrained shear strength is presented in Steps 6 to 9.
6. The effective stress at the depth of interest is computed.
7. The OCR is computed using the known value of maximum past consolidation pressure. (If the new effective stress exceeds the maximum past consolidation pressure, then the soil becomes normally consolidated and OCR = 1.0.)
8. The OCR is then used to obtain the strength ratio from the chart of strength ratio versus OCR logarithm.
9. The undrained shear strength is computed by multiplying by the vertical effective stress by the strength ratio.

When soil is to be placed on or excavated from the site, it is often necessary to estimate the changes in undrained shear strength resulting from consolidation or rebound due to changes in levels of effective stress. If this is the case, the effective stress computed for Step 1 may be computed both prior to and at the end of consolidation or rebound. The change in undrained shear strength is then the difference in the two values of computed strength.

If the stress-strain behavior is of interest, the OCR can be used to estimate the stress-strain curve.

Noted that if one is interested in the undrained shear strength under a foundation, one must compute the vertical distribution of stresses due to the foundation. A similar situation exists for stresses under an embankment.

3.6.8 Other Types of Shear Testing for Soils

Several other types of tests for measuring the shearing properties of soils have been developed in addition to the direct shear and triaxial shear tests. Short descriptions of several of these tests follow.

One type of test is the direct simple shear test. In this test, the soil specimen is distorted by rotation of the sides of the shear box. Rotation of the specimen is accomplished by confining the test specimen in a wire-reinforced membrane, an articulated shear box, or a stack of thin rings that can slide. In each of these configurations, the soil specimen is distorted to develop shear strain within it. This type of test has had limited use by consultants and is usually used in conjunction with triaxial shear tests and consolidation tests to develop constitutive parameters for advanced models of material behavior.

Another type of shear test is the ring shear test, in which the shear box is composed of two horizontal rings. The soil properties measured in this test are the same as those measured in the direct shear test. However, the ring shear test is not limited in its range of deformations because one ring is driven in shear using a worm gear system. Ring shear devices are very useful in measuring the residual shearing resistance on well-formed shear planes, such as those that might develop in a slope failure. These devices are commercially available, but have the limitation that a large-diameter soil sample must be used for testing undisturbed soils.

Other types of tests that are used to measure the dynamic shearing properties of soils are the cyclic triaxial test and the resonant column test. The results of these tests are used to solve problems in earthquake engineering or mechanical vibration for which dynamic soil properties are important.

In the cyclic triaxial test, the soil sample is prepared in the same manner as for a conventional triaxial test but is sheared using various levels of cyclic loading. The main property of interest measured in the cyclic triaxial test is the number of cycles of loading needed to cause the soil specimen to fail.

In the resonant column test, a cylindrical soil specimen is placed under a confining stress and permitted to consolidate. After consolidation is complete, the specimen is vibrated in torsion. The resonant frequency of the specimen is measured as a function of the shearing strain amplitude developed in the specimen, and after vibration ceases, the decay of vibrations is measured. The principal measurements made using the resonant column test are shear wave velocity, shear modulus, and material damping ratio.

Lastly, the performance of stress-path triaxial tests is possible. In this type of test, computer-controlled feedback testing is used to force the stress in the test specimen to follow a prescribed path. A stress path is the trajectory or path of stress states on a Mohr-Coulomb diagram. Stress-path triaxial testing is available at only a few commercial testing laboratories and is usually used only for major projects for which benefits might be realized. Modern computer-controlled triaxial equipment like that shown in Figure 3.63 is capable of performing stress-path testing. However, the foundation engineer must first perform advanced modeling of a problem to determine the desired stress path to be followed.

3.6.9 Selection of the Appropriate Testing Method

The foundation engineer must tell the testing laboratory how to test the soil so that the necessary soil properties are measured for the problem at hand. Often the engineer must select soil tests while considering how much money is available for testing. Additionally, the engineer must recognize that meaningful testing can be accomplished only on soil samples of good quality and

Figure 3.63 Computer-controlled triaxial testing equipment (photograph courtesy of Trautwein Geotac).

that testing of disturbed soil samples may obtain results that are not representative of field conditions.

In light of these considerations, the foundation engineer understands the following limitations and capabilities of the consolidation test, direct shear test, and triaxial test:

1. The consolidation test measures the consolidation properties of the soil needed for calculations of total settlement and time rate of settlement.
2. The direct shear test can be used economically to measure the angle of internal friction of sand. The shortcoming of this test is that it is not possible to measure a stress-strain curve for sand.
3. The direct shear test on clayey soils can be used to measure the drained friction angle of clays more economically than *S*-type triaxial tests. Again, as with direct shear tests on sand, the shortcoming of this test is that it is not possible to measure a stress-strain curve for the soil.
4. *UU*-type triaxial tests are the best tests to measure the undrained shear strength of cohesive soils. While these tests can also be used to test sands, this requires that a specimen be built for testing. This usually involves more work to set up the specimen for testing than for a specimen of cohesive soil. One limitation of the *UU*-type triaxial test is that the effects of consolidation or rebound of vertical stress cannot be adjusted for in the test because the soil specimen is not subjected to consolidation pressures. If the effects of consolidation are important, a shearing test that permits consolidation must be used.
5. The *CD* triaxial test is the best and most expensive test used to measure the fully drained shearing properties of soils. The test results include the consolidation properties of the soil, the angle of friction under fully drained conditions, and the stress-strain curve. Unfortunately, time and budget constraints often eliminate the use of *CD* triaxial test on many projects. If only the fully drained angle of friction is required, a drained direct shear test is a practical alternative.
6. The *CU* triaxial test without measurement of porewater pressures during shear is used when undrained shear strength values are needed for soils that are subject to consolidation in the field.
7. The *CU* triaxial test with measurement of porewater pressures during shear can be used to measure the effective stress shearing parameters of cohesive soils more economically than the *CD* triaxial test.

PROBLEMS

Problem Set 1

P3.1. Settlements of saturated clays are often modeled with a piston and spring, as shown below. If the load *P* is suddenly increased, describe the model's reaction at time zero, with the passage of time, and at infinite time.

P3.2. The piston shown in Problem P3.1 weighs 128 lb. The spring constant is 400 lb per foot of compression. The spring-piston assembly is initially in equilibrium (no change in the system as a function of time so long as no loads are applied to or removed from the piston), with the valve open and with the dimensions shown. The force P is suddenly increased from zero to 100 lb.

a. What was the reading on the pressure gage just before the 100-lb load was added? Use units of psf. The gage reads the pressure at the center of the gage, and the line to the gage is full of water.

b. What will the reading on the pressure gage be immediately after applying the 100 lb?

c. When enough fluid has leaked out so the piston has settled 0.1 ft, what will the gage read?

d. When equilibrium is reestablished, what will the gage reading be and how much will the piston have settled relative to its position when the 100-lb load was added?

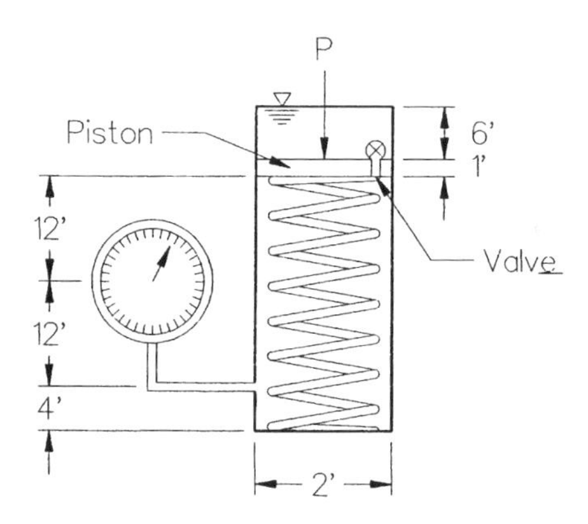

Problem Set 2

P3.3. The wet and dry weights of a lump of soil are 86.0 and 61.2 g, respectively. Calculate the water content (in percent).

P3.4. A wet soil sample weighs 110.0 g. Its water content (w) is 16.7%. Calculate the dry weight in grams.

P3.5. The total weight of a chunk of moist soil is 330 lb. Its volume is 3 ft^3. The water content was found to be 27%. Find e, γ, γ_d, and S. State any assumptions made.

P3.6. A specimen of saturated clay has a water content of 46.5% and a specific gravity of solids (G_s) of 2.7. Calculate (a) the total unit weight in pcf and kN/m^3 and (b) the dry unit weight in pcf and kN/m^3

P3.7. For a particular soil, $\gamma = 126.0$ pcf and $\gamma_d = 102.3$ pcf. Calculate the water content (w).

P3.8. A sand has a dry density of 101.0 pcf and $G_s = 2.67$. What water content is required to obtain 85% saturation?

P.8.9. A sample of dry clay weighing 485 g is sealed with 10 g of paraffin (the density of paraffin is 0.90 g/cc). The volume of the soil and paraffin, found by immersion in water, is 310 cc. The density of solids is 2.65 g/cc. What is the void ratio of the soil?

Problem Set 3

A soil boring has the following profile:

0 to 40 ft sand, dry density = 105 pcf, S_r = 100%, G_s = 2.72
40 to 60 ft silt, w = 35%, S_r = 100% G_s = 2.80

The water table is at the surface. Calculate the total stress, the porewater pressure, and the effective stress at the top of the silt layer and then at the bottom of the silt layer. Enter your answers in the following table.

Location	Total Stress	Porewater Pressure	Effective Stress
top	________psf	________psf	________psf
bottom	________psf	________psf	________psf

P3.10. Draw total effective and pore pressure diagrams.

P3.11. A steel mold that weighs 4.10 lb and has an inside volume of 1/30 of a cubic foot is filled with a compacted clay. The weight of the mold full of clay is 8.41 lb. The water content is 14.1%, and G_s = 2.75. Calculate the void ratio and the degree of saturation.

P3.12. Calculate the effective stress at a depth of 40 ft in a submerged clay deposit using a single direct calculation; that is, do not calculate either the total stress or the porewater pressure. The water content of the submerged clay is 70%, and G_s is assumed to be 2.65.

P3.13. Calculate the change in total stress at a depth of 30 ft in a clay layer if the water table is dropped from the surface to a depth of 10 ft. The clay has a water content of 40%, S_r = 100%, and G_s = 2.80. No change in degree of saturation or water content occurs when the water table is changed.

Problem Set 4

A soil profile consists of 19 ft of sand (submerged unit weight of 60 pcf) on top of 20 ft of normally consolidated clayey silt on top of an impervious rock. A 15-ft-thick layer of sand fill (total unit weight of 110 pcf) is placed on top of this soil profile. For simplicity, we take the water table at the ground surface (prior to placement of the fill), and we assume that the water table remains at the interface between the sand and the fill during settlement, so no submergence correction needs to be made. The average water content of the clayey silt is 70% and G_s is 2.65.

A one-dimensional consolidation test was performed on the clayey silt, and the following stress-strain data were obtained:

Effective Stress, psf	Strain	Effective Stress, psf	Strain
100	0.000	4,000	0.090
250	0.008	8,000	0.137
500	0.015	16,000	0.190
1,000	0.030	32,000	0.246
2,000	0.055	64,000	0.305

P3.14. Compute the changes in the void ratio in the consolidation test and plot the e versus log σ' curve on three-cycle semi-logarithmic paper.

P3.15. Compute the compression index C_c (the slope of the virgin curve).

P3.16. Calculate the total settlement of the embankment using single drainage. Subdivide the soil into four layers. Assume that the settlement caused by the compression of the sand layers is negligible. Calculate the settlement of the clayey silt layer by computing average stresses at the center of the layers. The curve you have plotted is not the same as the true field curve because the soil structure was disturbed by the boring and sampling operation. In calculating the settlement in the field, use the field consolidation curve. To estimate the location of the curve, draw a line from the point $e = e_0$, $\sigma' = \sigma'_0$, tangent to the laboratory virgin curve and use this line as the field curve.

P3.17. The one-dimensional consolidation apparatus used in the laboratory for this test allows water to drain out at both the top and bottom of the specimen, whereas the bottom of the clayey silt layer in the field is impervious. Does this difference in drainage conditions cause the design engineer to predict higher settlement, the same settlement, or less settlement than will occur in the field? Explain.

Problem Set 5

A corrected (field) consolidation curve is defined at the following points:

Effective Stress, psf	Void Ratio	Effective Stress, psf	Void Ratio
250	0.755	4,000	0.740
500	0.754	8,000	0.724
1,000	0.753	16,000	0.704
2,000	0.750	32,000	0.684

P3.18. Plot the curve of effective stress versus void ratio using three-cycle semi-logarithmic paper.

P3.19. Compute the compression index C_c.

P3.20. If the soil stratum in the field is 8 ft thick and the initial effective stress at the center of the layer is 1400 psf, how much additional stress can the stratum support before 0.75 in. of settlement occurs? Remember that the curve you plotted in Problem P3.18 is already the correct field curve.

Problem Set 6

A soil profile consists of 5 ft of sand overlying 40 ft of normally consolidated clay underlain by an incompressible sand, gravel, and then bedrock. The properties of the upper sand are not known, but we estimate that it has a dry density of 95 pcf and a G_s of 2.72. Tests on the clay indicate that it has the following average properties: water content = 40%, G_s = 2.85, and compression index = 0.35. The water table is at the surface.

A wide embankment 20 ft thick is placed at the surface. The dry density of the fill averages 115 pcf, with a water content of 14%.

Calculate the settlement of the surface:

P3.21. By dividing the clay layer into two sublayers, each 20 ft thick.

P3.22. By using the entire clay layer as a single layer and calculating average stresses at the mid-depth.

Problem Set 7

A light industrial building is designed using a reinforced concrete slab resting directly on a 10-ft-thick underlying clay layer. We lack adequate funds for a comprehensive soils investigation, but we do collect the following data: water content of the soil prior to construction averages 25%, liquid limit about 75%, plastic limit about 25%, shrinkage limit about 13%, and water content of a sample allowed to soak under a low applied stress about 45%. We assume that the soil is saturated prior to construction.

P3.23. How much would the building settle if the clay became air-dried and we assume that all shrinkage is in the vertical direction (no cracking)?

P3.24. How much would the building be lifted if the natural clay became soaked instead of being dried?

Problem Set 8

A soil deposit consists of 10 ft of miscellaneous "old fill" over 30 ft of normally consolidated clay over sand and, finally, bedrock. The clay has fully consolidated under the old fill. The water table is at the interface of the fill and clay. The total unit weight of the fill is 122 pcf. The water content of the clay is 49%, the degree of saturation is 100%, and G_s is 2.70. Assume that the fill and lower sand are freely draining. From experience with this saturated clay, the compression index is assumed to be 0.35 and the coefficient of consolidation 0.10 ft²/day. Sixteen feet of additional fill, at a total unit weight of 125 pcf, are placed on the surface.

P3.25. Calculate the settlement caused by compression of the clay layer. Use two layers for analysis of the clay.

P3.26. Calculate the time required to achieve half of the ultimate settlement.

Problem Set 9

A mud flat next to a bay is to be developed for industrial purposes. Sand will be sucked up from out in the bay and pumped as a slurry to the site, where it will be dumped and allowed to drain. This is called a *hydraulic fill.* The soils at the site are tolerably uniform and consist of about 30 feet of very soft, slightly organic mud (clay) overlying sand and pervious bedrock. Sufficient fill must be pumped in so that the top of the fill will be 5 ft above the original ground surface after settlement has ceased. The water table is at the original ground surface at the start of filling and stays at the same elevation throughout the settlement time, so that some of the fill becomes immersed as settlement progresses (in which case it loads the subsoil with its submerged unit weight instead of its total unit weight).

Calculate the total thickness of fill that must be pumped onto the site.

Calculate (as best you can) the time-settlement curve of the ground surface and present it in a plot. Show any assumptions made in the analysis.

The soil properties to be used in the analysis include:

Very soft clay—water content = 70%, G_s = 2.65 normally consolidated, compression index = 0.54, coefficient of consolidation = 0.05 ft²/day.

Fill—G_s = 2.70, dry density = 95 pcf; assume that the degree of saturation is 100% when the sand is below the water table and 30% when it is above the water table.

Problem Set 10

Many borings were made in San Francisco Bay during the design of the Bay Area Rapid Transit (BART) tunnel. The data you will be given came from a one-dimensional consolidation test on a sample from boring DH-21. The samples were taken using a Swedish foil sampler. The soil profile at the site was as follows:

0–21 ft	Water
21–32 ft	Clay: silty, black organic, very soft, many sand lenses
32–37.6 ft	Clay: silty, blue gray, soft, some organic matter, many lenses of clayey fine to medium sand
37.6–38.3 ft	Shells: silty, sandy, gray, loose
38.3–52 ft	Sand: silty, blue gray, loose, many lenses of soft silty clay 1/8 to 1/2 in. thick, some shell fragments
52–53.3 ft	Clay: very sandy, silty, blue gray, soft, numerous sand lenses
53.3–55.4 ft	Clay: silty, blue-gray, soft, many lenses of silty sand 1/4 to 1/2 in. thick
55.4–56.4 ft	Sand: silty, some silty clay layers
56.4–145 ft	Clay: silty, blue-gray, soft, many lenses of fine to medium sand, occasional lenses of shells (struck oil at 107.3 ft)
145–169 ft	Clay: silty, soft to stiff, greenish-gray, some fine gravel

A sample from a depth of 62 ft was used. The overburden effective stress at this depth is estimated to be 1850 psf. The soil was trimmed into a small ring (1.768-in. diameter) and loaded up to a peak pressure of 128,000 psf. The relevant data are summarized on the attached data forms. The recorded total settlements are the final settlements under each load, not S_{100}.

In addition, the time-settlement data for consolidation under the 4000-psf load are given.

Calculate void rations at each pressure.

P3.27. Plot void ratio versus effective stress (log scale) using five-cycle paper.

P3.28. Estimate the position of the field curve. Is the deposit normally consolidated or overconsolidated?

P3.29. Plot settlement versus time (log scale) using five-cycle paper. Determine S_0, S_{100}, S_{50}, and t_{50}. Calculate c_v (use units of in^2/min to match the units used in the plot).

P3.30. Plot the theoretical curve of settlement versus time on top of the experimental curve of Problem P3.29.

P3.31. Repeat Problems P3.27 to P3.29 using strain versus a log p plot and compare the results.

Problem Set 11

P3.32. Samples of compacted, clean, dry sand were tested in a large direct shear machine with a box 254 by 254 mm in area. The following results were obtained:

	Test 1	Test 2	Test 3
Normal load (kg)	500	1,000	1,500
Peak shear load (kN)	4.92	9.80	14.62
Residual shear load (kN)	3.04	6.23	9.36

Determine the angle of shearing resistance ø for the compacted state (dense) and for the loose state (large displacement or residual).

P3.33. Three specimens of nearly (not totally) saturated clay were tested in a direct shear machine. Shear loading was started immediately after the application of the normal load, and testing was completed within 10 minutes of the start of the test. The following results were obtained:

	Test 1	Test 2	Test 3
Normal stress (psi)	145	241	337
Shear stress at failure (psi)	103	117	132

What are the values for the apparent cohesion and angle of internal friction for the clay? What cohesion value would be obtained from an unconfined compression test of the same soil?

P3.34. The following results were obtained from undrained triaxial compression tests on three identical specimens of saturated soil:

	Test 1	Test 2	Test 3
Confining pressure (psi)	70	140	210
σ_1 at failure (psi)	217	294	357
Angle between failure plane and σ_1-plane (deg.)	51°	53°	52°

Determine c_u and ϕ_u for the soil. Estimate the drained (or effective stress) friction angle. From what type of soil would results such as these be expected?

P3.35. The following results were obtained from tests on a saturated clay soil:

a. Undrained triaxial tests:

	Test 1	Test 2	Test 3
Cell pressure (psi)	100	170	240
$(\sigma_1 - \sigma_3)$ at failure (psi)	136	142	134

b. Direct shear tests in which the soil was allowed to consolidate fully under the influence of both normal and shear loads:

	Test 1	Test 2	Test 3
Normal stress (psi)	62	123	185
Shear stress at failure (psi)	73	99	128

Determine values of apparent cohesion and ø for both undrained (total stress) shear strength and drained (effective stress) shear strength.

P3.36. A cohesive soil has an undrained friction angle of 15° and an undrained cohesion of 30 psi. If a specimen of this soil is subjected to an undrained triaxial compression test, find the value of confining pressure required for failure to occur at a σ_1 value of 200 psi.

P3.37. The results of undrained triaxial tests with pore water pressure measurements on a saturated soil at failure are as follows:

	Test 1	Test 2
Confining pressure (psi)	70	350
σ_1 at failure (psi)	304	895
Pore water pressure at failure (psi)	−30	+95

Determine the cohesion and friction angle for failure with respect to both total and effective stresses.

P3.38. Consolidated-undrained triaxial tests with pore water pressure measurements were performed on specimens of a saturated clay. Test results are given below. Plot the Mohr circles at failure and determine the c and ϕ values with respect to both total and effective stresses.

	Test 1	Test 2	Test 3
Cell pressure (psi)	200	400	600
Max. prin. stress dif. (psi)	120	230	356
Pore pressure at failure (psi)	102	200	299

P3.39. A sample of saturated clay has a pore water pressure of −800 psf when unconfined. A total stress equal to 1000 psf is then applied to all surfaces of the sample, and no drainage is allowed to occur. What is the effective stress in the sample? Fully explain your answer.

P3.40. For a saturated clay, $q_u = 2$ tsf. A confining pressure of 100 psi is added under undrained conditions, and a Q-type triaxial compression test is performed. What will c_Q be? Fully explain your answer.

P3.41. A sample of clay was consolidated to a cell pressure of 20 psi. If $c' = 0$, and $\phi' = 22°$, what will the principal stress difference be at failure in a fully drained test?

CHAPTER 4

INVESTIGATION OF SUBSURFACE CONDITIONS

4.1 INTRODUCTION

An adequate and appropriate investigation of the subsurface is critical in the design of the foundation for most projects. On occasion, information on the site can lead to a simple and straightforward approach. In parts of the world, soft soil overlies the founding stratum. Thus, the necessary investigation involves sounding to determine the thickness of the weak deposit so that the length of axially loaded piles can be selected. The situation changes, however, if the piles must sustain lateral loading, requiring detailed information on the soft soil.

For most subsurface investigations, three preliminary activities are proper: (1) gaining information on the geology at the site, as discussed in Chapter 2; (2) a field trip to the site to get specific information related to the design and construction of the foundations (this could be combined with the field trip to obtain geologic information); and (3) meetings with the architect and structural engineers to gain information on the requirements of the foundation.

Meetings with the owner and relevant professionals may lead to an early definition of the general nature of the foundation at the site, whether the foundation is shallow or deep. If the soil is soft, and if the design will require settlement and stability analyses, as for shallow foundations, Ladd (2003) presents a detailed discussion on necessary procedures for subsurface investigation and laboratory testing. If saturated clay exists at the site, the time-dependent behavior of all foundations must be considered. Soil properties are strongly influenced by the installation of deep foundations, but the prediction of the effects must start with a well-designed and effective soil investigation.

Chapter 2 discusses the desirability of a geologic investigation and indicates the availability of geologic information. Information on the geology at the site will be valuable in planning and executing the subsurface investigation.

A field trip will determine the nature of other structures near the proposed site. If possible, information should be gained on the kinds of foundations employed in nearby buildings, any problems encountered during construction, response of the foundations since construction, and if any undesired movements have occurred. On occasion, the results will be available for nearby structures from previous subsurface investigations. Municipal agencies will provide drawings showing the location of underground lines, and power lines should be noted, along with any obstructions that would limit access by soil-boring equipment.

The condition of the site with respect to the operation of machinery for the investigation of the subsurface is important. With regard to movement of construction equipment, countries in Europe are providing guidelines for what is termed the *working platform* (European Foundations, 2004). The purpose of such guidelines is to provide a clear statement on the safety of personnel and machinery on the site. If the site is unsuitable for the operation of boring machines for soil sampling, and later for the operation of construction equipment, such guidelines will inform the owner of the site about improvements required. For example, drainage may be needed, as well as treatment of the surface of the site by an appropriate form of soil stabilization.

Meetings with other professionals on the project are essential. The tolerance of the proposed structure to movement, both vertical and lateral, should be established. The requirements of the requisite building code should be reviewed. Of most importance is the magnitude and nature of the expected loadings, whether short-term, sustained, cyclic, and/or seismic. In some instances, the probability of certain loads may be considered. The maximum loads on the foundations of many offshore structures occur during storms, whose frequency must be estimated on the basis of historical information. Discussions among the principals should address the possible effects of a foundation failure, whether a minor monetary loss, a major monetary loss, or a catastrophic failure with loss of life.

The details of the site investigation should be addressed in meetings of the principals for the project. Field and laboratory testing can be done that will have little relevance to the success of the foundation. On the other hand, evidence is clear that a thorough and proper soil investigation will affect favorably the initial and final costs of the structure. Some contracts include a clause placing responsibility on the general contractor for the subsurface investigation, possibly leading to problems as construction progresses. A better solution is to employ the phased method, noted below, and then to require the geotechnical contractor to comply with specifications tailored for the structure and site conditions.

The importance of investing appropriately in the soil investigation is illustrated in Figure 4.1. If no money is spent, the structure may collapse. Spending a small amount could lead to later expenditures to correct for unequal settlement of the foundation. As shown in Figure 4.1, an optimum amount

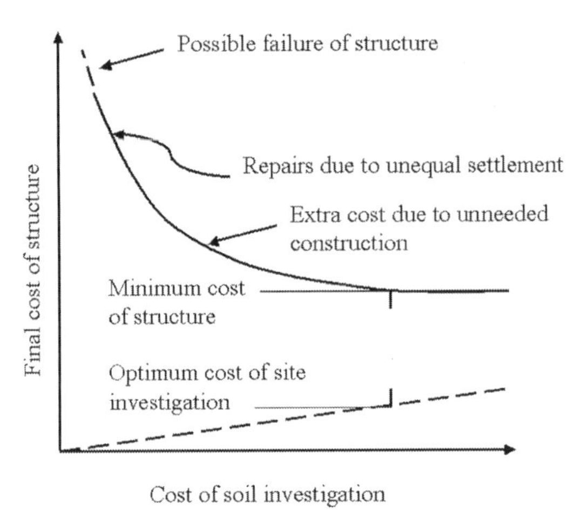

Figure 4.1 The effect of soil investigation on the final cost of structures.

spent on the soil investigation leads to a minimum cost of the structure. If more than the optimum amount is spent, the cost of the structure increases by the cost of the soil investigation, but because the cost of the soil study normally is minor compared to the cost of the structure, the final cost of the structure increases only slightly. If owners and their representatives are aware of the facts presented in the figure, less emphasis will be put on price competition for making a soil investigation.

The ideal procedure for the subsurface investigation for a major structure is to perform exploratory borings to identify the various strata at the site and to determine whether or not the strata are tilted. The final design would then become available, and the nature of the foundation system would be evident. Borings could then be undertaken to obtain the required specimens for laboratory testing and to perform in situ tests if needed. Field loading tests could also be done if needed. This two-stage process would result in the acquisition of precise data for the design of the foundation.

The two-stage process is not possible in some instances—for example, in performing borings for the design of piles for a fixed offshore platform (see Section 4.6). Also, price competition on many projects can lead to a single-stage investigation with a limited number of borings and reduced laboratory testing. Price competition is plainly unwise when specifying a soil investigation.

4.2 METHODS OF ADVANCING BORINGS

4.2.1 Wash-Boring Technique

The use of wash borings is the most common method for advancing a boring because the technique is applicable to any soil, the depth is limited only by the equipment employed, samples can be taken with a variety of tools, and in situ tests can be performed as the borehole is advanced. A typical drilling machine and its associated equipment are shown in Figure 4.2. The derrick is for handling the hollow drill pile that passes through a rotary table, powered by an engine with the necessary power. The hollow drill pipe carries an appropriate cutting tool. A surface casing is set with a T-section above the ground to direct the drilling water to a holding tank. A pump will drive water down the drill pile to raise the cuttings, which may be examined to gain an idea of the formation being drilled. The water is pumped from the top of the tank, with additional water provided as necessary. Drilling fluid can replace the water if caving occurs.

The system can be scaled up for drilling deeper holes and scaled down for hand operation. The Raymond Concrete Pile Company, now out of existence, distributed a movie for the classroom showing the use of a tripod, assembled on site, for raising and lowering the drill pipe by use of a pulley. A small gasoline engine rotated a capstan head used to apply tension to a rope that passed through the pulley to the drill pipe. A small gasoline pump picked up the drilling water from a tank or from an excavation on site. The entire system could be transported with a light truck and assembled and operated by two workmen. The components of such a system are shown in Figure 4.3.

4.2.2 Continuous-Flight Auger with Hollow Core

Borings to limited depths can be made with a continuous-flight auger driven by a powered rotary table. The stem of the auger is hollow, allowing samples to be taken through the stem without removing the auger. Boring and sampling are done rapidly but, even with a powerful rotary table, the penetration of the continuous-flight auger is limited.

Figure 4.2 A typical drilling machine and its associated equipment.

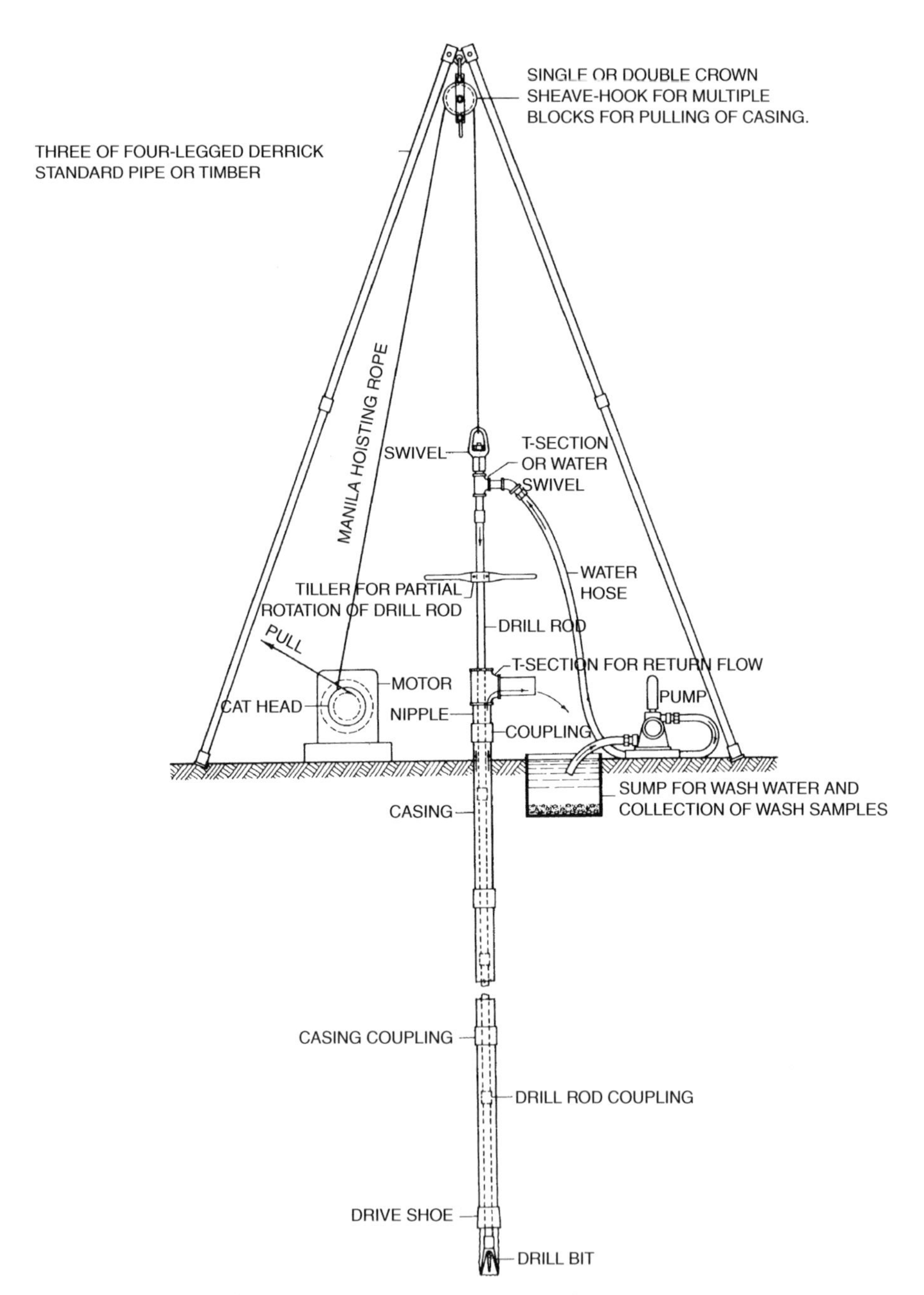

Figure 4.3 Portable wash boring system (from Hvorslev, 1949).

The drill rod fits inside the central pipe of the auger and includes a point that is in place while the auger is advanced. When the desired depth is achieved, the drill rod with the drilling point is removed, a sampling tube is put on the drill rod, the sample is taken by pushing or driving, and the desired sample is retrieved for testing. Alternatively, an in situ testing device, as described below, can be lowered with the drill rod. After testing at a particular depth, the drill rod with the drill point can be replaced and drilling to the desired depth can be done in preparation for the acquisition of the next sample or in situ data.

4.3 METHODS OF SAMPLING

4.3.1 Introduction

In a remarkable effort, the late Dr. M. Juul Hvorslev, working at the Waterways Experiment Station in Vicksburg, Mississippi, and supported by a number of other agencies, presented a comprehensive document on subsurface exploration and sampling of soils for purposes of civil engineering (Hvorslev, 1949). The document has been reprinted by the Waterways Experiment Station and remains a valuable reference.

The term *undisturbed* is used to designate samples of high quality. The ASTM uses the term *relatively undisturbed* in describing the use of sampling with thin-walled tubes (ASTM-D 1587). Disturbance of samples is due to a number of factors: change in the state of stress as the sample is retrieved from the soil, especially if the sample contains gas; disturbance due to resistance against the sides of the sample as it enters the sampling tube; disturbance during transportation to the laboratory; and disturbance as the sample is retrieved from the sampling tube.

Disturbance due to the presence of gas in soils is difficult to overcome. Methods must be implemented to prevent the specimen from expanding throughout the sampling, transportation, trimming, and testing periods. A most severe problem involves the sampling and testing of hydrates that occur in frozen layers at offshore sites.

Sampling disturbance due to resistance against the sample as it is pushed into the sampling tube can be significant. Hvorslev (1949) includes some remarkable photographs of samples that have been distorted in the sampling process (pp. 95, 96, 98, 104, 106, 112, 115, 116). In accounting for the effects of sampling disturbance, Ladd and Foott (1974) proposed experimental procedures for use in the laboratory to determine the strength of most soft, saturated clays.

4.3.2 Sampling with Thin-Walled Tubes

The ASTM has published a detailed standard (1587) for thin-walled tube sampling of soils. The specified dimensions of sampling tubes are shown in Figure 4.4.

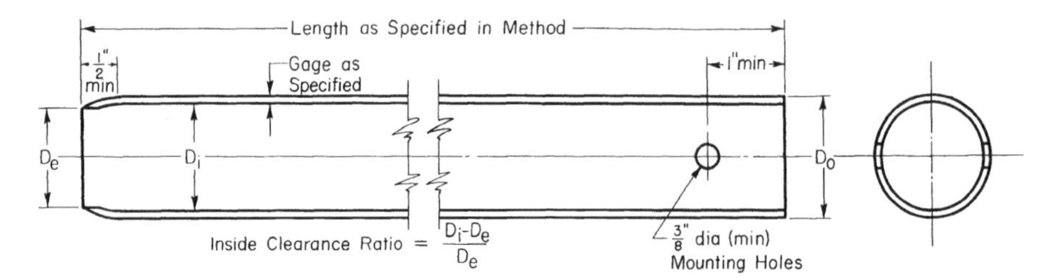

Figure 4.4 Thin-walled tube for sampling (from ASTM D 1587).

Two mounting holes are required for 2- to 3.5-in. samplers and four mounting holes for larger sizes. Hardened screws are required for the mounting. Dimensions for tubes of three diameters are given in Table 4.1, but tubes of intermediate or larger sizes are acceptable. The lengths shown are for illustration; the proper length is to be determined by field conditions.

The preparation of the tip of the sampler is specified, and the clearance of 1 percent is designed to minimize disturbance due to resistance to penetration of the sample. The interior of the sampling tube must be clean, and a coating is sometimes recommended. Procedures for the transportation of thin-walled tubes are specified in ASTM D 4220, Standard Practice for Preserving and Transporting Soil Samples.

Special tools have been developed to eliminate disturbance partially due to the interior resistance of the sampling tool. At the Swedish Geotechnical Institute, a sample more than 2 m long was laid out for examination. The sample had been taken by the Swedish Foil Sampler, which consisted of a short, thin-walled section followed by a thick-walled section in which was embedded a series of rolls of foil. As the sample was pushed into full penetration, the rolls of foil were simultaneously pulled back to eliminate completely resistance due to sample penetration. The description of a similar device is presented in ASTM D 3550, Standard Practice for Ring-Lined Barrel Sampling of Soils. A sketch of the sampling tool is shown in Figure 4.5.

The ASTM Book of Standards presents 13 standards, in addition to D1587, related to surface and subsurface characterization of soil and rock. A list of the relevant ASTM standards is presented in Appendix 4.1.

TABLE 4.1 Suitable Thin-Walled Steel Sampling Tubes

Outside diameter, in.	2	3	5
Wall thickness, in.	0.049	0.065	0.120
Tube length, in.	36	36	54
Clearance ratio, %	1	1	1

Source: ASTM D 1587-83

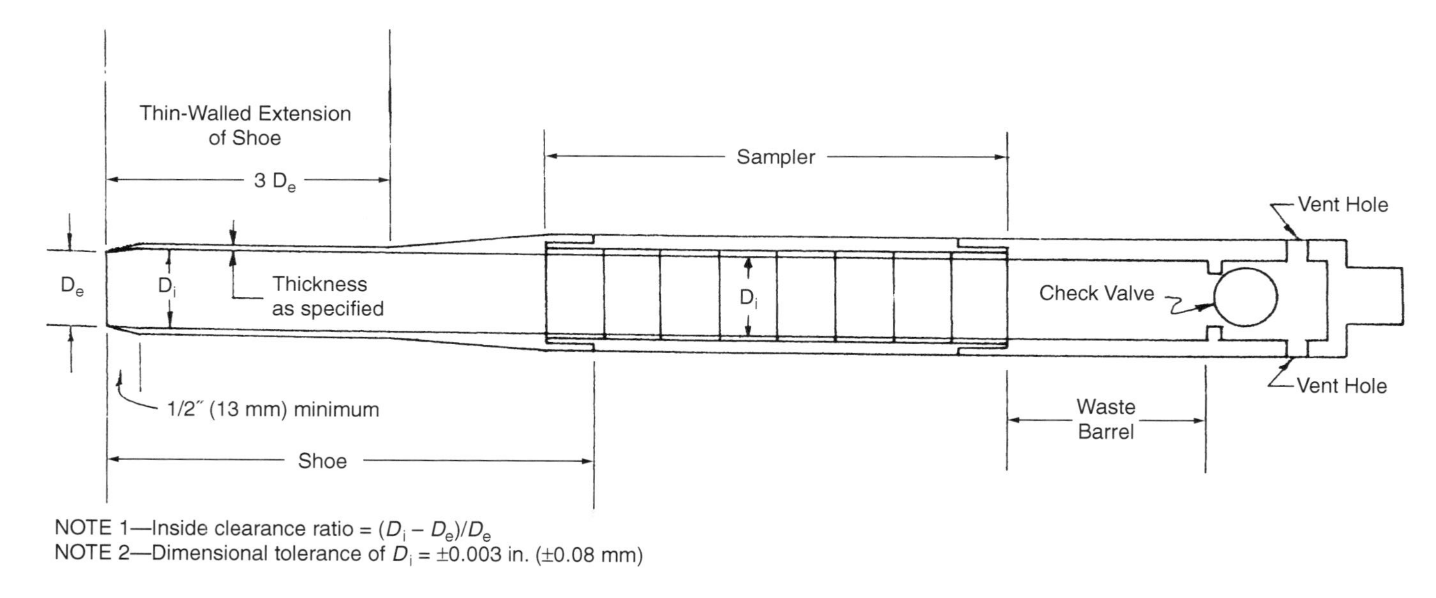

NOTE 1—Inside clearance ratio = $(D_i - D_e)/D_e$
NOTE 2—Dimensional tolerance of D_i = ±0.003 in. (±0.08 mm)

Figure 4.5 Ring-lined barrel sampling assembly (from ASTM D 3550).

4.3.3 Sampling with Thick-Walled Tubes

Exploratory investigations can be made using a thick-walled sampler that has a split barrel. The sampler can be opened in the field and the contents examined in order to log the stratum being bored. A sketch of the sampler is shown in Figure 4.6, taken from ASTM Standard D 1586, Standard Method for Penetration Test and Split-Barrel Sampling of Soils. After the boring has penetrated to the desired depth, the sampler is fixed to sampling rods and the Standard Penetration Test (SPT) can be performed.

The SPT is performed by dropping a 140-lb weight a distance of 30 in. to impact the top of the sampling rods. The blows required to drive the sampler for each of three 6-in. intervals are counted. The *N*-value is the sum of the number of blows required to drive the sampler through the second and third intervals. The SPT has been used for many years as an exploratory tool and sometimes to gain information for design. Undisturbed samples of sand cannot be taken except sometimes in the capillary zone or by freezing. As noted in Chapter 3, correlations have been proposed for values of the friction angle ϕ as a function of N. Even though the sample of sand is disturbed, it can be examined for grain shape and character, and grain-size distribution curves can be developed.

Samples of clay obtained by the split-spoon sampler can be examined in the field to gain information on the character of the deposit. Specimens can be taken to the laboratory for determination of water content and for determining Atterberg limits. However, correlations between shear strength and *N*-value for clay soils are not recommended. Most clays are either saturated or partially saturated, and the porewater pressures in the clay below the impacted sampling tool are certainly affected. Clays can be sampled with the thin-walled tube and tested in the laboratory for strength and other characteristics.

Soil sampling methods are not applicable if the blow count reaches 50 for a penetration of 1 inch. The material can then be sampled by core drilling.

4.3.4 Sampling Rock

Samples of rock can be taken by core drilling, as shown in Figure 4.7, and the standard practice for diamond core drilling for site investigation is given in ASTM D 2113, as shown in Figure 4.6. The *rock quality designation* (RQD) should be recorded, and is defined in percentage terms as determined by summing the length of the sound pieces of core that are at least 4 in. long and dividing that length by the length of core drilled. The RQD will vary, of course, through the thickness of a stratum, and the RQD should be recorded for a specified length of core.

Intact specimens of core can be tested in a triaxial device (ASTM D 2664) or in unconfined compression (ASTM D 2938). Tensile strength and elastic moduli may be tested using intact cores if the values are needed in design.

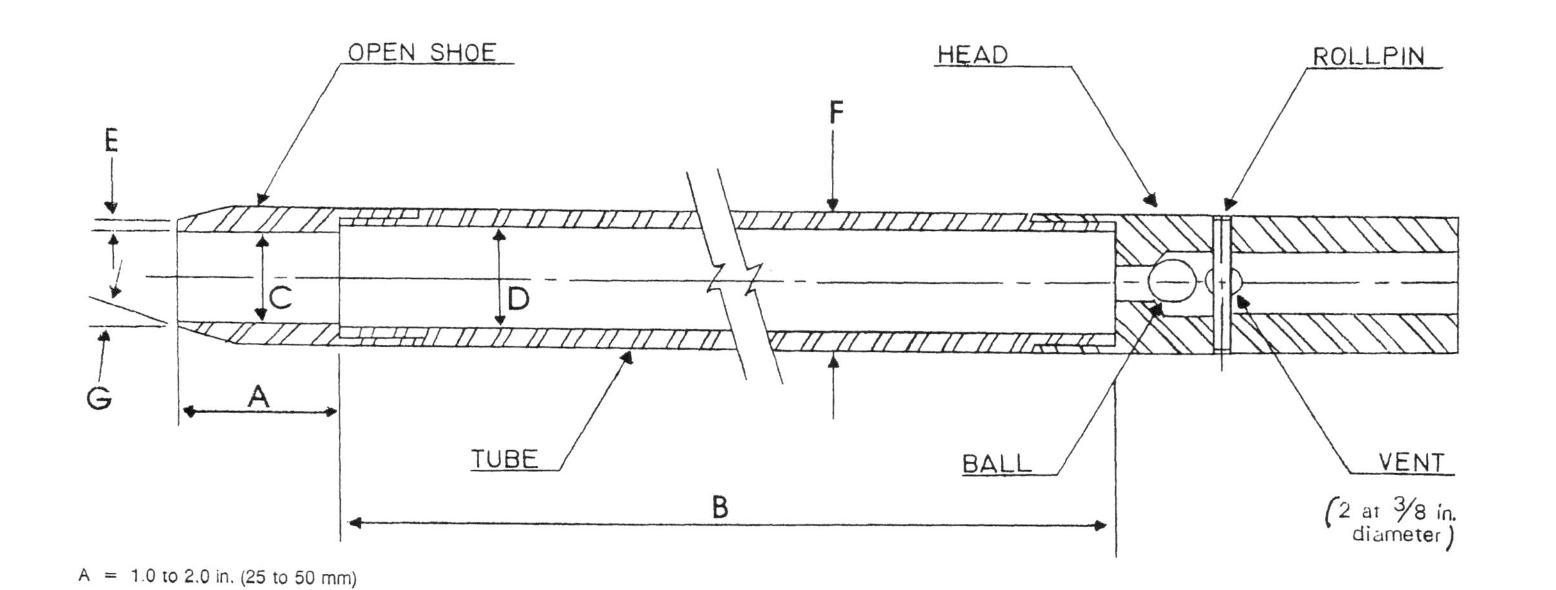

A = 1.0 to 2.0 in. (25 to 50 mm)
B = 18.0 to 30.0 in. (0.457 to 0.762 m)
C = 1.375 ± 0.005 in. (34.93 ± 0.13 mm)
D = 1.50 ± 0.05 − 0.00 in. (38.1 ± 1.3 − 0.0 mm)
E = 0.10 ± 0.02 in. (2.54 ± 0.25 mm)
F = 2.00 ± 0.05 − 0.00 in. (50.8 ± 1.3 − 0.0 mm)
G = 16.0° to 23.0°

The 1½ in. (38 mm) inside diameter split barrel may be used with a 16-gage wall thickness split liner. The penetrating end of the drive shoe may be slightly rounded. Metal or plastic retainers may be used to retain soil samples.

Figure 4.6 Split-barrel sampler (from ASTM D 1586).

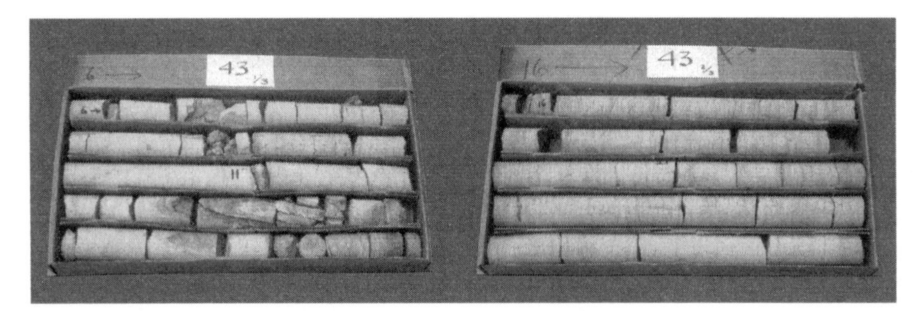

Figure 4.7 Rock-core samples.

During the investigation of the site, the face of the rock for the foundation may be exposed. If so, a comprehensive examination of the face of the rock is recommended (Gaich et al., 2003; Lemy and Hadjigeorgiou, 2003).

4.4 IN SITU TESTING OF SOIL

4.4.1 Cone Penetrometer and Piezometer-Cone Penetrometer

Several types of cone penetrometers are in use, including the mechanical cone, mechanical-friction cone, electric cone, electric-friction cone, and piezometer cone. The mechanical cone has a 60° point angle and a 1.406-in. base diameter. The cone is attached to hollow drill rods and may be pushed down about 3 in. by push rods inside the hollow drill rods while measuring the push force. The mechanical-friction cone has a point of the same size as the mechanical cone and, in addition, includes a sleeve that will be engaged and pulled down after the cone has been pushed down. Two resistances are measured, the resistance from the cone only and the resistance from the cone and the friction sleeve. Sketches of the friction-cone penetrometer are shown in Figure 4.8.

The electric-cone penetrometer has a cone the same size as the mechanical cone, but the resistance to penetration is measured by a load cell at the top of the penetrometer. Readings are taken by an electrical conduit coming to the ground surface. The electric-friction-cone penetrometer is similar to the electric-cone penetrometer, but two measurements on resistance are taken as the device is pushed into the soil. Load cells from strain gauges allow the cone resistance and the friction-sleeve resistance to be measured simultaneously. Sketches of the electric-cone penetrometer and the electric friction-cone penetrometer are shown in Figures 4.9 and 4.10, respectively.

The piezometer-cone penetrometer combines the electric cone with a piezometer that can be read electronically. In addition to yielding the cone resistance, the device gives information on the value of the porewater pressure at the depth of the test. The additional information gives the engineer a way to predict more accurately the characteristics of the stratum being investigated.

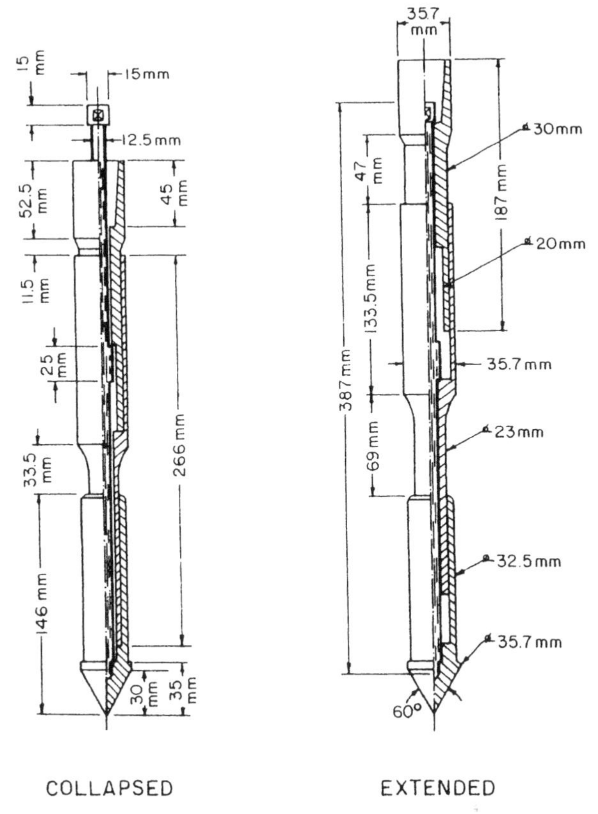

Figure 4.8 Example of a mechanical friction-cone penetrometer tip (Begemann friction-cone) (from ASTM D 3441).

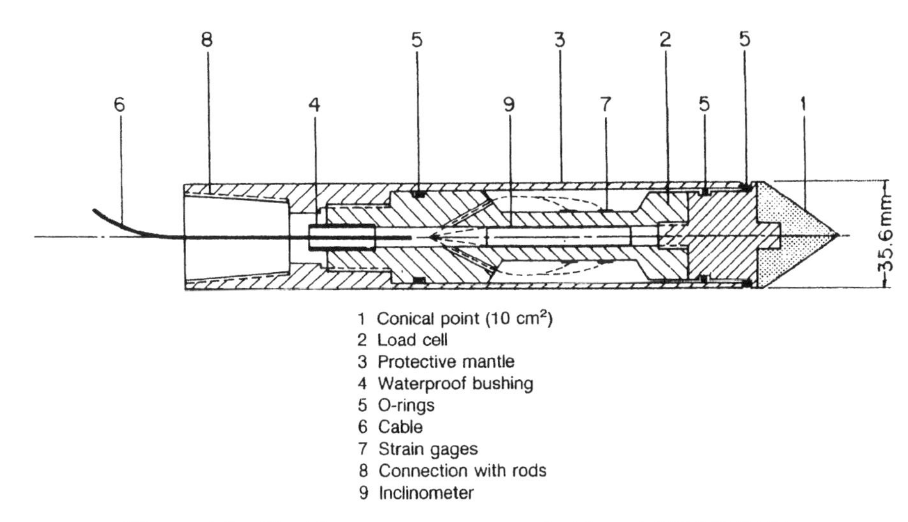

Figure 4.9 Electric-cone penetrometer tip (from ASTM D 3441).

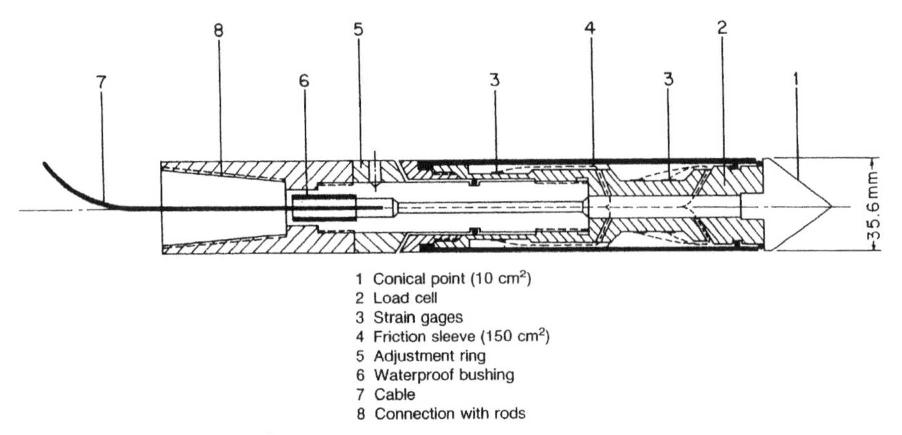

Figure 4.10 Electric friction-cone penetrometer tip (from ASTM D 3441).

ASTM provides the following statement about the quality of the data collected by the cone penetrometer (ASTM D 3441, p. 475): "Because of the many variables involved and the lack of a superior standard, engineers have no direct data to determine the bias of this method. Judging from its observed reproducibility in approximately uniform soil deposits, plus the q_c and f_c measurement effects of special equipment and operator care, persons familiar with this method estimate its precision as follows: *mechanical tips*—standard deviation of 10% in q_c and 20% in f_c; *electric* tips—standard deviation of 5% in q_c and 10% in f_c." If the shear strength of clay is to be determined, the engineer must divide the value of q_c by a bearing-capacity factor; opinions vary about what value of that factor should be employed.

4.4.2 Vane Shear Device

Field vane testing consists of inserting vanes at the ends of rods into soft, saturated soils at the bottom of a borehole and rotating the rods to find the torsion that causes the surface enclosing the vane to be sheared. The torsion is converted into a unit shearing resistance. Two views of typical vanes are shown in Figure 4.11. If the rod used to insert the vane is in contact with the soil, a correction must be made for the torsion on the rod.

With the vane in position, the first test is performed by rotating the rod attached to the vane at a rate not exceeding 0.1° per second, usually requiring 2 to 5 minutes to achieve the maximum torque, yielding the undisturbed shear strength. Then the vane is rotated rapidly through a minimum of 10 revolutions to remold the soil. Finally, the test is repeated to obtain the remolded shear strength of the soil.

The shear strength, s (lbf/ft^2), is found from the following equation:

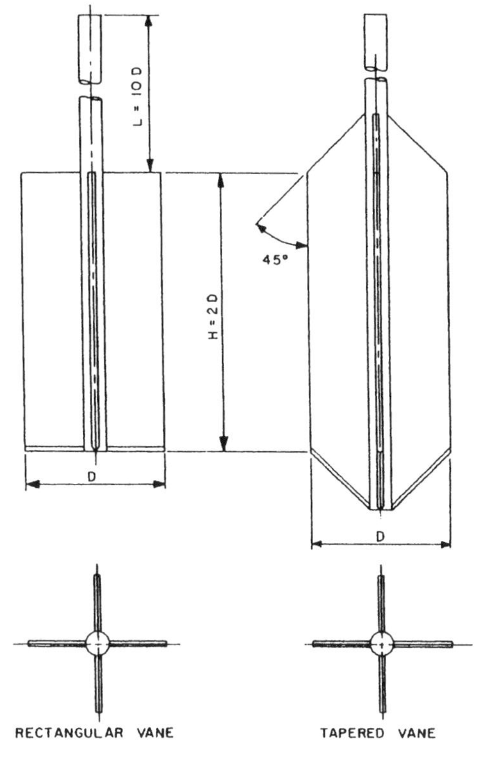

Figure 4.11 Geometry of a field vane (from ASTM D 2573).

$$s = \frac{T}{K} \tag{4.1}$$

where T is torque in lbf ft and K is a constant in ft^3, depending on the shape of the vane. As an example, the value of K for the rectangular vane shown in Figure 4.10 is as follows:

$$K = \left(\frac{\pi}{1{,}728}\right)\left(\frac{D^2H}{2}\right)\left(1 + \left(\frac{D}{3H}\right)\right) \tag{4.2}$$

where

D = measured diameter of the vane, in., and
H measured height of the vane, in.

A problem with no easy solution is one in which the soil has inclusions such as shells. The value of s from the vane would be higher than the actual shear strength and could lead to an unsafe design.

4.4.3 Pressuremeter

The pressuremeter test consists of placing an inflatable cylindrical probe into a predrilled hole and expanding the probe in increments of volume or pressure. A curve is obtained showing the reading of the volume as a function of the pressure at the wall of the borehole. Baguelin et al., (1978) state that Louis Ménard was the driving force behind the development of the pressuremeter. Early tests were carried out in 1955 with a pressuremeter designed by Ménard, a graduate student at the University of Illinois. Ménard later established a firm in France where the pressuremeter was used extensively in design.

The components of the pressuremeter system are shown in Figure 4.12. The probe consists of a measuring cell and two guard cells. The cells are filled with water, and gas is used to expand the cells against the wall of the borehole. The guard cells are subjected independently to the same pressure as the measuring cell, and all three cells are expanded at the same rate, ensuring two-dimensional behavior of the measuring cell. A borehole is dug slightly larger in diameter than the diameter of the pressuremeter probe, with the nature of the soil being noted at the depths of the tests. The test is per-

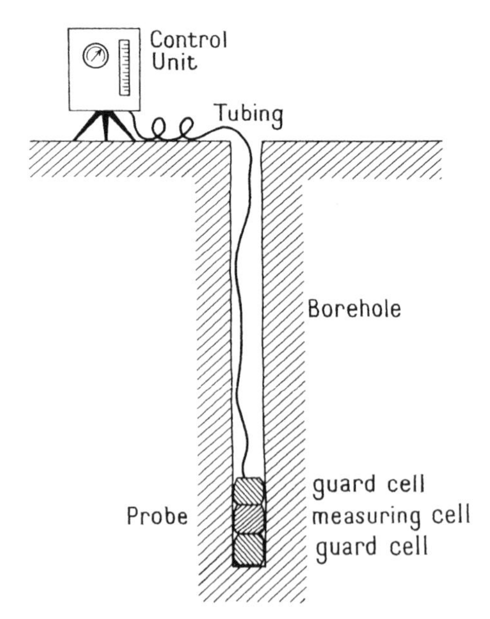

Figure 4.12 Components of the pressuremeter system (from Baguelin et al., 1978).

formed according to a standard procedure—for example, according to ASTM D 4719.

The results from a pressuremeter test are presented in Figure 4.13. The following data apply with respect to the curve: type of soil, silty clay; depth of test, 7 m, and depth to water table, 1.5 m. The following values are from the curve. The volume at the beginning of the straight-line portion of the curve, $v_0 = 170$ cm^3; the volume at the end of the straight-line portion of the curve, $v_f = 207$ cm^3; and the limit pressure, $p_l = 940$ kPa. The volume of the pressuremeter when the pressure was zero was 535 cm^3.

The pressuremeter modulus may be computed from the following equation (ASTM D 4719). Values from Figure 4.13 are shown where appropriate.

$$E_p = 2(1 + v)(V_0 + V_m) \frac{\Delta P}{\Delta V} \tag{4.3}$$

where

E_p = an arbitrary modulus of deformation as related to the pressuremeter, kPa,
v = Poisson's ratio, taken as 0.33,
V_0 = volume of the measuring portion of the probe at zero reading of the pressure (535 cm^3),
V_m = corrected volume reading at the center of the straight-line portion of the pressuremeter curve, measured as (170 + 207)/2), and
$\Delta P / \Delta V$ = slope of the straight-line portion of the pressuremeter curve, (measured as 13.7-kPa/cm^3).

$$E_p = 2(1 + 0.33)(535 + 188.5)(13.7) = 26{,}000 \text{ kPa}$$

Some error is inherent in the value of E_p due to the measurements and the value is an arbitrary measure of the stiffness of the soil, but the engineer can gain some useful information. The most useful information is presented by Baguelin et al. (1968), where correlations are given for the net limit pressure and properties of clay and sand (see Tables 4.2 and 4.3.)

The net limit pressure, p_1^*, is equal to the limit pressure minus the total initial horizontal pressure where the pressuremeter test was performed, as found from the following equation:

$$p_1^* = p_1 - [(z\gamma - u) K_0 + u] \tag{4.4}$$

where

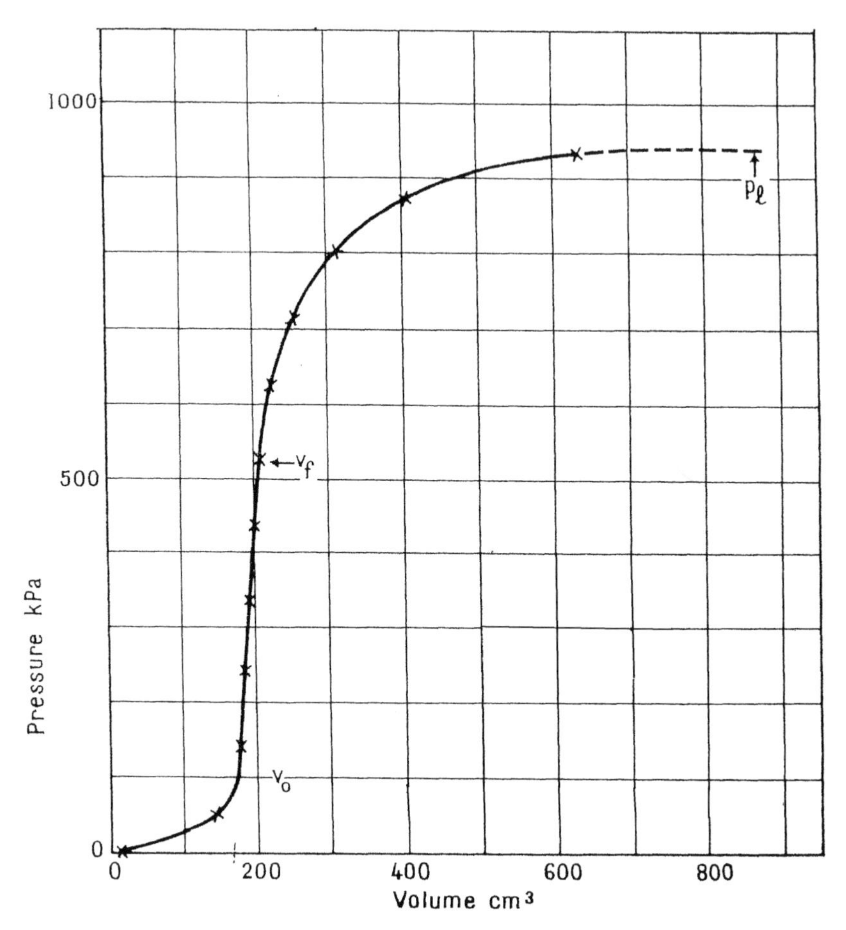

Figure 4.13 Corrected pressuremeter curve (from Baguelin et al., 1978).

z = depth below ground surface where test was performed,
γ = unit weight of soil,
u = porewater pressure, and
K_0 = coefficient of earth pressure at rest.

Use of the pressuremeter that is installed in a predrilled hole results in a pressure that is zero, or lower than the earth pressure at rest if the excavation is filled with water or drilling mud; thus, some creep of the soil could occur inside the borehole. In some instances, the test cannot be performed if the borehole collapses. A self-boring pressuremeter has been developed; it can be installed with a minimum of disturbance of the soil. The advantage of the self-boring pressuremeter is obvious; the disadvantages are that extra time is required to perform the test and, in some instances, the pressuremeter cannot be recovered.

TABLE 4.2 Correlations between Properties of Clay and Limit Pressure from the Pressuremeter Test

p_l^*, kPa	Description	Field Test	p_l^*, psi
0–75	Very soft	Penetrated by fist; squeezes easily between fingers	0–10
75–150	Soft	Penetrated easily by finger; easily molded	10–20
150–350	Firm	Penetrated with difficulty; molded by strong finger pressure	20–50
350–800	Stiff	Indented by strong finger pressure	50–110
800–1600	Very stiff	Indented only slightly by strong finger pressure	110–230
1600+	Hard	Cannot be indented by finger pressure; penetrated by fingernail or pencil point	230+

Source: Baguelin et al. (1968).

TABLE 4.3 Correlations between Properties of Sand and Limit Pressure from the Pressuremeter Test

p_l^*, kPa	Description	SPT N	p_l^*, psi
0–200	Very loose	0–4	0–30
200–500	Loose	4–10	30–75
500–1000	Compact	10–30	75–220
1500–2500	Dense	30–50	220–360
2500+	Very dense	50+	360+

Source: Baguelin et al. (1968).

4.5 BORING REPORT

The quality of a boring report is related to a number of critical features. Most important, a coordinate system must be established before the boring begins. Preferably, the coordinate system should be the one employed by the local governmental entity, but a site-specific system can be established if necessary. Horizontal and vertical controls are necessary. The horizontal coordinates and the elevation of each boring must be shown.

On occasion, the information from the field is analyzed and the boring report is combined with data from the analyses. There are strong reasons to believe that the information from the field should be presented in detail. Information of importance is the weather, personnel on the job and their responsibilities, time needed for each operation, equipment used to advance the boring, measures used to keep the borehole open if necessary, kinds of samples taken, any difficulty in sampling, depths of samples, in situ tests performed and depths, and description of the soils removed.

An engineer-in-training or a registered engineer should be on the job to log the information on the soils recovered from each borehole. As much as possible, the nature of the soils encountered and the grain sizes should be noted. Some field tests may be possible to ascertain the strength of the samples recovered. A hand-held penetrometer can be used to obtain and estimate the strength of clays, and in some cases the use of a miniature vane at the ends of a sample of clay in the sampling tube can reveal in situ shear strength. The engineer on site will have knowledge of the structure to be placed on the site and can acquire relevant information that may be difficult to list at the outset of the work. For example, the engineer will know if the structure is to be subjected to lateral loads and that the character of the soils near the final ground surface is quite important in performing analyses to determine the response of foundations. Special procedures may be necessary to find the strength and stiffness of the near-surface soils.

An open borehole should remain on the site of the exploration for the location of the water table, information that is critical in the design of foundations. Frequently, water or drilling mud is used in drilling the borehole, so drilling fluid must be dissipated in order to get the depth to the water table. If the water table is in granular soil, sounding of the borehole with a tape will reveal its location with little delay. However, if the water table is in fine-grained soil, the borehole must be left open for some time, requiring a workman to return to the site, perhaps on several occasions. After the water in the borehole achieves the same level for some time, the borehole may be filled with grout if required by the specifications.

The engineer must decide how deeply to drill the boreholes for the subsurface investigation. The problem has a ready solution if a founding stratum, such as bedrock, with substantial thickness exists at a reasonable depth. If no preliminary design has been made, the engineer charged with the subsurface investigation must decide on the probable type of foundation based on available information. For shallow foundations, computations should be made showing the distribution of pressure with depth. The borings should be extended to the depth below which the reduced pressure will cause no meaningful settlement. For deep foundations, computations should be made for the distribution of pressure below the tips of the piles to check for undesirable settlement. A problem would occur if the piles are designed as end bearing in a relatively thin stratum above compressible soil.

4.6 SUBSURFACE INVESTIGATIONS FOR OFFSHORE STRUCTURES

The initial step in an offshore investigation is to refer to a detailed map of the topography of the sea bed. Such a survey will be available from the owner of the site. In addition to the location of the structure to be constructed, the topography of the sea bed in the general area should be studied. There may be evidence of the existence of stream beds that have been refilled with soft material. A further step would be to review geophysical tests that have been performed at the site. In the North Sea, it is customary to use high-power sparker profiling with a spacing of 200 to 1000 m (Andresen et al., 1978).

An investigation of the character of the soils at an offshore site usually requires a special vessel, such as that shown in Figure 4.14. The daily cost for the drilling vessel is substantial, so double crews are necessary to allow the subsurface investigation to continue around the clock. A weather window is necessary to allow the vessel to stay on site for the few to several days required to complete the work on site.

Over the several decades that offshore borings have been made, procedures have been developed to allow the work to proceed with dispatch. Each rig includes a mechanism, as shown in Figure 4.14, to compensate for the heave of the ship. The ship goes to the drill site and is anchored in multiple directions; a casing is then placed on the ocean floor, and soil exploration proceeds. Drilling is accomplished through the drill string, and drilling fluid may be used to maintain the excavation if necessary. The common procedure is to lower a sampling tool with a wire line; the wire line can raise a weight, and

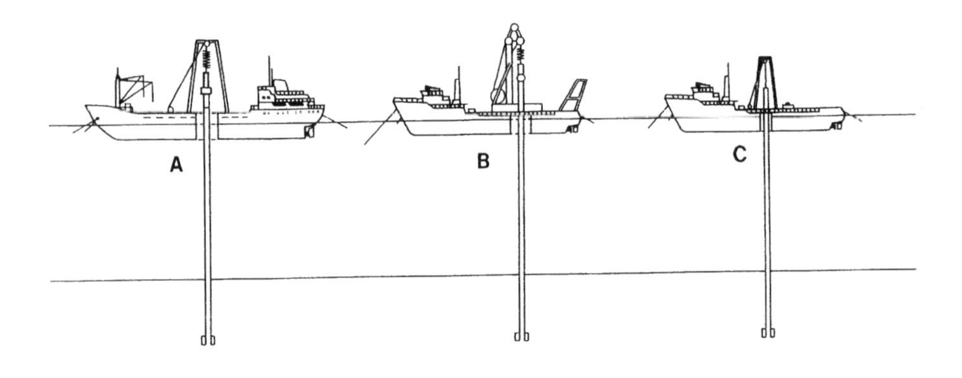

Figure 4.14 Sketches of different types of ships and drilling systems (from Richards and Zuidberg, 1983).

the tool is driven into position. The end of the drill string and the sampling tool are shown on the left-hand side of Figure 4.15 (Richards and Zuidberg, 1983). The sample is retrieved by lifting the wire line, and the borehole is advanced by drilling to the next position for sampling.

An alternative method of sampling is shown on the right-hand side of Figure 4.15. The wire line lowers a hydraulic piston and sampling tube. The tool can be activated from the drill ship, and the thin-walled sampler is pushed into place. A number of other tools may be lowered; however, the hammer sampler is the most common because of the extra work in latching more complex tools to the bottom of the drill string.

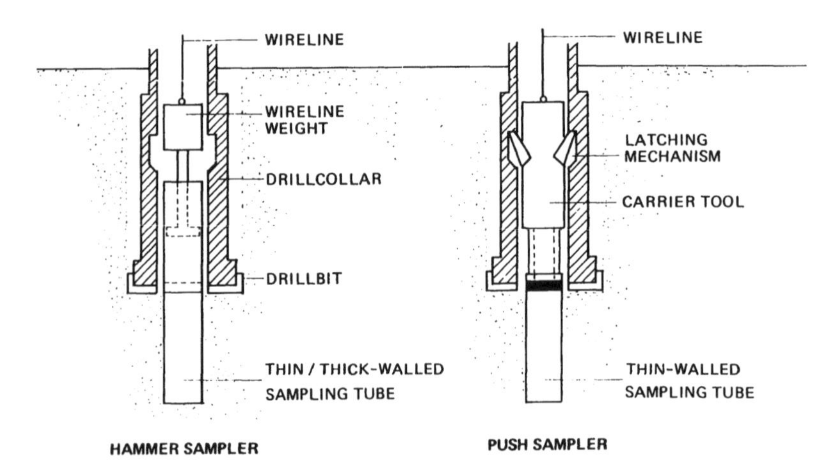

Figure 4.15 Diagrams of two types of wire line samplers (from Richards and Zuidberg, 1983).

Emrich (1971) reported on a series of studies aimed at comparing the results of the properties of clays obtained by various techniques in borings done to a depth of 91 m at an offshore site near the Mississippi River. Samples were taken using a hammer sampler with a diameter of 57 mm, with a hammer sampler with a diameter of 76 mm, with an open-push sampler with a diameter of 76 mm, and with a fixed-piston sampler with a diameter of 76 mm. Unconfined compression tests were performed on samples from each of the three methods of sampling. In addition, field vane tests were performed, and miniature vane tests were performed at the ends of some of the samples within the sampling tubes.

The data were analyzed by plotting the results on the same graph, with the results from the fixed-piston sampler being taken as the correct value. Compared to results from the fixed-piston sampler, the following results were obtained: 57-mm hammer sampler, 64%; 76-mm hammer sampler, 71%; open-push sampler, 95%. The results from the vane tests were scattered but were generally higher than those from the fixed-piston sampler. Sampling disturbance can explain the lower values of shear strength for some of the methods described above. To reduce the effect of sampling disturbance, some investigations subjected the specimens from the hammer sampler in a triaxial apparatus to a confining pressure equal to the overburden pressure. This procedure could be unwise if the soil at the offshore site is underconsolidated, as are many recent offshore deposits.

Many offshore investigations are performed at a site where the loads of the piles have been computed and the borings are to be carried deep enough so that information is available on the required penetration of the piles, employing the guidelines of the appropriate governing authority, such as the American Petroleum Institute (API, 1987). Because offshore platforms are frequently designed to sustain horizontal loads during a severe storm, some piles will be subjected to large tensile loads. The exploration must alert the construction managers if soils at the site may be impossible to penetrate, even with a very-heavy-impact hammer, so that proper procedures can be in place to allow construction to proceed without undue delay.

Factors involving soil properties that are critical to the design of piles under axial loading are numerous. The reader is referred to Chapters 10, 11, 12, 13, and 14 for further reading.

PROBLEMS

P4.1. Prepare a list of eight items you would investigate during a visit to the site of a new project.

P4.2. A meeting has been scheduled with the representatives of the owner of the project, the architect, and the structural engineer. List the questions you would ask to obtain needed information prior to going to the site and initiating the soil investigation.

P4.3. Soil investigations in the past have shown that the soil at the site of the project consists of soft clay over a founding stratum of dense sand. The decision has been made to use axially loaded piles with end bearing in the sand. Describe the techniques you would employ to obtain the thickness of the clay layer and the relative density of the sand.

P4.4. You are the owner of an established geotechnical firm with experience and are to be interviewed by representatives of the owner of a site where a high-rise building is to be constructed about performing a subsurface investigation at the site. (a) List the points you would make in stating that a preliminary investigation should be funded before reaching an agreement for the comprehensive investigation. (b) You need the work; if the owner insisted, would you give a fixed price for the comprehensive investigation?

P4.5. Discuss whether or not, as an owner of a geotechnical firm, you would want an engineer at the site of a subsurface investigation or would be comfortable having an experienced technician in charge.

P4.6. List the in situ methods mentioned in the text in the order of their complexity, with the least complex method first.

APPENDIX 4.1

1992 Annual Book of ASTM Standards, Volume 04.08
Soil and Rock, Surface and Subsurface Characterization,
Sampling and Related Field Testing for Soil Investigation

D 420-87	Guide for Investigating and Sampling Soil and Rock
D 1452-80 (1990	Practice for Soil Investigation and Sampling by Auger Borings
D 1586-84	Method for Penetration Test and Split-Barrel Sampling of Soils
D 1587-83	Practice for Thin-Walled Tube Sampling of Soils
D 2113-83 (1987)	Practice for Diamond Core Drilling for Site Investigation
D 2573-72 (1978)	Test Method for Vane Shear Test in Cohesive Soil
D 2664	Standard Test Method for Triaxial Compressive Strength of Undrained Rock Core Specimens Without Pore Pressure Measurements
D 2936	Standard Test Method for Direct Tensile Strength of Intact Rock Core Specimens
D 2938	Standard Test Method for Unconfined Compressive Strength of Intact Rock Core Specimens
D 3148	Standard Test Method for Elastic Moduli of Intact Rock Core Specimens in Uniaxial Compression
D 3441-86	Method for Deep, Quasi-Static, Cone and Friction-Cone Penetration Tests of Soils
D 3550-84 (1991)	Practice for Ring-Lined Barrel Testing of Soils
D 3967	Standard Test Method for Splitting Tensile Strength of Intact Rock Core Specimens
D 4220-89	Practice for Preserving and Transporting Soil Samples
D 4428/D 4428M–91	Test Method for Crosshole Seismic Testing
D 4633-86	Test Method for Stress Wave Energy Measurement for Dynamic Penetrometer Testing Systems
D 4719-87	Test Method for Pressuremeter Testing of Soils
D 4750-87	Test Method for Determining Subsurface Liquid Levels in a Borehole or Monitoring Well (Observation Well)
D 5195-91	Test Method for Density of Soil and Rock In-Place at Depths Below the Surface by Nuclear Methods

CHAPTER 5

PRINCIPAL TYPES OF FOUNDATIONS

5.1 SHALLOW FOUNDATIONS

A *shallow foundation* generally is defined as a foundation that bears at a depth less than about two times its width. There is a wide variety of shallow foundations. The most commonly used ones are isolated spread footings, continuous strip footings, and mat foundations.

Many shallow foundations are placed on reinforced concrete pads or mats, with the bottom of the foundation only a few feet below the ground surface. The engineer will select the relatively inexpensive shallow foundation for support of the applied loads if analyses show that the near-surface soils can sustain the loads with an appropriate factory of safety and with acceptable short-term and long-term movement. A shallow excavation can be made by earth-moving equipment, and many soils allow vertical cuts so that formwork is unnecessary. Construction in progress of a shallow foundation is shown in Figure 5.1. The steel seen in the figure may be dictated by the building code controlling construction in the local area.

Shallow foundations of moderate size will be so stiff that bending will not cause much internal deformation, and such foundations are considered rigid in analyses. The distribution of stress for eccentric loading is shown in Figure 5.2a, and bearing-capacity equations can be used to show that the bearing stress at failure, q_{ult}, provides an appropriate factor of safety with respect to q_{max}. The equations for the computation of bearing values are presented in Chapter 7.

Shallow foundations can also be designed to support horizontal loads, as shown in Figure 5.2b. Passive pressure on the resisting face of the footing and on the surface of a key, along with horizontal resistance along the base

Figure 5.1 Construction of a shallow foundation in progress.

of the footing, can be designed to resist the horizontal load. Active pressure would occur on faces moving away from the soil, but these may be ignored as being too small to make any difference in the solution.

Factors that influence the selection of a shallow foundation are discussed in Chapter 1. Usually shallow foundations are less expensive than deep foundations, but designs become more complicated as the foundation becomes larger in plan. Significant stress for a mat or larger shallow foundation reaches deeper soils, and the computation of deformation becomes more complicated than for the foundation of moderate size. Not only will the vertical movement

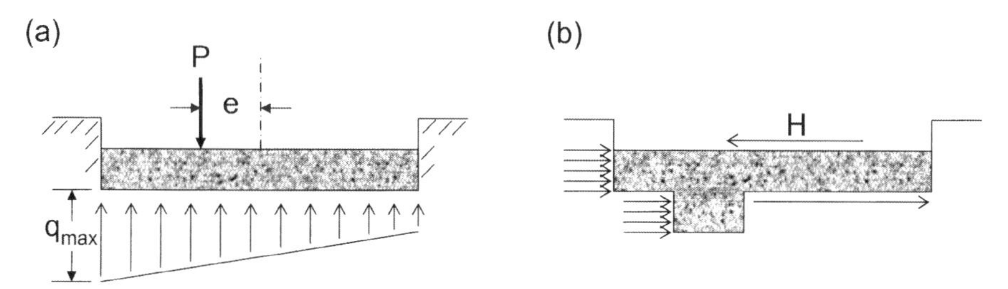

Figure 5.2 Elevation views of shallow foundations. (a) Bearing stress for a foundation under eccentric vertical loading. (b) Stress distribution for a shallow foundation under horizontal loading.

be larger than for a footing, but the deformation of the mat must be considered as well as the deformation of the supporting soil.

The principal problem with shallow foundations under light to moderate loading concerns expansive clay, discussed in Chapter 6. The problem is widespread and can be devastating to homeowners. Engineers must be especially diligent in identifying expansive clay at a building site and taking appropriate actions if such soil is present.

5.2 DEEP FOUNDATIONS

5.2.1 Introduction

Typical types of deep foundations are discussed in the following paragraphs. Entrepreneurs have developed several special and innovative types of deep foundations, and more will continue to be offered by the construction industry. Only one of the special types is discussed below.

Professionals associated with a particular project will make an appropriate study leading to the selection of the particular type of deep foundation. Such a study involves factors related to the structure, subsurface conditions, local practice, and special requirements. For example, the authors worked on a project where deep foundations were to be installed near an elementary school. The noise level was the overriding consideration. The contractor could work only when the school was not in session or had to use a drilled foundation with a noise level lower than would occur during pile driving. The engineers selected the drilled foundation.

Omitted from the presentation that follows is a consideration of deep foundations that may be used for soil improvement, such as sand piles or stone columns.

5.2.2 Driven Piles with Impact Hammer

The engineer frequently makes an extensive and thorough investigation prior to selecting of the type and configuration of a pile for a particular project. The pile must sustain the expected loadings with appropriate safety, and construction must be accomplished in a timely manner while complying with local regulations. In some cities, for example, noise is a major concern and regulations may preclude the use of many driven piles. The types of piles and the factors affecting the selection of a particular pile are noted in the following discussion.

Several later chapters address methods for computing the axial and lateral capacity of a pile. The piles may be timber, reinforced concrete, prestressed concrete, structural-steel shapes, steel pipe, or a tapered-steel pipe. The engineer selects the pile type and hammer on the basis of (1) loads to be supported, (2) tolerance of the superstructure to differential settlement, (3) expected life of the project, (4) availability of materials and construction machinery, (5) length of time required for installation, (6) difficulty of construction, (7) ability to make a proper inspection, (8) noise during construction, and (9) cost.

Environmental effects on the material composing the pile must be considered. Steel in some environments will be subjected to corrosion. In offshore practice, some piles may extend through the *splash* zone, where corrosion must be considered. Two procedures are common: extra-thick steel may be provided to account for progressive loss of the steel with time or a form of cathodic protection may be provided. Timber piles can be treated with creosote to prevent attack from insects, but this does not prevent damage from certain species of marine borers (Grand, 1970). And, as discussed below, special care must be employed in driving reinforced concrete piles to prevent cracking that could lead later to failure due to corrosion of the rebars.

The engineer may need to consider the *drift* of driven piles during installation. The tips of the piles in a group may move close to each other at full penetration and may actually touch each other during driving. The bending stiffness of the piles relative to the stiffness of the soil and the position of the head of the pile are factors that affect the drift. The closeness of the piles after installation may or may not affect their ability to sustain loading. If the final positions of the piles in a group are important, the engineer may stipulate that preboring be employed.

A large variety of impact hammers are available up to the size required for driving very long piles and down to the smallest that can be driven. Hammers may be drop-weight, diesel, single-acting steam, or double-acting steam

The structural engineer and the geotechnical engineer may cooperate in studying the relative advantage of a small number of larger piles and a large number of smaller piles. The selection may be dictated by local construction practice.

With the preliminary selection of a pile, an appropriate computer code should be employed in matching the hammer and cushioning to the pile. The cushioning material may be some form of plastic, or wood of various sorts. Plywood sheets are frequently used as cushioning. The computer code models a compressive wave yielding stress versus time, generated by the impact of the hammer, taking the cushioning into account. The wave travels down the pile and is reflected at the pile tip. For driving against bedrock, a compressive wave is reflected. For driving against very weak soil at the tip, the wave is reflected as a tensile wave. Tension in a reinforced concrete pile can cause cracking that could result in penetration of water, leading to corrosion. Some reinforced concrete piles under bridges have been so damaged due to tensile cracks as to need replacement.

The proper cushion is important. The authors were asked to review a project where the cushioning was provided by sheets of plywood. The plywood was replaced irregularly and actually caught fire on occasion. The result was cracking of some of the reinforced concrete piles due to compressive forces.

One of the authors participated in driving an instrumented, closed-ended steel pipe with a diameter of 6 in. A blow from a drop hammer caused the pile to move down in soft clay but with very little permanent set. Persistent driving was damaging the instrumentation, and the decision was made to use a coiled spring from a rail car as the cushioning. The pile drove readily, with no damage to the instrumentation. The decision to use the soft cushion was made by judgment, but the implementation of a computer code to model the pile driving would have been useful.

An impact hammer resting on a driven pile is shown in Figure 5.3. A continuous-flight auger is shown that can be used to predrill a small-diameter hole to a given distance to ease the driving into some formations. The photograph shows disturbance to the soil in the vicinity of the pile, which must be considered when computing pile capacity. Vibrations and soil movement due to pile driving can cause severe damage to existing structures and has been investigated in detail (Lacy and Gould, 1985; Lacy and Moskowitz, 1993; Lacy, et al, 1994; Lacy, 1998).

Some piles are driven with vibratory hammers, principally steel sheet piles used for retaining structures. Bearing piles are installed with vibratory techniques in some soils, and vibratory methods are frequently used to pull casings that were installed for the construction of drilled shafts. Vibratory hammers are less noisy than impact hammers and can result in less movement of the soil when driving into some sands.

5.2.3 Drilled Shafts

The design and construction of drilled shafts are discussed in detail in many publications (e.g., Reese and O'Neill, 1988). Drilled shafts (called *bored piles* in some countries) are constructed by predrilling a cylindrical excavation while keeping the excavation from collapsing by appropriate means, placing a rebar cage, and then filling the excavation with concrete. A typical drilling rig for constructing drilled shafts is shown in Figure 5.4.

The drilling machine may be mounted on one of three types of carriers: a truck, a crane, or a crawler. The smaller machines are truck-mounted and may move readily along a public road. The soil-filled auger is visible on the truck-mounted machine in Figure 5.4, and spinning will dislodge dry soil. Excavations for light loads may be made with diameters as small as 12 to 18 in. and to depths of a few feet. Excavations for massive loads may be made with diameters of 15 ft or more and to depths of 200 ft. The drilled-shaft industry provides technical information and training through the International Association of Foundation Drilling (ADSC) in Dallas, Texas.

The engineer develops an appropriate design by sizing the drilled shaft for a particular application, but significant effort is necessary in preparing of specifications for construction. The engineer who made the design and prepared the specifications should manage the field inspection. Drilled shafts are

Figure 5.3 Impact hammer atop a driven pile.

a popular type of deep foundation, but as with other deep foundations, special care must be taken by the engineer to ensure proper construction.

Drilling machines are fitted with powerful engines to drive a rotary table and kelly. A variety of drilling tools are available. The appropriate drilling tool operating with downward force from cables or with a weighted drill string can be used for drilling into rock. Rock sockets are common to accommodate design requirements, giving the drilled shaft an advantage over some other types of deep foundations.

The following sections describe three types of constructions in common use, but the details of each may vary with the contractor. The specifications for construction prepared by the engineer must be prescriptive in some in-

Figure 5.4 Construction of a drilled shaft with a truck-mounted unit.

stances (e.g., giving the required slump of concrete for the particular job) but should lean toward the performance desired from the drilled shaft. Prescribing a particular construction method is usually unwise because anomalies in the soil profile could lead to the use of a wet method even though the dry method appeared feasible at the outset.

Dry Method of Construction The dry method of construction may be employed in soils that will not cave, slump excessively, or deflect inward when the hole is drilled to the full depth. A type of soil that meets these requirements is stiff clay. The water table may be located in the stratum of clay, and above the water table the clay may be saturated by capillarity. A problem occurs if the clay below the water table contains fractures that allow water to flow into the excavation. If such fractures were not observed in the soil investigation, the dry method of construction may have been specified and then may have become impossible to achieve.

The steps in employing the dry method of construction are shown in Figure 5.5. A crane-mounted drilling machine is positioned as shown in Figure 5.5a. The location has been surveyed and staked to give the contractor precise information on location and on the final position of the top of the shaft. Specifications will inform the contractor about tolerance in placement and in deviation from the vertical as the shaft is advanced. A temporary surface

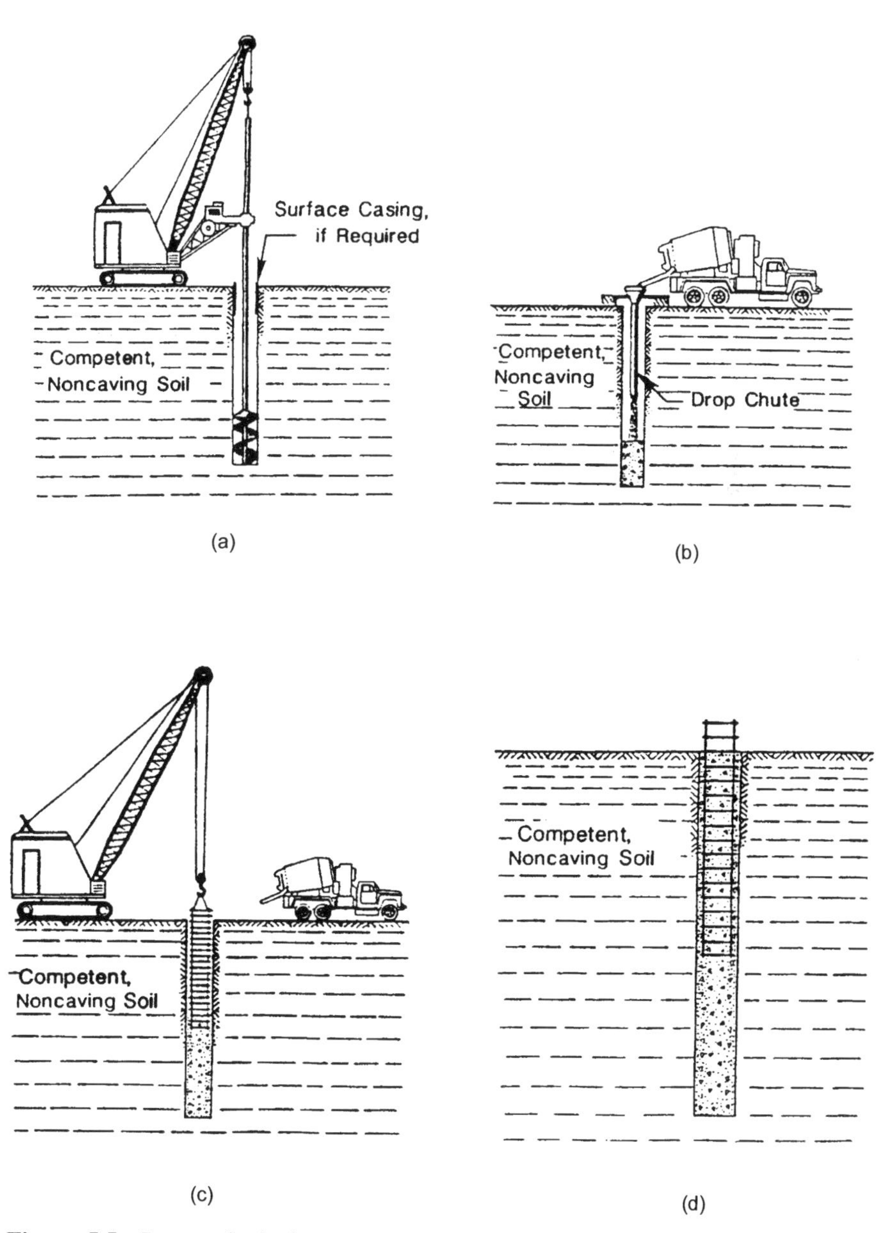

Figure 5.5 Dry method of construction of a drilled shaft: (a) start of drilling; (b) placing concrete of bottom of shaft with a drop chute; (c) placing partial length rebar cage; (d) completed shaft (from Reese and O'Neill, 1988).

casing is frequently placed after the excavation is advanced a few feet. The surface casing prevents raveling of the soil at the surface and provides a positive guide for inserting the auger as drilling proceeds.

Figure. 5.5b shows concrete being placed in the bottom of the excavation, where the computed stresses in the drilled shaft show that no reinforcing steel is needed. Specifications almost always state that the concrete must be poured without striking the sides of the excavation or any obstruction to prevent segregation during placement, and the drop chute serves to guide the concrete in free fall. Research has shown that concrete may fall great distances without segregation if no obstruction is encountered during falling.

The placement of the rebar cage is shown in Figure 5.5c, and the final concrete is placed by use of a tremie or by pumping. Guides are placed on the rebar cage to ensure centering. A service crane is required to hold the tremie or the pump line, as shown in the figure. Alternatively, the crane with the drilling machine could do the work, but further drilling would be delayed. Specifications frequently require that the concrete be placed the same day the excavation is completed to prevent time-dependent movement of the soil around the excavation. The completed shaft is shown in Figure 5.5d.

Casing Method of Construction The casing method of construction may be employed where caving soils are encountered, as shown in Figure 5.6a. Slurry, either from bentonite or polymer, is introduced when the caving soil is encountered and when drilling proceeds through the caving layer and into cohesive soil below. The casing is placed, and the bottom is sealed into the cohesive soil. Prior to placing the casing, the slurry is treated to remove excessive amounts of inclusions and to ensure that specifications are met for properties of slurry before placing concrete. The contractor twists and pushes the casing to make a seal that prevents the slurry from entering the excavation below the casing.

Figure 5.6b shows a crane-mounted drilling unit inserting a drill through the casing and drilling into the cohesive soil below. The excavation below the casing is smaller than that used in the initial drilling, and the difference in size, not as great as indicated in the figure, must be taken into account in computing the geotechnical capacity of the shaft. Figure 5.6c shows that an underream has been excavated at the base of the excavation and that the casing is in the process of being retracted. The fluidity of the concrete and the retained slurry are very important. The concrete must be sufficiently fluid that the excavation will be completely filled and the slurry will be ejected from the excavation. The cleaned slurry must be free of inclusions and easily displaced from the excavation by the fluid concrete. Specifications address the desirable slump of concrete and the characteristics of the slurry.

Figure 5.6c shows the slurry at the ground surface. Preferably the slurry is directed to a sump, where a pump sends the slurry to a tank for cleaning and reprocessing. The disposal of the slurry must meet environmental standards. Slurry from polymers is usually much more easily disposed of than slurry from bentonite. The completed shaft is shown in Figure 5.6d.

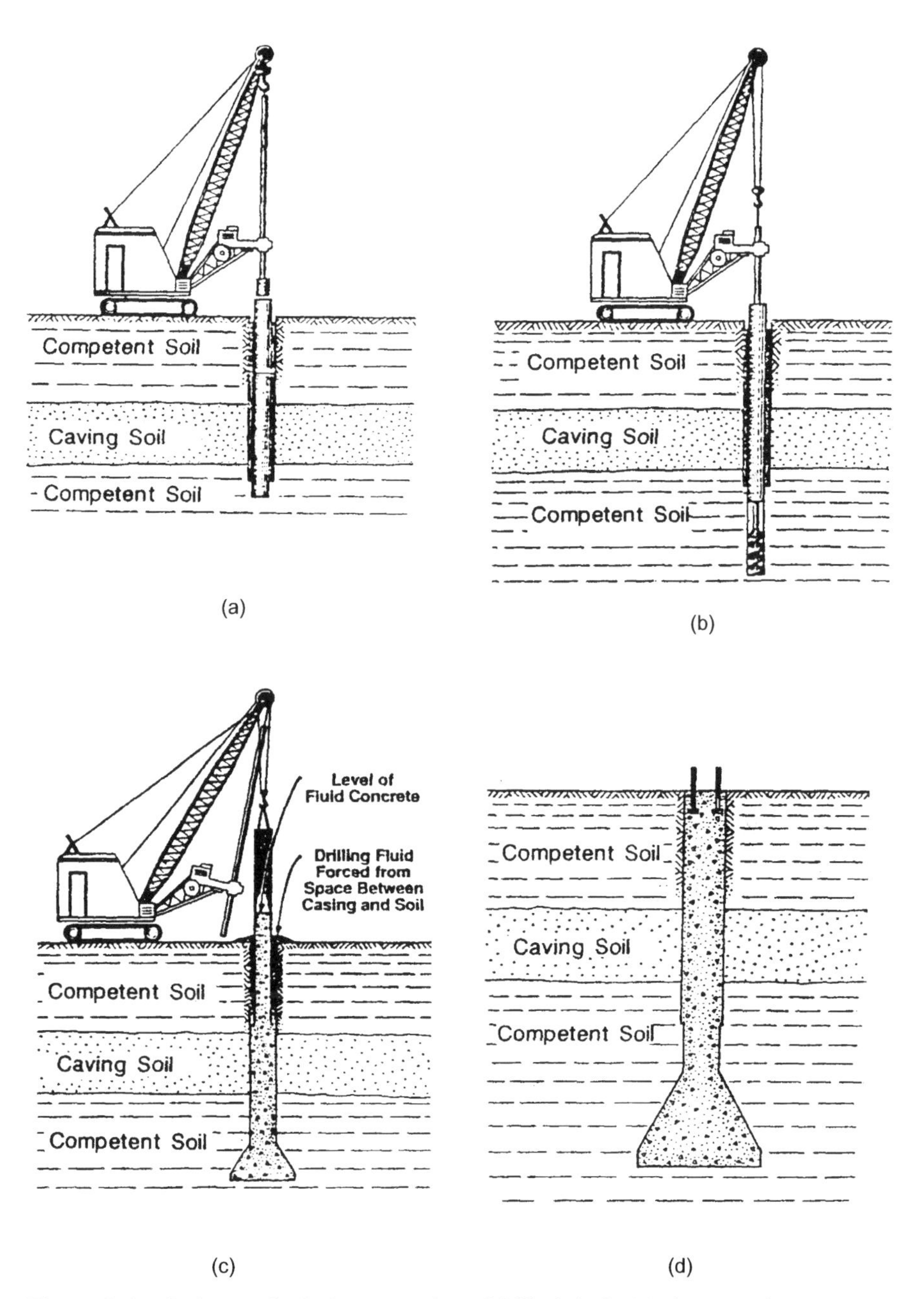

Figure 5.6 Casing method of construction of drilled shaft: (a) slurry used to penetrate caving soil and casing placed; (b) smaller auger used to drill through casing; (c) casing retrieved after drilling underream; (d) completed shaft (from Reese and O'Neill, 1988).

Wet Method of Construction The wet method of construction permits the rebar cage to be placed into a drilled hole filled with fluid. As with the casing method, the drilling fluid is slurry, as shown in Figure 5.7a. The excavation is made to the full depth with slurry, with the contractor exercising care to keep the height of the column of fluid in the excavation above the water table. The slurry acts to create a membrane at the wall of the excavation in the caving soil, usually a granular material, and any flow of fluid will be from the excavation into the natural soil.

The rebar cage may be placed directly into the fluid column, as shown in Figure 5.7b. Prior to placing the cage, samples of the slurry are taken from the excavation, with most samples taken from the bottom, where suspended particles collect. Not shown in the figure is the system for pumping the slurry from the excavation, directing the pumped fluid to a container, usually a tank, where the slurry can be cleaned with screens and centrifuges if necessary. Specifications are available from the owner of the project or from standard specifications regarding such characteristics as the sand content of the slurry, the pH, and the viscosity. The aim of the specifications is to ensure that no debris will collect at the bottom of the excavation to interfere with load transfer in end bearing and that no slurry remains along the sides of the excavation to interfere with load transfer in skin friction.

The placing of the concrete is shown in Figure 5.7c using a tremie and a concrete bucket. The bottom of the tremie is sealed with a plate that detaches when the tremie is charged fully with concrete. Alternatively, a plug is placed in the tremie to separate the concrete for the slurry and moves downward as the concrete is placed in the tremie. The plug, perhaps made of foam rubber, is compressed and remains in the concrete or floats upward in the column of fluid concrete. The completed shaft is shown in Figure 5.7d.

Plain water can sometimes be employed as the drilling fluid. One of the authors worked at a site in Puerto Rico where the founding stratum was a soft rock with joints and cracks. The water table was high, and drilling dry was impossible. Water was employed as the drilling fluid, and the level of water in the excavation was kept above the water table to prevent inward flow and possible weakening of the founding stratum.

5.2.4 Augercast Piles

Augercast piles (also known as *augered-cast-in-place* piles) are constructed by turning a continuous-flight auger with a hollow stem into the soil. On reaching the desired depth, grout is pumped through the hollow stem as the auger is withdrawn, forming a column of fluid grout. A short rebar cage can be set by gravity or sometimes by the use of vibration. In addition, if desired, a single rebar can be placed in the hollow stem before grout is pumped. These piles are termed *continuous-flight-auger* (CFA) piles in Europe and have received wide acceptance there and elsewhere.

Among the construction problems that have been encountered with the use of augercast piles are the sudden retrieval of the auger, allowing the soil to

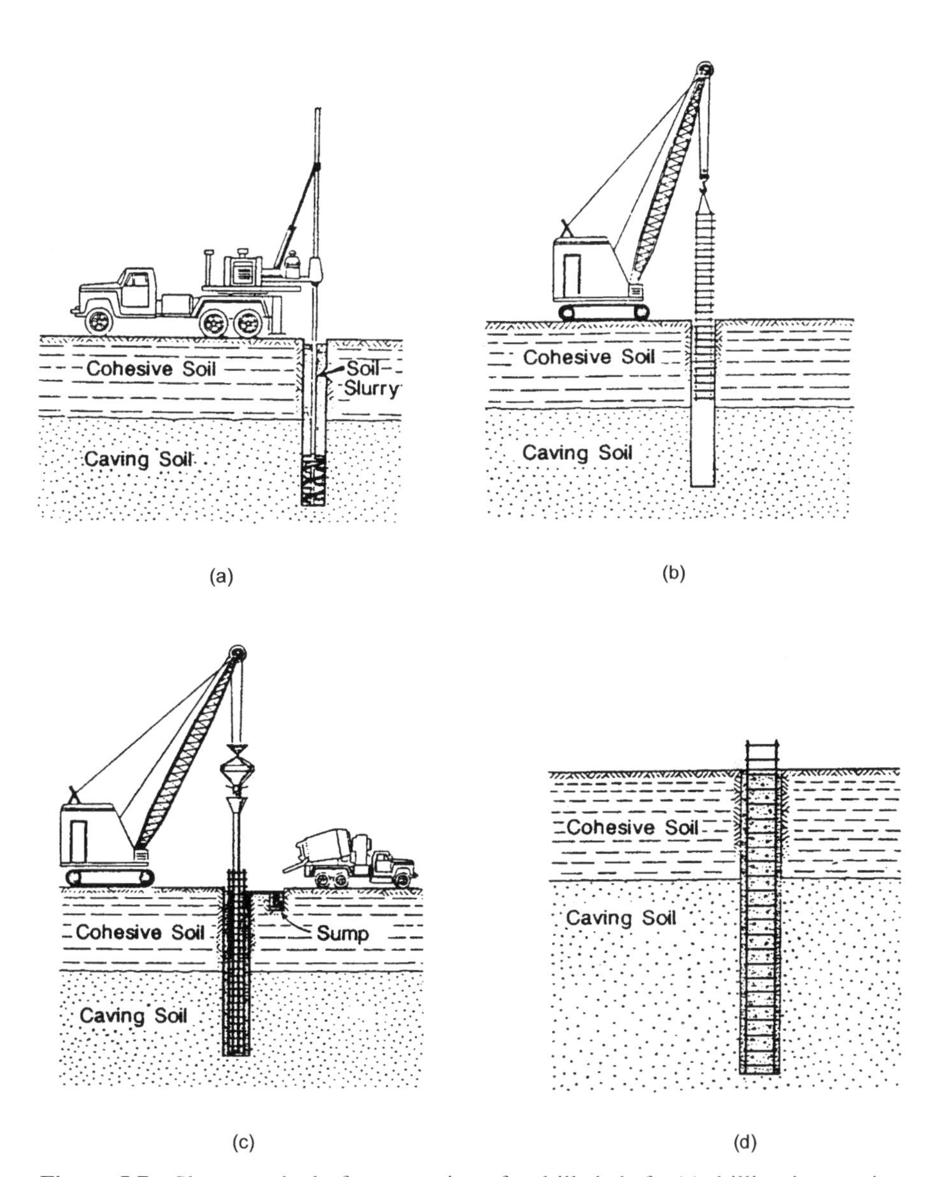

Figure 5.7 Slurry method of construction of a drilled shaft: (a) drilling into caving soil with slurry; (b) placing rebar cage; (c) displacing slurry with fluid concrete; (d) completed shaft (from Reese and O'Neill, 1988).

squeeze in and cause a neck in the pile; the continued rotation of the auger after it reaches the desired depth, with the result that soil is "mined"; and upward flow of water around the sides of a pile or through the center opening when installations occur where the water table is high. The mining of soil during drilling by an unqualified or careless workman can lead to a collapse of the ground surface in the area around the pile. Some problems can be eliminated by the use of instruments on the drilling machine. Such instruments produce a record showing the rate of rotation of the auger as a function of depth and the amount of grout that is pumped as a function of the depth of the auger.

Loss of ground due to mining of subsurface soils is more of a problem when constructing an augercast pile in sand than in clay. Some engineers are reluctant to specify the use of augercast piles in sand unless careful inspection is programmed.

A potential problem with augercast piles is possible eccentricity during the application of loading (Siegel and Mackiewitz, 2003). This problem may arise because of the lack of reinforcement except for the short cage near the top of the pile and a single bar along the axis.

Lacy et al. (1994) have discussed the use of augercast piles to reduce or eliminate the impact on adjacent structures. Driving piles will cause vibration that can affect nearby structures and can also cause the settlement of deposits of loose sand that can potentially cause great damage to structures in an urban setting. Augercast piles have proved to be very useful in some settings, but construction must be monitored carefully to ensure a sound and competent foundation.

One of the authors has observed the use of CFA piles in the overconsolidated clay near London. The drilling rigs are invariably instrumented, and axial-load tests are frequently performed at a site to prove the quality of the construction and to confirm the results of the analyses of axial capacity. The system for testing the pile is often controlled by a computer, and the working load can be confirmed by maintaining the load overnight or longer without the presence of a technician.

5.2.5 GeoJet Piles

The GeoJet pile is described as an example of a special kind of deep foundation. Numerous kinds of special deep foundations have been developed and employed in some projects. The engineer faces the problems of designing for the support of axial and lateral loading and of preparing specifications for construction that allow for inspection to ensure good quality. With regard to design, the properties of the soil will be affected by the method of installation. Therefore, the engineer must depend heavily on the results of load tests with the special foundation at the site or as reported in the technical literature where all relevant details are presented.

The GeoJet pile is constructed by rotating a special drilling head into the soil and introducing grout through the drilling head during withdrawal. Figure 5.8 shows a typical drilling head attached to a kelly prior to initiating construction at a site. If construction conditions are favorable, the column of soil cement can be constructed in less than 5 minutes. After the column of fluid soil-cement is created, a steel insert, such as a pipe, is placed to the full depth of the pile by gravity with minor driving. Tests of samples of soil-cement made with the soil from the construction site lead to the desirable percentage of cement to give the soil-cement the necessary strength and the desirable percentage of water so that the steel insert can fall under its own weight. The concept employed in construction is that sensors in the system will provide data to a computer on the specific gravity of the grout, the amount of grout being pumped, the measurements of pressures at points in the system, the

Figure 5.8 Drilling head used in construction of the GeoJet pile.

rate of rotation of the auger, and the force encountered in drilling. A printed record may be made showing relevant data as a function of depth. The record is valuable in evaluating the quality of the construction.

Some data have been collected on load tests with piles constructed by the GeoJet system, but the concept of design is that pile tests are performed at a site with designs of soil-cement based on the tests noted above. Some data have been collected on GeoJet piles that have been tested (Spear et al., 1994; Reavis et al., 1995).

5.2.6 Micropiles

Micropiles are deep foundations with a small diameter. They may be installed in a variety of ways and have several purposes. Bedenis et al. (2004) describe the use of micropiles as load-bearing elements to strengthen foundations that were to support a greater load. The piles were installed by drilling, and were employed because pile-installation equipment could not be operated in the available space inside the building. The piles consisted of a high-strength steel bar with a diameter ranging in size from 32 to 63 mm, grouted into place, and forming a pile 152 mm in diameter. The piles extended through overburden soils to sandstone at a depth of 24.7 m. The capacity was developed principally through side resistance in the sandstone.

Gómez et al. (2003) describe the testing of a micropile with a diameter of 219 mm and a length of 7.13 m, consisting of a central casing with a diameter of 178 mm that was embedded in grout. The soil at the test site consisted of 3.0 meters of residual soil above rock. The instrumentation revealed that the bond between the rock and the grout did not fail, but some loss of load transfer in skin friction was noted as the movement of the pile relative to the soil increased above 0.05 mm.

If micropiles are used to sustain compressive loads, the engineer should compute the buckling load by using a computer code. The unfactored lateral load should be applied, and the compressive load is increased in increments until failure occurs by excessive lateral deflection. Pile-head fixity must be modeled as well as possible or the analyses should be performed with a range of pile-head fixities that encompass the value to be anticipated.

5.3 CAISSONS

Caissons are often used as foundations for bridges, and may be large in cross section and seated deep beneath the mudline. While the sides of the excavation will be in contact with soil, the caissons may be designed as end-bearing units if the founding stratum is bedrock or strong soil.

Shuster (2004) gives an example of the construction of a caisson in discussing the foundations for the new Tacoma Narrows Bridge. Steel boxes, 18 ft high and 8 by 130 ft in plan, were lowered to the sea bed at a depth of

150 ft. Concrete lifts with a height of 15 ft were added to the top of the steel boxes as they were lowered. Dredging was done within the cells of the caisson until the bottom reached a depth of 60 ft.

5.4 HYBRID FOUNDATION

A hybrid foundation consists of both a soil-supported mat and piles and is used principally to support an axial load. In Europe it is commonly called a *piled raft* because engineers conceived the idea of designing the foundation for high-rise buildings using a mat (raft) resting on the ground with piles supporting the mat (Figure 5.9). The concept was that the combined foundation would be sufficient to support the applied axial loading with an appropriate factor of safety and that the settlement of the combined foundation at working load would be acceptable. The settlement of a mat foundation is

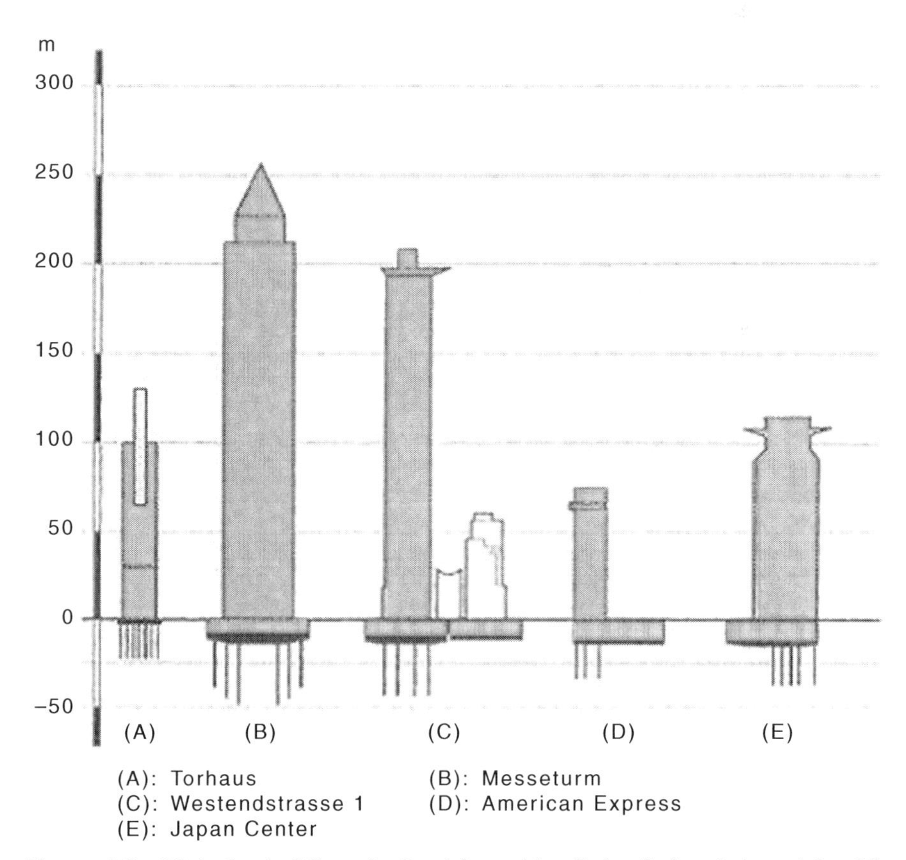

Figure 5.9 High-rise buildings in Frankfurt with piled raft foundations (after El-Mossallamy and Franke, 1997).

dish-shaped, with the largest settlement at the center of the mat. To achieve a more uniform settlement of a structure, it has been suggested that the piles be clustered near the center of the mat.

The analysis of such a system is complicated because the settlement of the raft is affected by the presence of the piles and because a piled raft foundation consists of conventional piles and a rigid raft, as shown in Figure 5.10. Considering each of these foundation elements separately leads to the conclusion that interaction is inevitable. The mat alone is certainly affected by the presence of the piles because the foundation is much stiffer than with the soil alone. The piles alone are affected by the earth pressure from the raft because the increased lateral stresses on the piles affect the capacity for side resistance. The problem can be solved by using the finite-element method, where appropriate plate or solid elements can be used for modeling the raft. Beam elements can be used for modeling piles. The soil around the piled raft system can be conveniently modeled as solid elements. Modeling the problem by the finite-element method, which is widely available, was found to be practical and yielded reasonable agreement between results from analyses and from experiments (Novak et al., 2005).

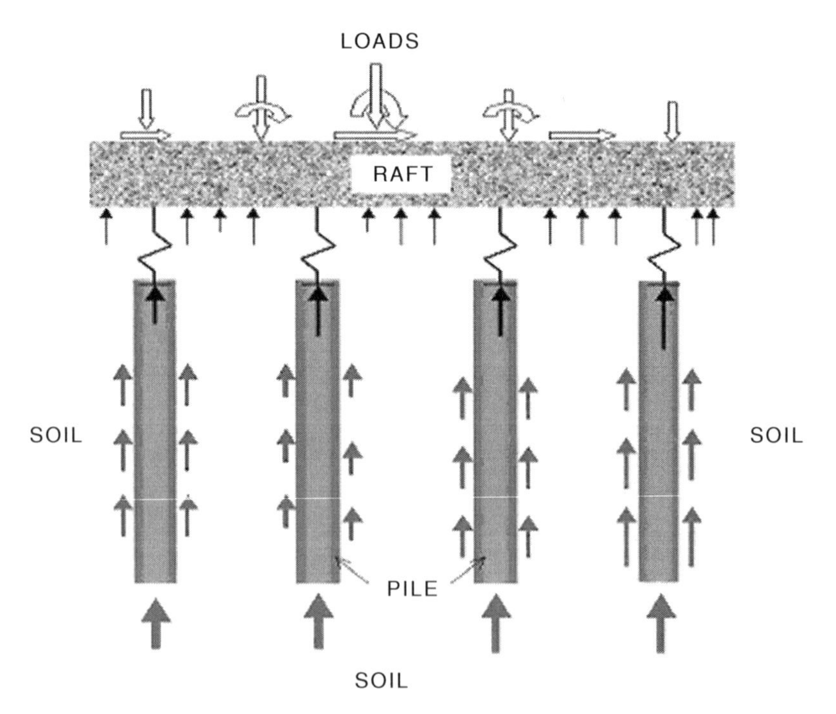

Figure 5.10 Soil–structure interaction of a piled raft structure.

PROBLEMS

P5.1. (a) Nine items were listed for the selection of the pile type and hammer for a particular project. You are the geotechnical engineer. List the items to be obtained from the owner, from the structural engineer, and by you. (b) Of the nine items, which one is most important to the geotechnical engineer and why?

P5.2. In addition to performing an appropriate subsurface investigation, if you are responsible for using a new and innovative type of deep foundation that would save the owner substantial money, what three steps in what order would you take to ensure construction of high quality?

P5.3. Name two factors that would lead you to select a deep foundation instead of a shallow foundation for a project even though strong soil exists at the ground surface

CHAPTER 6

DESIGNING STABLE FOUNDATIONS

6.1 INTRODUCTION

The design of every foundation presents a unique challenge to the engineer. Designs must take into account the nature and cost of the structure, geology and terrain, quality of subsurface investigation, kind of agreement with the owner, loadings over the life of the structure, effects of the proposed construction on buildings near the construction site, sensitivity of the proposed structure to total and differential settlement, requisite building codes, potential environmental effects due to the proposed construction including excessive noise, adequacy of the analytical tools available for making the design, result of a failure in terms of monetary loss or loss of life, availability of materials for foundations, and competent contractors in the area. A considerable effort will normally be necessary to gather and analyze the large amount of relevant data.

To create foundations that perform properly, the engineer must address carefully the topics in the above list and other relevant factors. Failures will occur if the foundation is not constructed according to plan and in a timely manner, if detrimental settlement occurs at any time after construction, if adjacent buildings are damaged, or if the design is wasteful.

The topic of designing wasteful foundations is of interest. An engineer could cause more than necessary resources to be expended on a foundation in the absence of peer review. An engineer increases the size of a foundation because the extra material gives a measure of "pillow comfort." An engineer in a firm was asked to describe the technique being used to design the piles for an important job and replied: "We use an approximate method and always add a few piles to the design." Such failures may never be seen, but they are failures nevertheless.

Casagrande (1964) wrote of *calculated risk* in earthwork and foundation engineering and gave several examples of projects where risk was used in the design and construction of major projects. Risk is always involved in geotechnical engineering because of the inability to get good information about the factors listed above and because of that sometimes unusual behavior of soil that cannot be anticipated in spite of diligence. Risk is minimized by prudent engineers who use available tools and techniques and who stay abreast of technological advances.

Peck (1967) presented a keynote address at a conference where the emphasis was on improved analytical procedures. He gave five sources of error with regard to the bearing capacity and settlement of foundations: (1) the assumed loading may be incorrect; (2) the soil conditions used in the design may differ from the actual soil conditions; (3) the theory used in the design may be wrong or may not apply; (4) the supporting structure may be more or less tolerant to differential movement; and (5) defects may occur during construction. The technical literature is replete with examples of foundations that have failed, many for reasons noted by Peck. The following sections present brief discussions of some examples, emphasizing the care to be taken by the engineer in planning, designing, and specifying methods of construction of foundations.

Lacy and Moskowitz (1993) made the following suggestions concerning deep foundations: understand the subsurface conditions in detail; select a qualified contractor; take care in preparing specifications for construction; provide full-time inspection by a knowledgeable person who has the necessary authority; and monitor adjacent structures as construction progresses.

6.2 TOTAL AND DIFFERENTIAL SETTLEMENT

The total settlement of a structure is the maximum amount the structure has settled with respect to its original position. The Palace of Fine Arts in Mexico has settled several feet but still remains usable because the differential settlement is tolerable. Differential settlement causes distortions in a structure, possible cracks in brittle materials, and discomfort to the occupants. The masonry building materials used in constructing the Palace of Fine Arts can tolerate the inevitable differential settlement, but some discomfort is inevitable. One can feel a definite tilting while sitting in a seat and watching a performance.

The subsequent chapters on computation of the settlement of a foundation show that the maximum settlement should occur under the center of the foundation and the minimum under the edges. Some proposals suggest that the difference between the maximum and minimum, that is, the differential, should be some fraction of the maximum. However, in all but exceptional cases, the engineer should create a design in which total settlement is moderate and differential settlement negligible.

Settlement as a result of subsidence can be important in the design of foundations. Subsidence occurs because water or perhaps oil is removed from the underlying formations and causes an increase in effective stress, triggering settlement that can amount to several feet. Subsidence has occurred in many cities because of dewatering. Houston, Texas, is an example. The movement of the surface usually is relatively uniform and foundation problems are minimal, but faults can occur suddenly and can pass through the foundation of a home, with serious and perhaps disastrous consequences. A less common cause of subsidence is the yielding of supports for tunnels used in mines that have been abandoned in some regions of the United States.

Areas of subsidence may be more frequent in the future as the need for water increases and possibly leads to more pumping of water from aquifers. The engineer must be aware of areas where subsidence is occurring or is expected to occur.

6.3 ALLOWABLE SETTLEMENT OF STRUCTURES

6.3.1 Tolerance of Buildings to Settlement

Recommendations have been published on total and differential settlement for various classes of buildings, starting with monumental structures and ending with temporary warehouses. Total settlement is important because connections to underground services can be broken and sidewalks can be distorted and out of place. Differential settlement leads to distortions in the building and cracking of brittle building materials such as glass and ceramics. Both total settlement and differential settlement must be considered by the engineer for each structure, and published recommendations should be avoided.

The modern approach to the tolerance of buildings to settlement is to perform analyses to predict immediate and time-related movement of the foundation, consistent with the stiffness of the superstructure. Procedures presented here, such as a deformable slab supported by bearing piles, provide guidance to a comprehensive solution. Tools are available for structural engineers to use in predicting the deformation of the superstructure, taking into account nonstructural elements such as partitions and non-load-bearing walls to conform to the time-related deformation of the foundation. Field observations are used to confirm the validity of the models, but analytical tools are at hand. Engineers and owners will be relieved of the burden of complying with prescriptive requirements.

The design of some structures may be very sensitive to differential settlement. Conferences between the structural engineer and the geotechnical engineer are useful in such cases. Extraordinary methods can be employed to limit differential settlement when necessary.

6.3.2 Exceptional Case of Settlement

The Valley of Mexico, the site of Mexico City, in some ways is a large-scale laboratory for demonstrating principles of soil mechanics. Pumping from deep-seated aquifers over many years caused several feet of subsidence in the Valley, and stresses from the foundations of major structures caused additional settlement due to the compressible nature of the soils. Zeevaert (1972) presents a profile to a depth of 80 m of the highly stratified subsurface soils in the Valley, with many meters having water contents averaging 200% and some strata with water contents as high as 400%. Settlement due to soil consolidation can be significant. Settlement in Mexico City is illustrated by the view of a government building (Figure 6.1). The building remained in use even though the settlement was much more than what might be tolerated at a site with less extreme soil conditions.

Figure 6.1 Photograph showing settlement of a government building in Mexico City.

Cummings (1941) noted that "There is no question but that Mexico City is one of the most interesting places in the world for field studies in the settlement of structures. Whereas it is often necessary to search carefully for structural settlements with precise leveling and other accurate methods, the observation of structural settlements in Mexico City can be done by the eye without the use of measuring instruments of any kind."

6.3.3 Problems in Proving Settlement

One of the authors was asked to express opinions on two sites where the failure of foundations was questionable. Structural steel was being erected for a high-rise structure in the southern United States when the contractor noticed that the steel members were not fitting together properly. Even though the settlement survey was inconclusive, the drilled-shaft foundations were underpinned without delay. More than ordinary skill is required to make precise measurements of building settlement. Even though settlement was claimed, the proof was lacking and the ensuing lawsuit was expensive, time-consuming, and perhaps unnecessary.

At a site in Hong Kong, a nearly horizontal crack was discovered in a thick pile cap during the erection of a reinforced-concrete warehouse. Settlement was postulated, even though no strong evidence was present from precise surveying. Coring of a multitude of bored piles was undertaken, and poor contact was found for some piles between their base and the founding stratum. Such piles were judged to be defective, even though the area of the core was only a fraction of the total area of the base of the pile. Hand-dug piles were being installed in some places as part of a huge program of underpinning when the decision was made to test some of the piles below the cracked cap. A slot was cut between a bored pile and the cap, flat jacks were installed, the height of the existing reinforced-concrete structure provided sufficient resistance, and load testing proceeded. Three piles were tested with the flat-jack method, and all showed ample capacity. Underpinning was discontinued, and the construction proceeded without a problem. Later, an investigation revealed that the crack in the pile cap was due to a cold joint resulting from a large concrete pour.

6.4 SOIL INVESTIGATIONS APPROPRIATE TO DESIGN

6.4.1 Planning

Many factors affect the plans for a proper investigation of the subsurface soils for a project. A cooperative effort is desirable in which the owner conveys to the architect and structural engineer the requirements for the proposed structure, the structural engineer and architect make a preliminary plan that dictates the foundation loads, and the geotechnical engineer describes the geology and suggests a type of foundation. Unfortunately, the geotechnical engineer is often selected later, after a plan is proposed for the subsurface investigation where price could be an important consideration. A cursory study of the soils could result, with the geotechnical engineer recommending design parameters on the basis of limited information. Such limited soil studies have resulted in a large number of claims by the contractor of a "change of conditions" when the soils were not as presented in the soils report.

The ideal plan, unless substantial information on subsurface conditions is available, is to perform exploratory borings for classification and to get data that will guide the borings for design. As noted in Chapter 4, a wide range of techniques are available to the geotechnical engineer. The data will allow the type of foundation to be selected and will provide numerical values for the relevant soil properties.

6.4.2 Favorable Profiles

At a number of locations in the United States and elsewhere, the use of bearing piles is dictated by the nature of the soil overlying the founding stratum. The soil investigation is aimed mainly at determining the thickness of the surface stratum in order to find the required length of the piles. Penetration tests can be used with confidence, and experience may show that the piles need only be driven to refusal into the bearing stratum.

For many low-rise buildings supported on a thick surface stratum of sand, the SPT (Chapter 4) can be used to determine that the sand is not loose or very loose and to ascertain the position of the water table. Spread footings or a raft can be designed with confidence.

Other soil profiles exist that are well known to local engineers, and the type of foundation can be selected without exploratory borings. The soil investigation can be made with methods that are locally acceptable and lead to a standard design. The geotechnical engineer must exercise caution in all cases to identify soils with special characteristics as discussed below. Nature is anything but predictable, and the engineer must be alert when characterizing soils.

6.4.3 Soils with Special Characteristics

Cambefort (1965) wrote of experiences with unusual soils and remarked: "There is no known 'recipe.' But the most dangerous thing is to think that the known formulas can explain everything. Only reasoned observation can lead to satisfactory results." The observational method should always be used, particularly on major projects, and some engineers have employed this method extensively. The thesis is that not all the features of the behavior of a foundation can be predicted but that strengthening can be done if excessive movement is observed.

The following sections describe several of types of soil that can cause problems in construction and deserve special attention from the engineer. Other sections of the book present methods of designing foundations for a number of soils noted below.

6.4.4 Calcareous Soil

The engineer may encounter an unusual soil, especially if working in an area where little is known about the soil. Several years ago, one of the authors attended a preconstruction meeting prior to building an offshore platform on the Northwest Shelf of Australia. A sample of sand was shown, and the results of laboratory testing were presented. The friction angle was consistent with the relative density, and the decision was made to design the piles using available equations even though the sand was calcareous due to the nature of the geologic deposition.

The template was set on the ocean floor, and piles were to be stabbed and driven with the legs of the template as guides (Jewell and Andrews, 1988). However, after the first open-ended pipe pile was placed into the template, the pile fell suddenly about 100 m and came to rest on a dense stratum. The calcareous grains had crushed under the walls of the pile, and natural cementation in the deposit prevented the calcareous sand from exerting any significant lateral stress against the walls of the pile. A stress-strain curve from the laboratory exhibited severe strain softening. Thus, skin friction was low to nonexistent as a result of the large deformation during installation, and the allowable end bearing on the piles was insufficient to provide adequate safety during the design storm. The result was that special strengthening was designed after an intensive study. The required construction was expensive, complex, and time-consuming.

Very Soft Clay The design of stable foundations must address the presence of soft clay at the construction site. Several of the chapters of this book deals with aspects of soft clay, including identification, strength, deformational characteristics, and design of foundations. Two procedures are noted below. In addition, the engineer must take special care if an excavation must be made in soft clay. Analytical techniques must be employed to investigate the possibility of sliding that would affect the site and possibly nearby structures as well.

Preloading. If the construction can be delayed for a period of time, the site may be preloaded with a temporary fill. A drainage layer can be placed on the surface of the soft clay, and boring can be used to install vertical drains at appropriate spacing. Analysis can predict the time required for the clay to drain to an appropriate amount, with a consequent increase in shear strength. Settlement plates can be installed to provide data to confirm the analytical predictions or to allow modification of the predictions.

Load-Bearing Piles. Another method of providing foundations where there is a stratum of very soft clay is to install piles through the clay to a bearing stratum below. Usually piles are driven, causing modification of the properties of the clay. The engineer must be aware that soft clay can settle, subjecting the piles to downdrag. Possible problems related to the buckling of piles in soft clay can be investigated by methods presented in Chapters 12 and 14.

Expansive Clay Expansive clay at a construction site, if not recognized, can sometimes lead to disastrous results. Chapters 2, 3, 6, and 9 will discuss the identification of expansive clay and the design of shallow and deep foundations. Expansive clay increases the cost of the foundation for a low-rise structure, and inadequate foundations are being built in spite of current knowledge and some building codes. Thus, the problem is faced more by homeowners than by the owners of commercial buildings.

Clay expands as it becomes wet and shrinks as it dries. Furthermore, moisture will collect when evaporation is cut off. In addition to swelling of the clay, other important factors are the nature of the foundation, weather, and transmission of moisture through the clay. A foundation on expansive clay may show no distress for perhaps years and then experience severe movement. On the other hand, one of the authors was asked to visit a site where a church building had cracks in the walls so wide that people in the congregation could see children in a playground. However, a rain had occurred the night before, and when the author arrived, the cracks were virtually closed! Figure 6.2 shows the cracking of a structure on expansive clay.

The severe differential movement of the foundations of homes on expansive clay can sometimes be devastating. In a home belonging to an assistant sports coach at a major university, doors did not close properly, the wallpaper was wrinkled, and the floors were very uneven. A portion of the home was on a slab, and another portion was supported by piers and beams. Repairs could be made, but the expense would be heavy. Later, the coach was divorced, and he moved away.

Some years ago, a nonprofit agency was asked to host a series of seminars on expansive clay. The plan was to encourage potential homeowners to look for cracks in the ground, damage to nearby homes, and elementary methods of identifying expansive clay. Geologists, geotechnical engineers, structural engineers, developers, and local officials were to be invited, along with the potential homeowners. The agency declined to host the seminars, perhaps due to fear of being sued by builders or developers who had tracts where expansive clay existed.

In addition to presenting information on identification of expansive clay, this book will give recommendations for the construction of shallow and deep foundations.

Loess In the Mississippi Valley of the United States, in Romania, in Russia, and in many other parts of the world, a soil exists called *loess*. It was created

Figure 6.2 Damage to a masonry structure on expansive clay.

by the transport by wind over long periods of fine grains ranging in size from about 0.01 to 0.05 mm. Grass or other vegetation grew during the deposition, so loess has a pronounced vertical structure with cementation associated with the vegetation.

Cuts in loess will stand almost vertically to considerable heights, but the soil will collapse under load when saturated. Loess is capable of sustaining a considerable load from a spread footing, but the design of foundations must consider the possibility of saturation. For relatively light structures, loess may be treated to a depth of up to 2 m by prewetting, compaction, and/or chemicals. Appropriate drainage is critical. Pile foundations extending through the loess are frequently recommended for major structures.

Loose Sand Terzaghi (1951) described the design of foundations for a factory building in Denver, Colorado, where the assumption was made of an *allowable bearing value* of 2 tons/ft^2 for the underlying sand. The dead load from the building was 0.9 ton/ft^2, but when a heavy snowfall increased the loading to 1.4 tons/ft^2, the building experienced settlement of the columns of up to 3.5 in. Tests performed later showed the sand to be loose to very loose and variable both vertically and horizontally.

A serious problem with loose sand is densification due to vibration. Vibration will cause the void ratio to decrease and settlement to occur. Problems have been reported with foundations of pumps at pipelines in South Texas, where unequal settlement occurred after operation for some time.

Pinnacle Limestone and Embedded Boulders Pinnacle limestone, where a deposit of limestone is riddled with solution cavities, and embedded boulders present similar problems. Both kinds of sites are extremely difficult to investigate by subsurface drilling or probing.

Pinnacle limestone is prevalent in the southeastern United States and elsewhere and is generally known by the geology of the area. Each site poses a different problem, and no straightforward method of determining an appropriate foundation is evident. Drilled shafts are usually recommended, with the depth of the foundation depending on the result of drilling.

A vertical surface exhibiting pinnacle limestone at the site of the foundation for the Bill Emerson Bridge in Missouri is shown in Figure 6.3. The existence of solution cavities is apparent. About 360 boreholes were made at the site of one of the foundations with a surface area of 90 by 120 ft. The subsurface condition was revealed in a three-dimensional plot generated by a computer. The computer depiction allowed slices to be taken through the formation, and a program of grouting was undertaken to eliminate zones of weakness. A foundation of drilled shafts was then executed.

Extraordinary solutions are sometimes required in constructing the foundations when pinnacle limestone exists at the site. At a construction site in Birmingham, Alabama, with dimensions of about 200 by 280 ft, three types of deep foundations were required: drilled shafts for about 40% of the site, micropiles for about 27%, and driven pipe piles for the remaining 33%. One

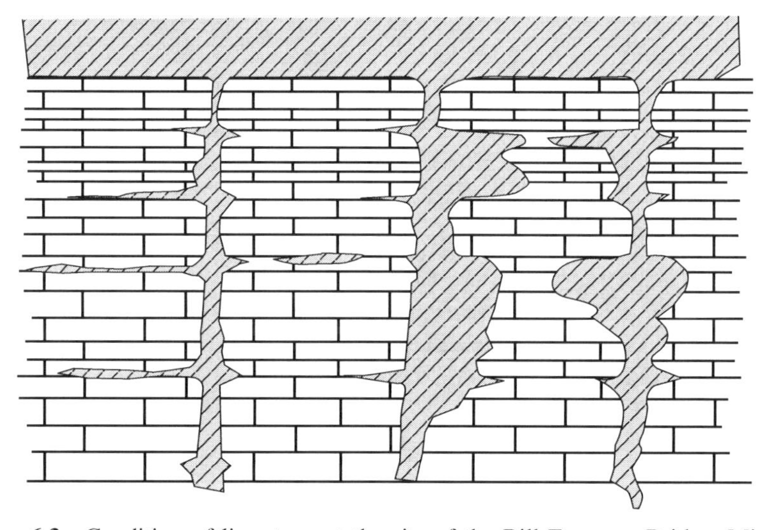

Figure 6.3 Condition of limestone at the site of the Bill Emerson Bridge, Missouri (from Miller, 2003).

of the micropiles extended to a penetration of 37 ft, and a few feet away another extended to a depth of 129 ft. The driven piles are anticipated to penetrate to a depth of 150 ft. The design created challenges for the engineer and the contractor to ensure that the axial movements of the different types of piles would be within the allowable range for the superstructure.

Embedded boulders create severe difficulties in designing and constructing deep foundations. Boulders in weaker soil occur because of the action of glaciers, or because of uneven weathering, or possibly because of being left in fills in the past. D'Appolonia and Spanovich (1964) describe large settlements of an ore dock because of the different response of supporting piles in settlement under axial load. Some of the 6000 piles rested on boulders and others were founded in hardpan. The authors stated that the boulder-supported piles were more compressible because of the short-term settlement of the soil beneath a boulder, causing load to be transferred to piles founded in hardpan, which then became overloaded.

If boulders are known to exist at a site, the design may call for the use of drilled shafts (bored piles). If the boulders are smaller than the diameter of the drilled hole, "grab" tools can be used to lift the boulders from the excavation and drilling can proceed. If the boulders are larger than the diameter of the drilled hole, special techniques are required. The boulders may consist of soft rock and can be broken by the use of a chopping bit; if they consist of hard rock, the size of the drilled hole may be increased. Sometimes the hard rock of the boulder can be drilled and a steel bolt can be grouted into place to allow the boulder to be lifted by a crane. Extreme care must be used if workmen enter a drilled hole because of the danger that carbon monoxide gas has settled into the excavation from a nearby motorway.

6.5 USE OF VALID ANALYTICAL METHODS

Models of various kinds have been proposed for the solution of every foundation problem. For example, failure surfaces are shown in the soil below a footing, and soil-mechanics theory is used to predict the location and stresses along these surfaces. Integration is then used to compute the load on the footing that generates the failure surfaces. Such a model is difficult to apply when obtaining the bearing capacity of a footing on layered soils, particularly considering three-dimensional behavior. But such models can be used with confidence if proven by full-scale load tests.

While the ultimate load on a footing may be computed with the model described above, another type of model must be used to obtain the movement of the foundation. The goal of analytical techniques is to have a model that may be used both for ultimate capacity and for the nonlinear movement of the foundation. The development of such models is continuing.

6.5.1 Oil Tank in Norway

Bjerrum and Överland (1957) studied the failure of an oil tank in Norway. Soil borings were made at the site of the tank. Properties of the soil were obtained by the use of the in situ vane and by the unconfined compression test. The upper layer of soil was a silty clay to a depth of 7 m with a soft marine clay below. The undrained shear strength was reconstituted to obtain values for the construction time of the tanks. The strength of the soil was almost constant to a depth of 10 m with a value of about 3 tons/m^2, but the shear strength was over 4 tons/m^2 at a depth of 15 m. The diameter of the tank was 25 m. If a failure of the entire tank was assumed, the average shear strength along the failure surface yielded a factor of safety of 1.72. On the other hand, if failure was assumed at one edge of the tank, the average shear strength along the shallower failure surface was less and the factor of safety was computed to be 1.05.

The heave of the soil was near one edge of the tank and indicated a local failure in which the weaker soil was mobilized. The error in the original design was due to the selection of a model for general failure where the average shear strength along an assumed failure surface was significantly larger than the average shear strength of the near-surface soil related to local failure. Thus, the careful selection of a model for a failure surface that will yield the lowest factor of safety is essential.

6.5.2 Transcona Elevator in Canada

This grain elevator located near Winnipeg, Canada, was constructed in 1913 on a raft foundation. Failure of the bin house by gradual tilting occurred after 875,000 bushels of wheat had been stored yielding a load of 20,000 tons, distributed uniformly. The bin house had plan dimensions of 77 by 195 ft and was 92 ft high. The elevator was founded on clay that had been deposited in a glacial lake and failed by tilting when the uniform load reached 3.06 tsf. The raft was 12 ft below the ground surface, giving a net increase in load of $3.06 - (120 \times 12) = 2.34$ tsf.

Soil borings were made in 1953, and properties were determined by testing of tube samples (Peck and Bryant, 1953). The stratum below the raft was 27 ft thick, and had an unconfined compressive strength (q_u) of 1.13 tsf and a sensitivity of 2. The next stratum was 8 ft thick, with a q_u of 0.65 tsf and a sensitivity of 2. The clay was underlain by a granular stratum with refusal to the penetration test at 20 ft below the clay.

The depth of the raft and its plan dimensions yielded a bearing capacity factor N_c of 5.56. Using a weighted average for the 35 ft of clay beneath the raft of 0.93 tsf, the allowable loading was computed to be $(0.5)(0.93)(5.56) = 2.57$ tsf. Thus, the factor of safety against collapse was slightly greater than 1. Many codes of practice, along with guidelines used by practicing engineers, suggest using a factor of safety of up to 3 to account for errors in obtaining soil properties and limitations of the theory.

6.5.3 Bearing Piles in China

The model for the design of a pile under axial loading (see Chapter 10) is fairly simple. For transfer of load in side resistance (skin friction) the model employs the distribution of stresses on the pile–soil interface of elements along the length of the pile. The axial load sustained by an element can be computed by integrating the vertical stresses along the face of the element. For transfer of load at the base of the pile (end bearing), the model employed is similar to that for a footing. If piles are driven close to each other, an allowance must be made for pile–soil–pile interaction.

While the models described above have been in use for many years, a design in 1959 used piles of different lengths along a pier (Figure 6.4) where the axial loading on the deck of the pier was presumably uniform. The shorter piles settled more than the longer ones and caused an unacceptable and uneven settlement of the pier. Long et al. (1983) describe the repair of the pier

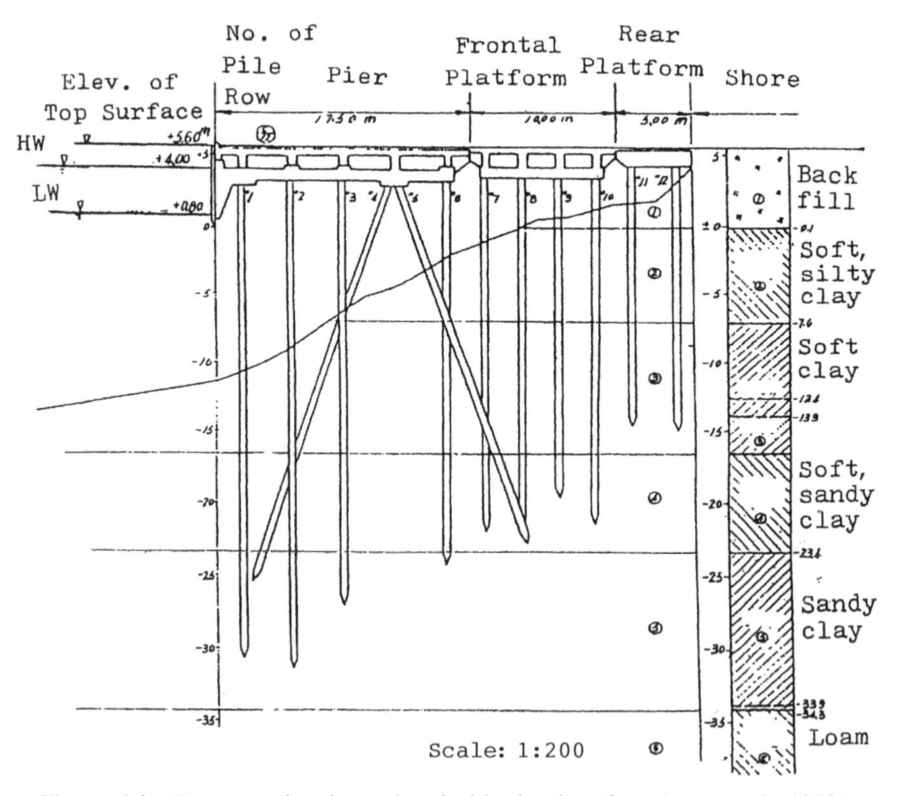

Figure 6.4 Structure of a pier and typical boring log (from Long et al., 1983).

caused by the unequal settlement of the piles. The soil profile across the site was relatively uniform. The use of the simple models described above should have produced an acceptable design. However, the use of the models for an axially loaded pile is obviously more complex if the soil profile varies across the construction site.

6.6 FOUNDATIONS AT UNSTABLE SLOPES

Chapter 3 presents an equation for the shearing resistance of soils: $s = c + (p - u) \tan \phi$, where c = cohesion, p = total normal stress on the plane of failure, u = porewater pressure, and ϕ = friction angle for the soil. The equation shows that if the value of u becomes equal to p, the shearing resistance of the soil becomes equal to c, which could be quite small for a clay or equal to zero for a sand. The value of the porewater pressure can increase as surface loading is applied, say from a building, and if the formation does not allow rapid drainage, the strength of the soil will be reduced. The following examples illustrate the importance of predicting the porewater pressure.

6.6.1 Pendleton Levee

A failure occurred as described by Fields and Wells (1944). The Pendleton Levee was built along the Mississippi River above a formation consisting of 4 to 9 ft of fine sand, underlain by 12 ft of soft clay, and underlain in turn by a relatively thick stratum of fine to coarse sand. Specimens of the clay were tested in the laboratory without drainage and were found to have a c value of 80 psf and a ϕ value of 19°. The clay exhibited strain softening with the ϕ value of 12°, corresponding to the ultimate strength of the clay.

The soil for the levee was moved and distributed by earth-moving equipment. When the fill reached the maximum height of 32 ft, a slide caused the levee to fail on the land side. The levee had been instrumented extensively to gain information on present and future construction.

Terzaghi (1944) studied the results of the soil investigation and the data from the instruments and concluded that failure had occurred in a thin horizontal stratum of fine sand or coarse silt in the soft clay that was undetected in the soil investigation. The layer of granular soil could reasonably have occurred during the process of deposition over the centuries as flooding deposited soil in layers. Variations in climatic conditions could have led to the deposition of the cohesionless soil. The fill caused the pore pressures to increase in the thin, cohesionless stratum, and failure occurred along that stratum as its strength was reduced.

Peck (1944) discussed the paper and noted that the identification of such a stratum of fine sand in the clay would require continuous sampling with proper tools. Kjellman et al. (1951) described the use of the Swedish sampler with thin strips of foil along the interior of a sampling tube. The foil is retracted as the sampler penetrates the soil, eliminating the resistance between the soil and the interior of the tube and allowing undisturbed samples to be obtained with lengths of several meters. Such a tool would have been ideal for sampling the soil at the Pendleton Levee but the method has found little use outside of Sweden.

6.6.2 Fort Peck Dam

A slide occurred at Fort Peck Dam that allowed impounded water to rush out, with consequent damage and public outcry. A comprehensive investigation followed, and Middlebrooks (1940) reported that the slide occurred in shale with seams of bentonite at the hydraulic-fill dam. The investigation prior to design failed to indicate the extent of weathering in the shale and bentonitic seams; further, the investigation failed to predict the high hydrostatic pressure in the bentonite due to the load from the hydraulic fill.

6.7 EFFECTS OF INSTALLATION ON THE QUALITY OF DEEP FOUNDATIONS

6.7.1 Introduction

All types of deep foundations can be damaged by improper methods of construction. As noted by Lacy and Moskowitz (1993), the engineer has several responsibilities prior to and during construction. Specifications for construction must be prepared that give the methods to be used to achieve foundations of good quality. Also, it is critical for qualified inspectors to be present during construction to ensure compliance with the specifications. The selection of a qualified contractor with experience with the proposed system and with proper equipment is vital, and the owners of the project should allow procedures that ensure a qualified contractor, such as prequalification of bidders. Preconstruction meetings with the successful contractor are almost always useful.

Driven Piles Piles are sometimes damaged by overdriving. The wave-equation method is useful to select an appropriate pile hammer for the pile to be driven. Attention to detail is important concerning such things as the driving equipment, the quality of the pile being driven, and the arrangements of the components of the system. Even with the use of an appropriate hammer and appurtenances, tough piles such as H-piles can be damaged at their ends by overdriving.

Damage of steel piles in unusual ways can involve great expense and loss of time. Open-ended steel pipes were being driven to support an offshore platform when the lower ends were distorted by being banged against the template structure. The distortion caused the lower portion of the piles to buckle inward during driving due to lateral forces from the soil. The collapse prevented drilling through the piles to allow the installation of grouted inserts. Many engineers worked for several months to design a remedial foundation.

Damage to precast concrete piles during driving can be vexing and expensive. The authors are aware of a bridge along the coast of Texas where many of the precast piles had been damaged due to overdriving, causing tensile cracks at intervals along the piles. Salt water entered the cracks, corroded the reinforcing steel, and destroyed the integrity of the piles. The contractor worked beneath the deck of the bridge to install a special system of drilled piles.

Selection and proper replacement of cushioning material when driving concrete piles is important (Womack et al., 2003). The authors were asked to review the installation of concrete piles where plywood was used as cushioning material. Driving continued until the plywood actually collapsed and burned due to continued use. The concrete piles exhibited a number of cracks in the portions above the ground.

Drilled Shafts Drilled shafts or bored piles are becoming increasingly popular for several reasons. The noise of installation is less than that of driven piles, drilling can penetrate soft rock to provide resistance in side resistance and end bearing, and diameters and penetrations are almost without limit. Drilled shafts are discussed in Chapter 11, and construction methods are described in detail in Chapter 5.

As noted earlier, preparation of detailed specifications for construction of drilled shafts is essential, and a qualified, knowledgeable engineer must provide inspection. Preconstruction meetings are highly desirable to ensure that the specified details of construction are consistent with the contractor's ability to construct the project.

CFA Piles As noted earlier, CFA piles have been used for many years and are competitive with respect to the cost per ton of load to be supported when the subsurface conditions are suitable. The CFA pile is constructed by rotating a hollow-stem CFA into the soil to the depth selected in the design. Cement grout is then injected through the hollow stem as the auger is withdrawn. Two errors in construction with CFA piles must be avoided: (1) the mining of soil when an obstruction is encountered and the auger continues to be rotated and (2) abrupt withdrawal of the auger, allowing the supporting soil to collapse, causing a neck in the pile. To assist in preventing these two construction problems, a data-acquisition system can be utilized to record rotations, grout pressure, and grout volume, all as a function of penetration of the auger. Alternatively, the inspector can obtain such data on each pile as construction progresses.

Other Types of Piles Many other types of piles for deep foundations are described in Chapter 5. The engineer must understand the function of such piles and the details of the construction methods. Preparation of appropriate specifications for construction and inspection of construction are essential. Field load tests of full-sized piles often must be recommended if such experimental data are unavailable from sites where the subsurface conditions are similar.

6.8 EFFECTS OF INSTALLATION OF DEEP FOUNDATIONS ON NEARBY STRUCTURES

6.8.1 Driving Piles

The installation of deep foundations obviously affects the properties of the nearby soils. Such effects may be considered in design. In addition, movements of soil from installation of piles must be considered. The driving of a pile will displace an amount of soil that can affect nearby construction. The displacement is greater if a solid pile is driven, such as a reinforced-concrete section, and less if an H-pile or an open-ended-pipe pile is driven. However, in some soils, the pipe pile will plug and the soil between the flanges will move with the pile; these piles can become *displacement* piles.

When a displacement pile is driven into clay, the ground surface will move upward, or heave, and the heave can cause previously driven piles to move up. Often the heaved piles must be retapped to restore end bearing. The heave and lateral movement could also affect existing structures, depending on the distance to the structures. The volume of the heaved soil has been measured and is a percentage of the volume of the driven piles.

If a pile is driven into loose sand, the vibration will cause the sand to become denser and the ground surface will frequently settle. Settlement occurs even though a volume of soil is displaced by the placement of the pile. As noted earlier, settlement will occur if mats or spread footings on loose sand are subjected to vibratory loadings.

Lacy and Gould (1985) describe a case where fine sand and varved silt from glacial outwash overlie bouldery till at Foley Square, New York City. H-piles were driven for a high-rise structure on 3-ft centers through 80 ft of sand and silt. The vibrations from pile driving caused settlement of adjacent buildings founded on footings above the glacial sand. Even though the construction procedures were changed, the vibration of the sand at the adjacent buildings caused settlement of the footings to continue. The adjacent buildings, 6 and 16 stories in height, had to be demolished.

The driving of any type of pile will cause vibrations to be transmitted, with the magnitude of the vibrations dependent on the distance from the construction site. Predicting whether or not such driving will damage an existing structure requires careful attention. The engineer must be cognizant of the possibility of such damage and take necessary precautions. Mohan et al. (1970) present the details of damage to existing buildings in Calcutta due to nearby pile driving.

The near-surface soils in Calcutta consist of silts and clays to a depth of about 15 m. Foundations of relatively small buildings can be placed in the top 5 m, where the clay has medium stiffness. Larger buildings are founded on piles that penetrate the 15 m into stronger soil below. Piles were being driven at a site where the central portion of the site was about 30 m from two existing buildings that were founded on spread footings. After 145 reinforced-concrete piles had been driven, cracks were observed in the existing buildings. While there was no danger of failure of the spread footings, the cracks were unsightly and reduced the quality of the buildings. A study was undertaken before driving the last 96 piles for the new building. Driving at more than 30 m from the site caused no damage to existing buildings, but as the driving moved closer, the height of fall of the pile hammer had to be reduced from 1 m to 15–30 cm.

Construction of a large federal building in New York City was halted when only a third of the piles had been driven. The building had plan dimensions of 165 by 150 ft and was to be 45 stories high. It was to be supported on 2,700 14BP73 piles. The water table was 15 ft below street level, and much of the underlying soil was medium to fine sand. The subsidence due to the pile driving had lowered a nearby street more than 12 in., and sewer lines had been broken on two occasions (*ENR,* 1963b, p. 20).

Lacy et al. (1994) recommend the use of CFA piles to reduce the impact on adjacent structures—for example, by eliminating densification ("loss of ground") of granular soil as a result of pile driving.

6.9 EFFECTS OF EXCAVATIONS ON NEARBY STRUCTURES

Excavation near existing structures can cause two problems if the excavation is carried out below the water table. First, the cut, either open or braced, can allow the soil to move toward the excavation, resulting in lateral movement of the foundations of an existing structure. Second, the lowering of the water table will result in the possible drop of the water table at some distance away from the cut. The increase in the effective stress due to the lowered water table can cause settlement in some soils.

With respect to the latter case, an excavation was made in Renton, Washington, for a sewer line. The excavation traversed an old river channel and an aquifer that existed below the bottom of the trench. Dewatering was done with large-diameter deep-well pumps. The Boeing Company claimed that the dewatering of the excavation caused significant damage to a pile-supported two-story steel-frame structure built at a site that was formerly a peat bog. Even though the building was 1,300 ft away from the dewatered excavation, settling of the ground floor was observed and cracks appeared in the interior walls (*ENR,* 1962).

6.10 DELETERIOUS EFFECTS OF THE ENVIRONMENT ON FOUNDATIONS

Steel pipes exposed to a corrosive environment can be damaged severely. Designs must address two conditions: corrosive water in natural soil deposits and structures in sea water. A soil investigation must determine the character of the water. If the water is found to be corrosive, the engineer may provide extra wall thickness to allow for an amount of loss of metal throughout the life of the structure or may provide a coating for the piles. Several types of coatings or wraps may be used but, in any case, the engineer must be assured by preparing appropriate specifications that the installation of the piles does not damage the coating or wrap.

Steel piles that support waterfront or offshore structures in the oceans must be protected against corrosion by the use of extra metal in the splash zone or by coatings. Alternatively, the engineer may specify the employment of cathodic protection, in which sacrificial metal ingots are installed in connection with appropriate electrical circuits.

6.11 SCOUR OF SOIL AT FOUNDATIONS

The scour or erosion of soils along streams or at offshore locations can be catastrophic. If soil is eroded around the piles supporting bridge bents, the possible collapse of the structure will cause inconvenience, may be very expensive, and could result in loss of life. Unfortunately, technical literature contains many examples of such failures (*ENR,* 1962).

Predicting the amount of scour is a complex undertaking. Predictions can be made of the velocity of a stream to put in suspension a particle of granular soil, up to the velocity of a mountain stream that moves boulders. Moore and Masch (1962) developed a laboratory apparatus that could be used to study the scour of cohesive soil. It is vital to predict the increase in velocity when an obstruction, such as a foundation pile, is in the stream.

Simply placing large stones around the foundations to prevent scour will not always suffice because in time the stones can settle into the fine soil beneath. One solution is to employ a reverse filter, in which layers of granular soil of increasing size are placed on the fine natural soil so that the particles at the surface are sufficiently large to remain in place during swift stream flow. The design of the reverse filter follows recommendations of Terzaghi on research at the Waterways Experiment Station at Vicksburg, the so-called TV grading (Posey, 1971).

PROBLEMS

P6.1. Develop a table identifying the failures mentioned in this chapter and state whether the incident (a) caused a minor monetary loss, (b) caused a major monetary loss, or (c) could have been a catastrophe in which loss of life was possible. Note: a particular failure could have involved more than one category.

P6.2. Using one of the cases in Problem P.6.1, write a short statement giving your opinion on the reasons for the noted failure. Some of the reasons may be an inadequate soil investigation, including inadequate interpretation; lack of the proper analytical procedure; failure to use the available analysis; failure in the construction operations; or other reasons.

CHAPTER 7

THEORIES OF BEARING CAPACITY AND SETTLEMENT

7.1 INTRODUCTION

Testing of soil in the laboratory, and perhaps in the field, to obtain properties is required to allow the computation of bearing capacity and settlement. The reader is referred to Chapter 3 for details on the determination of the required parameters of the soil at a site.

The problems to be addressed in this chapter can be illustrated by the shallow foundation of given lateral dimensions resting on soil, as shown in Figure 7.1a. The first of two problems facing the engineer is to find the unit vertical load on a shallow foundation, or on the base of a deep foundation, that will cause the foundation to settle precipitously or to collapse. The unit load at failure is termed the *bearing capacity.* With regard to bearing capacity, the following comments are pertinent:

1. Any deformation of the foundation itself is negligible or disregarded.
2. The stress-strain curve for the soil is as shown in Figure 7.1b.
3. The base of the foundation may be smooth or rough.
4. The soil is homogeneous through the semi-infinite region along and below the foundation.
5. The loading is increased slowly with no vibration.
6. There is no interaction with nearby foundations. In spite of the constraints on computing the values, the concept of bearing capacity has been used for the design of foundations for decades and presently remains in extensive use.

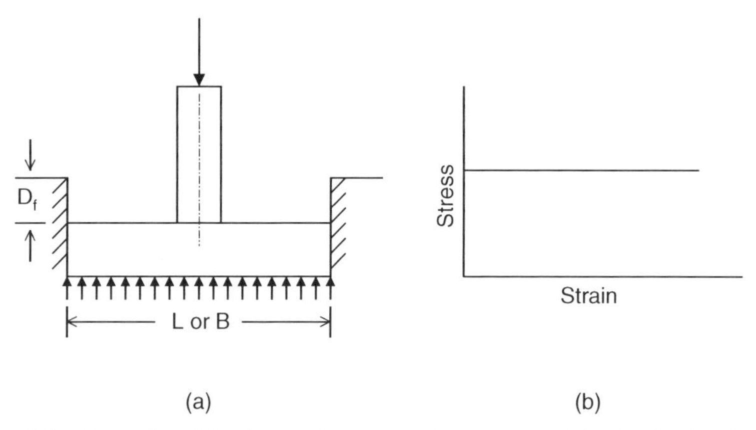

Figure 7.1 Example of footing and stress-strain curve used in developing bearing-capacity equations.

Investigators have noted that the bearing capacity equations will vary if the failure is *general* where symmetrical failure surfaces develop below the base of the foundation; is *local* where a failure occurs due to excessive settlement of the foundation; or is a *punch-through* where the foundation punches through a strong surface layer and causes the weak soil below to fail. The equations for general shear failure are most important.

The second of the two problems addressed here is the computation of settlement of a foundation such as that shown in Figure 7.1. Two types of settlement are noted: immediate or short-term settlement and long-term settlement due to the consolidation of saturated clays. The immediate settlement of foundations on sands of loose or medium density is relatively so large that settlement controls, termed a *local shear failure,* as noted above, and a general bearing capacity failure does not occur. Immediate settlement of foundations on sand and clay is discussed in Chapter 9.

Equations and procedures for dealing with long-term settlement due to consolidation are presented here. The settlement of deep foundations is discussed in chapters that deal specifically with piles and drilled shafts.

The finite-element method (FEM) discussed in Chapter 5 provides valuable information to the engineer on both bearing capacity and settlement, as demonstrated by the example solution presented. FEM can now be implemented on most personal computers, rather than on large mainframes, and will play a much greater role in geotechnical engineering as methods for modeling the behavior of soil are perfected. Leshchinsky and Marcozzi (1990) performed small-scale experiments with flexible and rigid footings and noted that the flexible footings performed better than the rigid ones. Rather than use the performance of expensive tests with full-sized footings, FEM can be used to study the flexibility of footings.

7.2 TERZAGHI'S EQUATIONS FOR BEARING CAPACITY

The equations developed by Terzaghi (1943) have been in use for a long time and continue to be used by some engineers. His development made use of studies by Rankine (1857), Prandtl (1920), and Reissner (1924). The two-dimensional model employed by Terzaghi (Figure 7.2) was a long strip footing with a width of $2B$ (later, authors simply use B for the width of footing) and a depth of the base of the footing below the ground surface of D_f. The base of the footing was either smooth or rough, with the shape of the solution for a rough footing shown in the figure. The wedge a, b, d is assumed to move down with the footing, the surface d, e is defined by a log spiral and the soil along that surface is assumed to have a failure in shear, and the zone a, e, f is assumed to be in a plastic state defined by the Rankine equation.

The equations developed by Terzaghi for bearing capacity factors are as follows:

$$Q_D = 2B(cN_c + \gamma D_f N_q + \gamma BN_\gamma) \quad \text{(general shear failure)} \tag{7.1}$$

$$Q'_D = 2B(\tfrac{2}{3}\, cN'_c + \gamma BN'_\gamma + \gamma BN'_\gamma) \quad \text{(local shear failure)} \tag{7.2}$$

where

Q_D = load at failure of the strip footing,
c = cohesion intercept,
γ = unit weight of soil, and
N_c, N'_c, N_q, N'_q, N_γ, N'_γ = bearing capacity factors shown graphically in Figure 7.3.

The differences in settlement for a general shear failure and a local shear failure are shown in Figure 7.4, where the general shear failure is depicted by the solid line. The footing on loose soil is expected to settle a large amount compared to the footing on dense soil, which Terzaghi elected to reflect by

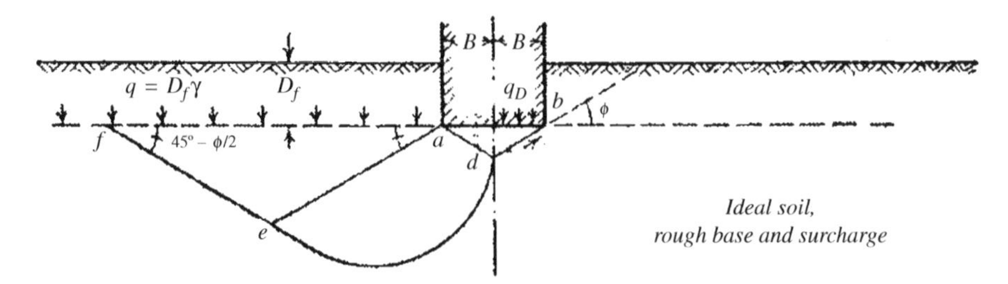

Figure 7.2 Model used by Terzaghi in developing valves for bearing-capacity factors (from Terzaghi, 1943, p. 121).

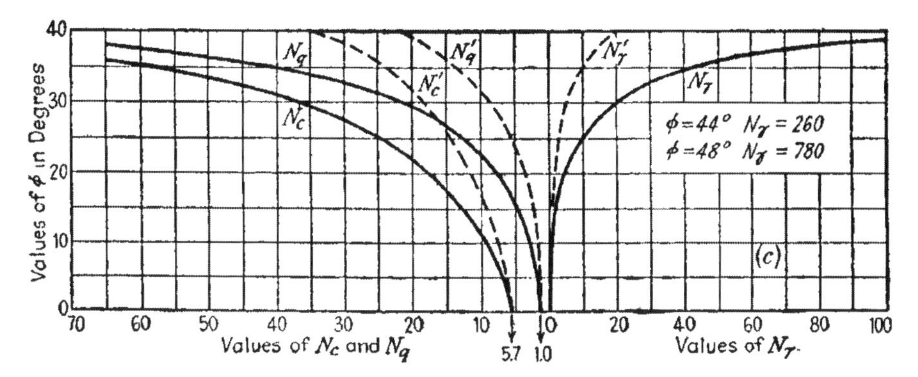

Figure 7.3 Terzaghi's bearing-capacity factors (from Terzaghi, 1943, p. 125).

reducing the values of the bearing capacity factors for loose soil. A proposed reduction in bearing capacity to deal with excessive settlement is disregarded here because Chapter 9 deals with short-term settlement of shallow foundations. The engineer may reduce the load on the foundation if the immediate settlement is deemed excessive.

7.3 REVISED EQUATIONS FOR BEARING CAPACITY

A number of authors have made proposals for bearing capacity factors, including Caquot and Kérisel (1953), Meyerhof (1963), Hansen and Christensen (1969), Hansen (1970), and Vesić (1973). The basic form of the bearing-capacity equation (Eq. 7.1), proposed by Terzaghi, has been accepted by most subsequent investigators; however, two modifications have been suggested: (1) improved analysis of the model proposed by Terzaghi (Figure 7.2), and

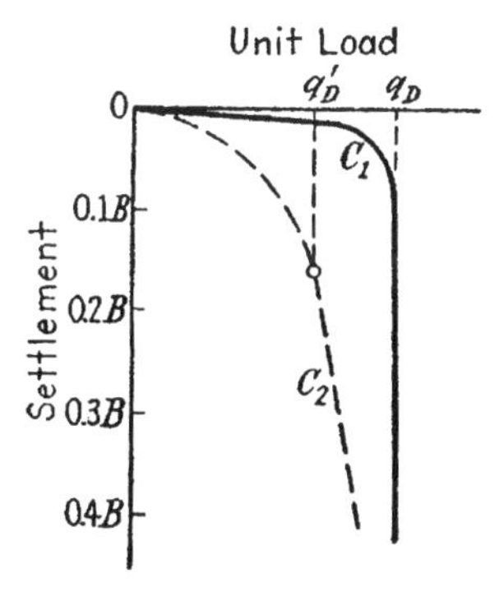

Figure 7.4 Assumed curves for unit load versus settlement for a footing on dense soil, C_1 and on loose soil, C_2 (from Terzaghi, 1943, p. 118).

(2) an extension of the method to include a number of factors such as inclined loading and a rectangular footing rather than a strip.

In addition, a number of other studies have been undertaken, such as those using FEM, to study the upper- and lower-bound values of the bearing capacity (Ukritchon et al., 2003). Hansen's method has been selected and is presented in detail because it is comprehensive and because the analytical results agree reasonably well with experimental results.

7.4 EXTENDED FORMULAS FOR BEARING CAPACITY BY J. BRINCH HANSEN

Hansen employed the same basic equation as Terzaghi except that the width of the footing is B instead of $2B$ as employed by Terzaghi. Hansen's basic equation is

$$\frac{Q_d}{B} = \frac{1}{2}\gamma B N_\gamma + \gamma D_f N_q + c N_c \tag{7.3}$$

where

Q_d = ultimate bearing capacity,
B = width of foundation, and
γ = unit weight of soil (use γ' for submerged unit weight).

The equations below are for the bearing-capacity factors N_q and N_c as expressed by exact formulas developed by Prandtl (1920):

$$N_q = e^{\pi \tan\phi} \tan^2\left(45 + \frac{\phi}{2}\right) \tag{7.4}$$

$$N_c = (N_q - 1)\cot\phi \tag{7.5}$$

(Note: if $\phi = 0$, $N_c = \pi + 2$.)

Hansen and Christensen (1969) presented a graph for N_γ as a function of the friction angle, δ, between the base of the footing and the sand (Figure 7.5). If the base is rough, $\delta = \phi$, and the values can be read from the indicated curve in Figure 7.5. Footings with a rough base, as would occur with concrete poured on a base of sand, are usually assumed in design. One of the authors witnessed construction in Moscow where concrete footings were in a plant and were being trucked to the job site. Extremely low temperatures for many months prevent casting of concrete in the field. If a designer wishes to account for low friction between the base of a footing and the sand, the curves in Figure 7.5 may be used, with the values for $\delta = 0$ indicating a perfectly

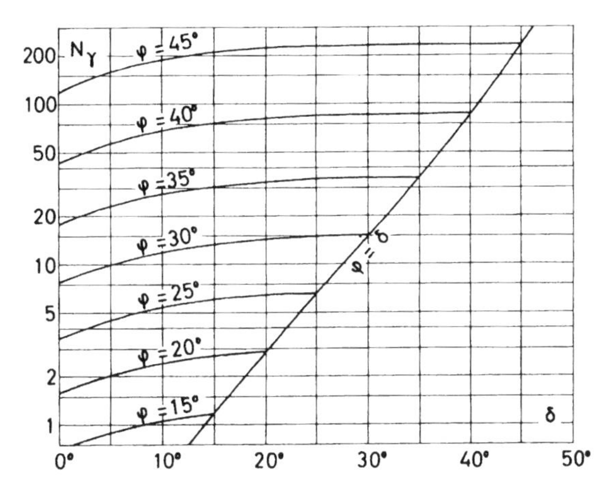

Figure 7.5 Bearing-capacity factor N_γ found for strip footings as a function of friction angles φ for sand and δ for subsurface of footing (from Hansen and Christensen, 1969).

smooth footing. The following equation is for a perfectly rough footing and is commonly used in design:

$$N_\gamma = 1.5(N_q - 1)\cot \phi \tag{7.6}$$

Hansen's bearing-capacity factors for the strip footing are presented in graphical form in Figure 7.6 and tabulated in Table 7.1. Perhaps Hansen's most important contribution was to extend the equations for bearing capacity

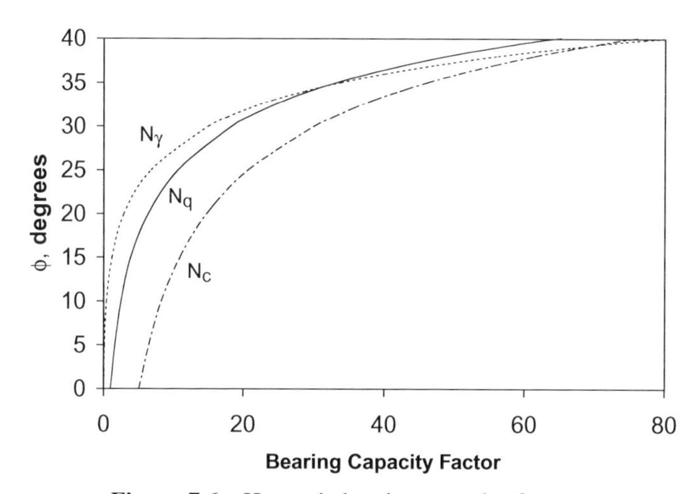

Figure 7.6 Hansen's bearing-capacity factors.

TABLE 7.1 Hansen's Bearing-Capacity Factors

ϕ (degrees)	N_q	N_c	N_γ
0	1.0	1.5	0.0
5	1.6	6.5	0.1
10	2.5	8.3	0.4
15	3.9	11.0	1.2
20	6.4	14.8	2.9
25	10.7	20.7	6.8
30	18.4	30.1	15.1
31	20.6	32.7	17.7
32	23.2	35.5	20.8
33	26.1	38.6	24.4
34	29.4	42.2	28.8
35	33.3	46.1	33.9
36	37.8	50.6	40.1
37	42.9	55.6	47.4
38	48.9	61.4	56.2
39	56.0	67.9	66.8
40	64.2	75.3	79.5
42	85.4	93.7	114.0
44	115.3	118.4	165.6
46	158.5	152.1	244.6
48	222.3	199.3	368.7
50	319.1	266.9	568.6

to deal with deviations from the simple case of a strip footing. The load may be eccentric, inclined, or both. The base of the foundation is usually placed at depth D_f below the ground surface. The foundation always has a limited length L and its shape may not be rectangular. Finally, the base of the foundation and the ground surface may be inclined. The equation for Hansen's extended formula is

$$\frac{Q_d}{A} = \frac{1}{2}\,\gamma B N_\gamma s_\gamma d_\gamma i_\gamma b_\gamma g_\gamma + \gamma D_f N_q s_q d_q i_q b_q g_q + c N_c s_c d_c i_c b_c g_c \tag{7.7}$$

where

A = area of footing,
s = shape factors,
d = depth factors,
i = inclination factors,
b = base inclination factors, and
g = ground inclination factors.

In the special case where $\phi = 0$, Hansen writes that it is theoretically more correct to introduce additive factors, and Eq. 7.8 presents the result:

$$\frac{Q_d}{A} = (\pi + 2)s_u(1 + s_c^a + d_c^a - i_c^a - b_c^a - g_c^a) \tag{7.8}$$

where s_u = undrained shear strength of the clay. The definitions of the modification factors are shown in the following sections. Hansen stated that when the modifications occur one at a time, a simple analytical solution or results from experiments can be used; however, when all of the factors are used together for more complicated cases, the resulting computation will be an approximation.

All of the loads acting above the base of the foundation are to be combined into one resultant, with a vertical component, V, acting normal to the base and a horizontal component H acting in the base. To account for eccentric loading, the foundation is re-configured so that the resultant intersects the base at a point called the *load center.* If the foundation has an irregular shape, a rectangular foundation meeting the above conditions is employed.

7.4.1 Eccentricity

With regard to eccentric loading on a shallow foundation, Hansen recommended that the foundation be reconfigured so that the eccentricity is eliminated on the foundation to be analyzed. The scheme proposed by Hansen is shown in Figure 7.7, where the original foundation is indicated by the solid lines. The point of application of the loading on the base of the foundation

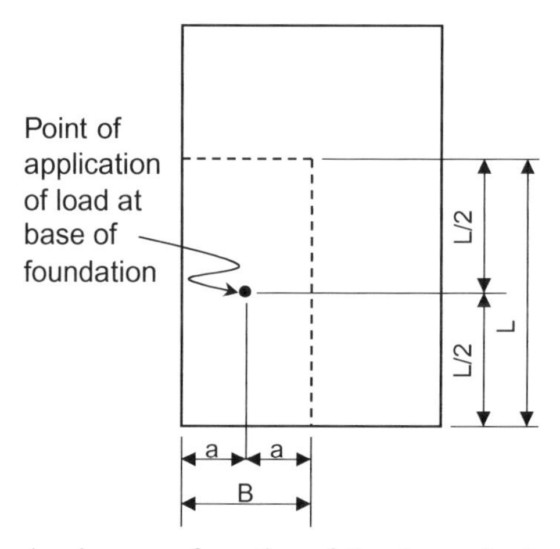

Figure 7.7 Sketch showing reconfiguration of the shape of a foundation to account for eccentric loading.

is shown by the dot, and the foundation to be analyzed is shown by the dashed lines. The eccentricity is exaggerated for purposes of illustration.

For foundations of other shapes with eccentric loads, the best possible rectangular foundation is configured with no eccentricity to replace the particular shape.

7.4.2 Load Inclination Factors

An inclination of the load will always mean a reduced bearing capacity, and solutions can be developed by revising the model shown in Figure 7.2. The following simple empirical formulas were proposed by Hansen:

$$i_q = \left[1 - 0.5\frac{H}{(V + Ac\cot\phi)}\right]^5 \tag{7.9}$$

$$i_\gamma = \left[1 - 0.7\frac{H}{(V + Ac\cot\phi)}\right]^5 \tag{7.10}$$

For the case where $\phi = 0$, the following equation may be used:

$$i_c^a = 0.5 - 0.5\left(1 - \frac{H}{As_u}\right)^{0.5} \tag{7.11}$$

where

H = component of the load parallel to the base and
V = component of the load perpendicular to the base.

Equations 7.9 and 7.10 may not be used if the quantity inside the brackets becomes negative.

The above equations are valid for a horizontal force where H is equal to H_B that acts parallel to the short sides B of the equivalent effective rectangle. If H_B is substituted for H, the computed factors may be termed i_{qB}, $i_{\gamma B}$, and i_{cB}^a. In the more general case, a force component H_L also exists, acting parallel with the long sides L. Another set of factors may be computed by substituting H_L for H, and factors will result that may be termed i_{qL}, $i_{\gamma L}$, and i_{cL}^a. The first set of factors with the second subscript B may be used to investigate possible failure along the long sides L, and the second set of factors with the second subscript L may be used to investigate possible failure along the short sides B. One of the sets of analyses will control the amount of load that may be sustained by the foundation.

7.4.3 Base and Ground Inclination

The sketches in Figure 7.8 defined the inclination υ of the base of the foundation and the inclination β of the ground. In the case of $\phi = 0$, the following equations were found:

$$b_c^a = \frac{2\upsilon}{\pi + 2} = \frac{\upsilon}{147} \tag{7.12}$$

$$g_c^a = \frac{2\beta}{\pi + 2} = \frac{\beta}{147} \tag{7.13}$$

where υ, β, and 147 are in degrees.

For values of ϕ other than zero, the following equation applies:

$$b_q = e^{-2\upsilon\tan\phi} \tag{7.14}$$

Equations 7.11 through 7.13 may be used for only positive values of υ and β, with β smaller than ϕ, and υ plus β must not exceed 90°.

7.4.4 Shape Factors

Hansen recommended the following equations to deal with footings with rectangular shapes:

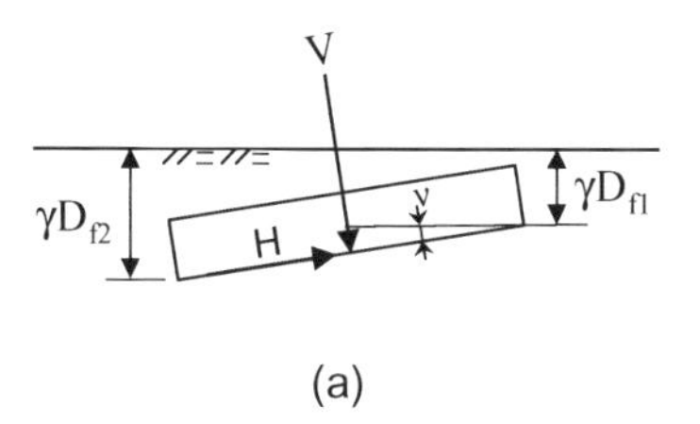

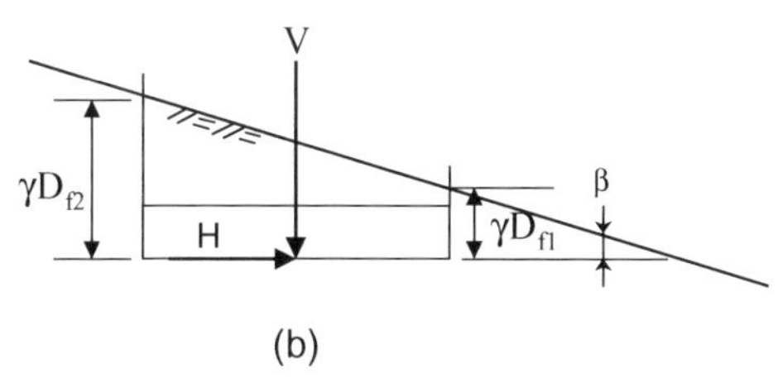

Figure 7.8 Sketches defining inclinations (a) of base υ and (b) of ground β (from Hansen, 1970).

$$s_\gamma = 1 - 0.4\frac{B}{L} \quad (7.15)$$

$$s_q = 1 + \sin\phi\frac{B}{L} \quad (7.16)$$

$$s_c^a = 0.2\frac{B}{L} \quad (7.17)$$

The above shape factors are only for vertical loads. For inclined loads, formulas must be introduced to account for the inclination. Failure may occur along the long sides or along the short sides; thus, two sets of equations are required.

$$s_{cB}^a = 0.2 i_{cB}^a \frac{B}{L} \quad (7.18)$$

$$s_{cL}^a = 0.2 i_{cL}^a \frac{L}{B} \quad (7.19)$$

$$s_{qB} = 1 + \sin\phi\frac{Bi_{qB}}{L} \quad (7.20)$$

$$s_{qL} = 1 + \sin\phi\frac{Li_{qL}}{B} \quad (7.21)$$

$$s_{\gamma B} = 1 - 0.4\frac{Bi_{\gamma B}}{Li_{\gamma L}} \quad (7.22)$$

$$s_{\gamma L} = 1 - 0.4\frac{Li_{\gamma L}}{Bi_{\gamma B}} \quad (7.23)$$

For Eqs. 7.22 and 7.23, the special rule must be followed that a value exceeding 0.6 must always be used.

7.4.5 Depth Effect

Foundations are always placed at depth D_f below the ground surface, so the term γD_f is used to indicate the soil above the base of the foundation. The shear strength of the overburden soil is assumed to be the same as that of the soil below the base of the footing. If the overburden soil is weaker than the foundation soil, the depth effect may be reduced or ignored entirely.

7.4.6 Depth Factors

The depth factor for the effect of the weight of the soil is always equal to unity:

$$d_\gamma = 1 \tag{7.24}$$

For values of D_f that are small with respect to B, the following equations may be used:

$$d_c^a = 0.4 \frac{D_f}{B} \tag{7.25}$$

$$d_q = 1 + 2 \tan \phi(1 - \sin \phi)^2 \frac{D_f}{B} \tag{7.26}$$

When D_f is large with respect to B, the following equations were proposed by Hansen:

$$d_c^a = 0.4 \arctan \frac{D_f}{B} \tag{7.27}$$

$$d_q = 1 + 2 \tan \phi(1 - \sin \phi)^2 \arctan \frac{D_f}{B} \tag{7.28}$$

To test the formulas for large values of D_f/B, the following result was found:

$$\frac{Q_d}{A} = (\pi + 2)c_u\left(1 + 0.02 + 0.4 \frac{\pi}{2}\right) = 9.4c_u \tag{7.29}$$

The result in Eq. 7.29 agrees with the well-known value found for the resistance of the tip for piles founded in clay. The following formula applies for other values of friction angles:

$$\frac{Q_d}{A} = \gamma D_f N_q (1 + \sin \phi)[1 + \pi \tan \varphi(1 - \sin \phi)^2] \tag{7.30}$$

Hansen noted that for friction angles between 30° and 40° the above equation yields results in good agreement with the Danish experience for point resistance for piles in sand, provided that ϕ is taken as the friction angle in plane strain.

For the usual case of failure along the long sides L on the base, Eqs. 7.29 and 7.30 may be used with Eqs. 7.25 and 7.26, yielding values of the depth factors d_{cB}^a and d_{qB}. For the investigation of a possible failure along the short sides, B, equations for another set of depth factors must be used:

$$d_{cL}^{a} = 0.4 \arctan \frac{D_f}{L} \tag{7.31}$$

$$d_{qL} = 1 + 2 \tan \varphi (1 - \sin \phi)^2 \arctan \frac{D_f}{L} \tag{7.32}$$

7.4.7 General Formulas

In many instances, the horizontal force will have a component H_B parallel with the short sides B and a component H_L parallel with the long sides L, and the following formulas must be used:

$$\frac{Q_d}{A} \leq \frac{1}{2} \gamma N_\gamma B s_{\gamma B} i_{\gamma B} b_\gamma + (\gamma D_f + c \cot \phi) N_q d_{qB} s_{qB} i_{qB} b_q - c \cot \phi \tag{7.33a}$$

or

$$\frac{Q_d}{A} \leq \frac{1}{2} \gamma N_\gamma B s_{\gamma L} i_{\gamma L} b_\gamma + (\gamma D_f + c \cot \phi) N_q d_{qL} s_{qL} i_{qL} b_q - c \cot \phi \tag{7.33b}$$

The above equations may be used as follows. Of the two possibilities for the terms including γ, the upper term should be used when $Bi_{\gamma B} \leq Li_{\gamma L}$ and the lower term should be used when $Bi_{\gamma B} \geq Li_{\gamma L}$. A check on the proper choice is that $s_\gamma \geq 0.6$. Of the two possibilities for the terms including γD_f, the one giving the smallest value must always be chosen.

In the special case where $\phi = 0$, the smallest of the values from the following two equations may be used:

$$\frac{Q_d}{A} \leq (\pi + 2) c_u (1 + s_{cB}^a + d_{cB}^a - i_{cB}^a - b_c^a - g_c^a) \tag{7.34a}$$

or

$$\frac{Q_d}{A} \leq (\pi + 2) c_u (1 + s_{cL}^a + d_{cL}^a - i_{cL}^a - b_c^a - g_c^a) \tag{7.34b}$$

7.4.8 Passive Earth Pressure

If sufficient lateral movements of a footing occur to develop passive earth pressure on the edge of the footing, the bearing-capacity equations will be affected. Hansen's studies showed that the movements necessary to develop bearing capacity and passive earth pressure are of the same order of magnitude. The movement associated with bearing capacity is discussed in Chapter 9. In special cases, the movements required to develop bearing capacity and passive earth pressure can be computed by FEM.

With respect to design, for mainly vertical loads no passive pressure is employed. For mainly horizontal loads, passive pressure must be employed; however, the details of construction play an important role. If an oversized excavation is made for the foundation, the nearby soil may have been disturbed and/or the backfill may have been placed poorly so that passive earth pressure does not develop.

7.4.9 Soil Parameters

In the case of saturated clays, the relevant value c_u of undrained shear strength must be employed for short-term behavior; for long-term behavior, the parameters ϕ and c must be employed as determined from drained triaxial tests.

For sand, c can be assumed to equal zero. The equations shown in the sections above are based on the friction angle in plane strain. The following identity is recommended:

$$\phi_{pl} = 1.1\phi_{tr} \tag{7.35}$$

where

ϕ_{pl} = friction angle for plane strain conditions, and
ϕ_{tr} = friction angle determined from triaxial tests.

7.4.10 Example Computations

Examples from Hansen (1970) The results from two tests of a foundation on sand were reported by Muhs and Weib (1969). The shallow foundation had a length of 2 m and a width of 0.5 m. The water table was at the ground surface, the submerged unit weight of the sand was 0.95 t/m^3, the friction angle measured by the triaxial test was 40° to 42°, and the base of the footing was 0.5 m below the ground surface. In the first test the foundation was loaded centrally and vertically, and failure occurred as a load of 190 t. Hansen suggested the use of ϕ_{tr} of 40° in his discussion of the Muhs-Weib tests.

$$\phi_{pl} = 1.1(40) = 44°$$

Eq. 7.4: $$N_q = (20.78)(5.55) = 115.3$$

Eq. 7.6: $$N_\gamma = (1.5)(114.3)(0.9657) = 165.6$$

Using the first two terms of the Eq. 7.33, failure along the short sides, with no inclined loading, inclined soil surface, or cohesion, and noting that $d_\gamma = 1$.

$$\frac{D_d}{A} = \frac{1}{2}\gamma N_\gamma B s_{\gamma B} + \gamma D_f N_q d_{qB} s_{qB}$$

Eq. 7.16: $$s_{\gamma B} = 1 - 0.4\,\frac{0.5}{2} = 0.90$$

Eq. 7.17: $$s_{qB} = 1 + (0.695)\,\frac{0.5}{2} = 1.174$$

Eq. 7.29: $$d_{qB} = 1 + 2(0.966)(0.0932)(0.785) = 1.141$$

$$\begin{aligned} Q_d &= 1\left(\frac{1}{2}(0.95)(0.5)(165.6)(0.90)(1) + (0.95)(0.5)(115.3)(1.174)(1.141)\right) \\ &= 35.4 + 73.4 = 108.8t \end{aligned}$$

The second test was performed with a foundation of the same size, in the same soil and position as the water table. The foundation was loaded centrally to failure with a vertical component $V = 108$ t and a horizontal component in the L direction $H_L = 39$ t. The revised Eq. 7.7 for the loading indicated is

$$\frac{Q_d}{A} = \frac{1}{2}\gamma L N_\gamma s_{\gamma L} d_{\gamma L} i_{\gamma L} + \gamma D_f N_q s_{qL} d_{qL} i_{qL};$$

because $H_B = 0$; $d_{qB} = 1.414$; $s_{qB} = 1.174$; $i_{\gamma B} = i_{qB} = 1$; $Bi_{\gamma B} = 0.5$; then $d_{qB} s_{qB} i_{qB} = (1.141)(1.174)(1) = 1.340$.

In the L direction, the following quantities are found:

Eq. 7.29: $$d_{qL} = 1 + 2(0.9657)(1 - 0.6947)^2(0.2450) = 1.044$$

Eq. 7.11: $$i_{\gamma L} = \left[1 - 0.7\left(\frac{39}{108}\right)\right]^5 = 0.233$$

$$Li_{\gamma L} = 0.466$$

Eq. 7.10: $$i_{qL} = \left[1 - 0.5\left(\frac{39}{108}\right)\right]^5 = 0.3695$$

$$Li_{qL} = 0.739$$

Because $Bi_{\gamma B} > Li_{\gamma L}$, use Eq. 7.33 and the following computations.

Eq. 7.23: $$s_{\gamma L} = 1 - 0.49 \left(\frac{0.466}{0.500}\right) = 0.627$$

Eq. 7.21: $$s_{qL} = 1 + 0.695 \left(\frac{0.739}{0.500}\right) = 2.027$$

$$d_{qL}s_{qL}i_{qL} = (1.044)(2.027)(0.3695) = 0.782$$

$$d_{qB}s_{qB}i_{qB} = (1.141)(1.174)(1) = 1.340$$

Because 0.782 < 1.340, use the lower q-term in Eq. 7.33.

$$Q_d = 1\left[\frac{1}{2}(0.95)(165.6)(2)(1)(0.627)(0.233) + (0.95)(0.5)(115.3)(1.044)(2.027)(0.3695)\right]$$
$$= 23.0 + 42.8 = 65.8 \text{ t}$$

In the first test noted above, the computed failure load was 108.8 t versus an experimental load of 180 t, yielding a factor of safety of 1.65 based on theory alone. In the second test, the computed failure load was 65.8 t versus an experimental load of 108 t, yielding a factor of safety of 1.64 based on theory alone.

Of interest is that Hansen employed a value of ϕ of 47°, instead of the 44° suggested by his equations, and obtained excellent agreement between the computed and experimental values of the failure load. Muhs and Weib (1969), as noted earlier, reported the measured friction angle to be between 40° and 42°. Had Hansen elected to use 42° instead of 40° for ϕ_{tr}, the value of ϕ_{pl} for use in the analytical computations would have been (42)(1.1) or 46.2 degrees, fairly close to the value that Hansen found to yield good agreement between experiment and analysis.

Ingra and Baecher (1983) discussed uncertainty in the bearing capacity of sands and concluded: "When the friction angle is imprecisely known (e.g., having a standard deviation greater than 1°), the effect of uncertainty in ϕ predominates other sources."

Examples from Selig and McKee (1961) In contrast to the footings tested by Muhs and Weib (1969), Selig and McKee (1961) tested footings that ranged in size from 2 by 2 in. to 3 by 21 in. The soil employed was an Ottawa sand that was carefully placed to achieve a uniformity of 112.3 lb/ft^3 in a box that was 48 in. square by 36 in. deep. The friction angle of the sand was measured in the triaxial apparatus and was found to range from 38° to 41°. An average value of the friction angle was taken as 39.5; thus, using the

Hansen recommendation, the following value was computed for use in the analysis:

$$\phi_{pl} = 1.1\phi_{tr} = 43.45°$$

Some small footings were tested with no embedment, which is not consistent with practice. The experimental results greatly exceeded the results from analysis using the Hansen equations. Selig and McKee performed three tests where the base of the footings was placed below the ground surface. The following equation is solved in the table below, with the following constant values: $N_q = 106.0$; $N_\gamma = 149.2$; $\gamma = 0.0650$ lb/in.^{3}; $s_\gamma = 0.6$; $d_\gamma = 1.0$; $s_q = 1.688$; and d_q is as shown in the table.

$$\frac{Q_d}{A} = \frac{1}{2}\gamma B N_\gamma s_\gamma d_\gamma + \gamma D_f N_q s_q d_q$$

For the three experiments that were evaluated, the factors of safety in the equation that were computed in Table 7.2 were 1.50, 1.38, and 1.23, yielding an average value for a fairly small sample of 1.37. The selection of a value of ϕ_{pl} is important, and Hansen's correlation with ϕ_{tr} may be used unless a more precise correlation becomes available. A factor of safety appears in the Hansen procedure, as noted from the small sample, and additional analyses of experimental data of good quality, particularly with full-sized footings, will be useful.

The settlement at which failure is assumed to occur is of interest when performing experiments. With regard to the bearing capacity of the base of drilled shafts in sand, several experiments showed that the load kept increasing with increasing settlement, and it was decided that the ultimate experimental load could be taken at a settlement of 10% of the base diameter. A possible reason for the shape of the load-settlement curve was that the sand kept increasing in density with increasing settlement.

Muhs and Weib (1969) apparently adopted 25 mm as the settlement at which failure was defined. Only a few of the load-settlement curves were reported by Selig and McKee (1961), but those that were given showed failure to occur at a relatively small settlement, with the load remaining constant or decreasing after the maximum value was achieved. In both sets of experiments discussed above, the relative density of the sand was quite high, probably

TABLE 7.2 Computation of Q_d to Compare with Experimental Value

Footing size, in.	Area, in.^{2}	D_f, in.	d_q	Comp. Q_d, lb	Exp. Q_d, lb	Factor of Safety in Equation
3 × 3	9	2	1.123	261	391	1.50
3 × 3	9	4	1.171	516	711	1.38
3 × 3	9	6	1.205	782	964	1.23

leading to the maximum load at a relatively small settlement and a settlement-softening curve for greater settlement.

The small number of experiments where full-sized footings were loaded to failure is likely due to the cost of performing the tests. Furthermore, the cost of footings for a structure is likely to be small with respect to the cost of the entire structure, so the financial benefits of load tests would likely be small in most cases.

7.5 EQUATIONS FOR COMPUTING CONSOLIDATION SETTLEMENT OF SHALLOW FOUNDATIONS ON SATURATED CLAYS

7.5.1 Introduction

As noted earlier, the geotechnical design of shallow foundations is presented in Chapter 9. Two kinds of settlement are discussed: short-term or immediate settlement when the settlement of granular soils (sands) and cohesive soils (clays) will occur rapidly, and long-term settlement of clays that can occur over a period of years. Some long-term settlement of sands can occur. The most evident case is when the sands are subjected to vibration, causing the grains to move and increase the density of the soil. Large amounts of settlement can occur if very loose sands are subjected to vibration. Loose sands below the water table cannot rapidly increase in density when affected by an earthquake and are subject to liquefaction because the water cannot immediately flow out. When a stratum of sand is subjected to sustained loading, some settlement of the foundation may occur that is time dependent.

Long-term settlement of clay deposits below the water table is of concern to the geotechnical engineer. Two types of predictions must be made: the total settlement that can occur over a period of time and the time rate of settlement. The procedures available for making the two predictions are presented in the following sections.

The assumptions made to develop the equations for computing the total settlement of a stratum of saturated clay are as follows:

1. The voids in the clay are completely filled with water.
2. Both the water and the solids of the soil are incompressible.
3. Darcy's law is valid (see Chapter 3); the coefficient of permeability k is a constant.
4. The clay is confined laterally.
5. The effective and total normal stresses are the same for every point in any horizontal plane through the clay stratum during the process of consolidation (Terzaghi, 1943).

Further, Terzaghi noted that the process of consolidation, as water flows vertically though soil, is analogous to the heat flow from an infinite slab if the slab is at a constant temperature and the boundaries are maintained at a different temperature.

If the assumptions noted above are correct, the equations for total settlement and the time rate of settlement of a stratum of clay below the water table when subjected to load are perfectly valid. The assumptions made in developing the theory cannot all be satisfied; for example, the loaded area must be finite, and some water could flow from the stratum laterally rather than vertically. However, if care is used in sampling and testing the stratum of clay, the consolidation equations can be used, as evidenced by past experience, to make predictions for the total settlement with reasonable accuracy and prediction of the time rate of settlement with less accuracy.

7.5.2 Prediction of Total Settlement Due to Loading of Clay Below the Water Table

Figure 7.9a shows a stratum of sand below the ground surface above a stratum of clay. The water table is shown in the stratum of sand. The assumption is made that the clay has always been saturated. The line labeled σ'_0 shows the distribution of pressure from the self-weight of the soil prior to the imposition of a surface load, and the quantity $\Delta\sigma'$ shows the increase in σ' due to the surface loading. The curve in Figure 7.9b shows the results from a laboratory consolidation test of a sample of superior quality from the stratum of clay. The dashed line shows the direct results from the laboratory data, with the solid line showing the corrected laboratory curve to yield a field curve, as described in Chapter 3. The portion of the curves marked "rebound" indicates that the soil will swell when load is removed. The simple model for computing total settlement is shown in Figure 7.9c.

The determination of the change in void ratio Δe for use in computing settlement S is shown in Figure 7.9b The average value of $\Delta\sigma'$ shown in Figure 7.9a is added to σ'_0, and the value of Δe is obtained as shown. The total settlement can now be computed with the equation in Figure 7.9c.

The equation for settlement S in Figure 7.9c may be rewritten in a convenient way to compute settlement based on the plot in Figure 7.10. The curve in the figure has been corrected to offset remolding and other disturbance. The plot of void ratio e versus the log of effective stress σ' is for a sample of clay that had been preloaded. The first branch of the curve indicates the reloading of the sample after removal of previous stress, the *reloading curve.* The other branch represents the behavior of the sample after exceeding the magnitude of the previous loading, the *virgin compression curve.* The equation for the two branches of the curve is as follows:

$$S = \frac{H}{1 + e_0}\left[c_r \log \frac{\sigma'_p}{\sigma'_0} + c_c \frac{\sigma'_0 + \Delta\sigma'}{\sigma'_p}\right] \qquad (7.36)$$

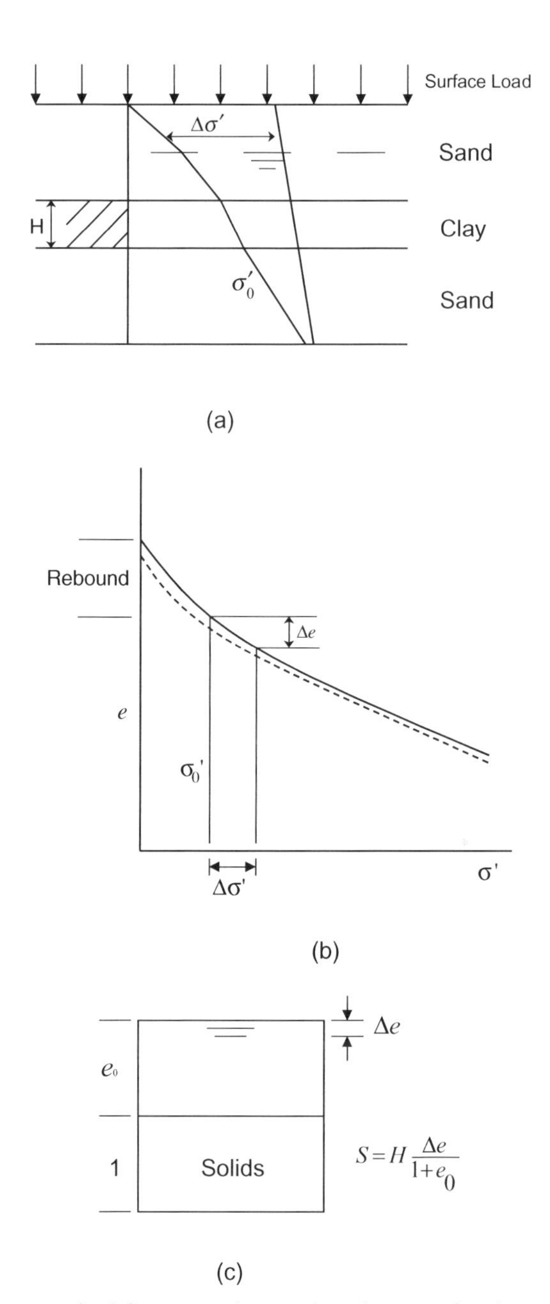

Figure 7.9 Data required for computing total settlement of a clay stratum due to an imposed surface load. (a) Solid profile showing initial effective stress σ'_0, surface load, and distribution of imposed load with depth $\Delta\sigma'$. (b) Field curve of void ratio e versus effective normal stress σ'. (c) Model for computing total settlement.

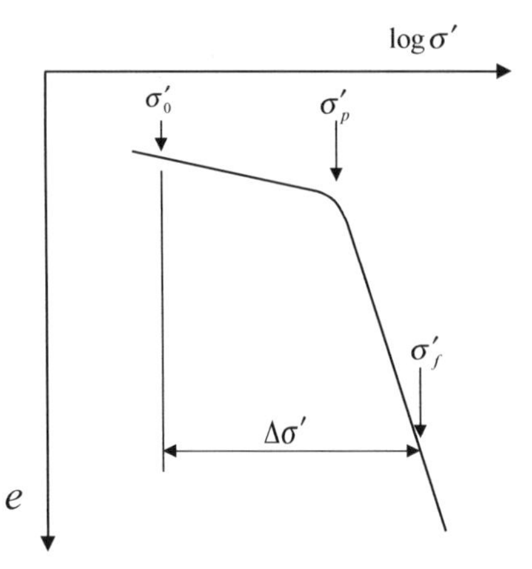

Figure 7.10 Plot of void ratio versus log of effective stress for an overconsolidated sample.

where

c_r = compression index (reloading), and
c_c = compression index (virgin loading), dimensionless.

In some instances, the settlement due to the reloading portion of the curves is deemed to be so small as to be negligible. For normally consolidated clay, only the virgin compression curve will be used.

To obtain the quantity $\Delta\sigma'$, a method for predicting the increase in load as a function of depth must be developed. An appropriate method has been proposed by Boussinesq (1885) as shown in the following equation (Terzaghi and Peck, 1948):

$$\sigma_v = \frac{3Q}{2\pi z^2}\left[\frac{1}{1 + (r/z)^2}\right]^{\frac{5}{2}} q \tag{7.37}$$

where

σ_v = stress in the vertical direction,
Q = concentrated vertical load,
z = vertical distance between a point N within an semi-infinite mass that is elastic, homogeneous, and isotropic, and
r = horizontal distance from point N to the line of action of the load.

The assumption is made that the applied load at the ground surface is perfectly flexible, with a unit load of q. The loaded area is divided into small parts, as expressed in the following equation, where load dq is assumed to be acting at the centroid of the area dA:

$$dq = qdA \tag{7.38}$$

Substitution can be made into Equation 7.36 with the following result:

$$d\sigma_v = \frac{3q}{2\pi z^2}\left[\frac{1}{1 + (r/z)^2}\right]^{\frac{5}{2}} dA \tag{7.39}$$

The magnitude of σ_v at point N due the entire loaded area can be found by integrating Eq. 7.39. Newmark (1942) produced a chart for easy use in the determination of σ_v (Figure 7.11). The following procedure is employed in using the chart to find σ_v at depth z: the distance to the depth is given by the length A-B in the chart; employing the scale, a sketch is made of the loaded area; the point in the sketch under which the value of σ_v is desired is placed over the center of the sketch; the number of subdivisions n, including fractions, covered by the sketch is counted; and $\sigma_v = 0.005n\ \sigma_0$, where σ_0 is the vertical stress at the ground surface due to the applied load. The number of subdivisions in the chart is 200; therefore, if all of the subdivisions are covered, $\sigma_v = \sigma_0$.

Sketches of the loaded area are made for increasing depths using the A-B scale, with the sketches of the loaded area becoming smaller and smaller. Thus, a plot of σ_v as a function of depth z beneath the ground surface can readily be made. As shown in Figure 7.12, the value of σ_v as a function of depth z can readily be made for a point under a particular footing if a number of closely spaced areas exist with differing values of σ_0. The scale A-B was set at 8 ft; thus, the value of σ_v was to be determined at a depth of 8 ft below the base of the footings. As shown in Figure 7.12, the scale A-B was used to draw sketches of the footings 6 by 6 ft and 8 by 8 ft, separated by a distance of 3 ft. The value of σ_v was desired below the center of the smaller footing. As shown in the figure, the unit load on the smaller footing was 2 k/ft^2 and the load on the larger footing was 3 k/ft^2. The number of values of n under the smaller footing was 42.4, and n for the larger footing was 10.4. The value of σ_v can now be computed:

$$(\sigma_v)_{8\,ft} = (0.005)[(42.4)(2.0) + (10.4)(3.0)] = 0.424 + 0.156 = 0.58\ \text{k/ft}^2$$

Examining the plotting of the footing sizes in the figure shows that there was a lack of precision that had only a minor effect on the result. The Newmark chart can be used to obtain the increase in effective stress as a function of depth beneath a given point under a loaded area due to various horizontal spacings of footings, considering the surface loading from each of them.

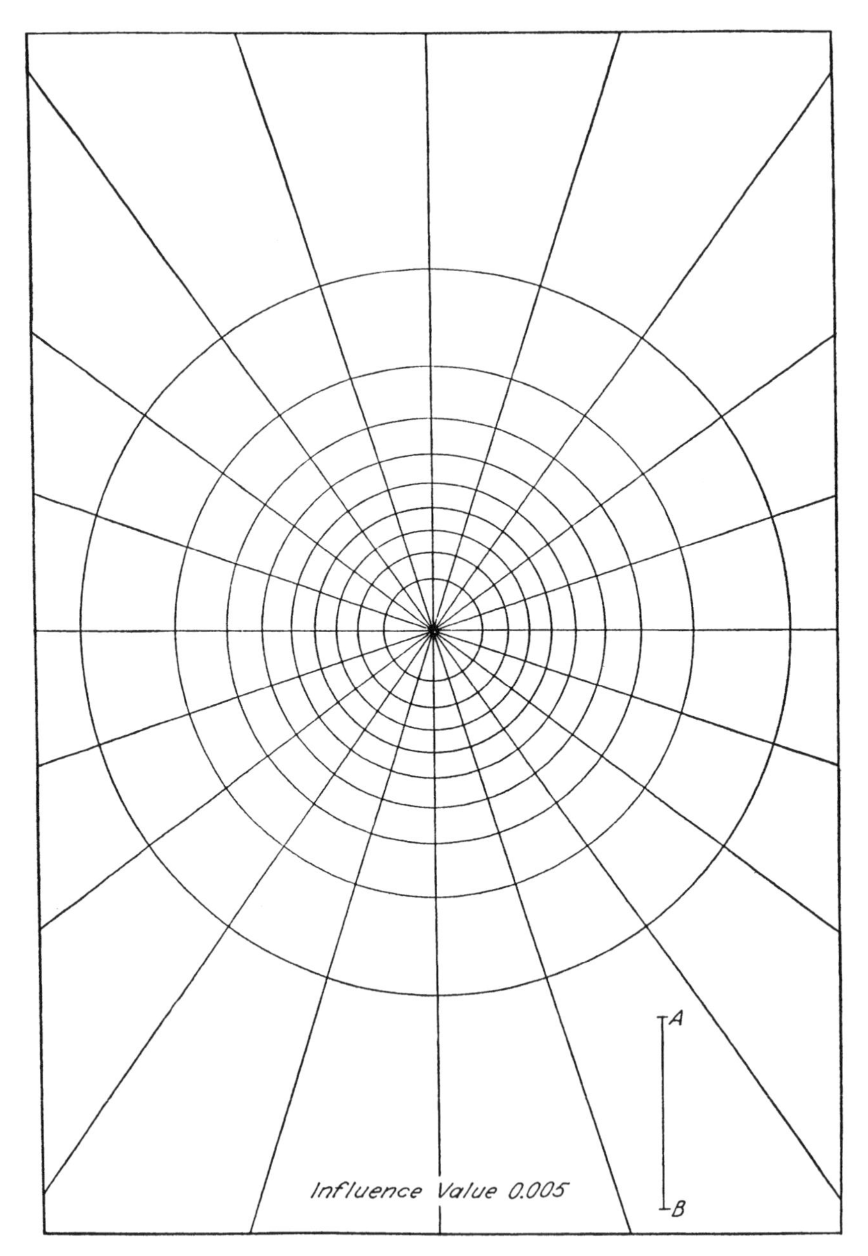

Figure 7.11 Newmark chart for computing vertical stress beneath a loaded area.

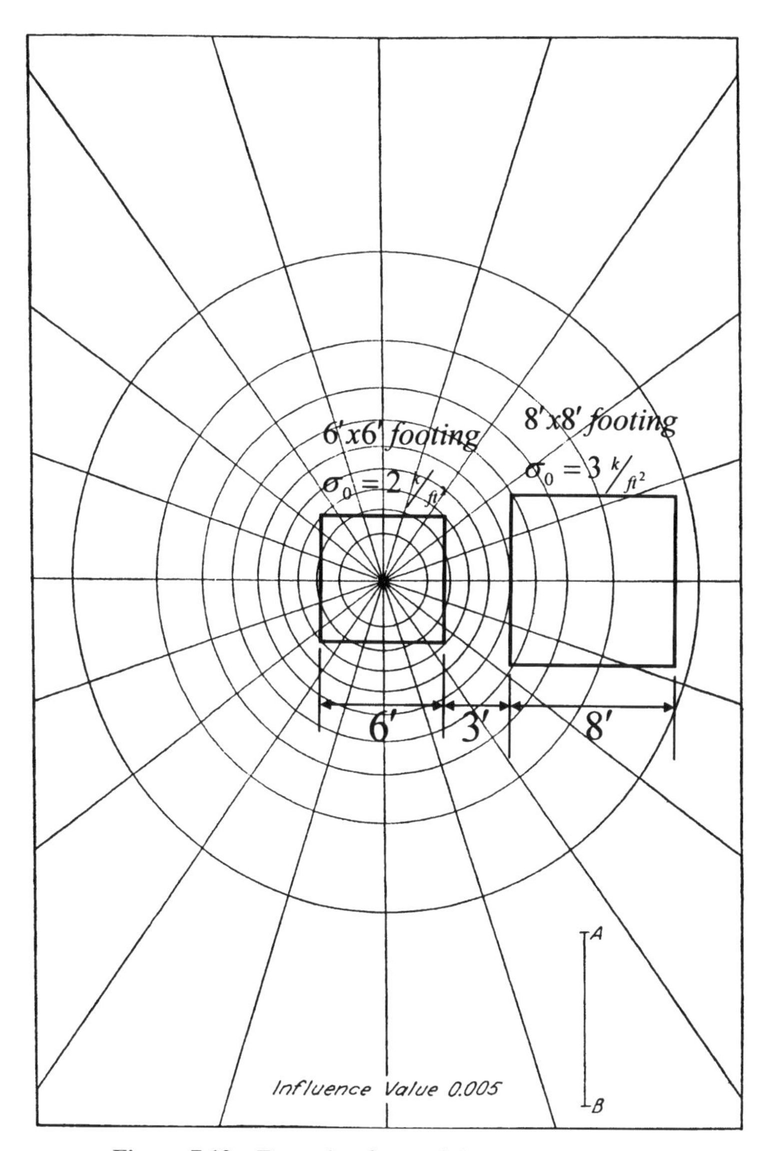

Figure 7.12 Example of use of the Newmark chart.

7.5.3 Prediction of Time Rate of Settlement Due to Loading of Clay Below the Water Table

Terzaghi (1943) reasoned that prediction of the time at which a certain amount of settlement occurs could be simplified by assuming that the surface loading occurred instantaneously, that excess porewater pressure would develop throughout the stratum of clay, and that drainage from the clay would occur instantly where the clay interfaced with a porous stratum of sand. Thus, water in the stratum of clay in Figure 7.9 would drain upward from the top of the clay and downward from the bottom of the clay. Terzaghi noted that the process was precisely analogous to the heat-flow problem, where a semi-infinite slab of material at an elevated temperature is suddenly subjected to zero temperature, or to a lower temperature, at both surfaces of the slab.

The differential equation for heat flow is shown in Eq. 7.40 (Carslaw and Jaeger, (1947)

$$\frac{\partial u}{\partial t} = \kappa \frac{\partial^2 u}{\partial z^2} \tag{7.40}$$

where

u = temperature (excess porewater pressure), and
κ = diffusivity reflecting the ability of the slab of material to transmit heat.

The problem is to find the properties of the soil, plainly involving the coefficient of permeability k, that will replace the diffusivity κ. The term κ is replaced by c_v in consolidation theory, where c_v is called the *coefficient of consolidation,* as defined by the following equation:

$$c_v = \frac{k}{m_v \gamma_w} \tag{7.41}$$

where the rate of flow of water from the clay is directly proportional to the coefficient of permeability k and inversely proportional to γ_w and m_v, where $m_v = a_v/1 + e$ and $a_v = -de/d\sigma'$. Thus, if the *coefficient of compressibility,* a_v, is larger, meaning that the *coefficient of volume compressibility,* m_v, is larger, the rate of flow will be less. Thus, the differential equation for the rate at which water will flow from a stratum of clay that has been subjected to an increase in effective stress, σ', due to the placement of a surface loading above the clay, is shown in Eq. 7.42:

$$\frac{\partial u}{\partial t} = c_v \frac{\partial^2 u}{\partial t^2} \tag{7.42}$$

The solution of Eq. 7.41 requires the implementation of the following boundary conditions: the excess porewater pressure, u, is equal to zero at the drainage faces; the hydraulic gradient i is equal to zero at the mid-height of the stratum of clay and at the boundary where flow is prevented, where $i = du/dz$; and after a very long time (infinite according to the differential equation), $u = 0$ at all depths.

The solution of Eq. 7.42 for the conditions given is as follows:

$$U_z(\%) = f\left(T_v, \frac{z}{H}\right) \tag{7.43}$$

where

$$T_v = \frac{c_v}{H^2}\,t \tag{7.44}$$

The dimensionless number T_v is called the *time factor.* The solution of Eq. 7.43 in terms of dimensionless coefficients is shown in Figure 7.13. The figure illustrates the process of consolidation graphically by showing the distribution of excess porewater pressure, u, with the passage of time. The figure readily shows that a very long time will be required for u to equal zero throughout the stratum. The hydraulic gradient causing the discharge of porewater becomes smaller and smaller over time.

The data in Figure 7.13 can be integrated to produce Figure 7.14. The time for a given percentage of the consolidation to take place in the laboratory for a representative sample is determined, usually at 50% of the laboratory consolidation. Using the time factor T_v of 0.197 (see Figure 7.14), the value of the coefficient of consolidation c_v can be found. And Eq. 7.43 can be used to plot a curve showing the amount of settlement of the clay stratum as a function of time.

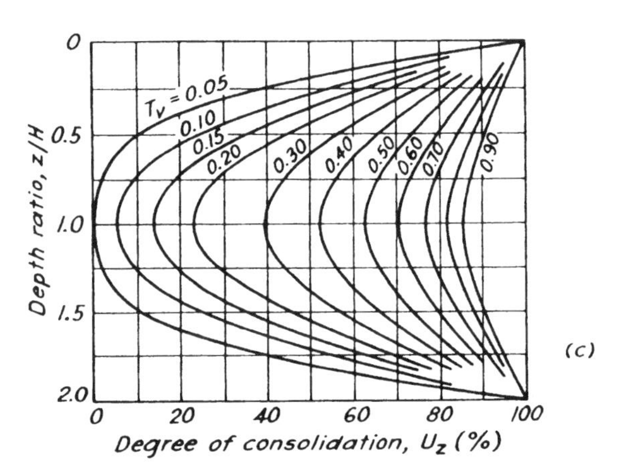

Figure 7.13 Percentage of consolidation U_z (%) as a function of relative depth Z/H and time factor T_v (note: thickness is 2H for the drainage top and bottom of the stratum) (from Peck et al., 1974).

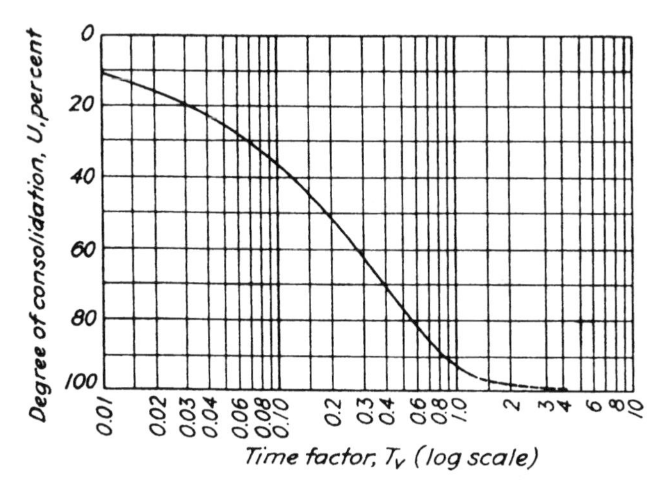

Figure 7.14 Theoretical relationship between degree of consolidation U and time factor T_v (from Peck et al., 1974).

PROBLEMS

P7.1. a. Compute the total load, Q_q, in kips, using the Hansen equations, for a footing 4 ft wide by 8 ft long founded on a stratum of sand. The vertical load is uniformly distributed; there is no horizontal load. The value of D_f is 3.5 ft, the water table is at the base of the footing, and the friction angle from the triaxial test is 30°. For the sand above the base of the footing, the water content is 20%, and the unit weight γ is 120 pcf. Compute a value of submerged unit weight γ' if needed, assuming specific gravity of the particles of sand to be 2.67.

b. The problem remains the same as in part a except that the footing rests on overconsolidated clay at a depth of 3.5 ft and the footing rests on the clay. Assume the same unit weights as before and determine the undrained shear strength of the clay to yield the value of Q_q found in part a.

P7.2. a. Under what condition might you wish to construct a footing with a base at some angle with the horizontal? Show a sketch.

b. What technique could you use to fine the most favorable angle for the base?

P7.3. Under what conditions would you wish to use the Hansen equations for a sloping ground? Show a sketch.

P7.4. The time to reach 55% consolidation for a laboratory sample of clay that was 0.6 in. thick and tested under double drainage was 24.5 seconds. How long would it take for a stratum of the same soil in the field that was 14 ft thick and drained on only one side to reach the same degree of consolidation?

CHAPTER 8

PRINCIPLES FOR THE DESIGN OF FOUNDATIONS

8.1 INTRODUCTION

The engineer is presumed to understand all of the methods presented here and to be able to make an appropriate decision about their applicability. What other things are necessary in developing a design? This chapter addresses that question.

The first section is closely related to the principles that must be employed in the design of foundations. To be an engineer is to be a member of a profession that has made innumerable contributions to the betterment of human life. The goal of an engineer is to act so as to add to such contributions.

8.2 STANDARDS OF PROFESSIONAL CONDUCT

Perhaps the most valuable characteristic of the engineer is integrity. The client expects the engineer to approach the proposed project with careful attention and relevant knowledge based on education and experience. The client understands that the engineer will comply fully with the standards of professional conduct set forth by the American Society of Civil Engineers (ASCE, 2000). The fundamental principles and the fundamental canons abstracted from the ASCE document are shown below.

8.2.1 Fundamental Principles

Engineers uphold and advance the integrity, honor, and dignity of the engineering profession by:

1. Using their knowledge and skill for the enhancement of human welfare and the environment;
2. Being honest and impartial and serving with fidelity the public, their employees, and clients;
3. Striving to increase the competence and prestige of the engineering profession; and
4. Supporting the professional and technical societies of their disciplines.

8.2.2 Fundamental Canons

1. Engineers shall hold paramount the safety, health, and welfare of the public and shall strive to comply with the principles of sustainable development in the performance of their professional duties.
2. Engineers shall perform services only in the areas of their competence.
3. Engineers shall issue public statements only in an objective and truthful manner.
4. Engineers shall act in professional matters for each employer or client as faithful agents or trustees, and shall avoid conflicts of interest.
5. Engineers shall build their professional reputation on the merit of their services and shall not compete unfairly with others.
6. Engineers shall act in such manner as to uphold and enhance the honor, integrity, and dignity of the engineering profession.
7. Engineers shall continue their professional development throughout their careers and shall provide opportunities for the professional development of those engineers under their supervision.

These 11 statements from ASCE are worthy of being framed and hung on the walls of university classrooms and engineering offices.

8.3 DESIGN TEAM

The geotechnical engineer becomes a member of the design team, where the owner and the architect provide information on the aims of the project, on special requirements, on an expected date for completion of construction, and on any requirements of governmental agencies with jurisdiction at the site. The structural engineer and other engineers are also members of the team. The team will be organized differently if a design-build contract is to be used.

A close working arrangement is developed between the geotechnical engineer and the structural engineer. As noted throughout this book, a number of problems must be solved that involve structural engineering, and the structural engineer may be the lead designer. However, the participation of the geotechnical engineer should continue through construction, and not end with the submission of *p-y* curves, *t-z* curves, and similar information. Questions to be answered by the geotechnical engineer will normally arise throughout the period of design and construction.

8.4 CODES AND STANDARDS

The governmental agency in the area where the project is located will specify a building code. Many agencies will specify a code such as the *Uniform Building Code* or the *Southern Building Code.* Some states, such as Florida and Oregon, have prepared their own building codes, as have some large cities, such as New York City. If the project is a bridge or another highway structure, most states have prepared specifications for highway construction, such as Standard Specifications, for the Department of Transportation of the State of California.

Many of the codes are silent on aspects of the design of foundations, but the engineer will study carefully any provisions that are given to prevent a violation. Most of the *Uniform Building Code* (1991 edition) contains requirements for the architect. Part V is entitled Engineering Regulations—Quality and the Design of Materials of Construction, and Part VI, Detailed Regulations, includes Chapter 29, Excavations, Foundations and Retaining Walls.

Detailed procedures for certain engineering operations have been written by professional societies, such as the ASTM. The ASTM standards are referenced extensively in Chapter 4, and the engineer will use them where appropriate.

8.5 DETAILS OF THE PROJECT

The engineer will develop a clear view of the nature of the project, whether a monumental structure for the ages or a temporary warehouse. The details of the design will, of course, vary with the project. Two types of failure can be identified with respect to the foundation: (1) failure due to excessive cost and (2) collapse of the structure. The first failure can be eliminated by careful work, by employing appropriate methods, and by using an appropriate factor of safety.

The results of a collapse need to be considered. Will the collapse be catastrophic, with loss of life, or will a large monetary loss be incurred or only a small one? While each possibility needs consideration, the collapse of even a minor structure should be avoided.

On occasion, what is thought to be a collapse is only excessive settlement. A theme of this book is computation of the movement of the foundation under any load. Therefore, with regard to all elements of the project being addressed, the engineer will require knowledge of the tolerance to total settlement and to differential settlement.

Determination of the loads to be employed in designing the foundation of a structure is sometimes no simple matter. The design of the foundations for an overhead sign depends on maximum wind velocity, which is a statistical problem. The pattern of the velocity with time is also important, no data may be available for use in the design. The loads to be used in designing the piles for an offshore platform are related to the maximum waves that will develop from a storm. Prediction of the maximum storm that will occur at some location in the ocean during the life of a structure is based on data from that area of the ocean, and such data may be scarce or nonexistent.

Professor Ralph Peck (1967) made some observations about the criteria for the design of a structure. He noted that the floor load was based on the weight of the goods to be stacked to a given height in a warehouse but the load on the foundations was computed by assuming that the entire of the warehouse floor was covered; however, the use of the warehouse required numerous aisles throughout the building. Another example given by Professor Peck was that a certain rotating machine had bearings whose differential movement could be only a small fraction of an inch. However, that same machine was used on a ship that rolled with the waves!

8.6 FACTOR OF SAFETY

A comprehensive discussion of factor of safety involves a multitude of factors. Only a brief presentation is given here to emphasize the importance of this topic in the design process. First, the engineer must refer to the building code covering the project for a list of requirements. Many building codes give some discretion to the engineer, depending on the details of the design.

Next, the engineer must make a study to determine the quality of the data related to the design. The loads to be sustained by the foundation were noted above. In many cases, the loads and tolerance to settlement will change as the design of the superstructure proceeds. The subsurface investigation was discussed in Chapter 4, and many factors can affect the quality of the information on the soil available for the design. If piles are to be used to support the structure, the engineer must examine carefully the available data on the results of load tests on the type of pile to be used in soils similar to those at the site. A critical factor, then, is whether or not load tests on a pile or piles are to be implemented for the project. If so, the tests should be performed as early in the design process as possible.

The idea of limit states was introduced 40 years ago and provides the engineer with a basis in considering the factor of safety. Table 8.1 presents a version of the limit states for a pile under axial and lateral loading. Two

TABLE 8.1 Limit States for a Pile Subjected to Axial and Lateral Loading

Ultimate Limit States	Most Probable Conditions
Sudden punching failure under axial loading of individual piles	Pile bearing on thin stratum of hard material
Progressive failure under axial loading of individual piles	Overloading of soil in side resistance and end bearing
Failure under lateral loading of individual piles	Development of a plastic hinge in the pile
Structural failure of individual piles	Overstressing due to a combination of loads; failure of buckling due principally to axial load
Sudden failure of the foundation of the structure	Extreme loading due to earthquake causing liquefaction or other large deformations; loading on a marine structure from a major storm or an underwater slide
Serviceability Limit States	
Excessive axial deformation	Design of large-diameter pile in end bearing on compressible soils
Excessive lateral deformation	Design with incorrect *p-y* curves; incorrect assumption about pile-head restraint
Excessive rotation of foundation	Failure to account for the effect of inclined and eccentric loading
Excessive vibration	Foundation too flexible for vibratory loads
Heave of foundation	Installation in expansive soils
Deterioraton of piles in the foundation	Failure to account for aggressive water; poor construction
Loss of esthetic characteristics	Failure to perform maintenance

Source: From Reese and Van Impe, 2001.

categories are presented: ultimate limit states and serviceability limit states. A collapse is considered in the first instance and the ability of the pile to perform acceptably in the second instance. While the table is not comprehensive, the engineer will find the concept of limit states useful in considering a particular design.

The factor of safety for the foundations of a project may be selected with a consideration of limit states, with the data at hand for use in the design, and after evaluation of the quality of the data. Constraints imposed by the building code must be taken into account. Two approaches are in use: selection of a global factor of safety and selection of a partial factor of safety. Both methods are discussed below.

8.6.1 Selection of a Global Factor of Safety

The first approach is to employ a global factor of safety that is normally used to augment the loads on a structure. For the design of a pile under lateral load, the working loads are augmented until the pile fails by the development of a plastic hinge or experiences excessive deflection. The global factor of safety is the load at failure divided by the working load, where the working load is that which is experienced by the structure during its regular use.

Is is convenient to consider the loads on the structure and the resistances supplied by the foundation to be time dependent, as shown in Figure 8.1.

The loads will vary with time as floor loads change and as the structure is subjected to wind and possibly waves. The resistance can vary as pore pressure dissipates around driven piles and as the water content changes in near-surface clays. The global factor of safety can be expressed in Eq. 8.1:

$$F_c = m_R / m_S \tag{8.1}$$

where

F_c = global factor of safety,
m_R = mean value of resistance R, and
m_S = mean value of loads S.

While producing the curves shown in Figure 8.1 may not be possible, the engineer considers all of the factors affecting the loads and resistances and arrives at values using best judgment to avoid the crosshatched area shown in the figure.

Values of the relevant soil parameters in a report on the subsurface investigation invariably reveal significant scatter as a function of depth. The usual

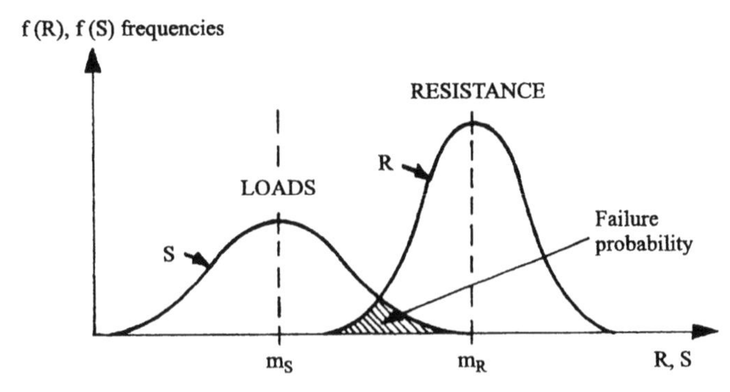

Figure 8.1 Probability frequencies of loads and resistance (from Reese and Van Impe, 2001).

approach is to select average values with depth, perhaps erring on the low side. The analyses are made with service loads and then augmented loads. The engineer then uses judgment to decide if the global factor of safety being employed is adequate for the project.

Another procedure is to select lower-bound values of soil resistance with the service loads augmented to obtain failure due to either overstressing a component of the foundation or due to excessive deflection. If the global factor of safety is adequate with the use of lower-bound soil properties, the engineer may have confidence in the design. However, for a large project, the potential savings on construction may be sufficient to justify full-scale testing of prototype foundations in the field.

8.6.2 Selection of Partial Factors of Safety

The second approach is implemented by the method of partial safety factors. The resistance R^* may be modified as shown in Eq. 8.2:

$$R^* = \frac{r_m}{\gamma_m \gamma_f \gamma_p} \tag{8.2}$$

where

r_m = mean resistance or strength,

γ_m = partial safety factor to reduce the strength of the material to a safe value,

γ_f = partial safety factor to account for deficiencies in fabrication or construction, and

γ_p = partial safety factor to account for inadequacies in the theory or model for design.

The loading S^* may be modified as shown in Eq. 8.3:

$$S^* = s_m \gamma_1 \gamma_2 \tag{8.3}$$

where

s_m = mean value of load,

γ_1 = partial load factor to ensure a safe level of loading, and

γ_2 = partial load factor to account for any modifications during construction, effects of temperature, and effects of creep.

Using the partial safety factors in Eqs. 8.2 and 8.3, a global factor of safety may be computed as shown in Eq. 8.4:

$$F_c = (\gamma_m \gamma_f \gamma_p)(\gamma_1 \gamma_2) \tag{8.4}$$

The method of partial safety factors may be formally implemented by selecting values from a set of recommendations by a building authority based on poor control, normal control, or good control. The engineer who uses the formal approach to development of a factor of safety can employ values for the factors shown in Eq. 8.4 but must also use judgment in giving specific values to the terms in the equation. If an informal approach to selecting the global factor of safety for a given foundation is employed, the factors in Eq. 8.4 provide a useful guide.

The American Association of State Highway and Transportation Officials (AASHTO) adopted a method in 1994 based on modification of load and resistance factors (LRFD), similar in concept to the method of partial safety factors. LRFD is discussed in numerous articles and reports in the current literature.

8.7 DESIGN PROCESS

A design can be made with the following information at hand: type of foundation; nature and magnitude of loads; constraints on movement, both initial and final; results from the soil investigation; quality of the soil investigation; analytical model along with computer code; nature of the structure and the effects of a failure; and all other factors affecting the selection of a factor of safety. Figure 8.2 shows an engineer at work with a computer. The analyses and design are carried out in such a manner that a review can be made with a full understanding of the assumptions and the results of all computations.

The computer has an important role. Computer codes exist for the solution of many problems, as noted in many of the chapters in this book. Two characteristics define the computer: (1) results can be obtained so rapidly, even for nonlinear problems where iteration is required, that the influence of many parameters on a solution can be investigated with little effort by the engineer; (2) the results from a computer solution may be incorrect. In regard to the first item, rarely does a soil report fail to show scatter in the results of soil properties. The usual procedure is to employ average values, with some leaning toward lower values. However, the engineer can easily compare results from selection of lower-bound values, upper-bound values, and average values.

However, in view of the second item, the engineer must be prepared to check the results of a computer solution. Usually, the results can be easily checked to see that the equations of static equilibrium are satisfied. Even for nonlinear problems, some checks can be made with hand computations to investigate correctness. Such checks of the computer results should be indicated in the written material that accompanies the report on design.

Figure 8.2 View of an engineer at work.

In the course of an analysis, close interaction between the various engineers is necessary, particularly between the geotechnical and structural engineers. Frequently, the final design is continuing and significant changes may have been made that affect the design of the foundations. Furthermore, the decision may be made to prepare alternate designs for submittal in the contract documents in order to achieve the lowest cost for the project.

Appendix 8.1 presents a list of considerations that can lead to improved designs. The aim is to create a design that is understood by the contractor and that is relatively easy to construct.

8.8 SPECIFICATIONS AND INSPECTION OF THE PROJECT

The preparation of specifications should be considered an important area of work for the engineer. Many firms will have a computer file containing specifications for the construction of various kinds of foundations. Using such information without careful consideration of the job at hand is unwise. Methods of construction change, and new ways of performing work are constantly being developed. The Internet can be used to pull up specifications by some agencies of contractors, and other sources may have specifications that can be examined. The writers have spoken to contractors who have discovered defective specifications in bid documents but refrained from comment for a variety of reasons. However, dealing with defective specifications after a job has begun is always difficult.

Details that are important in achieving good construction may be easy to overlook. For example, concrete for drilled shafts is poured with a tremie after the rebar cage has been placed. The design of the concrete is critical to ensure that it fills the entire excavation and develops an appropriate pressure value at the interface with the soil. A feature that is sometimes overlooked is that the spacing between rebars must be adequate to allow free passage of the concrete, depending on the maximum size of the coarse aggregate in the mix. Ample guidance exists here and in the literature on construction of drilled shafts to assist the engineer in sizing the spaces between rebars.

After construction has been completed, the engineer should take advantage of the contractor's knowledge to ask how the design and specifications could have been improved to achieve better construction. Such interaction between the engineer and the contractor is generally not possible during the bidding phase of the project, but it can be extremely valuable after the job has been completed.

Inspection of the construction of the foundations frequently is a gray area. Sometimes a special firm is employed by the owner to perform all of the inspection. However, the geotechnical engineer should take all desirable steps to ensure that the foundations are inspected by a firm that is fully knowledgeable about their design and about the specifications that have been prepared. Some geotechnical firm insist on inspecting the foundations they have designed.

8.9 OBSERVATION OF THE COMPLETED STRUCTURE

Far too few detailed observations have been made of completed structures, but the *observational method* is employed in some countries. With the permission of the owner, perhaps a governmental agency, a design is completed with the understanding that the behavior of the structure will be observed over time. If the structure shows some deficiency, such as total or differential settlement, provisions are made in the design to allow strengthening of the structure. Such a method is unusual.

One of the authors wished to install instruments at the base of the raft foundation for a high-rise building. The instruments would have allowed pressures and movements to be measured, providing insight into the response of the soil and the assumptions made in the design of the raft. A dinner was held in a luxurious club, and the architect and engineers came from a distant city. The research proposal was made, and a representative of the owner asked, "Is there any danger of failure of the foundation of the building?" "No, absolutely not!" "Then there is no reason for the company to allow the research." Unfortunately, this response was typical, but it was understandable. Engineers, however, should push for the opportunity to instrument candidate structures to gain information to be used in improving methods of design.

PROBLEMS

P8.1. Select no more than three of the items from the ASCE Standard for Professional Conduct of Civil Engineers and write a one-page, single-spaced essay using your word processor. The emphasis will be on your qualifications as a civil engineer.

P8.2. List the 11 items in Appendix 8.1 in the order of their importance to you and write a paragraph about one of them.

P8.3. Look at Chapter 11 and find the recommendation for the space between the rebars in a drilled shaft as related to the maximum size of the coarse aggregate.

APPENDIX 8.1

Considerations to Lead to Improved Designs

1. Communicate well. Written documents, computation sheets, and drawings should be clear and concise. If you have questions about presentation of information, ask a colleague to review your work.
2. Carefully consider complexities that are unusual. Designs are rarely "run of the mill," so deal with technical challenges to the best of your ability.
3. Use the computer as a tool and with judgment. Develop a healthy skepticism about results from computer runs until you have confirmed them by checking.
4. Design by considering tolerances. In practice, for example, piles cannot be installed precisely, so account for some eccentricity in vertical loads.
5. Mistakes are difficult to avoid. An in-house review of your work is essential, and you can encourage your firm to adopt a peer review process by outside designers.
6. Take advantage of the experienced engineers in your firm. Ask questions and solicit ideas on difficult problems.
7. Eagerly tackle jobs for which you are competent; ask for help when your knowledge is weak. Even the most reputable firms ask for help from outside experts when necessary.
8. Make certain that your data are accurate. Making assumptions about features of a problem is unacceptable.
9. View the construction of foundations you have designed. Experience is a great teacher.
10. Consider the ground to be riddled with previous construction. Take all possible steps to locate underground utilities or other obstructions before making final designs.
11. Make lifelong learning a goal in your engineering practice. Attending conferences and workshops and reading technical literature in a field are methods the engineer can use to learn. Such learning should never cease while the engineer continues to practice.

CHAPTER 9

GEOTECHNICAL DESIGN OF SHALLOW FOUNDATIONS

9.1 INTRODUCTION

The geotechnical design of shallow foundations requires specific computations for the following two purposes: (1) assurance that the foundation does not collapse by plunging into the soil or by rotating excessively and (2) assurance that the short-term and long-term total and differential settlement will be tolerable for the particular structure. The most frequent approach is to use the bearing-capacity equations in Chapter 7, or a similar set of equations, to compute the ultimate bearing stress q_d that will cause collapse; to apply a factor of safety to the ultimate bearing stress; to obtain the allowable soil bearing stress q_a; and to compute the short-term or long-term settlement with the value of q_a. These procedures are presented in detail in the following sections along with other features related to the design of shallow foundations.

9.2 PROBLEMS WITH SUBSIDENCE

Subsidence is the general settlement of the ground surface due principally to the extraction of water or perhaps oil. Subsidence can also result from the collapse of underground cavities (Gray and Meyers, 1970) or the vibration of deposits of sand (Brumund and Leonards, 1972). Lowering of the water table increases the vertical stress on water-bearing soil and on all lower strata. In simple terms, lowering of the water table means that vertical stress is increased by the total unit weight less the buoyant unit weight times the distance of the lowering. If the affected soil is compressible and fine-grained, the resulting settlement will be time-dependent and perhaps large.

Mexico City has been especially affected by subsidence (Zeevaert, 1980). Figure 9.1 shows a group of attendees at a conference in Mexico City standing next to a well casing whose top had originally been near the ground surface. The result of pumping water from strata below the city is graphically evident by comparing the exposed length of the casing to the 6-ft height of Dr. William R. Cox. If the water table is lowered uniformly and if the soil is uniform, subsidence could be uniform, but uniformity seldom exists. As noted earlier the Palace of Fine Arts has settled significantly but remains in heavy use. Some design guidelines may show that such a structure is intolerant of dif-

Figure 9.1 Photograph from Mexico City showing conference attendees standing next to a well casing showing subsidence over many years.

ferential settlement. The authors are reluctant to present such guidelines because of special circumstances that apply to many such structures.

Subsidence due to lowering of the water table has affected other places, including Ottawa, Canada (Bozuzuk and Penner, 1971), the Gulf Coast of Texas (Dawson, 1963), Venice (Berghinz, 1971), and London (Wilson and Grace, 1939). Surface and near-surface facilities such as roads, streets, and pipelines have been adversely affected (Zeevaert, 1980). In addition, fractures have been observed near the edges of the area of subsidence where the change in ground surface elevation was severe. Fracturing due to subsidence from the lowering of the water table is difficult and perhaps impossible to predict, but the geotechnical engineer must be aware of the chance of such occurrences.

Subsidence is a potential problem in areas of abandoned mines (Gray and Meyers, 1970) and in areas of karstic geology. Location of underground cavities emphasizes the need for a proper investigation of the soil at a site and suggests the possible use of geophysical techniques.

9.3 DESIGNS TO ACCOMMODATE CONSTRUCTION

9.3.1 Dewatering During Construction

Placing shallow foundations below the water table presents a series of challenges to the geotechnical engineer. Pumping from a sump in the excavation is frequently unacceptable because of the danger of collapse of the bottom of the excavation as a result of the lowered effective stress due to the rising water. The use of well points with good controls is usually acceptable. Furthermore, care must be taken that a nearby building is not affected by lowering the water table beneath the building.

Dr. Leonardo Zeevaert (1975) spoke of dewatering during construction of the Latino Americano Tower. Water that was removed from beneath the excavation was injected behind a sheet-pile wall to keep the water table at almost a constant elevation with respect to nearby structures. Also, records were made of a crack survey of the nearby buildings so that any new damage could be detected. No additional damage was found.

9.3.2 Dealing with Nearby Structures

Installation of foundations close to an existing structure can affect that structure. Most problems occur with the installation of deep foundations, where driving of piles or excavation for drilled shafts could cause vibrations or even the lateral movement of soil. However, an excavation with a substantial depth for a mat, for example, could create a severe problem. Extraordinary measures must sometimes be taken, including possible underpinning of the foundations of adjacent structures.

9.4 SHALLOW FOUNDATIONS ON SAND

9.4.1 Introduction

As noted in Chapter 6, designers of footings on sand must be aware of the possible settlement due to dynamic loads. Many investigators have studied the problem, with most concern focusing on the design of compaction equipment. Brumund and Leonards (1972) performed tests in the laboratory with a tank of Ottawa sand at a relative density of 70%. A plate with a diameter of 4 in. was vibrated and a settlement of approaching 1 in. was measured. The relative density of sand deposits and all relevant properties of the sand must be considered when designing the foundations for vibrating machinery. Data provided by the machine's manufacturer must be accessed in making such a design.

The discussion in Chapter 4 indicated that sampling deposits of sand without binder below the water table is not possible without extraordinary measures, such as freezing, and in situ techniques for investigating the sand are normally required. The two principal tools for obtaining data on the characteristics of sand are the SPT and the static cone, sometimes termed the *Dutch cone.* The cone consists of a 60° hardened steel point, a projected end area of 10 cm^3, and a rate of advancement of 2 cm/sec. Data from the static cone, if obtained in the recommended manner, are reproducible because the resistance along the push rods is eliminated, the force of penetration is measured with a calibrated load cell, and the rate of penetration is standard.

The SPT has been used widely in the United States for many years and is the usual technique for determining the in situ characteristics of sand. Testing with the static cone is popular in Europe but has been slow in gaining popularity in the United States; however, many U.S. geotechnical firms can now perform the cone test.

Data from the SPT, described in Chapter 4, can vary widely from one operator to another and during the performance of a particular test. Robertson and Campanella (1988) have referenced other authors and have noted that the energy delivered to the driving rods during an SPT can vary from about 20% to 90% of the theoretical maximum, with the variation related to the number of turns of rope around the cathead, the height of fall, the type of drill rig, and the operator. They concluded that an energy ratio of 55% to 60% is the average energy level employed in the field, suggesting that SPT values have some degree of uncertainty.

In the two sections that follow, two methods are presented for obtaining the immediate settlement of shallow foundations on sand. The first is based on the static-cone test and the second is based on the SPT.

9.4.2 Immediate Settlement of Shallow Foundations on Sand

Schmertmann Method A method for estimating the settlement of shallow foundations on sand has been proposed by Schmertmann (1970) and colleagues (1978). The resistance to penetration is termed the *static-cone bearing capacity* and is given by the symbol q_c. Schmertmann noted that the values of q_c can vary widely with depth, and constant values, termed $\overline{q}_c$, are used over a particular depth. Other static-cone systems can be used for measuring the resistance of penetration if correlations are obtained with the Dutch cone.

Schmertmann's concept for the settlement of foundations on sand was to integrate the strain of the sand beneath the loaded foundation where the unit load is substantially less than the failure load. FEM studies were performed, and the results revealed that the maximum strain did not occur at the base of the foundation but at some distance below the base. Further, after a certain distance below the loaded area, the amount of strain could be considered negligible. Schmertmann and his co-workers adopted two patterns for the influence factor for strain due to rigid footings on sand, as shown in Figure 9.2, one for axisymmetrical loading and the other for plane-strain loading, as from a continuous footing. Schmertmann includes recommendations for the stiffness of the sand E_s as a function of the value of q_c or $\overline{q}_c$. The primary computation, based on experimental values of $\overline{q}_c$ and on the location and magnitude of the applied load, is modified to account for strain relief due to embedment and for time-related creep. With regard to time-related properties of sand, the phenomena related to creep or stiffening are not as evident as is the drainage of water in the settlement of soft clay, but empirical evidence is strong (Mitchell and Solymar, 1984; Mesri et al., 1990). In his original paper (1970), Schmertmann showed good to excellent agreement between computed values and values obtained from a significant number of foundations where settlement was measured.

With respect to the examples that were analyzed and the presentation that follows, it is assumed that the net load imposed by the shallow foundation does not approach the value causing a collapse. Section 9.4.3 shows the computation of the magnitude of the pressure from the shallow foundation that causes a bearing-capacity failure in the cases where settlement has been computed.

Schmertmann noted that strain ε_z at any depth z beneath a loaded area may be computed as follows:

$$\varepsilon_z = \frac{p}{E}(1 + \nu)[(1 - 2\upsilon)A + F] \tag{9.1}$$

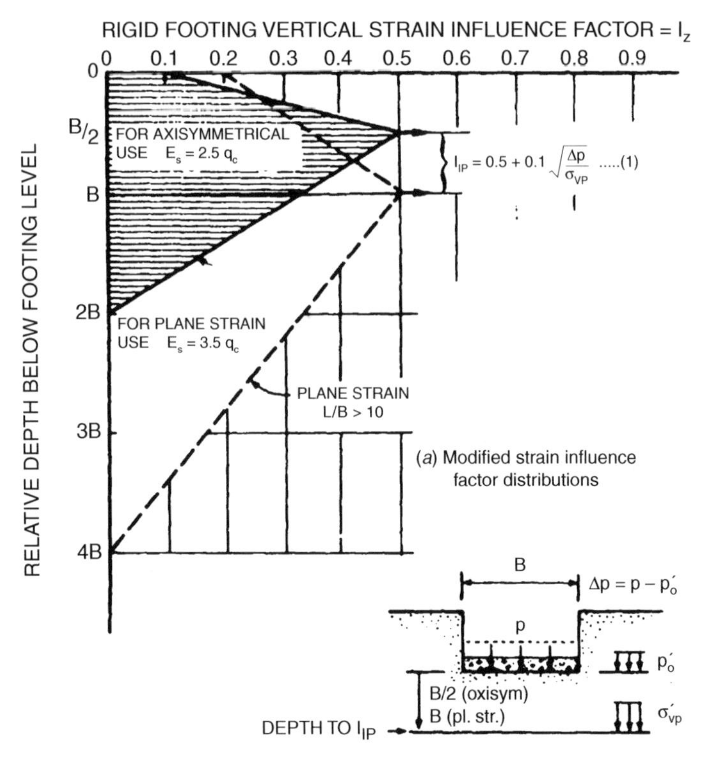

Figure 9.2 Recommended modified values for strain influence factor diagrams and matching sand moduli (from Schmertmann et al., 1978).

where

p = intensity of load on a homogeneous, isotropic, elastic half space,

E = modulus of elasticity,

ν = Poisson's ratio, and

A and F = dimensionless factors that depend on the location of the point considered.

If p and E remain constant, the vertical strain is dependent on the vertical strain factor, I_z:

$$I_z = (1 + \nu)[(1 - 2\nu)A + F] \tag{9.2}$$

The Schmertmann method for settlement is implemented by use of the following equations:

$$S_S = \int_0^\infty \varepsilon_z dz \cong \int_0^{f(B)} \left(\frac{I_z}{E_s}\right) dz \cong C_1 C_2 \Delta p \sum_0^{f(B)} \left(\frac{I_z}{E_s}\right) \Delta z \tag{9.3}$$

$$C_1 = 1 - 0.5\left(\frac{p_0}{\Delta p}\right) \tag{9.4}$$

$$C_2 = 1 + 0.2 \log\left(\frac{t_{yr}}{0.1}\right) \tag{9.5}$$

where

S_S = settlement according to the Schmertmann method,
$f(B)$ = $2B$ for the axisymmetric case and $4B$ for the plane-strain case (see Figure 9.2),
E_s = modulus of the sand,
C_1 = coefficient reflecting arching-compression relief,
C_2 = coefficient reflecting creep with time, and
t_{yr} = time in years.

Schmertmann and colleagues (1978) recommend that, for intermediate cases of footing shape, both diagrams in Figure 9.2 should be employed and interpolation employed for the appropriate solution.

The Schmertmann method is best demonstrated by the computation of an example that was presented in 1970 but is modified here to reflect changes presented in 1978. The footing in the example has a width of 2.60 m and a length of 23 m; therefore, the influence-factor diagram for the plane-strain case will be used. The base of the footing is 2 m below the ground surface with a unit load p of 180.4 kPa, and the overburden pressure p_0 is 31.4 kPa, yielding the net load Δp at the base of the footing of 149 kPa.

Schmertmann (1970) assumed that the cone-penetration test was performed and that values of $\overline{q}_c$ had been obtained for layers of sand of various thicknesses, starting at the base of the footing. The unit weight of the soil was assumed to be 15.7 kN/m^3 above the water table and 5.9 kN/m^3 below the water table, with the water table at the base of the footing. The diagram for plane strain may be used, but the value of I_{zp} must be increased as shown by the equation in the Figure 9.2; therefore, σ'_{vp} = (2.0)(15.7)+(2.0)(5.9) = 43.2 kN/m^3.

$$I_{zp} = 0.5 + 0.1\left(\frac{149}{43.2}\right)^{0.5} = 0.5 + 0.1(1.85) = 0.69$$

The values of I_z, shown in Figure 9.2, must be modified by increasing the value of I_{zp} from 0.5 to 0.69. The revised diagram for I_z is shown in Figure 9.3 along with assumed values of $\overline{q}_c$ from the cone test, following the pattern of values employed by Schmertmann in the original example.

The results from the computations for the revised example are shown in Table 9.1. The stratum was divided into layers, as shown in Figure 9.3, depending on values of $\overline{q}_c$, except where a layer was adjusted to fit the peak value of $\overline{q}_c$. The table extends below the footing to a depth of $4B$ or (4)(2.60) = 10.4 m. Eleven layers are shown in Table 9.1 with pertinent values of the relevant parameters. For a layer of a given thickness and with the corresponding value of $\overline{q}_c$, an average value of I_z was obtained by taking values at the top and bottom of each layer. The value of E_s for the plane-strain case was taken as 3.5 times $\overline{q}_c$. The values of C_1 and C_2, for a period of 5 years, were computed:

$$C_1 = 1 + 0.5\left(\frac{31.4}{149}\right) = 1.11$$

$$C_2 = 1 + 0.2 \log\left(\frac{5}{0.1}\right) = 1.34$$

The computations of settlement, layer by layer, shown in Table 9.1 yielded a total settlement of 54.1 mm, or 2.1 in. after 5 years. The immediate settle-

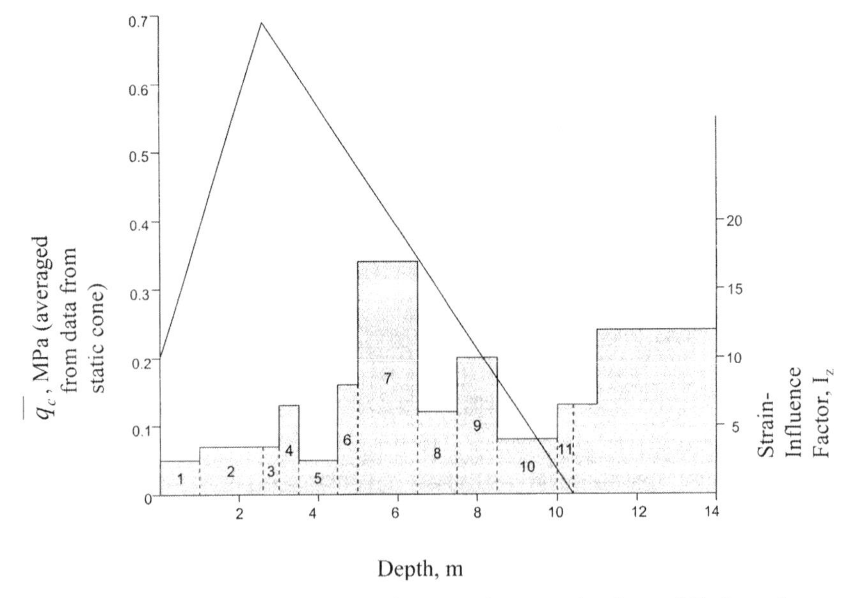

Figure 9.3 Modified strain-influence factor and assumed values of $\overline{q}_c$ from the cone test.

TABLE 9.1

Layer	ΔZ (m)	$\bar{q}_c$ (MPa)	E_s (MPa)	Z_L (m)	I_z	C_1	C_2	Δp (MPa)	Δz (m × 10^{-3})
1	1.0	2.50	8.75	0.50	0.295	1.11	1.34	0.149	7.5
2	1.6	3.50	12.25	1.80	0.540	1.11	1.34	0.149	15.6
3	0.4	3.50	12.25	2.80	0.673	1.11	1.34	0.149	4.9
4	0.5	7.00	24.50	3.25	0.633	1.11	1.34	0.149	2.9
5	1.0	3.00	10.50	4.00	0.568	1.11	1.34	0.149	12.0
6	0.5	8.50	29.75	4.75	0.500	1.11	1.34	0.149	1.9
7	1.5	17.00	59.50	5.75	0.410	1.11	1.34	0.149	2.3
8	1.0	6.00	21.00	7.00	0.300	1.11	1.34	0.149	3.2
9	1.0	10.00	35.00	8.00	0.213	1.11	1.34	0.149	1.3
10	1.5	4.00	14.00	9.25	0.103	1.11	1.34	0.149	2.4
11	0.4	6.50	22.75	10.20	0.018	1.11	1.34	0.149	0.1
									Sum = 54.1 mm

ment was computed to be (54.1)/(1.34) = 40.4 mm or 1.6 in. Schmertmann and his co-workers present a formal, step-by-step procedure for estimating the immediate and time-related settlement of shallow foundations on sand. The method is largely empirical but is founded on sound principles of mechanics and is validated by yielding reasonable comparisons with experimental values of settlement.

With regard to the magnitude of the settlement after 5 years, the geotechnical and structural engineers must decide what settlement is tolerable for the structure being designed. The informal guideline for the design of shallow foundations on sand is that settlement will control the allowable load. Frequently, a settlement of 1 in. (25.4 mm) is selected as the allowable settlement. The authors believe that the allowable settlement and the factor of safety to prevent a bearing-capacity collapse are matters to be considered for each structure, except as dictated by building codes.

An example of the bearing capacity of a shallow foundation on sand is presented in Section 9.4.3.

Peck, Hanson, and Thornburn Method The following equations present a method based on use of the SPT (Peck et al., 1974; Dunn et al., 1980). As noted, appropriate modifications are included to account for the position of water table and overburden pressure.

$$q_a = C_w(0.41)N_{SPT}S \tag{9.6}$$

$$C_w = 0.5 + 0.5\frac{D_w}{D_f + B} \tag{9.7}$$

where

q_a = net bearing pressure (kPa) causing a settlement S in millimeters,
C_w = term to adjust for the position of the water table,
N_{SPT} = corrected blow count from the SPT,
D_w = depth to water table from ground surface,
D_f = depth of overburden, and
B = width of footing (D_w, D_f, and B are in consistent units).

The authors included Figure 9.4 to correct N_{SPT} to account for pressure from the overburden, σ_0' (the dependence on units should be noted).

The example of a footing on sand is shown in Figure 9.5. The footing is 3 ft by 3 ft^2 and placed 1.2 m below the ground surface. The unit weights of the sand above and below the water table are shown. The corrected value of N_{SPT} is given as 20. The above equations are used to find the value of q_a where the value of S is 25 mm.

$$C_w = 0.5 + 0.5\frac{1.2}{1.2 + 3} = 0.64$$

$$q_a = 0.64(0.41)(20)(25) = 131 \text{ kPa}$$

The bearing capacity of the footing in Figure 9.5 will be studied in the next section as a means of evaluating the answer shown above.

9.4.3 Bearing Capacity of Footings on Sand

The procedures presented in Chapter 7 may be used to compute the bearing capacity of the footing shown in Figure 9.5, where the corrected value of

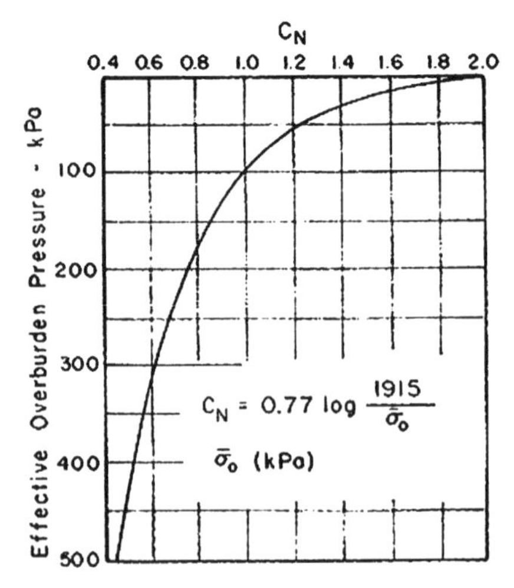

Figure 9.4 Factor used to multiply by N_{SPT} to obtain the corrected value accounting for pressure from overburden (from Peck et al., 1974).

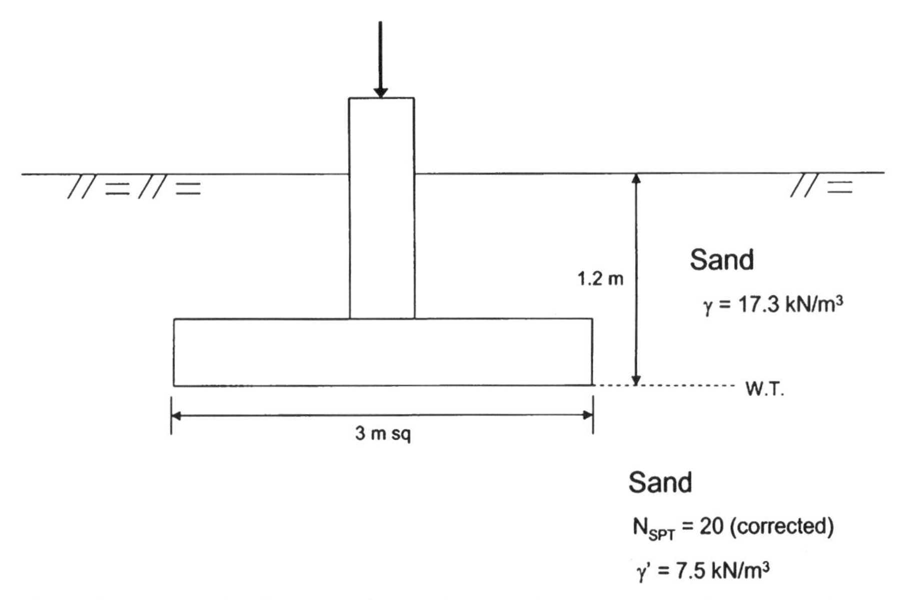

Figure 9.5 Example of a computing settlement where the value of N_{SPT} is available.

N_{SPT} is given as 20. Figure 9.6 shows a correlation between the corrected value of N_{SPT} and ϕ', where ϕ' is the friction angle from effective stress analysis. If data are available from the cone test, Figure 9.7 shows the correlation between the value of q_c from the cone test and ϕ' for uncemented quartz sand, taking the vertical effective stress into account.

For a corrected value of N_{SPT} of 20, Figure 9.6 shows the value of the friction angle ϕ' to be 33°. Table 7.1 shows the value of N_q to be 26.1 and

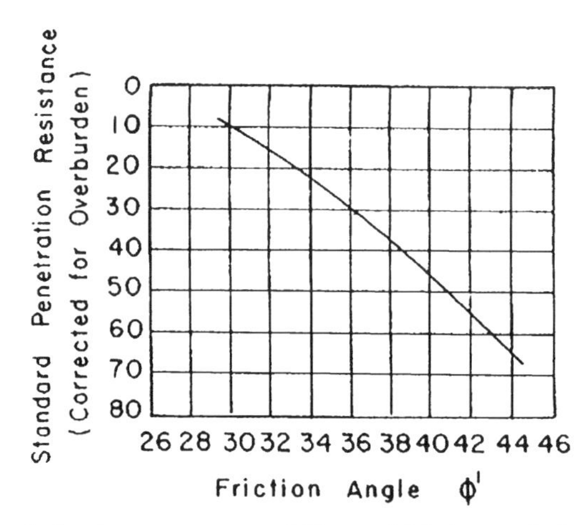

Figure 9.6 Correlation between corrected N_{SPT} and friction angle from effective stress (from Peck et al., 1974).

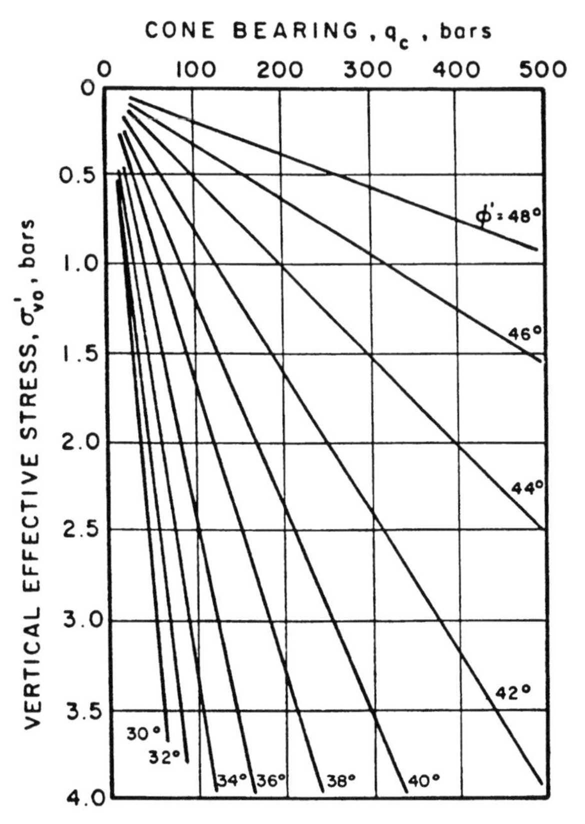

Figure 9.7 Suggested correlation between q_c from the cone test and peak friction angle Φ' for uncemented quartz sands considering vertical effective stress (from Robertson and Campanella, 1983).

the value of N_γ to be 24.4. The requisite formula is Eq. 7.7 for the case in Figure 9.5:

$$\frac{Q_d}{A} = q_d = \frac{1}{2}\,\gamma B N_\gamma s_\gamma d_\gamma + \gamma D_f N_q s_q d_q$$

where $s_\gamma = 0.6$, $s_q = 1.54$, $d_\gamma = 1$, and $d_q = 1.37$. The following values are computed:

$$q_d = \frac{1}{2}(7.5)(3)(24.4)(0.6)(1) + (17.3)(1.2)(26.1)(1.54)(1.37)$$

$$= 165 + 1143 = 1308 \text{ kPa}$$

If a factor of safety of 3 is employed, the value of q_a is 436 kPa. Employing the correlation shown in the previous section, the settlement for a q_a of 436 kPa would be 83 mm, a value that is too large for most structures.

This brief exercise shows that settlement will control the design of many shallow foundations on sand. However, the authors recommend that each case be considered in detail and that the decision on the value of q_a be reserved until all factors are considered.

If the cone test had been performed for uncemented quartz sand, the value of ϕ' could have been obtained from Figure 9.7. The bearing capacity q_d could have been computed by using Eq. 7.7, as shown above.

9.4.4 Design of Rafts on Sand

The procedures for computation of bearing capacity and settlement shown earlier may be extended to the design of a raft supported by sand. The principal problems facing the geotechnical engineer are (1) characterizing the soil below the raft and (2) selecting the properties to be used in the design. The geotechnical engineer will evaluate all of the available data gathered from field trips, geologic studies, and soil borings. The depth of the relevant soil beneath the footing may be selected by the following rule of thumb: use the net load at the base of the raft, use the theory of stress distribution with depth (Chapter 7), and find the depth where the stress increase under the area of heaviest load is equal to 10% or less of the net load.

Use of the rule of thumb will normally result in a depth encompassing various kinds of soil. With specific regard to a raft on sand, the properties of the sand may vary widely, perhaps as shown by the value of q_c in Figure 9.4. For the computation of settlement the sand may be considered layer by layer, but for bearing capacity, the engineer must select parameters that reflect the overall behavior of the stratum of sand. Modifying the equations for bearing capacity is not feasible, and no rule of thumb exists; instead, the engineer uses judgment, taking into account properties of the sand in the zones of maximum stress and adopting conservative values of soil properties.

9.5 SHALLOW FOUNDATIONS ON CLAY

9.5.1 Settlement from Consolidation

The procedure for predicting settlement of a shallow foundation on saturated clay was presented in Chapter 7. The theories for total settlement and for time rate of settlement are one-dimensional; that is, the flow of water from the soil is vertical, with no lateral flow or lateral strain. The theories are undoubtedly deficient for a number of reasons. Some of the deficiencies are as follows: excess porewater pressure can be dissipated by the lateral flow of water; the stratum of clay will likely have a degree of nonuniformity, and average values must be used in the analyses; the stiffness of a structure will affect the distribution of stress and the differential settlement; and settlement can be shown to depend to some extent on the value of the pore-pressure coefficient A (Skempton and Bjerrum, 1957). In spite of the limitations noted above, the equations for total settlement and time rate of settlement of shallow foundations on saturated clay can be used with some confidence.

Skempton and Bjerrum (1957) include data on the final settlement of four structures on normally consolidated clay where settlement was significant and where the computed settlement can be compared with the observed settlement (Table 9.2). The data are undoubtedly for the portion of the structure where the settlement was largest. The computed settlement includes an estimate of the initial settlement, as discussed in the following section.

The data in Figure 9.8 are remarkable where the time rate of settlement is shown for three of the structures listed in Table 9.2 and where measurements were made over many years. Excellent agreement is indicated between observed and computed settlements over much of the life of each structure.

The procedures presented in Chapter 7 for the computations of total settlement and time rate of settlement can be used with confidence in many instances. Limitations on the theory are noted above, and implementation of the procedures can be expensive and time-consuming. Foundation investigation and laboratory studies must be of high quality, and even then the geotechnical engineer is faced with difficult decisions about relevant soil properties.

The settlement due to consolidation of clay will be illustrated using Figure 7.9a. The following assumptions are made with regard to depths and soil properties: the top of the clay layer is at 8 ft below the bottom of the footings, the clay layer has a thickness of 4 ft, the footings are embedded 2 ft into the upper sand, the upper 2 ft of the sand are saturated by capillarity, the water table is at the base of the footings, the total unit weight of the sand below the water table is 120 lb/ft^3, and the total unit weight of the clay is 110 lb/ft^3.

The effective stress at the mid-height of the clay layer prior to construction the footing may be computed as follows:

TABLE 9.2 Comparison of Observed and Computed Settlement for Four Structures

Structure	Observed Settlement (in.)	Computed Settlement (in.)
Oil tank, Isle of Grain	21	21.5
Masonic Temple, Chicago	10	12
Monadnock Block, Chicago	22	22
Auditorium tower, Chicago	24	28.5

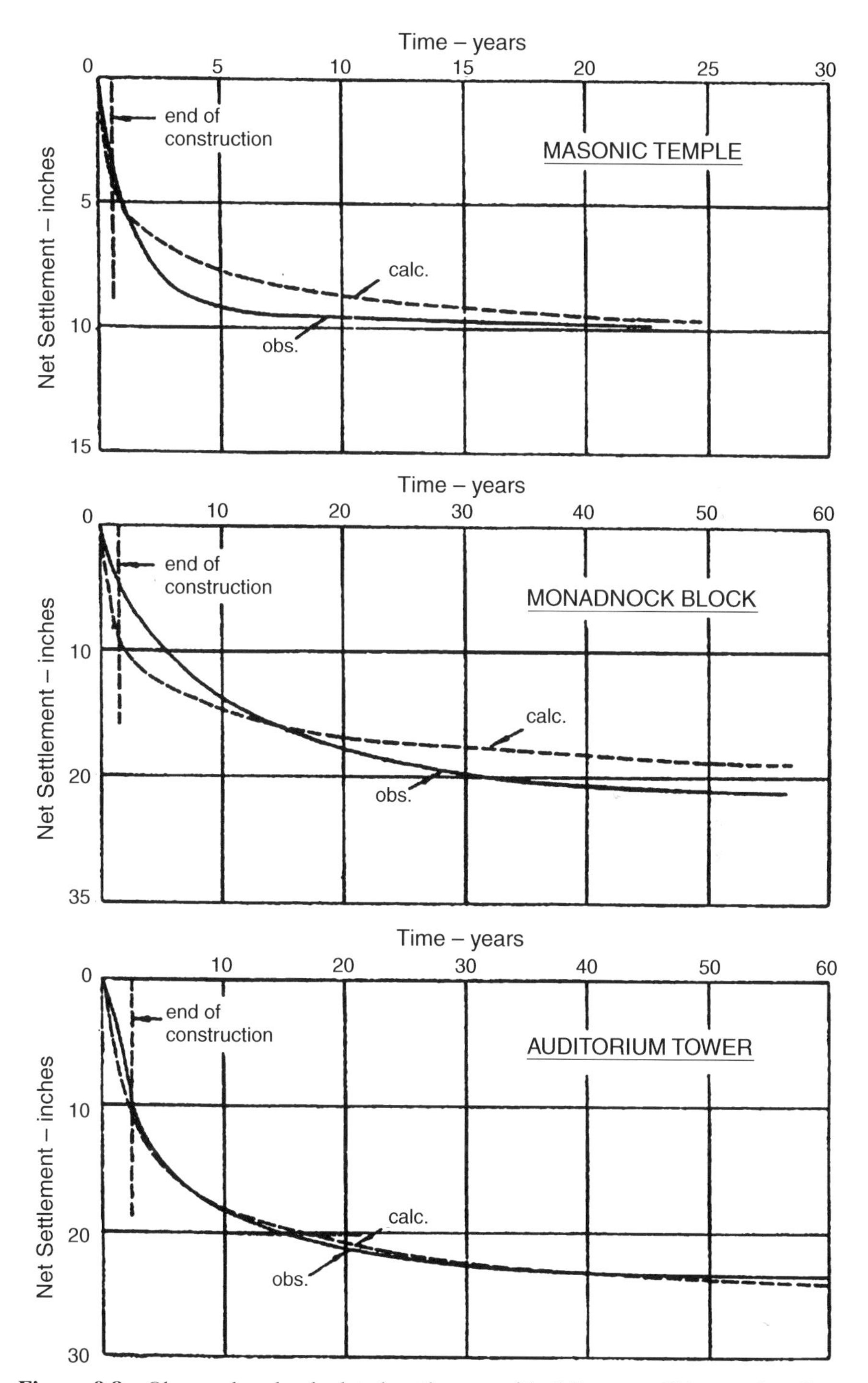

Figure 9.8 Observed and calculated settlement of buildings on Chicago clay (from Skemptom and Bjerrum, 1957).

$$\sigma'_0 = (2)(120) + (8)(120 - 62.4) + (2)(110 - 62.4) = 796 \text{ lb/ft}^2$$
$$= 0.796 \text{ k/ft}^2$$

Figure 7.12 shows that the value of σ_v was 0.580 k/ft^2 at the top of the clay layer due to the imposition of the footing loads. A similar exercise used the Newmark chart, and the value of σ_v at the bottom of the clay layer was 0.423 k/ft^2. The value of σ_v at the mid-height of the clay layer was 0.502 k/ft^2.

It is assumed that consolidation tests have been performed for the clay in the layer between the sand strata. The clay was found to be normally consolidated, and the following values were obtained: $e_0 = 0.7$, $c_c = 0.36$, and $c_r = 0.04$ ft^2/day.

From Eq. 7.36,

$$S = \frac{4}{1 + 0.7}\left[0.36 \log \frac{0.796 + 0.502}{0.796}\right] = 0.286 \text{ ft}$$

Referring to Figure 7.14, data can be obtained for preparing a table showing settlement as a function of time, as shown below. The time for a specific amount of settlement to occur may be obtained from Eq. 7.44, as shown for time for 50% of the settlement to occur.

$$T_v = \frac{c_v}{H^2}t \qquad 0.197 = \frac{0.04 \text{ ft}^2/\text{day}}{4 \text{ ft}^2}t \qquad t = 19.7 \text{ days}$$

U, %	T_v	S, ft	t, days
11	0.01	0.032	1.0
20	0.03	0.057	3.0
30	0.066	0.086	6.6
40	0.12	0.114	12.0
50	0.197	0.143	19.7
60	0.28	0.172	28.0
70	0.38	0.200	38.0
80	0.48	0.229	48.0
90	0.85	0.243	85.0
100		0.289	Infinite

Information such as that in the table above can be of great value to the engineer and other professionals in planning the design of a structure of saturated clay.

9.5.2 Immediate Settlement of Shallow Foundations on Clay

Skempton (1951) developed an equation using of some approximations of the theory of elasticity and showed the application of the methods to the immediate settlement of shallow foundations. The theory of elasticity yields the following expression for the mean settlement ρ of a foundation of width B on the surface of a semi-infinite solid:

$$\rho = qBI_\rho \frac{1 - \nu^2}{E} \tag{9.8}$$

where

q = bearing pressure on the foundation,
I_p = influence value, depending on the shape and rigidity of the foundation,
ν = Poisson's ratio of the soil, and
E = modulus of elasticity of the soil.

The above equation may be rewritten in the following convenient form:

$$\frac{\rho}{B} = \frac{q}{q_d}\frac{q_d}{s_u} I_\rho \frac{1 - \nu^2}{E/s_u} \tag{9.9}$$

where

q_d = ultimate bearing capacity, ν for undrained condition of clay = ½,
I_ρ = $\pi/4$ for a rigid circular footing on the surface of the clay, and
q_d/s_u = 6.8 for the circular footing on the surface of clay as found by Skempton, where s_u is the undrained shear strength of the clay.

Making the substitutions, the following equation was obtained for a circular footing on the surface of saturated clay:

$$\frac{\rho_1}{B} = \frac{4}{E/s_u}\frac{q}{q_d} \tag{9.10}$$

Skempton noted that for footings at some depth below the ground surface, the influence value of I_ρ decreases but the bearing capacity factor N_c increases; therefore, to a first approximation, the value of $I_\rho N_c$ is constant, and Eq, (9.10) applies to all circular footings. In the axial compression test for undrained specimens of clay, the axial strain for the deviator stress of $(\sigma_1 - \sigma_3)$ is given by the following expression:

$$\varepsilon = \frac{(\sigma_1 - \sigma_3)}{E} \tag{9.11}$$

where E is the secant Young's modulus at the stress $(\sigma_1 - \sigma_3)$. The above equation may be written more conveniently as follows:

$$\varepsilon = \frac{\sigma_1 - \sigma_3}{(\sigma_1 - \sigma_3)_t} \frac{(\sigma_1 - \sigma_3)_t}{s_u} \frac{1}{E/s_u} \tag{9.12}$$

But in saturated clay with no change in water content under the applied stress, $\frac{(\sigma_1 - \sigma_3)_t}{s_u} = 2.0$; thus, the following equation may be written:

$$\varepsilon = \frac{2}{E/s_u} \frac{(\sigma_1 - \sigma_3)}{(\sigma_1 - \sigma_3)_t} \tag{9.13}$$

Comparing Eq. 9.10 with Eq. 9.13 shows that, for the same ratio of applied stress to ultimate stress, the strain in the loading test is related to that in the compression test by the following equation:

$$\frac{\rho_1}{B} = 2\varepsilon \tag{9.14}$$

The equation states that the settlement of a circular footing resting on the surface of a stratum of saturated clay can be found from the stress-strain curve of the clay. Skempton performed some tests with remolded London clay and found that Eq. 9.14 could be used to predict the results of the footing tests except for high values of $\frac{q}{q_d}$. He further noted that the values of I_ρ increased from 0.73 for a circle to 1.26 for a footing with a ratio of length to width of 10:1, while the bearing capacity factor N_c decreased from 6.2 to 5.3 for the same range of dimensions. Skempton concluded that the proposed formulation could be used with a degree of approximation of $\pm 20\%$ for any shape of footing at any depth.

9.5.3 Design of Shallow Foundations on Clay

The equations presented in Chapter 7 for computing the bearing capacity for footings of a variety of shapes, depths below ground surface, at some slope with the ground surface and in sloping soil, may be used in design. The proposal by Skempton (1951) shown above may be used for computing the immediate settlement of a shallow foundation. The applicability of the Skempton procedure will be demonstrated by analyzing experiments performed by O'Neill (1970).

O'Neill placed instruments along the length of four drilled shafts to measure the magnitude of axial load in the shafts during the performance of axial-load tests. The downward movement along each of the shafts was found by subtracting computed values of the shortening of points along the shafts from the observed movement at the top of the shafts. The instrumentation with interpretation allowed load-settlement curves to be produced for each of the four tests.

Beaumont clay was deposited by the alluvial process in the first Wisconsin Ice Age and is somewhat heterogeneous, with inclusions of sand and silt. The liquid limit is around 70 and the plastic limit is around 20, although wide variations in both indices are common. The formation was preconsolidated by desiccation, with the indicated preconsolidation pressure at about 4 tons/ft^2. Numerous cycles of wetting and drying produced a network of randomly oriented and closely spaced fissures. The fissures led to considerable variations in measured shear strength, as indicated by the shear-strength profile shown in Figure 9.9. Confining pressure in the performance of the tests was equal to the overburden pressure. The scatter in the test results is not unusual and indicates practical problems in sampling and testing of many overconsolidated clays. The solid line in the figure was used in interpreting the results of testing of the foundation.

A stress-strain curve is shown in Figure 9.10 for the upper stratum at O'Neill's site. The shear strength was the average of two diameters below the base of each of the shafts. Significant data for the four test shafts are shown in Table 9.3. Computed values of the bearing capacity factor for three of the tests are very close to the Hansen value of 9.0. The overburden pressure and

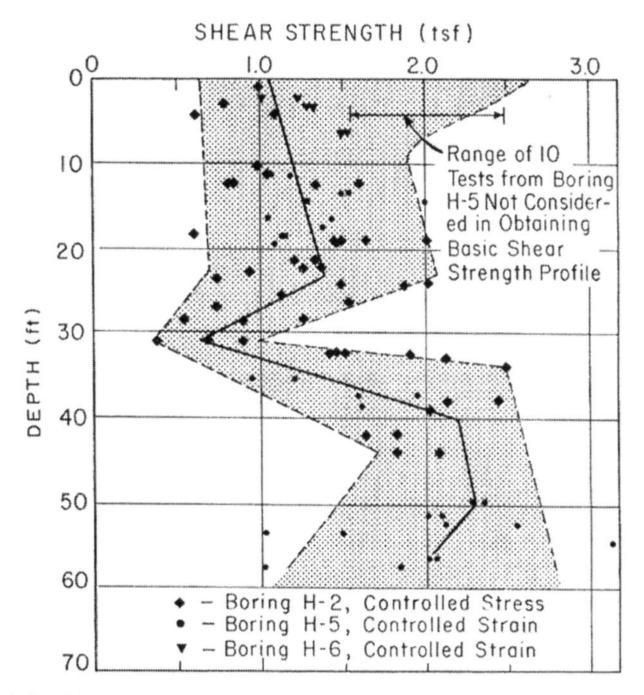

Figure 9.9 Shear strength profile from triaxial testing (from O'Neill, 1970).

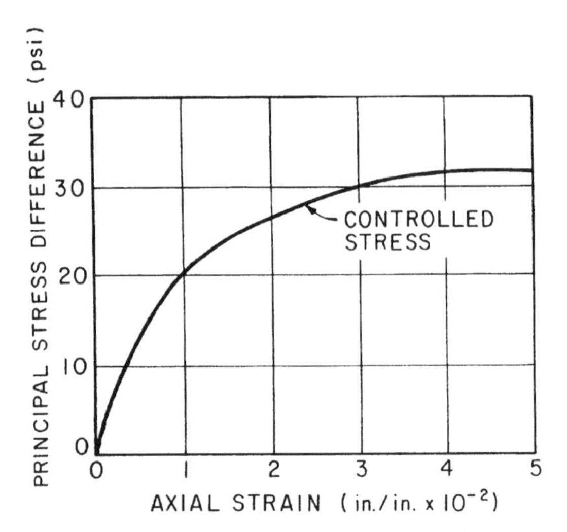

Figure 9.10 Stress-strain curves for beaumont clay (from O'Neill, 1970).

the pressure from the weight of the concrete in the drilled shaft were assumed to be equal in computing the bearing capacity.

Load-settlement curves from the experiments performed by O'Neill are plotted in Figures 9.11 through 9.14. The stress-strain curve in Figure 9.10 was used in making the Skempton computations. The lack of much better agreement between the measured and computed values is not surprising in view of the nature of the soil at the test site. The presence of randomly oriented and closely spaced fissures means that the soil in the triaxial tests was less stiff than the soil at the base of the test shafts during the testing.

As shown in Figures 9.11 to 9.14, the values for pile settlement are indicated for a factor of safety of 3 based on the data from the stress-strain curves. Assuming that the data from load tests were not available, the stress-strain curves yielded an average settlement value of 0.24 in. for the shafts with a base diameter of 30 in. and 0.27 in. for the shaft with a base diameter of 90 in. The numbers should not have caused the designer much concern, even

TABLE 9.3 Values for Each Shaft Tested by O'Neill

Shaft Designation	Penetration (ft)	Diameter of Base (in.)	Avg. S_u at 2D Below Base (tsf)	Ultimate Load on Base (tons)	Computed Value of N_c
S1	23	30	1.16	52	8.91
S2	23	90	1.16	445	8.68
S3	24	30	1.05	47	9.12
S4	45	30	2.28	142	12.68

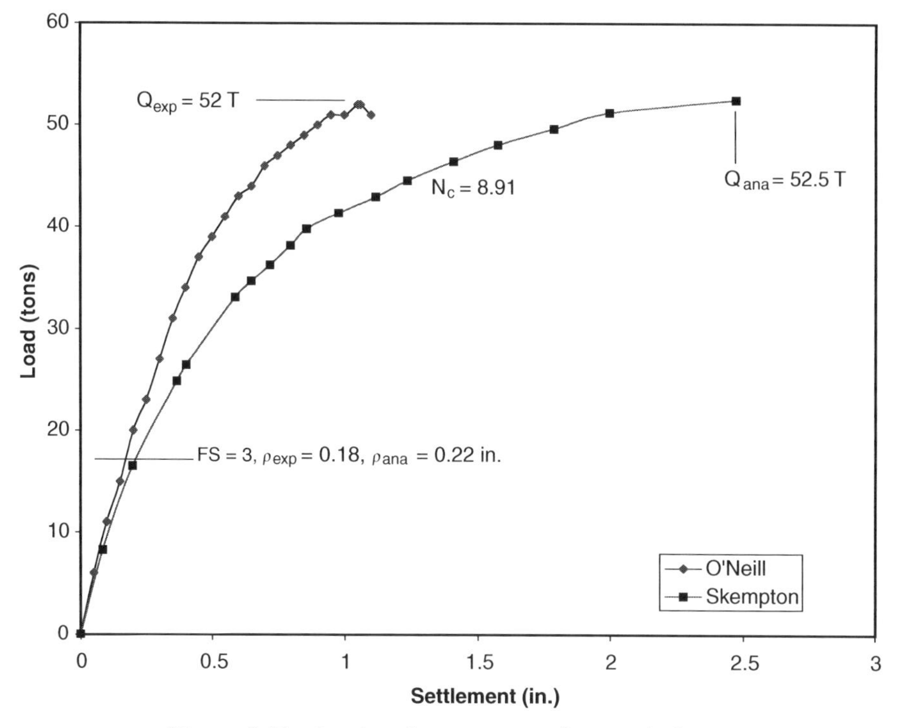

Figure 9.11 Load-settlement curves for test shaft S1.

though the corresponding numbers from the load tests were 0.18 in., 0.05 in., and 0.09 in. (shafts 1, 3 and 4), average of 0.11 in. and 0.21 in.

9.5.4 Design of Rafts

The comments on the design of rafts on sand apply as well to the design of rafts on clay. The equations for bearing capacity and settlement of shallow foundations presented in Chapter 7 may be used. Emphasis is placed on a soil investigation and a laboratory study of high quality. The engineer must exercise extra care in selecting parameters for use in design. If differential settlement is revealed to be a problem, the geotechnical engineer may work closely with the architect and the structural engineer in placing and sizing the shallow foundations to achieve an optimal solution.

9.6 SHALLOW FOUNDATIONS SUBJECTED TO VIBRATORY LOADING

The response of foundations under dynamic loads is significantly different from that under static loads. The analysis and design of foundations subjected

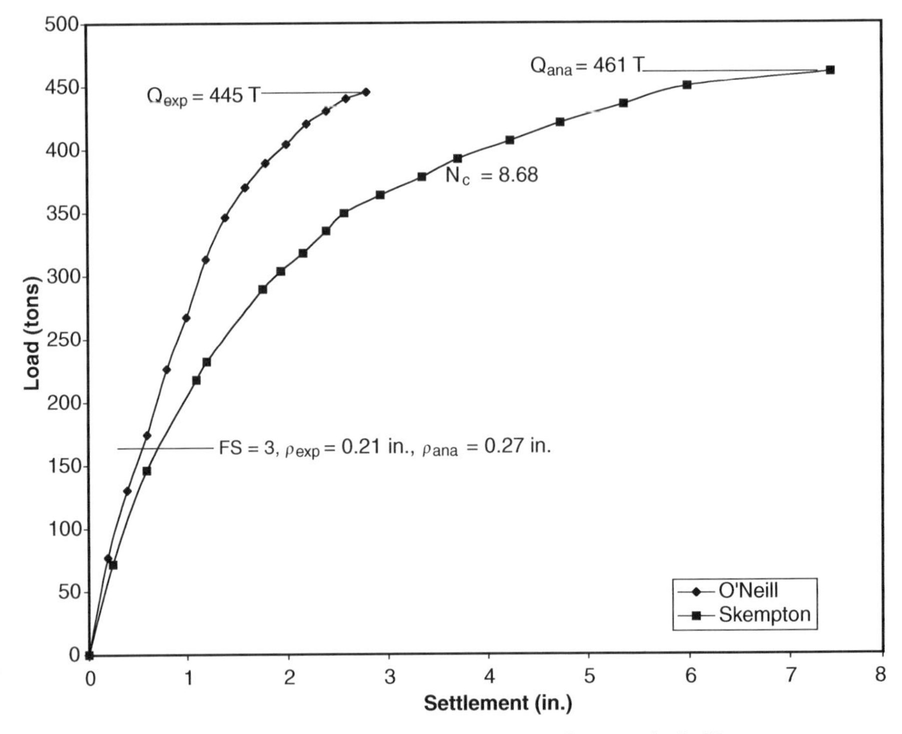

Figure 9.12 Load-settlement curves for test shaft S2.

to machine vibration or impact from earthquake is a difficult problem because of the complex interaction between the structural system and supporting soil/rock stratum. The proper solution of the problem requires an accurate prediction of the foundation response to these loads. A complete and rigorous analysis must account for the following: the three-dimensional nature of the problem, material and radiation damping of the soil, variation of soil properties with depth, embedment effects, and interaction effects between multiple foundations through the soil. In the last 30 years a number of solutions have been developed for design of machine foundations. Extensive reviews of these developments were presented by Richart et al (1970), Lysmer (1978), Roessett (1980), Gazetas (1983, 1991), and Novak (1987). These are beyond the current scope of this book.

The topic is discussed here to emphasize the fact that the vibration of sand can cause densification of the sand with consequent settlement of the foundation. Therefore, the geotechnical engineer must consider the relative density of the sand at the site. If it is close to unity, vibration is likely not to be a problem. Otherwise, settlement will occur and provisions must be made to deal with it.

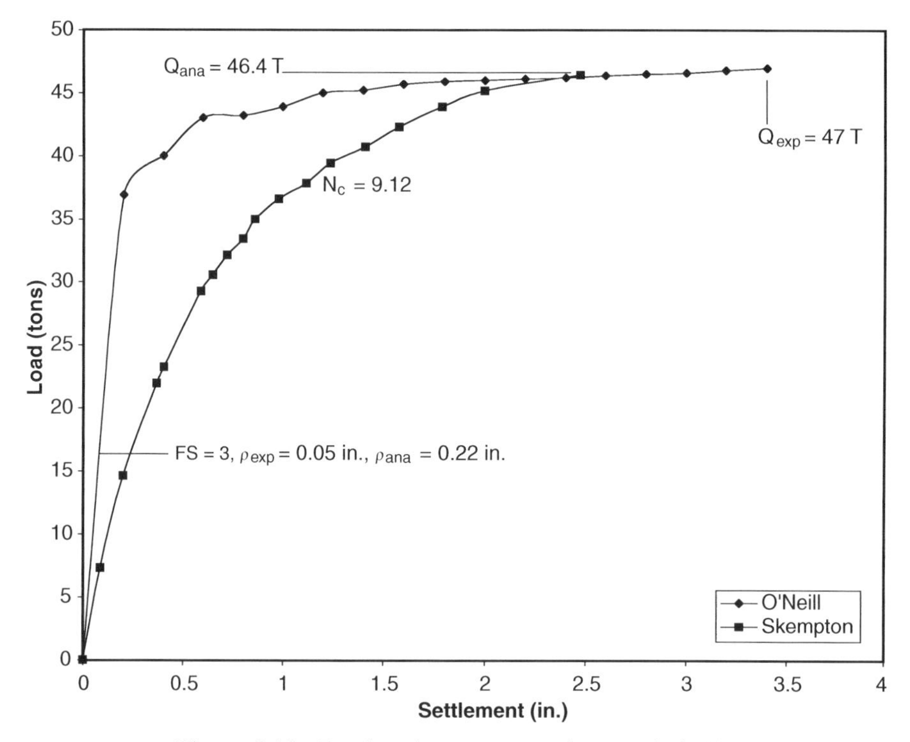

Figure 9.13 Load-settlement curves for test shaft S3.

Two procedures may be considered: (1) the use of deep foundations driven into stable soil with respect to vibration and (2) the use of soil-improvement methods to make the sand resistant to vibration.

9.7 DESIGNS IN SPECIAL CIRCUMSTANCES

9.7.1 Freezing Weather

Frost Action The depth of freezing of the soil in regions of the United States and areas to the south of Canada is shown in Figure 9.15. Ice lenses may form if the frozen surface soils include significant moisture, and the expansion due to the ice could cause some heave of the foundation. Engineers in cold climates must consult local building codes on the required depth of shallow foundations, taking into account the nature of the soils. A particular problem occurs if the soil at and above the water table consists of silt and fine sand, as shown in Figure 9.16.

The capillary rise in the granular soil between the water table and the shallow foundation allows water to accumulate in quantities, creating ice

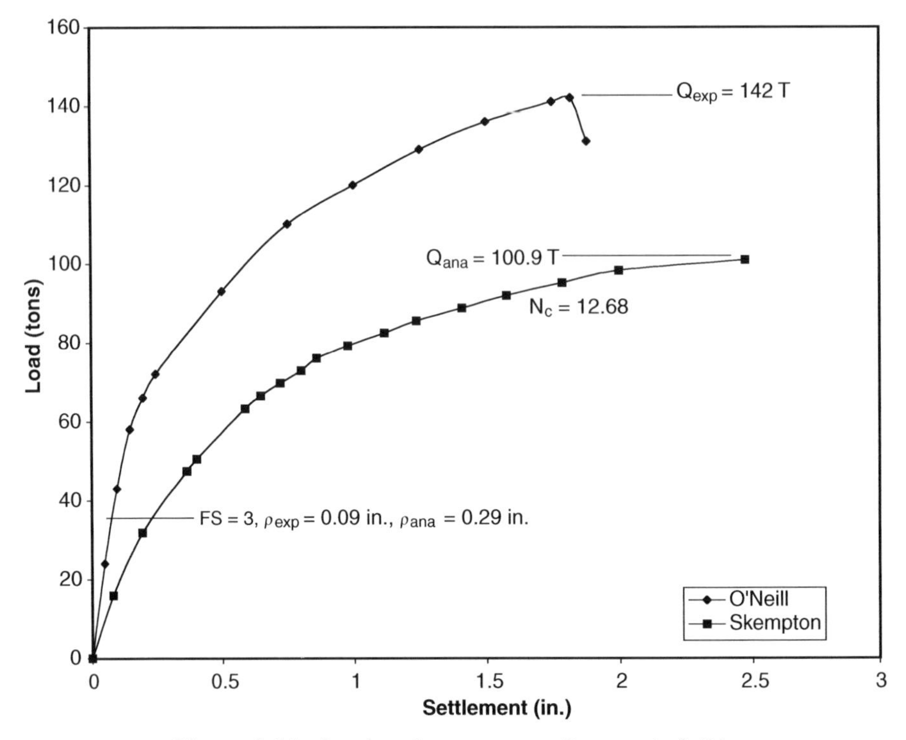

Figure 9.14 Load-settlement curves for test shaft S4.

lenses. With time, more water rises and additional ice lenses are formed. On some occasions, a large quantity of ice can collect in a restricted area. The increased thickness of the ice causes a heave of the shallow foundation, as shown in Figure 9.16. When warm weather occurs the ice in the soil melts, leaving the soil with high water content and a low value of bearing capacity. The conditions shown in Figure 9.15 can lead to failures of roadway pavement that can be dangerous to vehicular traffic. A remedy to the problem illustrated in Figure 9.16 is to place a layer of coarse granular soil in the zone of capillary rise to prevent the upper flow of capillary water.

In northern climates, the soil may remain permanently frozen, a condition termed *permafrost.* The construction of light structures and other facilities in such regions is aimed at preventing the thawing of the permafrost. Structures are supported on posts or short piles so that air circulation beneath the structure and above the frozen soil is allowed to continue.

Placing Concrete Freezing weather increases the cost of construction because the concrete must be protected. The protection against freezing must last until the concrete "takes a set" or hardens so that free water is not present. In some climates where freezing weather continues for many months, foun-

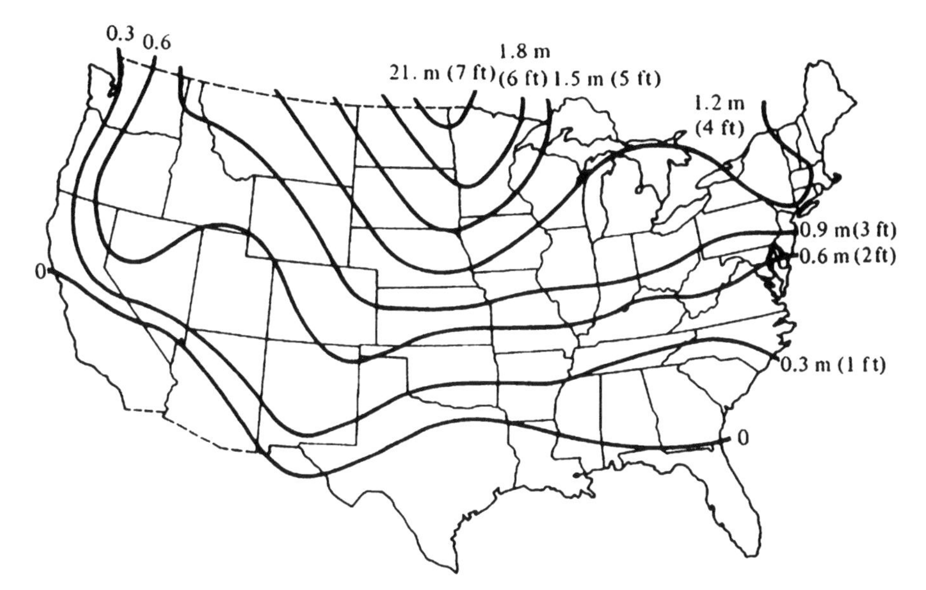

Figure 9.15 Maximum anticipated depths of freezing as inferred from city building codes. Actual depths may vary considerably, depending on cover, soil, soil moisture, topography, and weather (from Spangler and Handy, 1982).

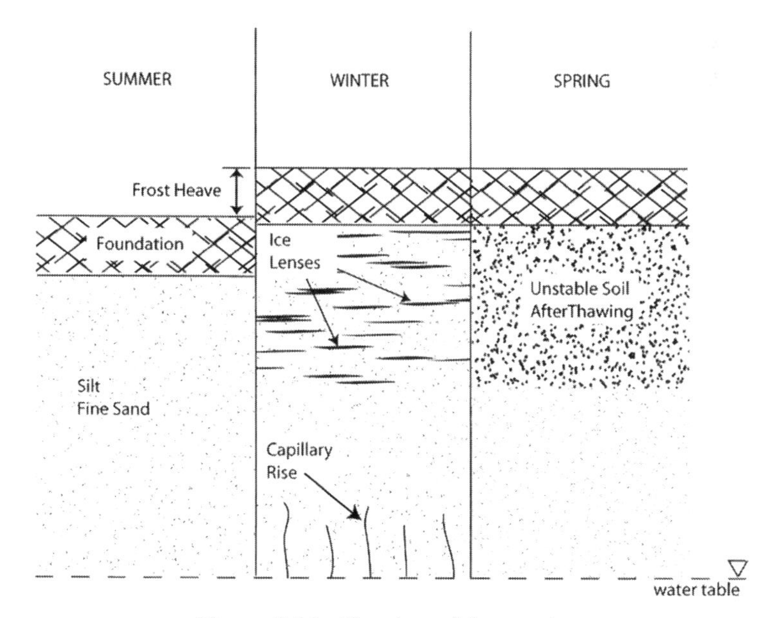

Figure 9.16 Sketches of frost action.

dation elements have been prefabricated in workshops and placed in prepared areas at the job site. Short-term settlement is more likely to occur with prefabricated foundations because the contact between the soil and the foundation is less than perfect.

9.7.2 Design of Shallow Foundations on Collapsible Soil

Collapsible soils were described briefly in Chapter 2. Such soils consist of thick strata of windblown fine grains, deposited over long periods of time, and reinforced by remains of vegetation or by cementation. The soil will remain stable under moderate loads as determined by the appropriate testing procedures; however, on becoming saturated, the soils will collapse. Preventing a rising water table or saturation from precipitation is essential. Therefore, if drainage can be assured or a rising water table can be prevented, shallow foundations can be installed and will remain stable. If saturation of the collapsible soil is judged to be possible, some form of deep foundation will be necessary.

9.7.3 Design of Shallow Foundations on Expansive Clay

Expansive clay is an extremely troublesome foundation material and exists widely in the United States, as shown in Figure 9.17. Chapters 2 and 6 present some of the factors related to expansive clay. Owners of homes and small businesses must pay a premium for building or repairing the foundation. Otherwise, they will be left with a substandard and unsightly structure.

The soil at a particular site may appear to be firm and substantial, and potential foundation problems may not be evident. Unfortunately, the local building codes in many cities may not address the requirements for identifying expansive clay and then specifying appropriate procedures for constructing a foundation if expansive clay is present. Furthermore, years may pass before the damage from an improper foundation becomes evident.

Expansive clay is readily seen if a site is visited during dry weather. Cracks in the surface soil are visible and may extend several feet below ground surface. Geotechnical investigations to identify the problem may involve several kinds of laboratory tests. These tests include liquid limit, plastic limit, shrinkage limit, swelling pressure under confinement, free-swell test, and identification of clay minerals. Tables 9.4 and 9.5 give recommendations for indicating the swelling potential of clay on the basis of laboratory tests.

The value of the plasticity index is used by many engineers to evaluate the character of a clay soil. Tables 9.4 and 9.5 suggest that a plasticity index of 15 or less means that the clay is expected to swell very little, if at all.

If a clay soil is judged to be expansive, the thickness of the zone is considered. If it is shallow, the troublesome soil may be removed. Alternatively, the expansive clay may be treated with chemicals such as lime. The lime must be mixed with the soil by tilling, not by the injection of a slurry of lime and

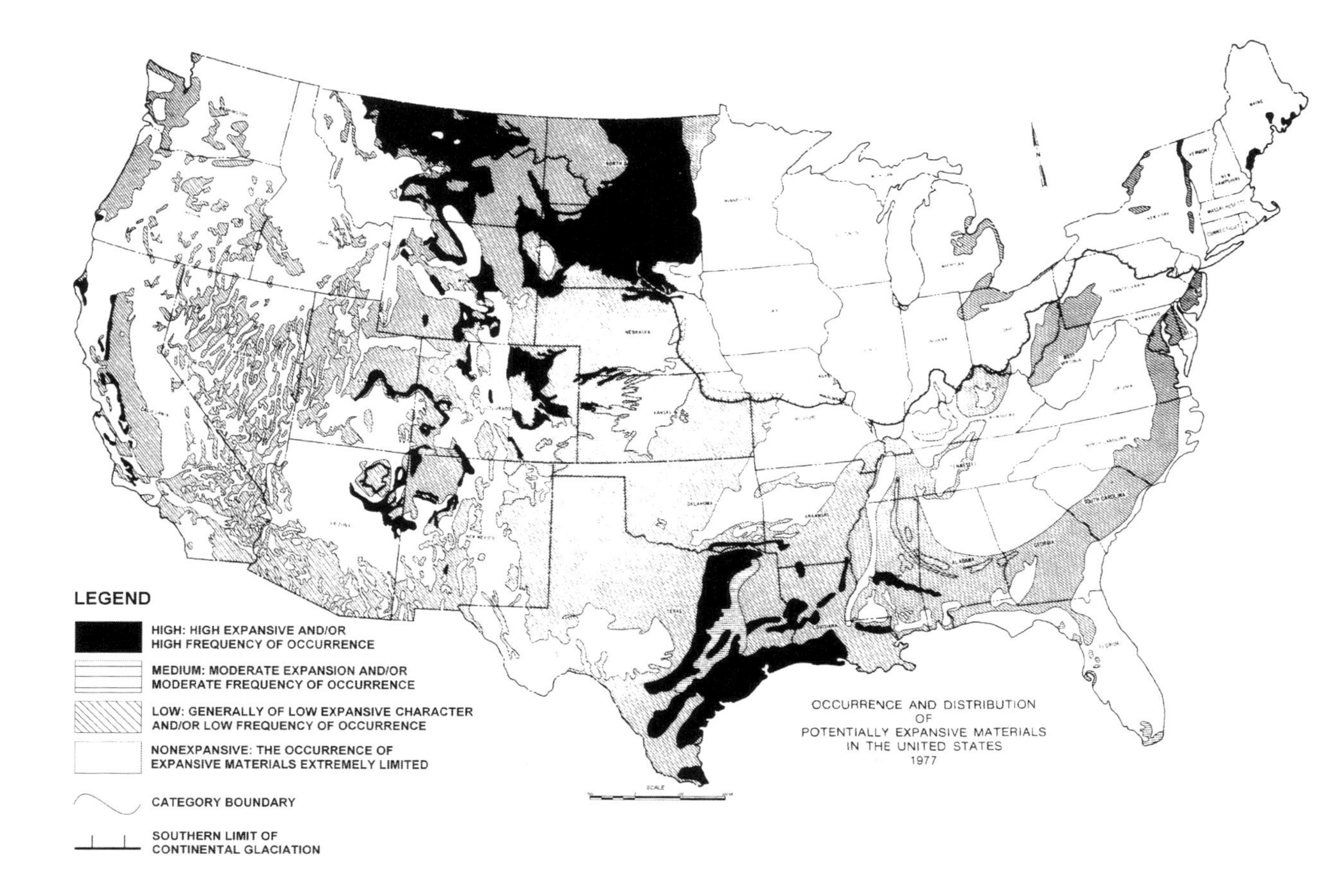

Figure 9.17 Distribution of potentially expansive soils in the United States (from Nelson and Miller, 1992).

TABLE 9.4 Swelling Potential from Laboratory Tests

Percentage of Sample Less Than 1 μm	Plasticity Index	Shrinkage Limit	Percentage of Volume Expansion from Dry to Saturation
>28	>35	<11	>30
20–31	25–41	7–12	20–30
13–23	15–28	10–16	10–20
<15	<18	>15	<10

Source: Holtz and Gibbs (1954).

water. The permeability of clay is so low that the slurry will not readily penetrate it.

If the stratum of clay is so thick that it cannot be removed or treated, then the engineer usually considers the use of a stiffened slab on grade, such as the BRAB slab (Building Research Advisory Board, 1968). The site is graded with deepened sections at relatively close spacing where reinforcing steel is placed to form beams into the foundation. The depth of the beams and the spacing depend on the judgment about shrinkage potential. The purpose of the beams is to make the slab so stiff that differential settlement is minimal. Over time, moisture will collect under the center of the slab because of restricted evaporation, but the edges of the slab will respond to changes in weather. Some homeowners have used plantings or other means to prevent shrinkage of the soil around a stiffened slab.

In some cases, a founding stratum may occur below the expansive clay. Drilled shafts can be installed with bells and designed to sustain the uplift from the expanding clay. If drilled shafts are used, a beam system should be employed to provide a space above the surface of the clay to allow swelling to occur without contact with the beams and floor system. Further, the bond between the expanding clay and the shaft should be eliminated or the shaft should be heavily reinforced to sustain uplift forces.

9.7.4 Design of Shallow Foundations on Layered Soil

Rarely does the geotechnical engineer encounter a design where the founding stratum is uniform and homogeneous. The reverse is usually the case; either

TABLE 9.5 Swelling Potential from the Plasticity Index

Swelling Potential	Plasticity Index
Low	0–15
Medium	10–15
High	35–55
Very high	≥55

Source: Chen (1988).

the soil is of the same sort but with widely varying properties, or two or more layers exist in the zone beneath the foundation. Settlement may be computed with procedures that have been presented, but questions arise when selecting the bearing capacity. If the additional cost is minor, the bearing capacity may be based on the weakest stratum in the founding stratum. The second possibility is that settlement will control the design, as shown earlier in this chapter, rather than bearing capacity.

In some studies, the shapes of the failure surfaces have been modified to reflect the presence of layers with different characteristics (Meyerhof, 1978; Meyerhof and Hanna, 1974). A more favorable approach is to employ FEM, as demonstrated by the material in the following section. A number of codes are available. Such codes will continue to be important in the analytical work of engineering offices.

9.7.5 Analysis of Response of a Strip Footing by Finite Element Method

As noted in the previous section, the FEM may be used to a considerable advantage in the analysis of shallow foundations. The application to the following example is an elementary use of FEM by serves to indicate the power of the method. A strip footing with a width, B, of 120 in. is resting on the surface of a soil, Figure 9.18, with the following characteristics: $E = 30{,}000$ psi, $\upsilon = 0.3$, $c = 10$ psi, and $\phi = 20°$. The unit weight of the soil was

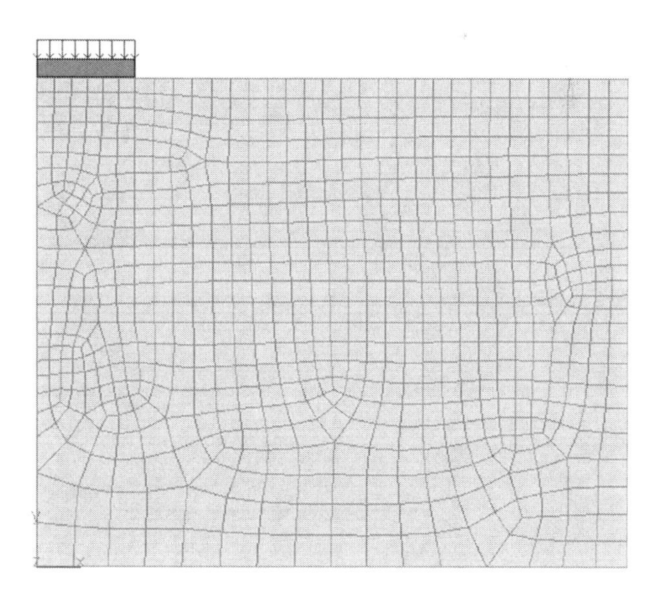

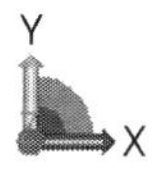

Figure 9.18 Model of a strip footing by FEM.

assumed to be 120 pcf. The Mohr-Coulomb model was used to represent the nonlinear behavior of the soil.

The result from the FEM computation, where a series of deformations of the footing was used as input, is shown in Figure 9.19. The analysis assumed a rough footing because the soil was not allowed to spread at the base of the footing. As may be seen, the unit value of the ultimate bearing capacity (q_u) of the footing was predicted as 185 psi. The Terzaghi equation for bearing capacity, Eq. 7.3, is re-written to obtain the unit value of ultimate bearing capacity and shown in Eq. 9.15.

$$q_u = \frac{1}{2}\gamma B N_\gamma + \gamma D_f N_q + cN_c \tag{9.15}$$

The footing is resting on the soil surface, so the value of D_f is 0. The properties of the soil show above were employed, along with the following values of the bearing capacity factors (see Chap 7): $N_\gamma = 3.64$, $N_c = 17.69$, and the value of q_u was computed as follows

$$q_u = \frac{1}{2}(0.0694)(120)(3.64) + (10)(17.69) = 15.2 + 176.9 = 192.1 \text{ psi.}$$

The FEM yielded a value quite close to the result from the Terzaghi equations. The methods presented herein do not allow a check to be made of the initial slope of the load-settlement curve in Fig. 9.19. The deformation field of soil in the vicinity of the strip footing generated by the FEM output is presented in Fig. 9.20 which is consistent with the theory introduced in Chapter 7.

The excellent comparison between results from the FEM for a solution of a straightforward problem indicates that the method can be applied to complex problems in bearing capacity including response to eccentric loading and behavior of shallow foundations on layered soils. However, in the solution of

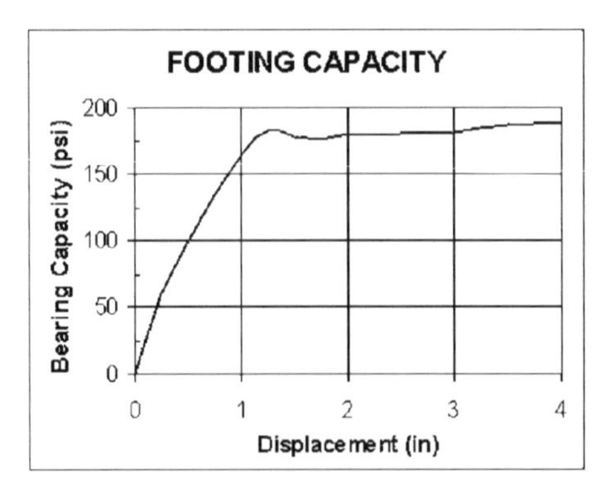

Figure 9.19 Load-settlement curve for a strip footing from analysis by FEM.

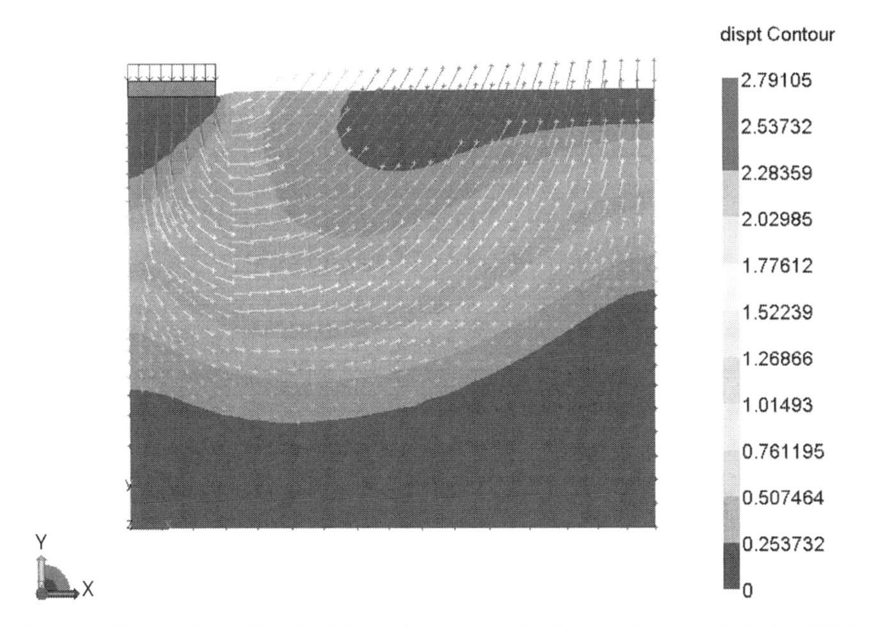

Figure 9.20 Deformation field for soil near a strip footing from analysis by FEM.

the example, the engineer faced the problem of selecting a constitutive model for the soil, the size and number of the elements, the distance to each of the boundaries, and appropriate restraint at the boundaries. Experience in using the method is valuable, but different trials by varying the parameters in the solution may be necessary. Many engineering offices are routinely employing codes for the FEM and more use is expected in the future.

PROBLEMS

P9.1. Cummings (1947) described the nature of the clay in the Valley of Mexico and showed the results of consolidation tests performed at Harvard, where the void ratios were extremely high and unusual.

The e-log-p plot in Figure 9.21 is patterned after the test results reported by Cummings. Using that plot and estimating the settlement at a location in Mexico City from the photograph in Figure 9.1, compute the amount the water table must be lowered to cause the settlement indicated. Assume the clay stratum to be 150 ft thick, the initial void ratio to be 6.60, the initial compressive stress to be 0.9 tsf, and that the clay remains saturated by capillarity as the water table is lowered. The answer may be surprising.

P9.2. The plot in Figure 9.22 shows values of q_c from the cone penetration test at a site where a footing is to be constructed. Use the Schmert-

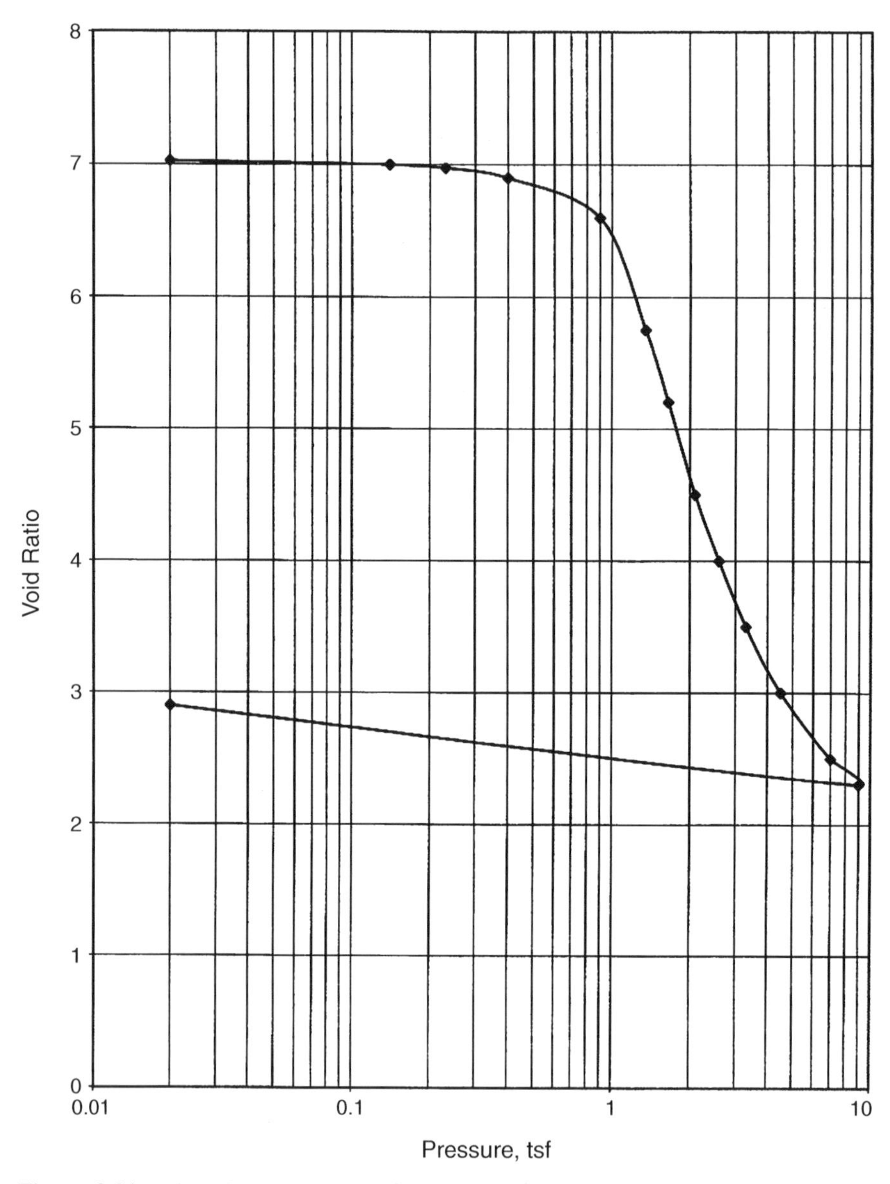

Figure 9.21 Plot of e versus log-p for soil similar to that in the Valley of Mexico.

mann method to compute the settlement of a square footing, 2 by 2 m in plan, with its base at 1 m below the ground surface. Assume that the soil is sand with an average grain size of 0.5 mm and a submerged unit weight of 9.35 kN/m^3. The load at the base of the footing is 80 kPa. Use three layers below the base of the footing in making settlement computations for 1 year and 10 years.

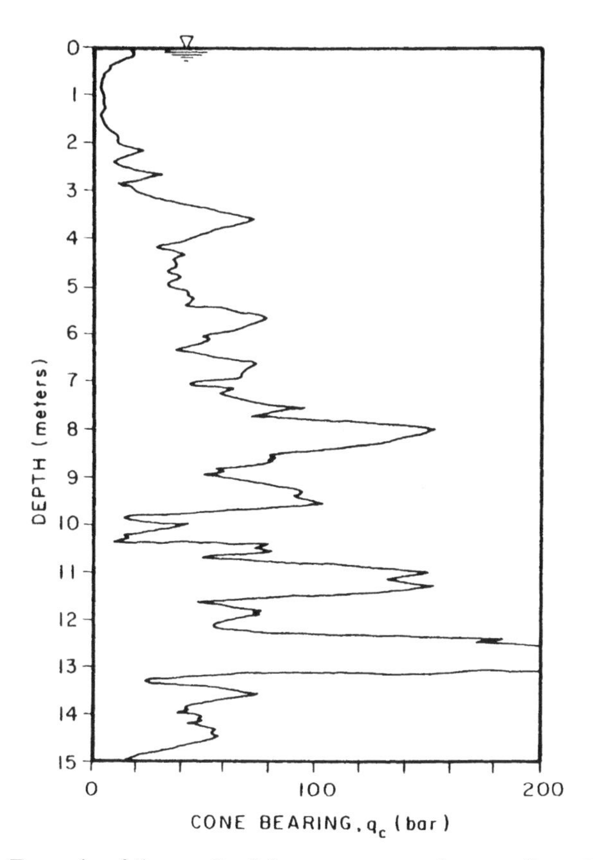

Figure 9.22 Example of the result of the cone penetration test (from Robertson and Campanella, 1988).

P9.3. Use the data from Problem P9.2 to compute the bearing capacity for the footing. Use Figure 9.23 in selecting a value of ϕ for use in bearing-capacity computations, and consider the low value q_c at the base of the footing and the significant increase with depth below the base. Assume a factor of safety of 3 and compute the value of the safe bearing capacity. Compare the safe bearing capacity with the unit load used in Problem P9.2. Discuss the allowable load or safe load you would select. (Check construction on figure and modify as necessary.)

P9.4. Estimate the settlement of a footing 10 ft^2 resting at a depth of 2 ft below the surface of a coarse sand and subjected to a total load of 400 kips. The water table is at a depth of 2 ft, and the corrected value of N_{SPT} is 30.

P9.5. Compute the unit bearing capacity in kips per square foot of the footing in Problem P9.4, assuming that the total unit weight of the sand is 122 pcf. Compute the factor of safety for the load of 400 kips.

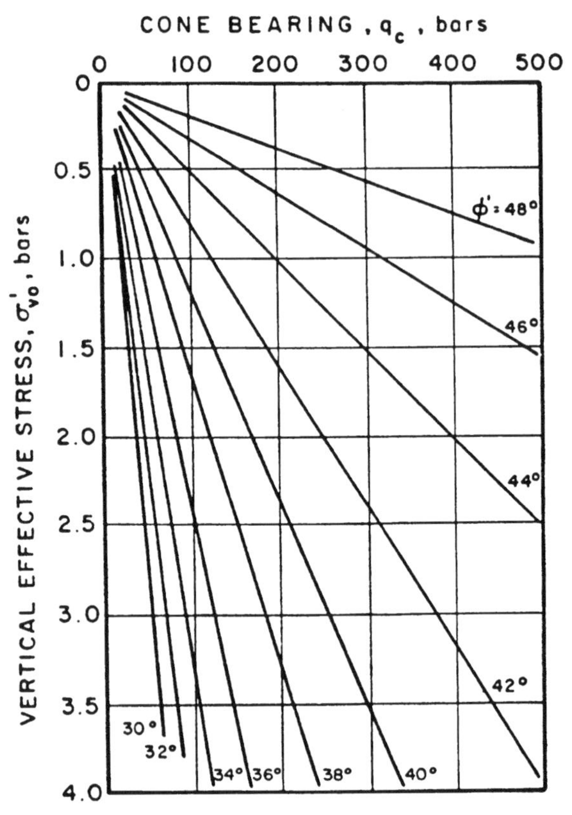

Figure 9.23 Suggested correlation between q_c from the cone test and peak friction angle Φ' for uncemented quartz sands considering vertical effective stress (from Robertson and Campanella, 1983).

P9.6. The unconfined compressive strength of a clay is 1.2 kips/ft^2, the water table is 4 ft below the ground surface, the total unit weight of the clay γ is 118 lb/ft^3, and a drilled shaft with a diameter of 4.25 ft placed at a depth of 10 ft supports a column load by end bearing only. (a) Use a factor of safety of 2.8 and compute the safe column load on the foundation. (b) An excavation near the drilled shaft removes the soil down to the water table. What is the effect on the factor of safety?

P9.7. A water tank with a height of 25 ft and a diameter of 50 ft rests on a stratum of sand with a thickness of 12 ft. A stratum of clay with a thickness of 20 ft exists below the sand, and a thin layer of gravel exists below the clay. The water table is at the surface of the clay. Plot a curve showing the stress increase in the clay as a function of depth for the center and edge of the tank. Assume the tank to be filled with water to a depth of 23 ft. Ignore the weight of the steel in the tank.

P9.8. Assume that consolidation tests were performed on the clay in Problem P9.7, that the soil is normally consolidated, and that the following values were obtained for the clay: $\gamma_{sat} = 120$ pcf; $e_0 = 1.10$; $c_c = 0.40$; $c_v = 0.025$ ft^2/day; sand, $\gamma = 125$ pcf. Compute the settlement in inches at the center and edge of the tank.

P9.9. Plot curves showing the settlement of the center of the tank as a function of time.

P9.10. Plot a curve showing the distribution of excess porewater pressure in the clay stratum for consolidation of about 50% (use $T_v = 0.2$).

P9.11. If the differential settlement of the tank is judged to be a problem, list three steps that may be taken to reduce it.

CHAPTER 10

GEOTECHNICAL DESIGN OF DRIVEN PILES UNDER AXIAL LOADS

10.1 COMMENT ON THE NATURE OF THE PROBLEM

Piles are generally used for two purposes: (1) to increase the load-carrying capacity of the foundation and (2) to reduce the settlement of the foundation. These purposes are accomplished by transferring loads through a soft stratum to a stiffer stratum at a greater depth, or by distributing loads through the stratum by friction along the pile shaft, or by some combination of the two. The manner in which the load is distributed from the pile to the supporting soil is of interest. A typical curve of the distribution of load along the length of an axially loaded pile is shown in Figure 10.1. This transfer of loads is extremely complex, highly indeterminate, and difficult to quantify analytically.

The method of pile installation is one of the important factors that affect load transfer between the pile and the supporting soil layers. Current construction practice for installation of piles may be divided into three basic categories: driving, boring, and jetting/vibrating. Driven piles are open-ended or closed-ended steel pipes, steel H-shapes, timber piles, or precast concrete piles driven by various types of driving hammers to the desired penetration for the required load-carrying capacity. This chapter focuses on the load-carrying capacity (also known as the *geotechnical capacity*) of driven piles only.

Computing the capacity of a single driven pile under axial loading is by no means a straightforward exercise. In practice, pile foundations have typically been designed based on either the pile-driving formula or the static formula. The static formula is considered the more reliable of the two. With this formula, the pile capacity is based on the static soil resistance determined

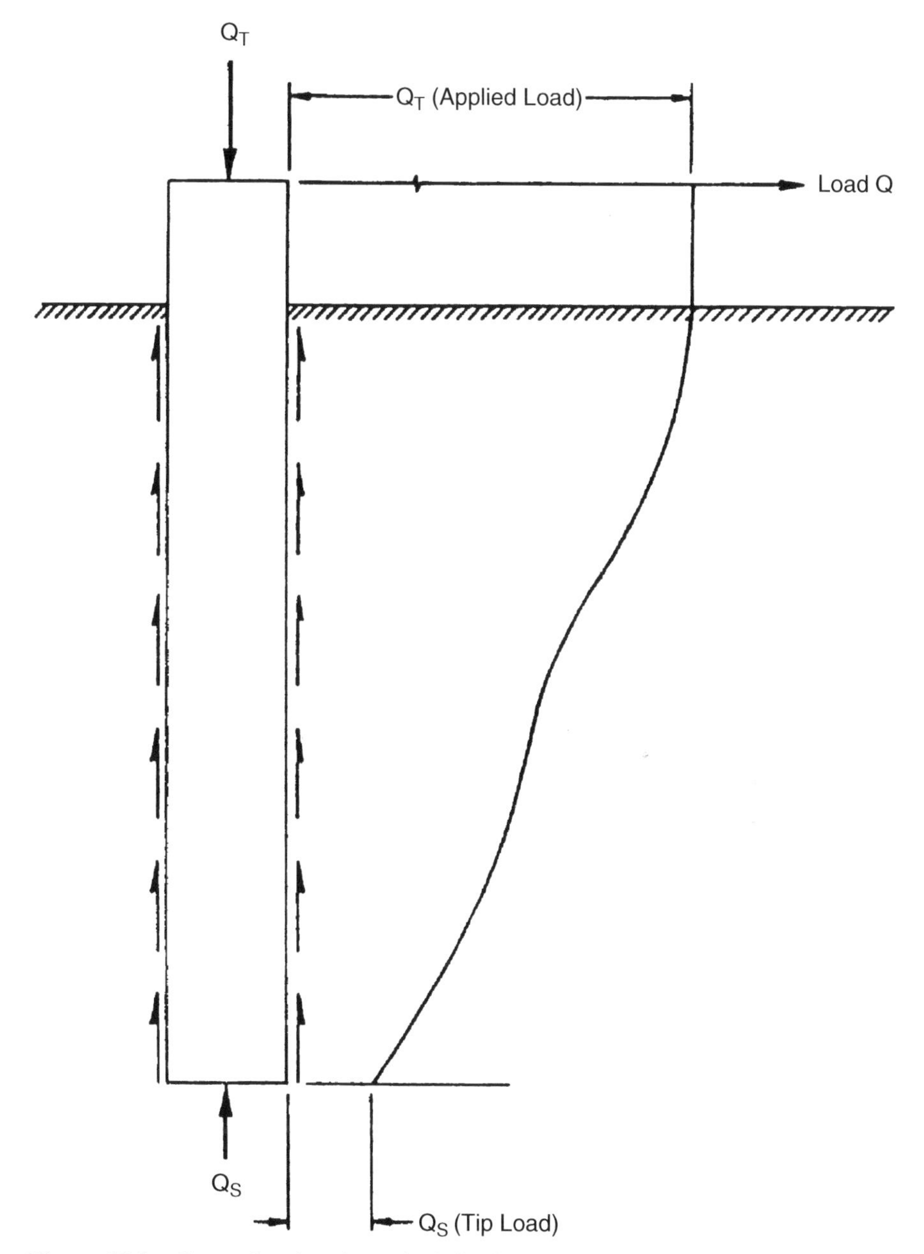

Figure 10.1 Curve showing the typical distribution of a load along the length of an axially loaded pile.

by the limit equilibrium theory for the tip of the pile and load transfer along the sides of the pile, termed *side resistance.* In discussing piles under axial load, the term *skin friction* has long been used to indicate the load along the sides of a pile. This term has a direct application to piles in granular soils but not to piles in cohesive soils, where *side resistance* is more descriptive. For piles in cohesive soils, because of traditional usage, *skin friction* is also used in some instances. The soil parameters used in the static formula are derived from field and/or laboratory tests. The procedures adopted in practice are revised or improved from time to time. There are two principal reasons for the necessary changes in methods of computation:

1. The lack of high-quality data from the results of full-scale testing of piles under axial loading. Various procedures are often used in tests reported in the technical literature, and the failure to use a recognized method leads to difficulty in evaluating results. Also, many techniques are used to obtain properties of in situ soil, and the results of soil tests often cannot be interpreted comparably. Furthermore, in only a few instances have measurements been made, using the necessary instrumentation along the length of a pile, that reveal the detailed manner in which the foundation interacts with the supporting soil.
2. The interaction between a pile and the supporting soils is complex. The soil is altered by pile installation, for example, and there are many other important factors that influence the response of a pile under axial loading. The reader is urged to read the *Proceedings, Workshop on Effects of Piles on Soil Properties,* U.S. Army Corps of Engineers, Waterways Experiment Station, Vicksburg, MS, August 1995.

Two kinds of computations are required for the analysis of pile capacity under axial loads: the computation of the ultimate axial capacity of piles under short-term loading and the computation of curves showing axial load versus settlement. More than one method is presented for obtaining axial capacity to reflect current practice among geotechnical engineers. The user may be somewhat confused because the methods produce differing results. However, the differences should not be so great as to cause concern. The reader is urged to study the basis of each analytical method in order to decide which one to use for a particular project.

In keeping with the theme of this book, soil–structure interaction, the reader is urged to use the material presented in Chapter 13 on the computation of load-settlement curves. Such a curve is shown in Figure 10.2, where the "plunging load" is the axial capacity. The value of the load-settlement curve, modified as necessary for long-term loading, is evident to the designers of various structures. Not only will such a curve provide a basis for the response of groups of piles, but it can also be used to determine the interaction between the foundation and the superstructure.

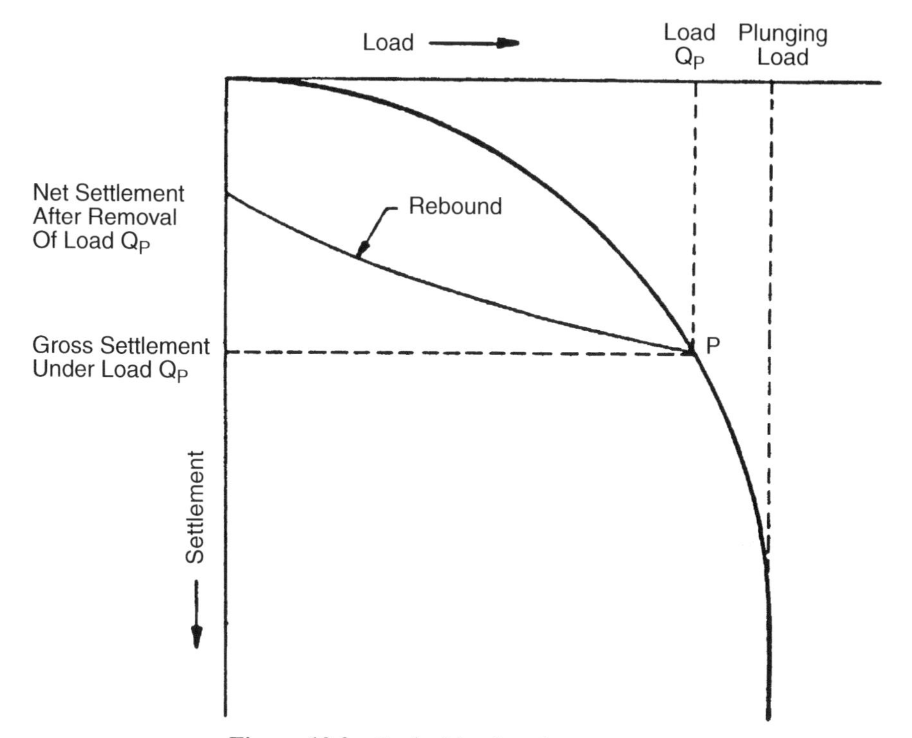

Figure 10.2 Typical load-settlement curve.

The engineer's confidence in computing the axial capacity of a pile and the load-settlement curve can be greatly enhanced by the results of field tests of piles under axial load in soils that exist at the site and installed with methods to be used in production. If there is no information in the technical literature, the engineer may be able to show that such tests, with piles of the size and penetration to be used in production, are cost effective and will greatly benefit to the proposed project.

Referring to Figure 10.2, if the load has a value of Q_p, shown by point P on the load-settlement curve, the gross settlement is indicated by the horizontal dashed line. If the load is released, some rebound will occur, as shown in the figure, with the net settlement shown for the load at zero.

10.2 METHODS OF COMPUTATION

10.2.1 Behavior of Axially Loaded Piles

The pile stiffness for axial loading in general is represented by the load-versus-settlement curve at the pile head. Some analytical methods are based on the theory of elasticity, specifically the Mindlin equation, and solutions have been proposed by D'Appolonia and Romualdi (1963), Thurman and D'Appolonia (1965), Poulos and Davis (1968), Poulos and Mattes (1969), Mattes and Poulos (1969), and Poulos and Davis (1980). The elasticity method presents the possibility of solving for the behavior of a group of closely spaced piles under axial loadings, (Poulos, 1968; Poulos and Davis, 1980), but the weakness of the approach is that the actual ground conditions are rarely satisfied by the assumptions that must be made for E and v, the modulus of elasticity and Poisson's ratio, respectively.

A second method for obtaining the response of a single pile to axial loading represents the soil with a set of nonlinear mechanisms. The method was first used by Seed and Reese (1957); other studies are reported by Coyle and Reese (1966), Coyle and Sulaiman (1967), and Kraft et al. (1981). The method is known as the *t-z method.* A model for it is shown in Figure 10.3. Figure 10.3a shows the free body of a pile in equilibrium where the applied load Q is balanced by a tip load Q_b plus side loads Q_s. As shown in Figure 10.3c, the pile is replaced by an elastic spring and the soil is replaced by a set of nonlinear mechanisms along the pile and at the pile tip (Figure 10.3d).

The nonlinear characteristics of a hypothetical set of mechanisms, representing the response of the soil, are shown in Figure 10.3c. With regard to

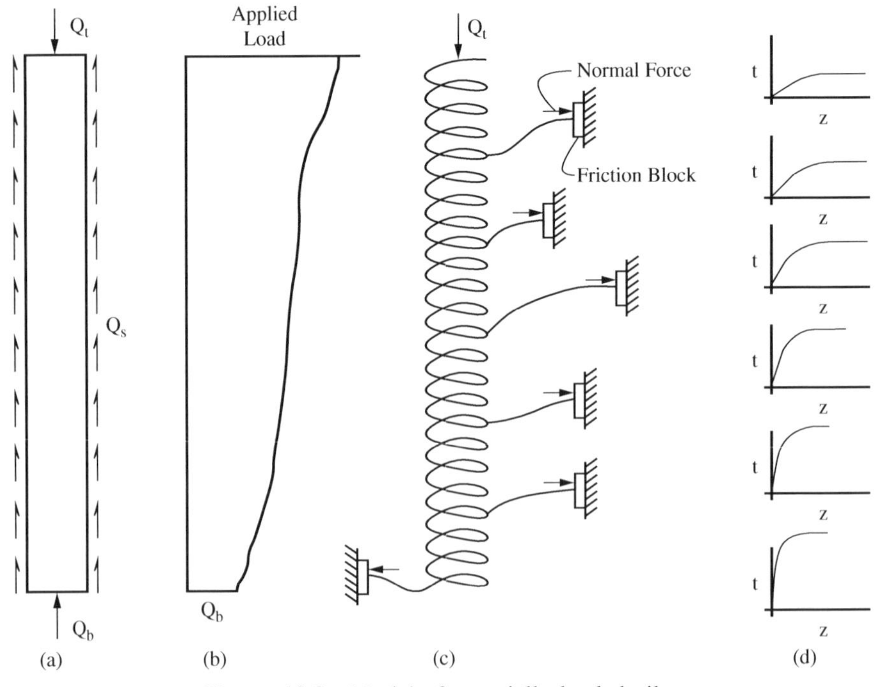

Figure 10.3 Model of an axially loaded pile.

side resistance, the ordinate t of the curves is load transfer and the abscissa z is shaft movement. The corresponding parameters for load transfer in end resistance are q and w_b. No load is transferred from pile to soil unless there is a relative movement between them. The movement is dependent on the applied load, on the position along the pile, on the stress-strain characteristics of the pile material, and on the load transfer curves.

Methods based on elasticity, while having limited applicability, satisfy the continuum effect, in contrast to the model shown in Figure 10.3, which employs the Winkler concept. That is to say, the nonlinear curves shown in Figure 10.3 are independent, while if the continuum is satisfied, all of the curves are mutually dependent. However, the recommendations for load-transfer curves are strongly based on results from full-scale loading tests where the continuum was satisfied. Further, the method has been applied with good success in the analysis of a number of field tests (Coyle and Reese, 1966; Coyle and Sulaiman, 1967), with the result that recommendations have been made for predicting the family of curves shown in Figure 10.3d.

10.2.2 Geotechnical Capacity of Axially Loaded Piles

The geotechnical capacity of an axially loaded pile is defined as the ultimate soil resistance at the point where the pile either plunges down into the ground without any further increase in load or the displacement at the pile head is too great for the superstructures. The use of static equations to compute the geotechnical capacity of piles is well established, and numerous procedures have been suggested. The first method introduced here is presented by the American Petroleum Institute (API) in their manual on recommended practice. The API procedures (see Section 10.3) have been adopted by a number of organizations. Thus, this method has some official recognition. The API procedures for clay are based essentially on the use of undrained shear strength and, of course, are largely empirical. The API procedure for sand is also strongly empirical, but effective-stress techniques are employed because no excess pore water pressures are assumed.

Three other methods, also presented here, are the method recommended by the U.S. Federal Highway Administration (FHWA), the method recommended by the U.S. Army Corps of Engineers, and the so-called Lambda method (Kraft et al., 1981). An earlier version of the Lambda method was published in 1972 (Vijayvergiya and Focht). It has been used in offshore practice because the influence of effective stresses is considered in computing skin-friction resistance.

The reader is urged to utilize all of the methods that apply to a particular soil profile. Furthermore, multiple computations should be made in which the soil parameters are varied through a range that reflects, as well as possible, the upper-bound and lower-bound values of the significant variables.

The performance of field-loading tests is strongly recommended. As noted earlier, the soil profile at a particular site can be "calibrated" by such field tests so that a method of computation can be used with more confidence. The construction procedures used for the test pile and those used for the production piles should be as nearly alike as possible.

Skin Friction and Tapered Piles If tapered piles are used, only the FHWA method has the option of taking into account the taper angle for computing skin friction. All the other methods simply use the straight section for computing skin friction. Field-loading tests are strongly recommended for tapered piles because limited field test results are reported in the technical literature and there are some uncertainties about tapered piles in layered soils.

In some cases, it may be desirable to discount completely the resistance in skin friction over a portion of the pile at the ground surface—for example, if lateral deflection is such that clay is molded away from the pile. To obtain the capacity of a pile in tension, the resistance in skin friction (side resistance) may be employed. Some methods provide specific recommendations for reduction of skin friction in tension. If no reduction factor is employed in the computation, a factor of safety must be applied to the loading.

If scour close to the pile is anticipated, allowance can be made by using zeros for the shear strength of clay and for the angle of internal friction of sand in the region where scour is expected. Other values should remain unchanged so that the overburden stress reflects the position of the original ground surface.

End Bearing End bearing is computed by using the weighted average of soil properties over a distance of two diameters below the tip of the pile. In using the results of such computations, however, the designer should employ judgment in those cases where the total capacity in compression shows a sharp decrease with depth.

Soil Plug If an open pipe is used as a pile, analysis may show that a plug of soil will be forced up the inside of the pile as the pile is driven. Therefore, the end-bearing capacity of the pile is computed by adding resistance in skin friction along the inner surface plus the end bearing of the material area only. In computing resistance in skin friction along the inner surface of the pile, only the area of the thick section (driving shoe) at the end of the pile will be considered. The load from the plug is compared to the load from end bearing over the full area of the base, and the smaller of the two values is used.

The length of the end section of the pile and the internal diameter of that section are important parameters to allow the designer to make appropriate decisions concerning the development of a plug in the pile as it is driven. The remolded strength may be used to compute the internal resistance to account for the fact that the internal plug is highly disturbed as it enters the pile. If there is a length of pipe at the end of the pile with a significantly greater thickness than the pipe above, the designer may decide that only that thicker section will resist the skin friction as the plug moves up.

The designer can also change the computation of the resistance from the plug by selecting the value of the remolded shear strength of the clay that is inside the end section. The same line of reasoning holds for a pile driven into sand, except that the unit skin friction for the interior of the pile is computed the same way as the unit skin friction for the exterior of the pile.

10.3 BASIC EQUATION FOR COMPUTING THE ULTIMATE GEOTECHNICAL CAPACITY OF A SINGLE PILE

The ultimate bearing capacities are computed by use of the following equation:

$$Q_d = Q_f + Q_p = fA_s + qA_p \tag{10.1}$$

where

Q_f = total skin-friction resistance, lb (kN),
Q_p = total end bearing, lb (kN),
f = unit load transfer in skin friction (normally varies with depth), lb/ft^2 (kPa),
q = unit load transfer in end bearing (normally varies with depth), lb/ft^2 (kPa),
A_p = gross end area of the pile, ft^2 (m^2), and
A_s = side surface area of the pile, ft^2 (m^2).

Equation 10.1 is used increment by increment as the pile is assumed to penetrate the ground.

While Eq. 10.1 is generally accepted, there is no general agreement on the methods of obtaining f and q. As noted earlier, the engineer is urged to use several recommendations in the literature and use judgment in making a final design. Alternatively, a special study can be made for a specific site.

10.3.1 API Methods

The API has been active in addressing problems related to the design of fixed offshore platforms. Recommendations have been presented in a document, entitled "API Recommended Practice for Planning, Designing, and Constructing Fixed Offshore Platforms," Report RP-2A. Two methods were given in RP-2A in 1986 for computing the side resistance (skin friction) in cohesive soils, Method 1 and Method 2. Method 1 was recommended for normally consolidated, highly plastic clays and Method 2 for all other types of clay. In RP-2A in1987, Method 1 and Method 2 were combined into a revised method. The revised method was included in RP-2A of 1993. The following sections will summarize Methods 1 and 2 and the revised method of 1987. The revised method has been widely used in the offshore industry.

Skin Friction in Cohesive Soil

API Method 1. The following brief statement from API RP2A (1986) presents Method 1:

> For highly plastic clays such as found in the Gulf of Mexico f may be equal to c for under-consolidated and normally-consolidated clays. For overconsolidated clays f should not exceed 1/2 ton per square foot (48 kPa) for shallow penetrations or c equivalent to normally consolidated clay for deeper penetrations, whichever is greater. [Note: c is shown as c_n in the presentation that follows.]

The detailed explanation of the application of Method 1 follows.

$$Q_f = fA_s \tag{10.2}$$

where

Q_f = axial load capacity in skin friction,
f = unit skin friction resistance (adhesion), and
A_s = side surface area of pile.

The above equation may be written more properly as

$$Q_f = \int_0^L f_x dA_s \tag{10.3}$$

where

L = penetration of pile below ground surface, and
f_x = unit resistance in clay at depth x, measured from ground surface.

The API provision is for highly plastic clay, such as that found in the Gulf of Mexico, and the method of obtaining f_x are shown in Figure 10.4.

The step-by-step computation procedure is:

1. Construct a diagram showing undrained shear strength as a function of depth. API suggests the use of unconfined compression or miniature vane tests, and judgment should be employed in regard to the use of other data.

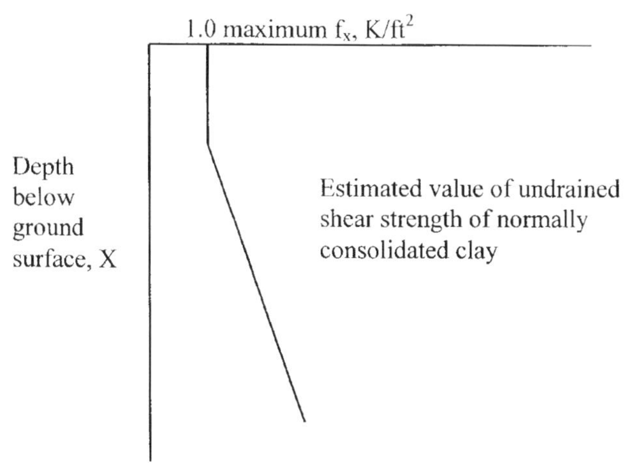

Figure 10.4 Method of obtaining f_x recommended by API RP2A (1986).

2. Construct a diagram showing the maximum f_x, as shown in Figure 10.4. The estimated value of the shear strength of normally consolidated clay may be obtained from the following expression:

$$c_n/\bar{p} = k_c = f\ (\mathrm{PI}) \tag{10.4}$$

where

c_n = undrained shear strength of normally consolidated clay,
$\bar{p}$ = effective stress = γx,
γ = effective unit weight of soil, and
k_c = constant.

The constant k_c has been shown by Skempton and others to be a function of the plasticity index, PI. Some engineers suggest that k_c should be taken as 0.25 for Gulf of Mexico soils based on their experience. The unit weight of the soil, either total or submerged, depending on the location of the water table, should be obtained from a study of data from the soil borings for the site.

3. Select a pile of a given geometry.
4. Select an incremental length of the pile, extending from the ground surface to some depth. The length to be used is selected on the basis of the variations in the diagrams prepared in Steps 1 and 2.
5. For the incremental depth being considered in Step 4, obtain a value of the average shear strength from the diagram prepared in Step 1.

6. Compare the value obtained in Step 5 with the value of $(f_x)_{max}$ for the same depth obtained from the diagram prepared in Step 2. The smaller of the values is to be used in computing the load capacity of the pile.
7. Use the value of f_x obtained in Step 6 with the appropriate surface area of the side of the pile to compute the load capacity of the first increment. Equation 10.2, appropriately modified, can be used.
8. Select the second increment of pile penetration, with the top of the second increment being the bottom of the first increment.
9. Repeat Steps 5 through 7.
10. The loads obtained for each of the two increments can be added to obtain the total axial loading capacity of the pile in skin friction for the two increments.
11. Select other increments until the total length of the pile is taken into account.

API Method 2. The following brief statement from API RP2A (1986) presents Method 2:

> For other types of clay, f should be taken equal to c for c less or equal to 1/4 ton per square foot (24 kPa). For c in excess of 1/4 ton per square foot (24 kPa) but less than or equal to 3/4 ton per square foot (72 kPa) the ratio of f to c should decrease linearly from unity at c equal to 1/4 ton per square foot (24 kPa) to 1/2 at c equal to 3/4 ton per square foot (72 kPa). For c in excess of 3/4 ton per square foot (72 kPa), f should be taken as 1/2 of c.

The detailed explanation of the application of Method 2 follows.

$$f_x = \alpha_x c_x \tag{10.5}$$

where

α_x = coefficient that is a function of c_x, and
c_x = undrained shear strength at depth x.

The recommendation for obtaining α is shown in Figure 10.5.

Figure 10.5 states that f may be considered equal to c for undrained shear strengths of 0.5 kip per square foot or less, and equal to $0.5c$ for values of c equal to or greater than 1.5 kips per square foot. For values of c between 0.5 and 1.5 kips per square foot, there is a small decrease linearly, as shown.

The step-by-step computation procedure is as follows:

1. Same as Step 1 in Method 1.
2. Same as Step 3 in Method 1.
3. Select an incremental length of pile (see Step 4 in Method 1).

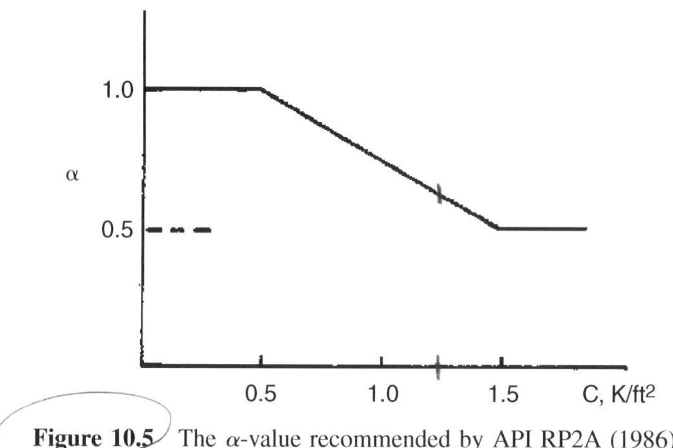

Figure 10.5 The α-value recommended by API RP2A (1986).

4. Obtain a value of c_x for the incremental length from the diagram prepared in Step 1.
5. Enter Figure 10.5 with value of c_x obtained in Step 4 and obtain α_x.
6. Compute the unit skin friction resistance (unit load transfer) f_x using Eq. 10.5.
7. Compute the load capacity of the first incremental length of the pile by using Eq. 10.2 as appropriately modified.
8. Select a second increment and repeat Steps 4 through 7.
9. Add the load computed for the first two increments.
10. Select other increments of length until the length of the pile is taken into account.

Revised API Method (1987). API RP2A has proposed a revised equation for the coefficient α. The factor, α, recommended by API can be computed by two equations:

$$\alpha = 0.5\psi^{-0.5} \quad \text{if } \psi \leq 1.0$$
$$\alpha = 0.5\psi^{-0.25} \quad \text{if } \psi > 1.0 \tag{10.6}$$

with the constraint that $\alpha \leq 1.0$

where

$\psi = c/\overline{p}$ for the depth of interest,
$\overline{p}$ = effective overburden pressure, and
c = undrained shear strength of soil.

Figure 10.6 presents a comparison between α-factors calculated from 1986 and 1987 recommendations. The comparison can be made only for a specific depth because the overburden stress will change with depth; hence, for a given value of c, the value of $c/\bar{p}$ will change. The comparisons shown in Figure 10.6 are for depth where $\bar{p}$ is 0.5, 1.0, and 2.0 T/ft^2. Careful study of Figure 10.6 indicates that for normally consolidated clay, where $c/\bar{p}$ is equal approximately to 0.25, the revised method yields higher values of α than does Method 2. The revised method yields higher values of α at the depth where $\bar{p}$ is equal to 2.0 T/ft^2 (80 to 100 ft) except for heavily overconsolidated clay. The revised method yields lower values of α where $\bar{p}$ is equal to 0.5 T/ft^2 (20 to 25 ft) than does Method 2.

The step-by-step computation procedure is identical to that for Method 2 except that Eq. 10.6 is used in Step 5 instead of Figure 10.5.

End Bearing in Cohesive Soil The recommendations of API RP2A (1987) for end bearing in clay are brief: "For piles end bearing in clay, q in lb/ft^2 (kPa) should be equal to $9c$. If the strength profile below the pile tip is not uniform, then the c utilized should reflect appropriate adjustment." This recommendation is consistent with that of many other investigators.

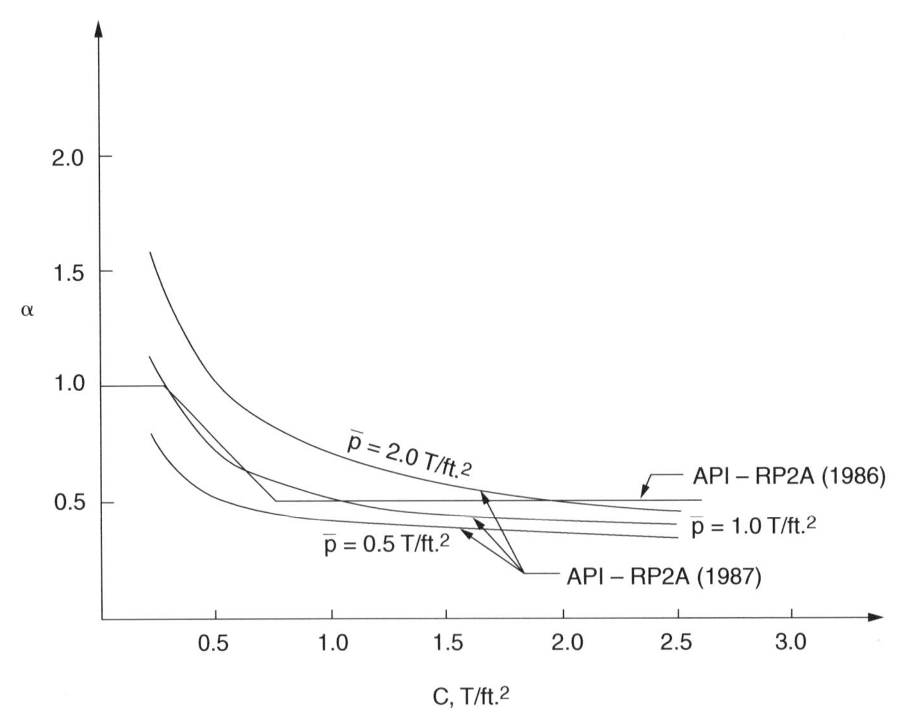

Figure 10.6 Comparison between α-factors calculated from 1986 and 1987 recommendations.

The API recommendation in equation form is as follows, along with a suggestion for modifying the value of c:

$$Q_p = qA_p \tag{10.7}$$

$$q = 9c \tag{10.8}$$

where

Q_p = axial load capacity in end bearing,
q = unit end-bearing resistance,
c = undrained shear strength at the tip of the pile, usually taken as the average over a distance of two diameters below the tip of the pile, and
A_p = cross-sectional area of the tip of the pile.

Skin Friction in Cohesionless Soil Experimental results for driven piles in sands show considerable scatter for values of skin friction and end bearing. The scatter is likely due to the influence of installation methods on soil properties and the state of stress. The following recommendations for computing unit values of skin friction and end bearing for piles in sand are consistent with the state of the practice; however, the user should use such values with caution. Pile-load tests are recommended if feasible.

The API RP2A (1987) recommendations for side resistance for pipe piles in cohesionless soil (sand) are given in the following sections.

$$f = K\bar{p}_o \tan \delta \tag{10.9}$$

where

K = coefficient of lateral earth (ratio of horizontal to vertical normal effective stress),
$\bar{p}_o$ = effective overburden pressure at the point in question, and
δ = friction angle between the soil and the pile wall.

A K value of 0.8 was recommended for open-ended pipe piles, which are driven unplugged, for loadings in both tension and compression. A K value of 1.0 was recommended for full-displacement piles. In the absence of data on δ, Table 10.1 was recommended as a guideline only for siliceous soil.

Equation 10.9 implies that the value of f increases without limit; however, Table 10.1 presents guidelines for limiting values.

End Bearing in Cohesionless Soil For end bearing in cohesionless soils, API recommends the following equation:

'eline for Side Friction in Siliceous Soil

	δ, degrees	Limiting f, kips/ft^2 (kPa)
..ose to medium, sand to silt	15	1.0 (47.8)
Loose to dense, sand to silt	20	1.4 (67.0)
Medium to dense, sand to sand-silt	25	1.7 (83.1)
Dense to very dense, sand to sand-silt	30	2.0 (95.5)
Dense to very dense, gravel to sand	35	2.4 (1110.8)

$$q = \bar{p}_0 N_q \tag{10.10}$$

where

$\bar{p}_0$ = effective overburden pressure at pile tip, and
N_q = bearing capacity factor.

Table 10.2 is recommended as a guideline only for siliceous soil.

The API publication points out that many soils do not fit the description of those in the tables and that the design parameters are not suitable for such soils. Examples are loose silts, soils containing large amounts of mica or volcanic grains, and calcareous sands. These latter soils are known to have substantially lower design parameters.

Drilled and grouted piles may have higher capacities than driven piles in calcareous soils.

10.3.2 Revised Lambda Method

Vijayvergiya and Focht (1972) proposed Eq. 10.11 for computing the skin-friction resistance:

$$Q_f = \lambda(p_m + 2c_m)A_s \tag{10.11}$$

TABLE 10.2 Guideline for Tip Resistance in Siliceous Soil

Soil	N_q	Limiting q, kips/ft^2 (MPa)
Very loose to medium, sand-silt	8	40 (1.9)
Loose to dense, sand to silt	12	60 (2.9)
Medium to dense, sand to sand-silt	20	100 (10.8)
Dense to very dense sand to sand-silt	40	200 (9.6)
Dense to very dense, gravel to sand	50	250 (12.0)

where

λ = a coefficient that is a function of pile penetration,
A_s = surface contact area,
p_m = mean vertical effective stress between the ground surface and the pile tip, and
c_m = mean undrained shear strength along the pile.

A curve, giving λ as a function of pile penetration, was introduced in 1972. Kraft et al. (1981) made a further study of data and revised the method of obtaining λ. The following equations were employed.

For normally consolidated soils:

$$\lambda = 0.178 - 0.016 \ln \pi_3 \tag{10.12a}$$

For overconsolidated soils:

$$\lambda = 0.232 - 0.032 \ln \pi_3 \tag{10.12b}$$

where

$$\pi_3 = (\pi B f_{\max}(L_e^2)/(AEU)), \tag{10.13}$$

B = pile diameter,
$f_{\max}$ = peak skin friction (taken as the mean undrained shearing strength),
L_e = embedded pile length,
A = cross-sectional area of pile material,
E = modulus of elasticity of pile material, and
U = pile displacement needed to develop the side shear (taken as 0.1 in.).

In the absence of data consolidation, Kraft et al. considered the clay to be overconsolidated if the undrained shear strength divided by the overburden stress was equal to or larger than 0.10. The value of the lower limit of λ was taken as 0.110.

The step-by-step computation procedure is as follows:

1. Construct a diagram showing undrained shear strength as a function of depth.
2. Construct a diagram showing vertical effective stress as a function of depth.

3. Select a pile of a given geometry.
4. Select a pile penetration for computation of the axial load capacity in skin friction.
5. Use the diagram constructed in Step 1 and compute c_m.
6. Use the diagram constructed in Step 2 and compute p_m.
7. Obtain the value of λ from Eqs. 10.12 through 10.13.
8. Substitute values of λ, p_m, and c_m, along with the value of A_s, into Eq. 10.11 and compute Q_f.
9. Select a different pile penetration and repeat Steps 5 through 8 to obtain the pile capacity Q_f as a function of pile penetration.

10.3.3 U.S. Army Corps Method

Piles in Cohesionless Soil

Skin Friction. For design purposes, the skin friction of piles in sand increases linearly to an assumed critical depth (D_c) and then remains constant below that depth. The critical depth varies between 10 to 20 pile diameters or widths (B), depending on the relative density of the sand. For cylindrical piles, B is equal to the pile diameter. The critical depth is assumed as:

$D_c = 10B$ for loose sands,
$D_c = 15B$ for medium dense sands, and
$D_c = 20B$ for dense sands.

The unit skin friction acting on the pile shaft may be determined by the following equations:

$$f_s = k\sigma'_v \tan\delta$$

$$\sigma'_v = \gamma' D \quad \text{for} \quad D < D_c$$

$$\sigma'_v = \gamma' D_c \quad \text{for} \quad D \geq D_c$$

$$Q_s = f_s A \tag{10.14}$$

where

K = lateral earth pressure coefficient (K_c for compression piles and K_t for tension piles),
σ'_v = effective overburden pressure,
δ = angle of friction between the soil and the pile,

γ' = effective unit weight of soil, and
D = depth along the pile at which the effective overbu
calculated.

Values of δ are given in Table 10.3.

Values of K for piles in compression (K_c) and piles in tension (K_t) are given in Table 10.4. Tables 10.3 and Table 10.4 present ranges of values of δ and K based upon experience in various soil deposits. These values should be selected for design based upon experience and pile load test results. The designer should not use the minimum reduction of the ϕ angle while using the upper-range K values.

For steel H-piles, A_s should be taken as the block perimeter of the pile and δ should be the average friction angle of steel against sand and sand against sand (ϕ.). Note that Table 10.4 provides general guidance to be used unless the long-term engineering practice in the area indicates otherwise. Under prediction of soil strength, parameters at load test sites have sometimes produced back-calculated values of K that exceed the values in Table 10.4. It has also been found, both theoretically and at some test sites, that the use of displacement piles produces higher values of K than does the use of nondisplacement piles. Values of K that have been used satisfactorily but with standard soil data in some locations are presented in Table 10.5.

End Bearing. For design purposes, the bearing capacity of the tip of the pile can be assumed to increase linearly to a critical depth (D_c) and then remain constant. The same critical depth relationship used for skin friction can be used for end bearing. The unit tip bearing capacity can be determined as follows:

$$q = \sigma'_v N_q \qquad (10.15)$$

where

$\sigma'_v = \gamma' D$ for $D < D_c$,
$\sigma'_v = \gamma' D_c$ for $D \geq D_c$, and
N_q = bearing capacity factor.

TABLE 10.3 Values of δ

Pile Material	δ
Steel	0.67 ϕ to 0.83 ϕ
Concrete	0.90 ϕ to 1.0 ϕ
Timber	0.80 ϕ to 1.0 ϕ

TABLE 10.4 Values of K

Soil Type	K_c	K_t
Sand	1.00 to 2.00	0.50 to 0.70
Silt	1.00	0.50 to 0.70
Clay	1.00	0.70 to 1.00

Note: The above values do not apply to piles that are prebored, jetted, or installed with a vibratory hammer. Selection of K values at the upper end of these ranges should be based on local experience. K, δ, and N_q values back-calculated from load tests may be used.

For steel H-piles, A_t should be taken as the area included within the block perimeter. A curve to obtain the Terzaghi-Peck (1967) bearing capacity factor N_q (among values from other theories) is shown in Figure 10.7. To use the curve, one must obtain measured values of the angle of internal friction (ϕ) which represents the soil mass.

Tensile Capacity. The tensile capacity of piles in sand, which should not include the end bearing, can be calculated as follows using the K_t values for tension from Table 10.4:

$$Q_{\text{ult}} = Q_{s,\text{tension}}$$

Piles in Cohesive Soil

Skin Friction. Although called skin friction, the resistance is due to the cohesion or adhesion of the clay to the pile shaft.

$$f_s = c_a$$

$$c_a = \alpha c$$

$$Q_s = f_s A_s \tag{10.16}$$

TABLE 10.5 Common Values for Corrected K

	Displacement Piles		Nondisplacement Piles	
Soil Type	Compression	Tension	Compression	Tension
Sand	2.00	0.67	1.50	0.50
Silt	1.25	0.50	1.00	0.35
Clay	1.25	0.90	1.00	0.70

Note: Although these values may be commonly used in some areas, they should not be used without experience and testing to validate them.

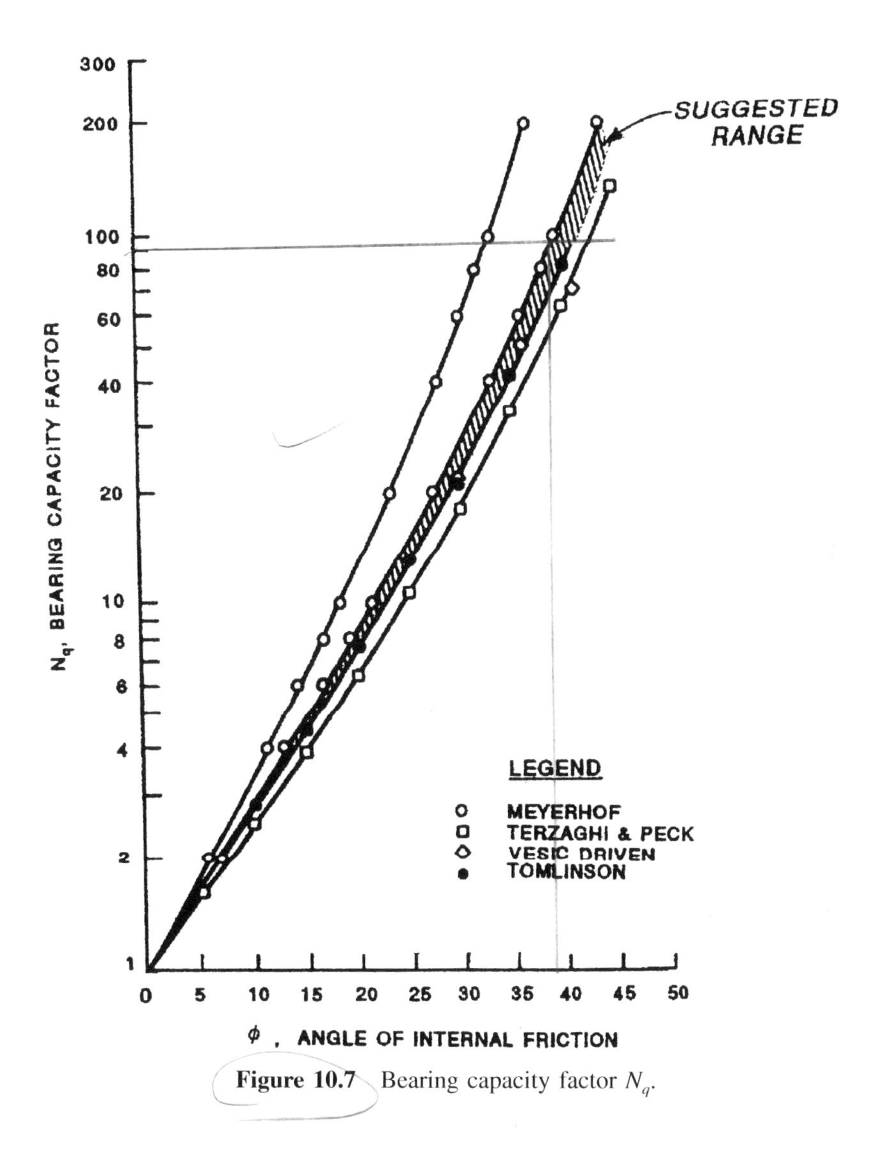

Figure 10.7 Bearing capacity factor N_q.

where

c_a = adhesion between the clay and the pile,
α = adhesion factor, and
c = undrained shear strength of the clay from a quick (Q) test.

The values of α as a function of undrained shear strength are given in Figure 10.8.

An alternative procedure developed by Semple and Rigden (1984) to obtain values of α that are especially applicable to very long piles is given in Figure 10.9, where

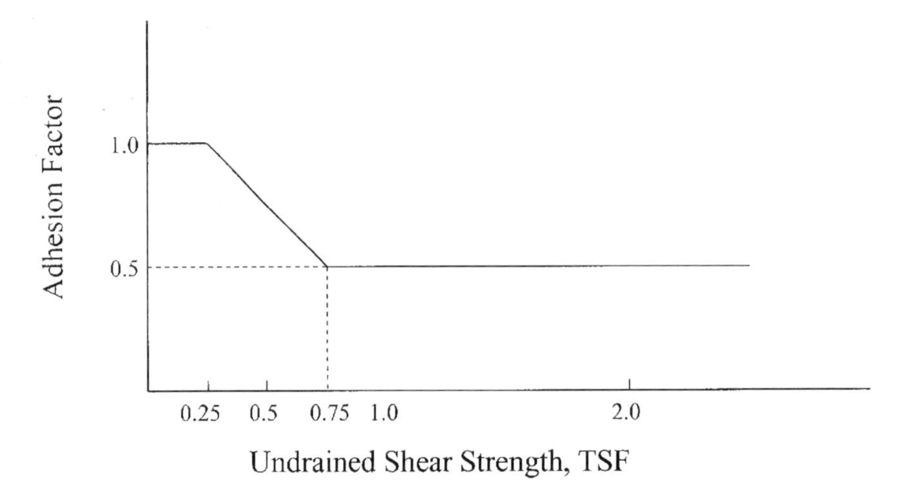

Figure 10.8 Value of α versus undrained shear strength recommended by the U.S. Army Corps method.

$$\alpha = \alpha_1 \alpha_2 \tag{10.17}$$

and

$$f_s = \alpha c \tag{10.18}$$

End Bearing. The pile unit-tip bearing capacity for piles in clay can be determined from the following equations:

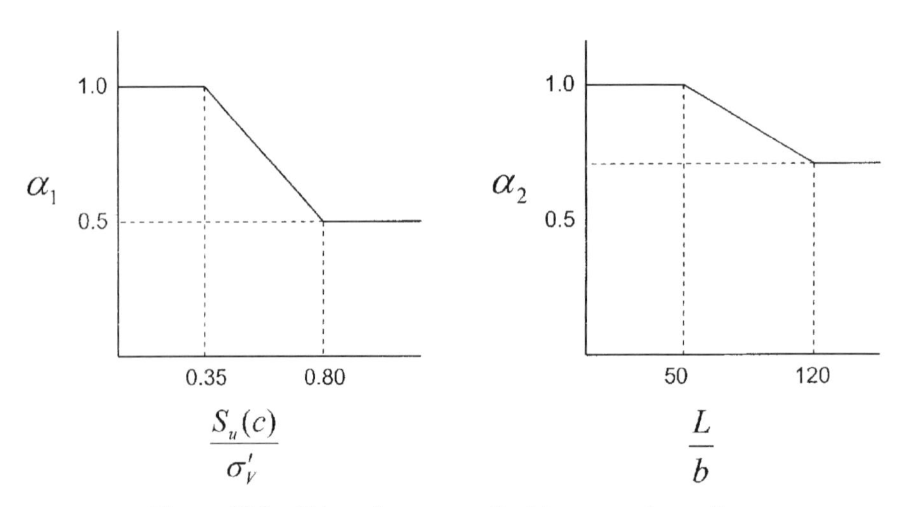

Figure 10.9 Value of α_1, α_2 applicable to very long piles.

$$q = 9c \tag{10.19}$$

$$Q_t = A_t q \tag{10.20}$$

However, the movement necessary to develop the tip resistance of piles in clay soils may be several times larger than that required to develop the skin friction resistance.

Compressive Capacity. By combining the skin friction and the top bearing capacity, the ultimate compression capacity may be found:

$$Q_{\text{ult}} = Q_s + Q_t \tag{10.21}$$

Tensile Capacity. The tensile capacity of piles in clay may be calculated as

$$Q_{\text{ult}} = Q_s \tag{10.22}$$

S-Case Shear Strength. The pile capacity in normally consolidated clays (cohesive soils) should also be computed in the long-term S shear strength case. That is, the engineer should develop an S-Case shear strength trend as discussed previously and proceed as if the soil is drained. The computational method is identical to that for piles in granular soils, and to present the computational methodology would be redundant. Note, however, that the shear strengths in clays in the S-Case are assumed to be $\phi > 0$ and $c = 0$. Some commonly used S-Case shear strengths in alluvial soils are reported in Table 10.6.

These general data ranges are from tests on specific soils in site-specific environments and may not represent the soil in question.

10.3.4 FHWA Method

Introduction Contrary to the methods presented above, the FHWA method uses the same general equations for computing the response of a pile in both sands and clays. A single pile derives its load-carrying ability from the fric-

TABLE 10.6 S-Case Shear Strength

Soil Type		Consistency	Angle of Internal Friction
Fat clay	(CH)	Very soft	13–17
Fat clay	(CH)	Soft	17–20
Fat clay	(CH)	Medium	20–21
Fat clay	(CH)	Stiff	21–23
Silt	(ML)		25–28

Note: The designer should perform testing and select shear strengths.

tional resistance of the soil around the shaft and the bearing capacity at the pile tip:

$$Q = Q_p + Q_s \tag{10.23}$$

where

$$Q_p = A_p q_p \tag{10.24}$$

and

$$Q_s = \int_0^L f_s C_d dz \tag{10.25}$$

where

A_p = area of pile tip,
q_p = unit bearing capacity at pile tip,
f_s = ultimate skin resistance per unit area of shaft C_d,
C_d = effective perimeter of pile,
L = length of pile in contact with soil, and
z = depth coordinate.

Point Resistance The point bearing capacity can be obtained from

$$q_p = cN_c + qN_q + \frac{B}{2} N_\gamma \tag{10.26}$$

where N_c, N_q, and N_γ are dimensionless parameters that depend on the soil friction angle ϕ. The term c is the cohesion of the soil, q is the vertical stress at the pile base level, B is the pile diameter (width), and γ is the unit weight of the soil.

For end bearing in cohesive soils, Eq. 10.27 is recommended:

$$Q_p = A_p \, c \, N_c \tag{10.27}$$

Values of N_c lie between 7 and 16. A value of $N_c = 9$ is usually used.

For end bearing in cohesionless soils, the FHWA method uses the following equation:

$$Q_p = A_p \, \overline{q} \, \alpha \, N_q' \tag{10.28}$$

where

N'_q = bearing capacity factor from Figure 10.10 an
α = dimensionless factor dependent on the depth
the pile in Figure 10.10.

Meyerhof (1976) recommends the limiting value for the tip resistance in cohesive soils as shown in Figure 10.11.

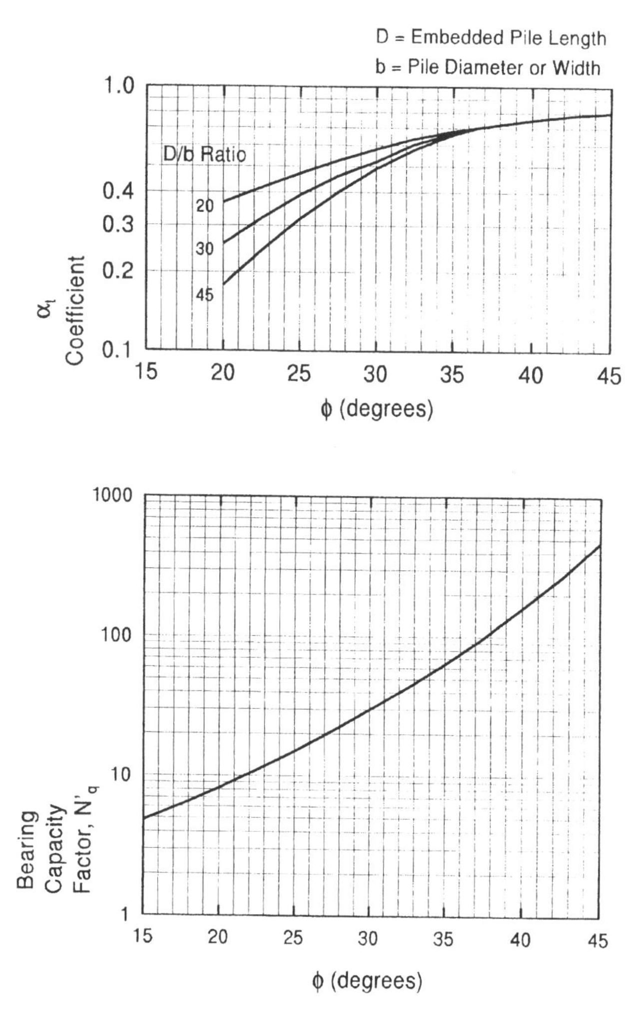

Figure 10.10 Chart for estimating the α coefficient and the bearing capacity factor N_q.

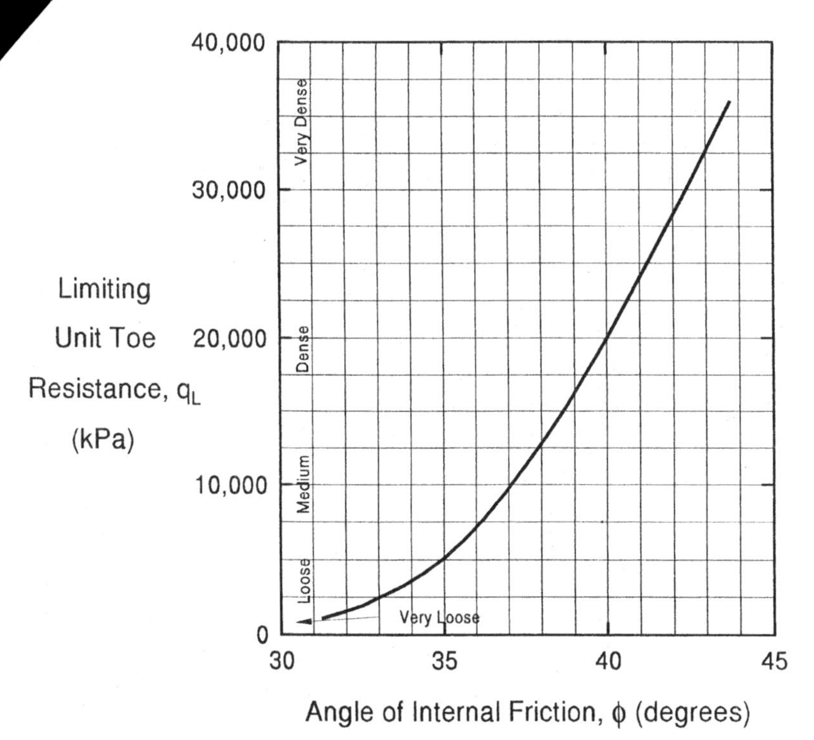

Figure 10.11 Relationship between maximum unit pile point resistance and friction angle for cohesionless soils recommended by Meyerhof (1976) (after FHWA, 1993).

Side Resistance The ultimate side resistance per unit area of shaft is calculated as follows:

$$f_s = c_a + \sigma_h \tan \delta \tag{10.29}$$

where

c_a = pile–soil adhesion,
σ_h = normal component of stress at pile–soil interface, and
δ = pile–soil friction angle.

The normal stress σ_h is related to the vertical stress σ_v as $\sigma_h = K\sigma_v$, where K is a coefficient of lateral stress. Substituting into Eq. (10.29) produces this result:

$$f_s = c_a + K\sigma_v \tan \delta \tag{10.30}$$

For cohesive soils, Eq. (10.30) reduces as follows:

$$f_s = c_a \tag{10.31}$$

where the adhesion c_a is usually related to the undrained shear strength c_u in the following way:

$$c_a = \alpha c_u \tag{10.32}$$

where α is an empirical adhesion coefficient that depends mainly upon the following factors: nature and strength of the soil, type of pile, method of installation, and time effects. Figure 10.12 presents the α-values used by the program as suggested by Tomlinson (1980).

For cohesionless soils, Eq. 10.30 reduces to

$$f_s = \bar{c}_a + K\bar{\sigma}_v \tan \delta \cong K\bar{\sigma}_v \tan \delta \tag{10.33}$$

because $\bar{c}_a$ is either zero or small compared to $K\bar{\sigma}_v \tan \delta$.

The main difficulty in applying the effective stress approach lies in having to predict the normal effective stress on the pile shaft $\bar{\sigma}_h = K\bar{\sigma}_v$.

Nordlund (1963, 1979) developed a method of calculating skin friction based on field observations and results of several pile load tests in cohesionless soils. Several pile types were used, including timber, H, pipe, monotube, and others. The method accounts for pile taper and for differences in pile materials.

Nordlund suggests the following equation for calculating the ultimate skin resistance per unit area:

$$f_s = K_\delta C_f \bar{p}_d \frac{\sin(\omega + \delta)}{\cos \omega} C_d dz \tag{10.34}$$

Combine Eq. (10.25) with Eq. (10.34) to calculate the frictional resistance of the soil around the pile shaft as follows:

$$Q_s = \int_0^L K_\delta C_f \bar{p}_d \frac{\sin(\omega + \delta)}{\cos \omega} C_d dz \tag{10.35}$$

which simplifies for nontapered piles ($\omega = 0$) as follows:

$$Q_s = \int_0^L k_\delta C_f \bar{p}_d \sin(\delta) C_d dz \tag{10.36}$$

where

Q_s = total skin friction capacity,
K_δ = coefficient of lateral stress at depth z,

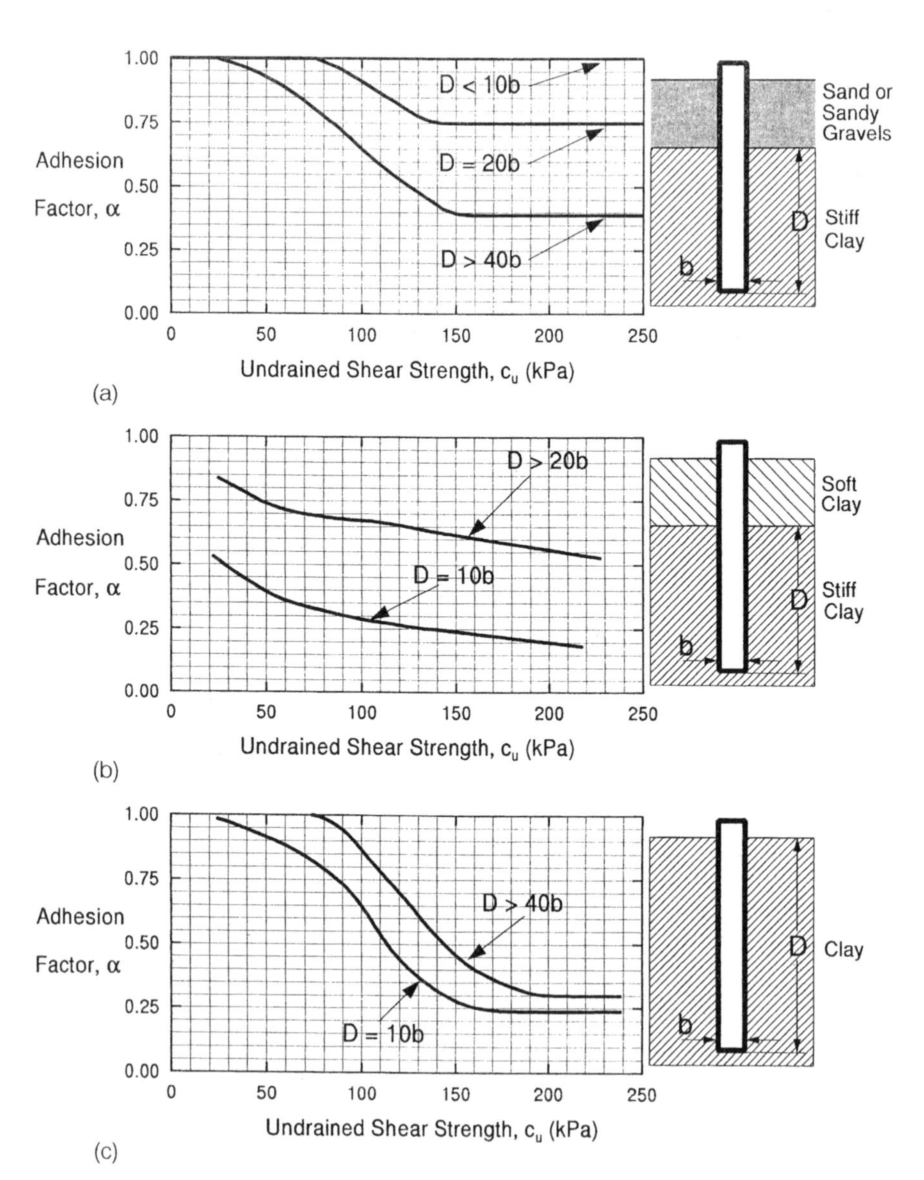

Figure 10.12 Adhesion factors for driven piles in clay recommended by Tomlinson (1980) (after FHWA, 1993).

$\bar{p}_d$ = effective overburden pressure,
ω = angle of pile taper from vertical,
δ = pile–soil friction angle,
C_d = effective pile perimeter, and
C_f = correction factor for K_δ when $\delta \neq \phi$.

To avoid numerical integration, computations are performed for pile segments within soil layers of the same effective unit weight and friction angle. Then Eq. 10.36 becomes

$$Q_s = \sum_{i=1}^{n} K_{\delta_i} C_{f_i} \bar{p}_{d_i} D_i \sin(\delta) \qquad (10.37)$$

where

n = number of segments, and
D_i = thickness of single segment.

Figures 10.13 to 10.16 give values of K_δ versus ϕ with δ equal to ϕ, and Figure 10.17 gives a correction factor to be applied to K_δ when δ is not equal to ϕ.

Nordlund (1963, 1979) states that the angle of friction (δ) between the soil and the pile depends on the pile material and the displaced volume per foot during pile installation. Figure 10.18 gives δ/ϕ for different pile types and sizes.

Figure 10.19 shows the correction factor of field SPT N-values for the influence of effective overburden pressure, and Figure 10.20 shows a correlation between SPT N-values and $\tilde{\phi}$.

10.4 ANALYZING THE LOAD–SETTLEMENT RELATIONSHIP OF AN AXIALLY LOADED PILE

10.4.1 Methods of Analysis

The foundation engineer must make a number of simplifying assumptions in the design procedure for axially loaded piles. These assumptions are necessary to quantify the behavior of the pile-soil system. One of the most grossly oversimplifying assumptions in the static formula is that the ultimate bearing capacity of the soil at the tip of the pile and the ultimate skin friction along the pile shaft are mobilized simultaneously, with no regard for the displacement compatibility of these separate components. Another oversimplifying assumption is that the soil conditions existing prior to installation are still present after placement of the pile.

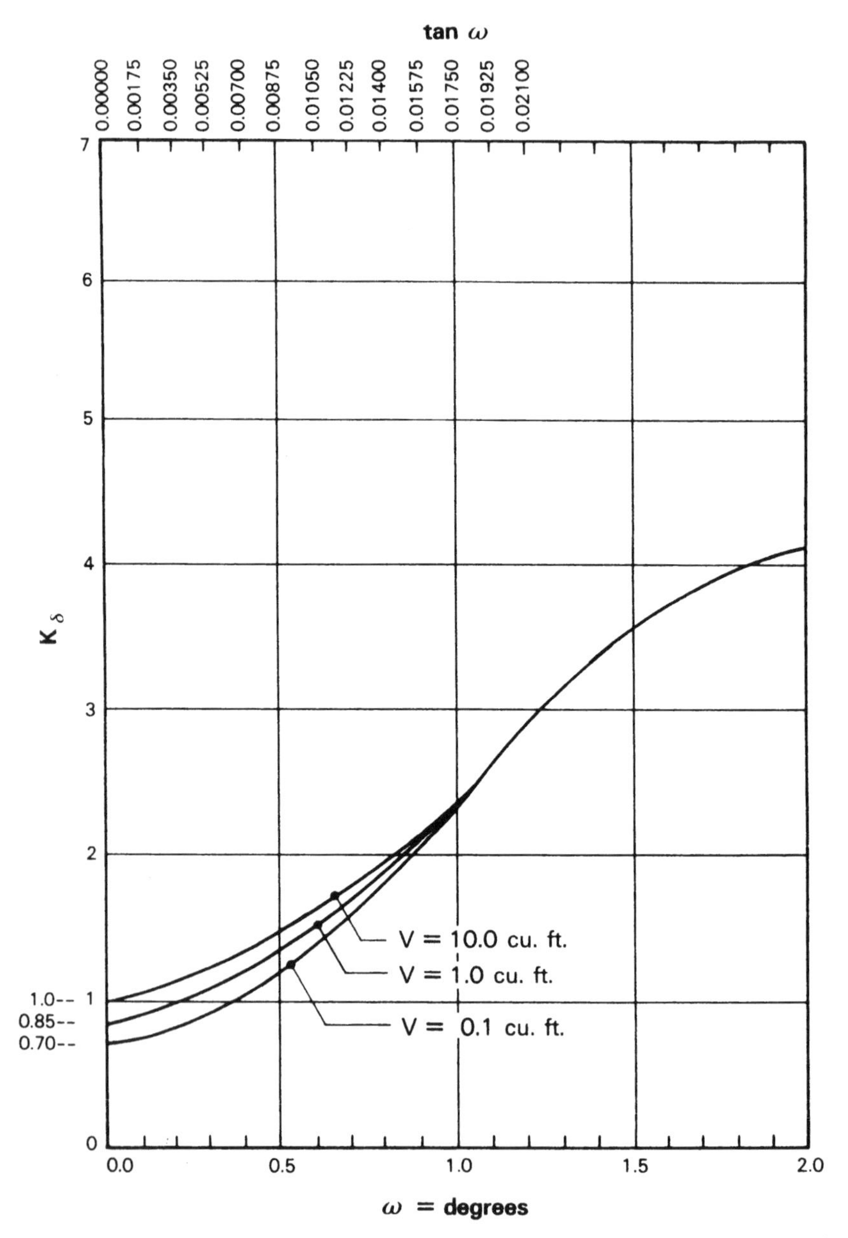

Figure 10.13 Design curves for evaluating K_δ for piles when $\phi = 25°$ recommended by Nordlund (1979) (after FHWA, 1993).

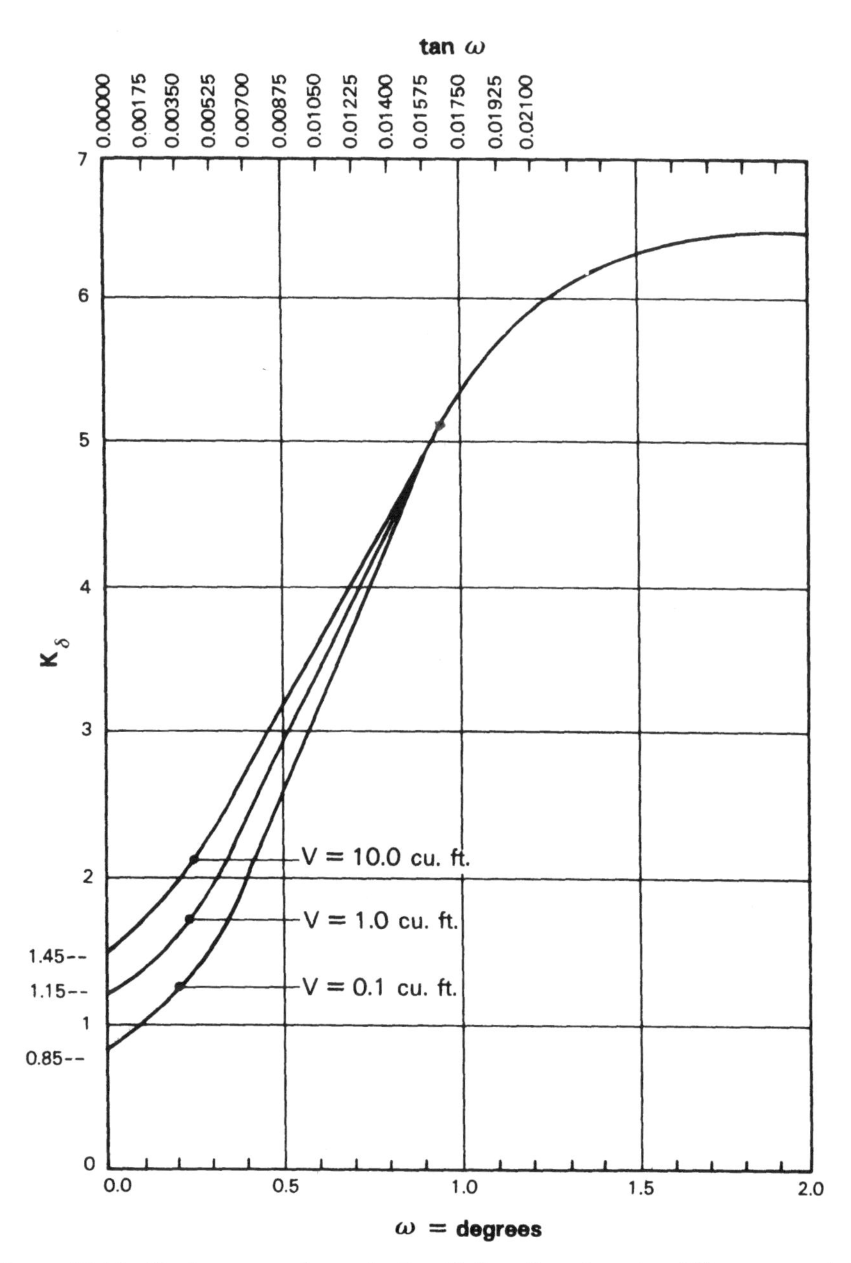

Figure 10.14 Design curves for evaluating K_δ for piles when $\phi = 30°$ recommended by Nordlund (1979) (after FHWA, 1993).

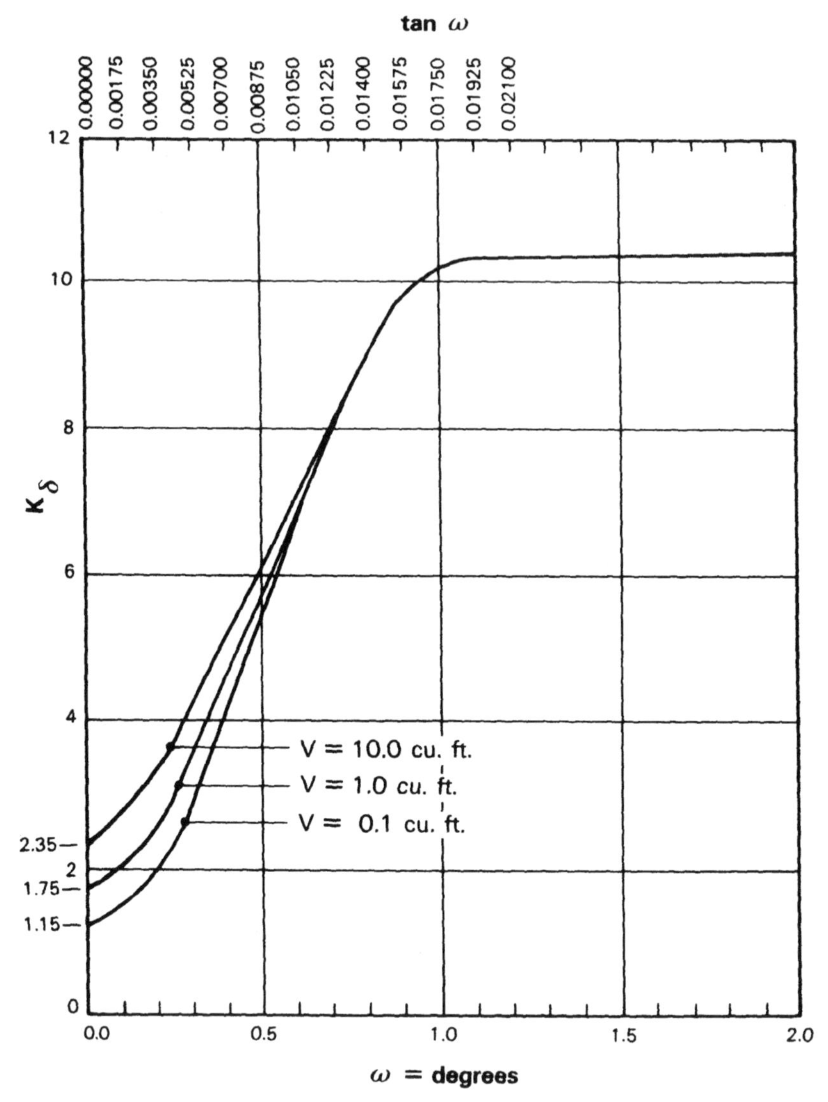

Figure 10.15 Design curves for evaluating K_δ for piles when $\phi = 35°$ recommended by Nordlund (1979) (after FHWA, 1993).

The drawback of the elasticity method lies in the basic assumptions that must be made. The actual ground condition rarely satisfies the assumption of uniform and isotropic material. In spite of the highly nonlinear stress-strain characteristics of soils, the only soil properties considered in the elasticity method are Young's modulus E and Poisson's ratio v. The use of only two constants, E and v, to represent soil characteristics is a gross oversimplification. In actual field conditions, v may be relatively constant, but E can vary through several orders of magnitude.

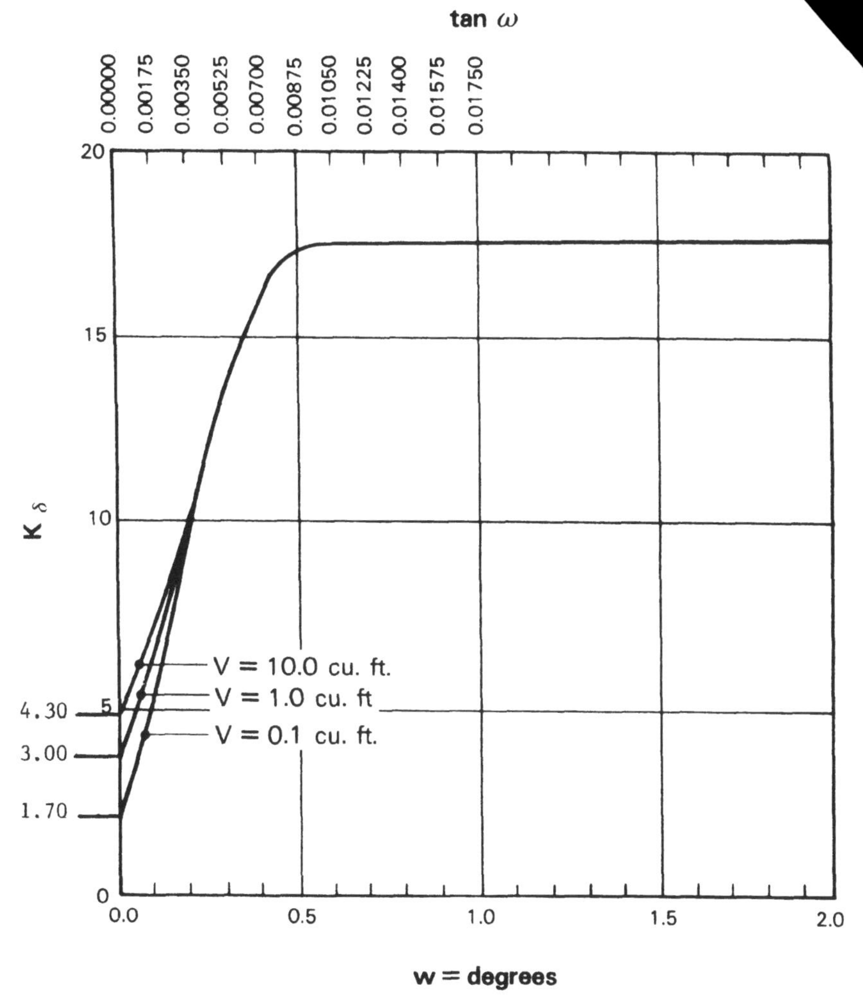

Figure 10.16 Design curves for evaluating K_δ for piles when $\phi = 40°$ recommended by Nordlund (1979) (after FHWA, 1993).

The other method used to calculate the load-versus-settlement curve for an axially loaded pile may be called the *t-z method* as presented in Section 10.2.1. Finite-difference equations are employed to achieve compatibility between the pile displacement and the load transfer along a pile and between displacement and resistance at the tip of the pile. The t-z method assumes the Winkler concept; that is, the load transfer at a certain pile section and the pile-tip resistance are independent of the pile displacement elsewhere. The close agreement between prediction and loading-test results in clays (Coyle and Reese, 1966) and the scattering of prediction for loading tests in sands (Coyle and Sulaiman, 1967) may possibly be explained by the relative sen-

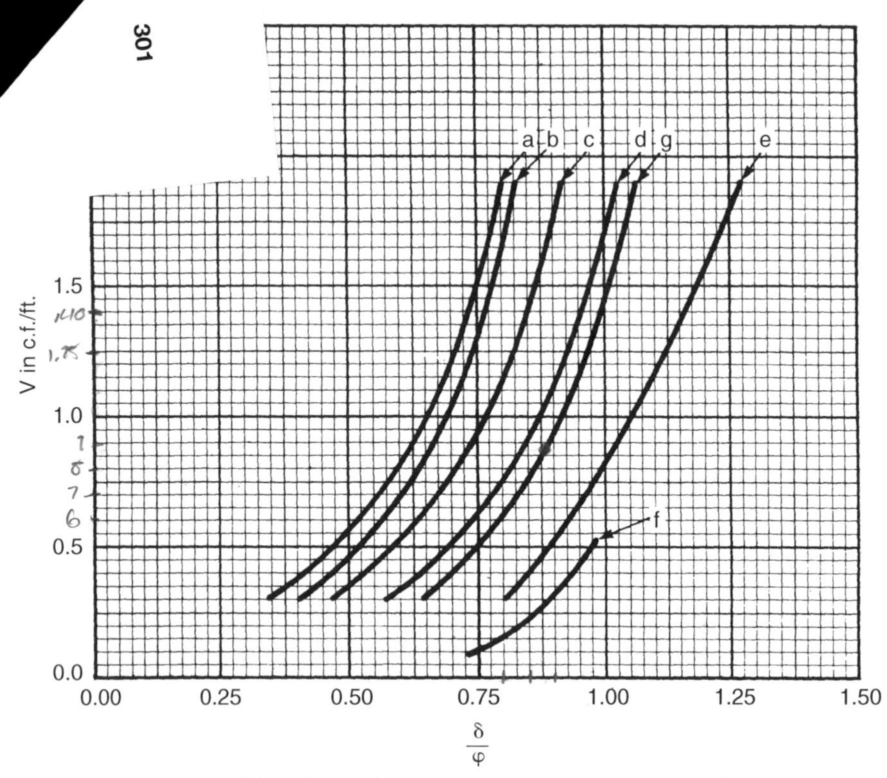

a. Pipe piles and nontapered portion of monotube piles
b. Timber piles
c. Precast concrete piles
d. Raymond step-taper piles
e. Raymond uniform taper piles
f. H-piles & augercast piles
g. Tapered portion of monotube piles

Figure 10.17 Correction factor for K_δ when δ and ϕ are not equal recommended by Nordlund (1979) (after FHWA, 1993).

sitivity of a soil to changes in patterns of stress. Admitting the deficiency in the displacement-shear-force criteria of sand, the t-z method is still a practical method because it can deal with any complex composition of soil layers with any nonlinear relationship of displacement versus shear force. Furthermore, the method can accommodate improvements in soil criteria with no modifications of the basic theory.

The fundamental problem in computing a load-settlement curve, and curves showing the distribution of axial load as a function of distance along the pile, is to select curves that show load transfer in side resistance (skin friction) as a function of pile movement at the point in question (commonly called *t-z curves*) and a curve showing end bearing as a function of movement of the pile tip (commonly called a *q-w curve*).

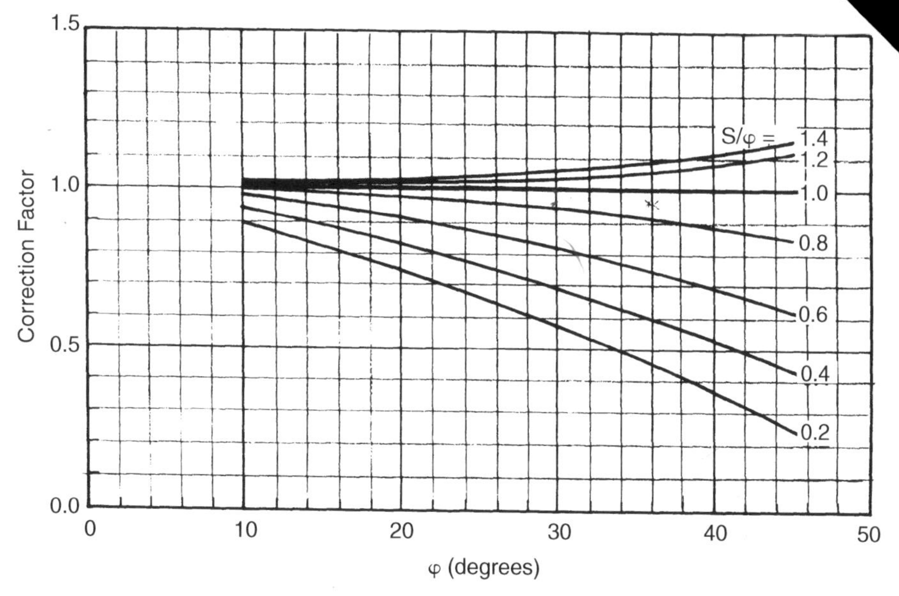

Figure 10.18 Relation δ/ϕ and pile displacement, V, for various types of piles recommended by Nordlund (1979) (after FHWA, 1993).

A computer program named APILEplus was developed based on the methods of computation presented in this chapter. The program utilizes two related codes to provide information on the behavior of driven piles under axial loading. The first code uses four different sets of recommendations for computing the axial capacity of piles as a function of depth. The second code employs *t-z* curves to compute the load versus settlement of the pile at the greatest length that is specified by the engineer. The program also has the flexibility to allow users to specify any types of values for the transfers in skin friction and end bearing as a function of depth. This feature is useful for cases when the site measurements were made from axial load tests.

10.4.2 Interpretation of Load-Settlement Curves

A typical load-settlement curve for a deep foundation was presented in Figure 10.2. The failure load of a driven pile can be interpreted in several different ways from the load settlement curves at the pile head. There is no doubt that the failure occurs when the pile plunges down into the ground without any further increase in load. However, from the point of view of the service requirement, the pile fails when its settlement reachs the stage at which unacceptable distortion might occur in the structure. The maximum allowable settlement at the top of the pile is typically specified by the designer of the structure.

The procedure recommended by Davisson (1972) for obtaining the failure load based on the load-settlement curve is widely accepted for driven piles.

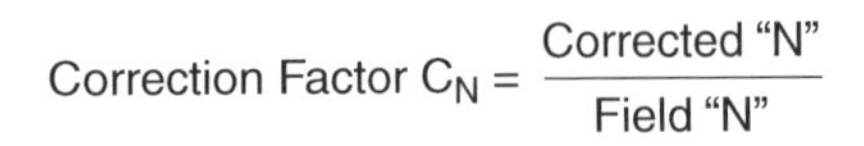

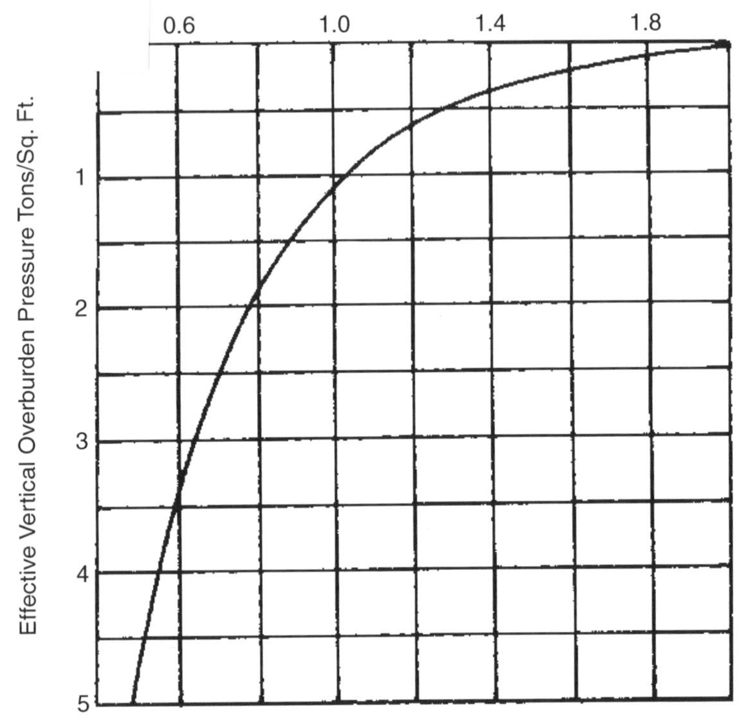

Figure 10.19 Chart for correction of *N*-values in sand for influence of effective overburden pressure (after FHWA, 1993).

Davisson's method takes into account the elastic shortening of the pile under the axial load, the required relative movement (0.15 in.) between the soil and the pile for full mobilization of side friction, and the amount of tip movement (1/120th of the pile diameter in inches) for mobilization of tip resistance. Therefore, the total movement under the failure load is as shown in Figure 10.21.

$$\Delta = (P_{ult})(L)/(AE) + 0.15 \text{ in.} + D/120 \text{ in.} \qquad (10.38)$$

where

P_{ult} = failure load,
L = pile length,
A = area of pile cross section,
E = elastic modulus of pile material, and
D = pile diameter.

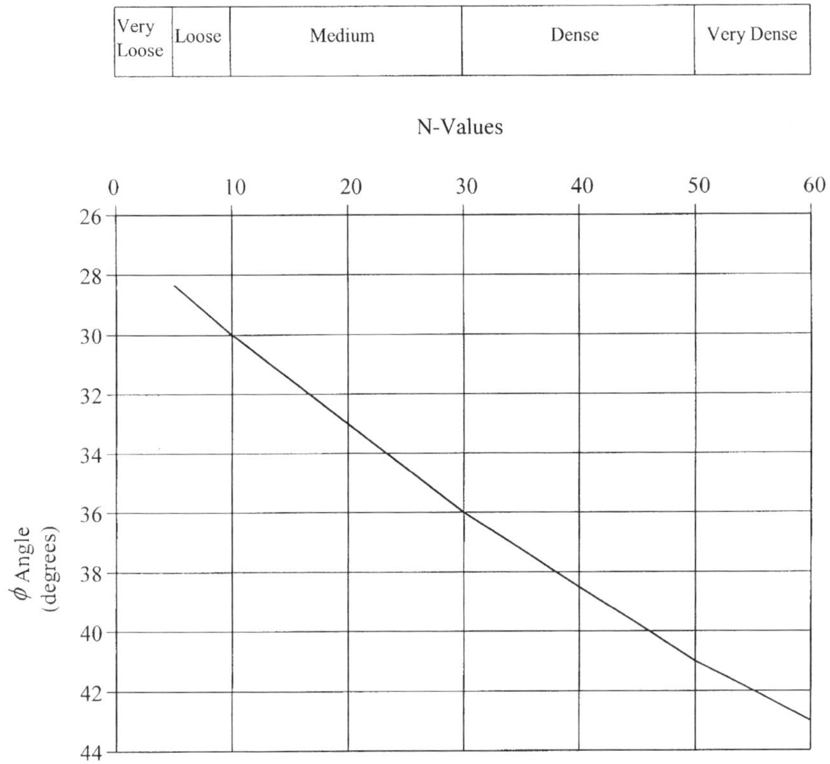

Very Loose
Loose
Medium
Dense
Very Dense
N-Values
0
10
20
30
40
50
60
26
28
30
32
34
36
38
40
42
44
φ Angle (degrees)

Figure 10.20 Relationship between SPT values and ϕ or relative density descriptions recommended by Peck et al. (1974) (after FHWA, 1993).

Axial Load at Pile Head
Q_ult
= Q_ult L/AE
(inches)
=0.15+D/120
(inches)
Movement

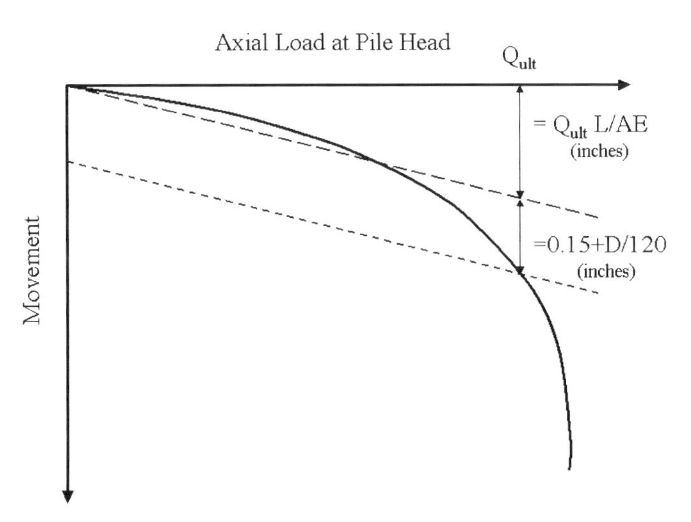

Figure 10.21 Interpretation of the load-settlement curve.

Other recommendations consider that the pile has failed when the pile head has moved 10% of the pile end diameter or a total gross settlement of 1.5 in. The definition of the failure load may involve judgment for some cases. However, the full range of the load-settlement curve at the pile head predicted by the *t-z* method is always beneficial to the designer, allowing better control of the movement of the structure components.

10.5 INVESTIGATION OF RESULTS BASED ON THE PROPOSED COMPUTATION METHOD

Minami (1983) made a study employing one set of recommendations for load-transfer curves in which the results from field-load tests of 10 piles were analyzed. The piles were tested in Japan, and the relevant data were available. Table 10.7 shows pertinent data about the various piles. The first step was to modify the soil properties as necessary to ensure that the ultimate capacity as computed was in agreement with the experiment. Then load-settlement curves were computed; Figure 10.22 shows comparisons of settlements for the 10 cases. The comparisons were made at one-half of the ultimate capacity, as found in the experiment.

Figure 10.22 shows that agreement was reasonable for this type of computation except for perhaps three cases where the predicted settlements were significantly greater that the measured ones. Some of the error may be due to experimental values such as load, settlement, pile geometry and stiffness, and soil properties. Most of the difference probably is due to the inability to predict correctly the load-transfer curves. Therefore, it is recommended that upper-bound and lower-bound solutions be developed by varying the input parameters through an appropriate range.

TABLE 10.7 Pertinent Data on the Various Piles Studied by Minami (1983)

No.	Type	Length (ft)	Outside Diameter (in.)	Soil	Experimental Load Capacity (kips)
1	SP	67.3	31.5	Sand	1058
2	SP	88.6	28.0	Clay, sand	1234
3	SP	105.0	31.5	Clay, sand	727
4	SP	67.3	31.5	Sand	1102
5	SP	137.8	31.5	Sand	661
6	PC	53.3	19.7	Clay, sand	661
7	PC	100.1	23.6	Clay, sand	881
8	CP	88.6	39.4	Sand	1432
9	PC	106.6	17.7	Clay, sand	705
10	SP	111.6	18.0	Sand	639

*Type of pile
SP = Steel pipe pile, PC = Precast concrete pile, CP = Cast-in place pile

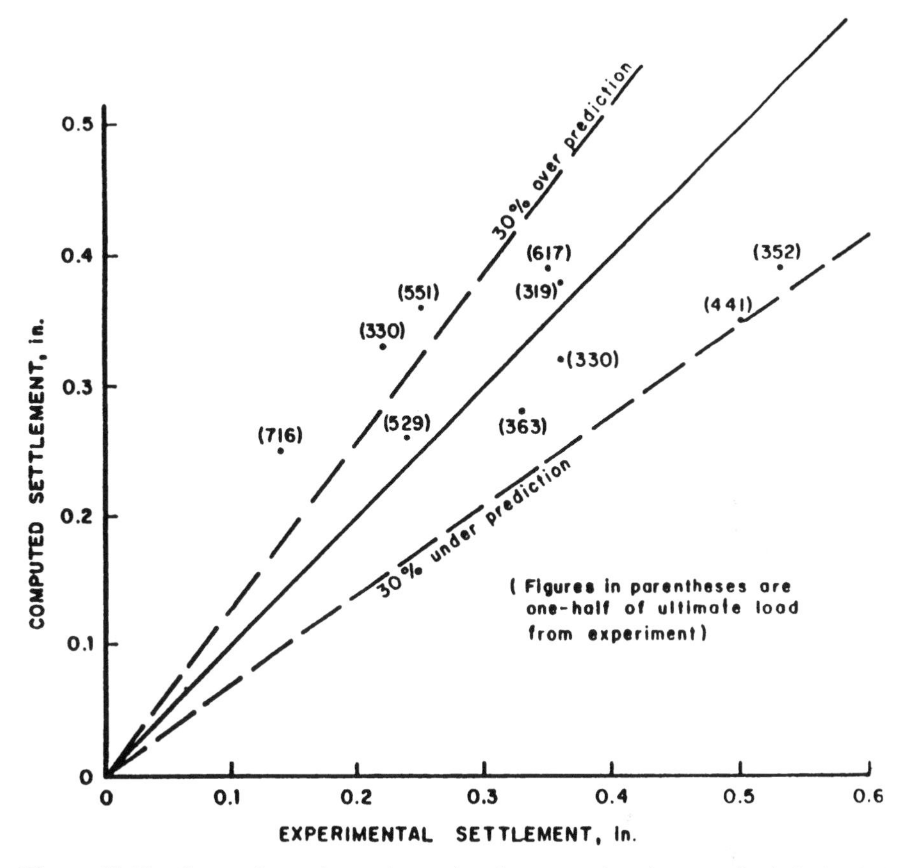

Figure 10.22 Comparison of experimental and computed settlement of axially loaded piles at one-half load capacity (after Minami, 1983).

The information presented here relates to short-term settlement. Settlement due to the consolidation of soft clay must be computed separately and added to the short-term settlement. Long-term settlement of cohesionless soil, perhaps due to vibration, must also be treated separately.

10.6 EXAMPLE PROBLEMS

This example is included to illustrate a common case for a bridge foundation. The soil deposit consists of two sublayers. The top layer is 5-m stiff clay underlain by a dense sand layer. The proposed pile for foundations is prestressed concrete pile with a 0.45-m (18-in.) outside diameter. The elastic modulus of the prestressed pile is 31,000,000 kPa. Each pile is required to be driven to 20 m below the ground surface. The soil properties interpreted from the geotechnical investigation report can be summarized as follows:

Layer I consists of tan to brown stiff clay; its thickness is about 5 m. Based on unconfined compressive tests conducted in the laboratory, the stiff clay has an unconfined-compressive strength of 95 kN/m^2. The unit weight of the soil is 18.9 kN/M^3. The PI for the clay is about 15, which indicates insignificant expansive properties.

Layer II consists of brown silty dense sand to sandy gravel; its total thickness is about 25 m. The fine to coarse sand has a blow counts ranging from 23 to 28. The estimated internal friction angle based on the blow counts is 36°. The unit weight of the soil is 19.1 kN/M^3. The water table is at the top of the sand layer.

10.6.1 Skin Friction

The general equation for skin friction is

$$Q_s = \Sigma f A_s \tag{10.39}$$

Skin Friction from Layer I—Stiff Clay (0 to 5 M)

API Method

$$fA_s \text{ (clay)} = \alpha c \pi B h \tag{10.40}$$

where

Avg. effective overburden pressure (p) is 18.9 * 2.5 = 47.3 kN/m^2,
$\alpha = (0.5)\ (\mathrm{c/p})^{-0.25} = (0.5)\ (95/47.3)^{-0.25} = 0.42$ (from Eq. 10.6)
$fA_s \text{ (clay)} = (0.42)\ (95) * \pi * (0.45) * (5) = 282$ kN

FHWA Method

$$\begin{aligned} fA_s\ (clay) &= \alpha c_u \pi B h \\ &= (0.72)(95)\pi(0.45)(5) \\ &= 483 \text{ kN} \end{aligned} \tag{10.41}$$

U.S. Army Corps Method

$$\begin{aligned} fA_s \text{ (clay)} &= \alpha c_u \pi B h \\ &= (0.5)(95)\pi(0.45)(5) \\ &= 335 \text{ kN} \end{aligned} \tag{10.42}$$

Skin Friction from Layer II—Dense Sand (5 to 20 M)

Avg. effective overburden pressure (p) from 0 to 5 m (18.9)(2.5) = 47.3 kN/m^2,

Avg. effective overburden pressure (p) from 5 to 10 m (18.9(5) + (9.1)(2.5) = 117.3 kN/m^2,

Avg. effective overburden pressure (p) from 10 to 15 m (18.9)(5) + (9.1)(7.5) = 162.8 kN/m^2, and

Avg. effective overburden pressure (p) from 15 to 20 m (18.9)(5) + (9.1)(12.5) = 208.3 kN/m^2.

API Method

$$fA_s\ (sand) = k_0 \sum (\gamma h)\tan(\phi - 5)\pi Bh \tag{10.43}$$

FHWA Method

ω (taper angle) = 0.0 (because of a uniform pile section)

$$fA_s\ (sand) = k_0 C_f \sum (\gamma h)\sin(\delta)\pi Bh \tag{10.44}$$

The average displaced volume per foot is $\pi(0.45)^2/4 = 0.159$ m^3. For a precast concrete pile with $V = 0.159$ m^3, $\delta = 0.9\phi$.

U.S. Army Corps Method

$$fA_s(\text{sand}) = k_0 \sum (\gamma h)\tan(\delta)\pi Bh \tag{10.45}$$

Skin friction in sand increases linearly to an assumed critical depth, D_c. $D_c = 20B = 20(0.45) = 9$ m for dense sand.

TABLE 10.8 Computation of Skin Friction from Each Sublayer Based on the API Method

(1) Depth Interval	(2) Total Area (πBh)	(3) Average Effective Stress	(4) $k_0 \tan(\phi - 5)$ $k_0 = 1$, $\phi = 36$	(5) $fA_s = (2)*(3)*(4)$
5–10 m	7.07 m^2	117.3 kN/m^2	0.6	497.6 kN
10–15 m	7.07 m^2	162.8 kN/m^2	0.6	690.6 kN
15–20 m	7.07 m^2	208.3 kN/m^2	0.6	883.6 kN
				$\Sigma fA_s = 2071.8$ kN

TABLE 10.9 Computation of Skin Friction from Each Sublayer Based on the FHWA Method

(1) Depth Interval	(2) Total Area (πBh)	(3) Average Effective Stress	(4) $k_0 C_f \sin(\delta)$ $k_0 = 2.1$, $\delta = 0.9\phi$, $C_f = 0.95$	(5) $fA_s = (2)*(3)*(4)$
5–10 m	7.07 m^2	117.3 kN/m^2	1.07	887.4 kN
10–15 m	7.07 m^2	162.8 kN/m^2	1.07	1231.6 kN
15–20 m	7.07 m^2	208.3 kN/m^2	1.07	1575.8 kN
				$\Sigma fA_s = 3694.8$ kN

Average effective overburden pressure p from 5 to 9 m (18.9)(5) + (9.1)(2) = 112.7 kN/m^2.

Effective overburden pressure p at 9 m (18.9)(5) + (9.1)(4) = 130.9 kN/m^2.

TABLE 10.10 Computation of Skin Friction from Each Sublayer Based on the U.S. Army Corps Method

(1) Depth Interval	(2) Total area (πBh)	(3) Average Effective Stress	(4) $k_0 \tan \delta$ $k_0 = 2$, $\delta = 0.9\phi$	(5) $fA_s = (2)*(3)*(4)$
5–9 m	5.66 m^2	112.7 kN/m^2	1.27	810.1 kN
9–20 m	15.55 m^2	Critical depth control 130.9 kN/m^2	1.27	2585.1kN
				$\Sigma fA_s = 3395.1$ kN

End Bearing

API Method. The general equation for end bearing is

$$Q_p = N_q \sigma_v A_p \tag{10.46}$$

where

$$A_p \text{ (tip area)} = (0.45)\,(0.45)\,(\pi/4) = 0.159 \text{ m}^2,$$
$$\sigma_v \text{ (effective stress at tip)} = (18.9)\,(5) + (9.1)\,(15) = 231 \text{ kN/m}^2, \text{ and}$$
$$N_q = 40.$$

$$Q_p = (40)(231)(0.159) = 1469$$

FHWA Method. The general equation for end bearing is

$$Q_p = \alpha N_q \sigma_v A_p \tag{10.47}$$

where

$$\begin{aligned} A_p \text{ (tip area)} &= (0.45)\ (0.45)\ (\pi/4) = 0.159 \text{ m}^2, \\ \sigma_v \text{ (effective stress at tip)} &= (18.9)\ (5) + (9.1)\ (15) = 231 \text{ kN/m}^2, \\ N_q &= 60, \\ \alpha &= 0.7, \text{ and} \end{aligned}$$

The limiting value from Meyerhof (1976) is 7258 kN/m^2.

$$Q_p = (0.7)\ (60)\ (231) = 9702 > 7258 \text{ kN/m}^2$$

Therefore, the limiting value should be used for the bearing capacity.

$$Q_p = (7258)(0.159) = 1154$$

U.S. Army Corps Method. The general equation for end bearing is

$$Q_p = N_q \sigma_v A_p \tag{10.48}$$

where

$$\begin{aligned} A_p \text{ (tip area)} &= (0.45)\ (0.45)\ (\pi/4) = 0.159 \text{ m}^2, \\ \sigma_v \text{ (effective stress at } D_c) &= (18.5)\ (5) + (9.1)\ (4) = 130.9 \text{ kN/m}^2, \text{ and} \\ N_q &= 45. \end{aligned}$$

$$Q_p = (45)\ (130.9)\ (0.159) = 936$$

Total Bearing Capacity

API Method. The total bearing capacity of the pile is given in Eq. 10.49. To determine the total bearing capacity of the pile, the value from the skin friction equations (Eqs. 10.40 and 10.43) and the end-bearing equation (Eq. 10.46) must be used in Eq. 10.49.

$$Q_{\text{total}} = Q_s + Q_p = (282 + 2{,}071) + 1{,}469 = 3{,}822 \text{ kN} \tag{10.49}$$

FHWA Method. The total bearing capacity of the pile is given in Eq. 10.50. To determine the total bearing capacity of the pile, the values from the skin friction equations (Eqs. 10.41 and 10.44) and the end-bearing equation (Eq. 10.47) must be used in Eq. 10.49.

$$Q_{\text{total}} = Q_s + Q_p = (483 + 3695) + 1154 = 5332 \text{ kN} \tag{10.50}$$

U.S. Army Corps Method. The total bearing capacity of the pile is given in Eq. 10.51. To determine the total bearing capacity of the pile, the value from the skin friction equations (Eqs. 10.42 and 10.45) and the end-bearing equation (Eq. 10.48) must be used in Eq. 10.49.

$$Q_{\text{total}} = Q_s + Q_p = (335 + 3395) + 936 = 4666 \text{ kN} \qquad (10.51)$$

The values of pile capacity shown above, computed by three different methods, vary significantly. The engineer may select an average value, or the lowest value, or make a decision based on the perceived validity of the three methods. Recommending a field load test may be justified for a large job.

10.7 ANALYSIS OF PILE DRIVING

10.7.1 Introduction

Ever since engineers began using piles to support structures, attempts have been made to find rational methods for determining the load a pile can carry. Methods for predicting capacities were proposed using dynamic data obtained during driving. The only realistic measurement that could be obtained during driving was pile set (blow count); thus, concepts equating the energy delivered by the hammer to the work done by the pile as it penetrates the soil were used to obtain pile-capacity expressions, known as *pile formulas.*

Up to the late 19th century, the design of piling was based totally on prior experience and rule-of-thumb criteria. One of the first attempts at theoretical evaluation of pile capacity was published in *Engineering News* under "Piles and Pile Driving," edited by Wellington in 1893. This approach is still known today as the *Engineering News pile-driving formula.* Since Wellington proposed the well-known Engineering News Formula in 1893, a number of other attempts have been made to predict the capacity of piles, and many reports of field experience have become available. Because of their simplicity, pile formulas have been widely used for many years. However, statistical comparisons with load tests have shown poor correlations and wide scatter. As a result, except where well-supported empirical correlations under a given set of physical and geologic conditions are available, pile formulas should be used with caution.

Another approach to modeling a pile during driving involves the use of the wave-equation method. The application of wave-equation analysis in pile driving has been of great interest to engineers for many years. Not only does this method provides better information on load capacity than other dynamic methods, but the wave equation will yield stresses in the pile during driving. The wave-equation method is a semitheoretical method whose reliability in predicting bearing capacity and driving stress depends primarily on the accuracy of the parameters defining pile and soil properties. However, the wave-equation method has been compared with a number of results of actual field tests performed throughout the world, and the results show that this method is rational and reliable.

The following sections provide a brief review of the conventional dynamic formulas and discuss the problems associated with them. The wave-equation method will also be described briefly.

10.7.2 Dynamic Formulas

The simplest form of dynamic formulas in use today is based on equating the energy used to the work done when a pile is moved a certain distance (set) against soil resistance. Various methods have been used to evaluate the many variables that could affect the results of a dynamic formula. In general, there is no indication that the more elaborate formulas are more reliable or that any one dynamic formula is equally applicable to all types of piles under all soil conditions. The driving formulas were all derived using grossly simplified assumptions.

Pile-driving formulas are based on the assumption that the bearing capacity of a driven pile is a direct function of the energy delivered to it during the last blows of the driving process, and that the energy from the hammer to the pile and soil is transmitted instantaneously on impact. These two assumptions have been proven wrong by many investigators. Researchers have clearly demonstrated that the bearing capacity of a pile is related less to the total hammer energy (per blow) than to the distribution of this energy with time during the driving process. Data on pile driving using wave-propagation theory indicate that time effects related to the propagation of the impact forces in the pile govern the behavior of piles during driving. The lack of confidence in dynamic formulas is demonstrated by the fact that the safety factors applied to determine the allowable loads are always very large. For example, a nominal value of 6 is used for the safety factor in the Engineering News Formula.

Most dynamic formulas yield the ultimate pile capacity rather than the allowable or design capacity. Thus, the specified design load should be multiplied by a factor of safety to obtain the ultimate pile capacity that is inputted in to the formula to determine the required *set* (amount of pile penetration per blow).

The equations used in two of the dynamic formulas are shown below. If the formula is in terms of allowable (design) pile capacity, then a factor of safety is already included in the formula and the application of an additional safety factor is not required. Equations 10.51 to 10.53 show several variations of the commonly used Engineering News Formula. Any consistent set of units may be used for formulas unless noted otherwise.

Engineering News Record:

$$P_u = \frac{W_r \cdot h}{s + 1.0} \quad \text{for drop hammers (} h \text{ and } s \text{ both in ft)} \tag{10.52}$$

$$P_u = \frac{W_r \cdot h}{s + 0.1} \quad \text{for steam hammers (} h \text{ and } s \text{ both in ft)} \tag{10.53}$$

$$P_a = \frac{2 \cdot W_r \cdot h}{s + 0.1} \quad \text{for steam hammers (} h \text{ in ft, } s \text{ in inches)} \tag{10.54}$$

AASHTO (1983)

$$P_a = \frac{2 \cdot h \cdot (W_r + A_r \cdot P)}{s + 0.1} \quad \text{for double-acting steam hammers (} h \text{ in ft, } s \text{ in inches)} \tag{10.55}$$

A factor of safety of 6 is included in Eqs. 10.54 and 10.55. The following is a list of symbols used for the pile formulas.

A_r = ram cross-sectional area,
h = height of ram fall,
P_u = ultimate pile capacity,
P_a = allowable pile capacity (with a factor of safety = 6),
p = steam (or air) pressure,
s = amount of point penetration per blow, and
W_r = weight of ram (for double-acting hammers, includes weight of casing).

10.7.3 Reasons for the Problems with Dynamic Formulas

The problems associated with pile-driving formulas can be related to each component within the pile-driving process: the driving system, the soil, and the pile. An investigation of these parameters indicates three reasons why the formulas lack validity.

First, the derivation of most formulas is not based on a realistic treatment of the driving system because they fail to consider the variability of equipment performance. Typical driving systems can include many elements in addition to the ram, such as the helmet (drive cap), cap block (hammer cushion), pile cushion, and anvil. These components affect the distribution of the hammer energy with time, both at and after impact, which influences the magnitude and duration of peak force. The peak force determines the ability of the hammer system to advance piles into the soil.

Second, the soil resistance is very crudely treated by assuming that it is a constant force. This assumption neglects even the most obvious characteristics of real soil performance. Dynamic soil resistance, or resistance of the soil to rapid pile penetration produced by a hammer blow, is by no means identical to the static load required to produce very slow penetration of the pile point. Rapid penetration of the pile point into the soil is resisted not only by static friction and cohesion, but also by soil viscosity, which is comparable to the viscous resistance of liquids against rapid displacement under an applied force. The net effect is that the driving process creates resistance forces along the pile shaft and at the pile tip, which are substantially increased because of the high shear rate. The viscous effect tends to produce an apparent increase in soil resistance, which is evident only under dynamic conditions.

Third, the pile is assumed to be rigid, and its length is not considered. This assumption completely neglects the flexibility of the pile, which reduces its ability to penetrate soil. The energy delivered by the hammer sets up time-dependent stresses and displacements in the pile-head assembly, in the pile, and in the surrounding soil. In addition, the pile behaves not as a concentrated mass, but as a long elastic bar in which stresses travel longitudinally as waves. Compressive waves that travel to the pile tip are responsible for advancing the pile into the ground.

10.7.4 Dynamic Analysis by the Wave Equation

As discussed in the previous section, dynamic formulas do not yield acceptably accurate predictions of actual pile capacities and provide no information on stresses in the piles. The one-dimensional wave-equation analysis has eliminated many shortcomings associated with pile formulas by providing a more realistic analysis of the pile-driving process. The use of the wave-equation method for predicting the behavior of driven piles was employed in 1938 (Glanville et al.). A model was proposed by Smith (1960) as shown in Figure 10.23, and has been discussed and utilized by a number of writers (Samson et al., 1963; Goble et al., 1972; Holloway and Dover, 1978). It appears to describe reasonably well the behavior of the pile during driving if soil parameters are properly selected. The selection of soil parameters has advanced only slightly since the time of Smith's proposal. The existence of residual stresses in a pile after its installation can be explained by the wave-equation model.

Wave equation is a term applied to several computer programs, including TTI by Hirsch et al. (1976), WEAP86 by Goble and Rausche (1986), and GRLWEAP by Pile Dynamics, Inc. (2005), that use a finite-difference method to model dynamic pile behavior. The wave-equation analysis uses wave-propagation theory to model the longitudinal wave transmitted along the pile axis when it is struck by a hammer. As the ram impact occurs, a force pulse is developed in the pile that travels downward toward the pile tip at a constant velocity that depends on the properties of the pile material. When the force pulse reaches the portion of the pile that is embedded, it is attenuated by frictional resistance from the soil along the pile. If the attenuation is incomplete, the force pulse will reach the pile tip and a reflected force pulse, which

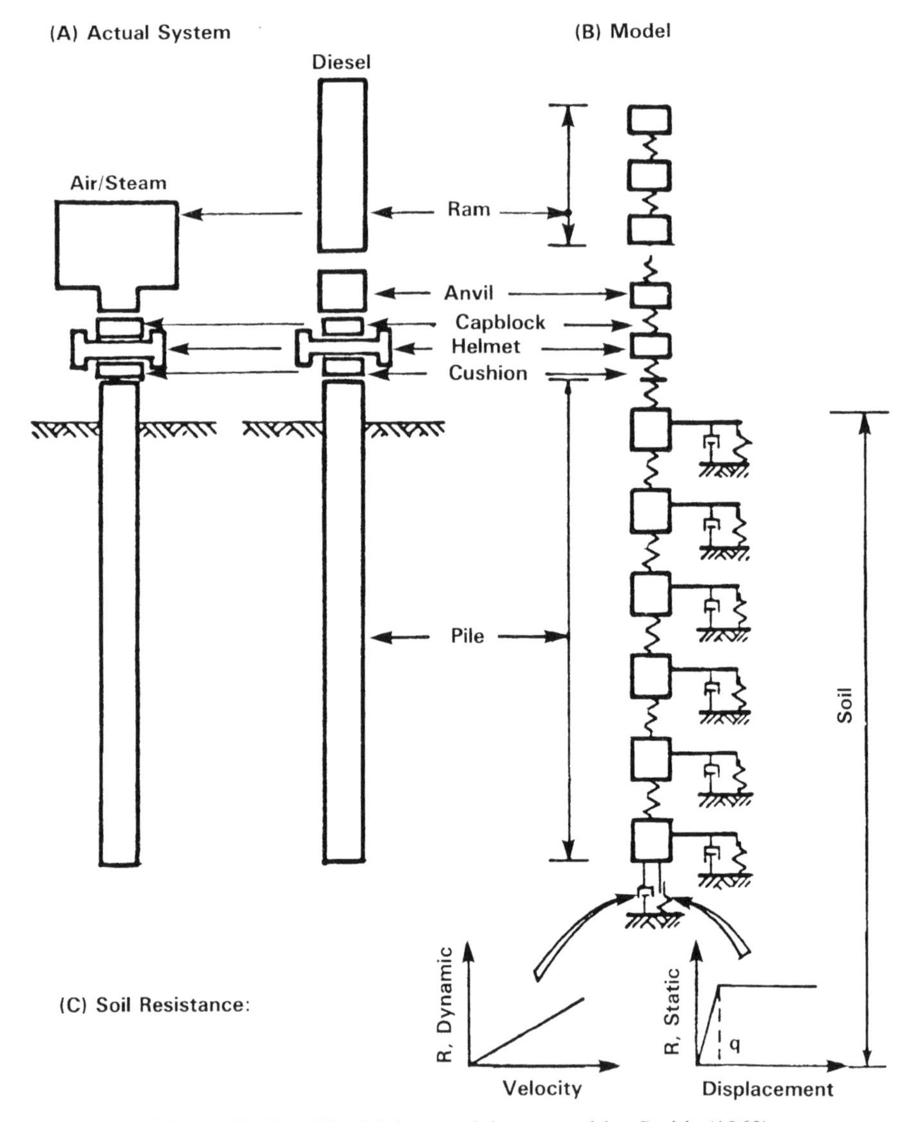

Figure 10.23 Pile-driving model proposed by Smith (1960).

is governed by the tip resistance, is generated. The pile will penetrate into the soil when the peak force generated by the ram impact exceeds the ultimate soil resistance at the pile tip and along the pile. The wave-equation analysis provides two types of information:

1. It provides information on the relationship between pile capacity and driving resistance. The user inputs data on soil side shear and end bearing, and the analysis provides an estimate of the set (inches/blow) under

one blow of the specified hammer. By specifying a range of ultimate pile capacities, the user obtains a relationship between ultimate pile capacity and penetration resistance (blows per inch or blows per foot), as shown in Figure 10.24.

2. The analysis also provides relationships between driving stresses in the pile for a specific soil and penetration resistance, as shown in Figure 10.25.

In summary, the wave-equation analysis enables the user to develop curves of capacity versus blow count for different pile lengths. This information may be used in the field to determine when the pile has been driven sufficiently to develop the required capacity. The analysis is used to (1) select the right combination of driving equipment to ensure that the piles can be driven to the required depth and capacity; (2) design the minimum required pile section for driving; (3) minimize the chances of overstressing the pile; and (4) minimize driving costs.

10.7.5 Effects of Pile Driving

The first and most obvious effect of installing a pile is the displacement of a volume of soil nearly equal to the volume of the pile. While some speculation has occurred about the patterns of such displacement as a pile is progressively driven into the soil, there is no accepted solution that would allow plotting of displacement contours. The volume changes cause distortion of clay and changes in the structure of clay, and usually result in loss of strength of clay. The volume changes during pile driving result in lateral deformations and uplift, and lateral movement can occur in piles previously driven.

If an open-ended-pipe pile is driven into clay, a soil plug may form during the pile driving after a certain pile penetration and may be carried down with the pile as the pile is driven farther. If the plug does not form, the pile will behave as a nondisplacement pile, with heave and lateral displacement being minimized. Pile driving in clay causes increases in total pressure and porewater pressure above the at-rest pressures in the vicinity of the pile wall as the pile tip passes a point. Such increases have been observed in the soil at several diameters away from the pile, with the effect being analogous to the increase in temperature of a slab due to the imposition of a line source of heat. Observations indicate that a significant dynamic effect occurs in the porewater continuum during impact, with sudden and sizable increases in these quantities occurring as a result of a hammer blow. A porewater-pressure gauge was embedded in the soil about 15 diameters away from the point where a 6-in. (15-cm) pile was being driven. There was a sudden jump in the pressure-gauge reading with each blow of a drop hammer.

An obvious effect of pile driving is a shearing deformation of large magnitude as the pile wall moves past a particular point in the soil. At present, it is not clear whether all of the shearing deformation occurs at the interface

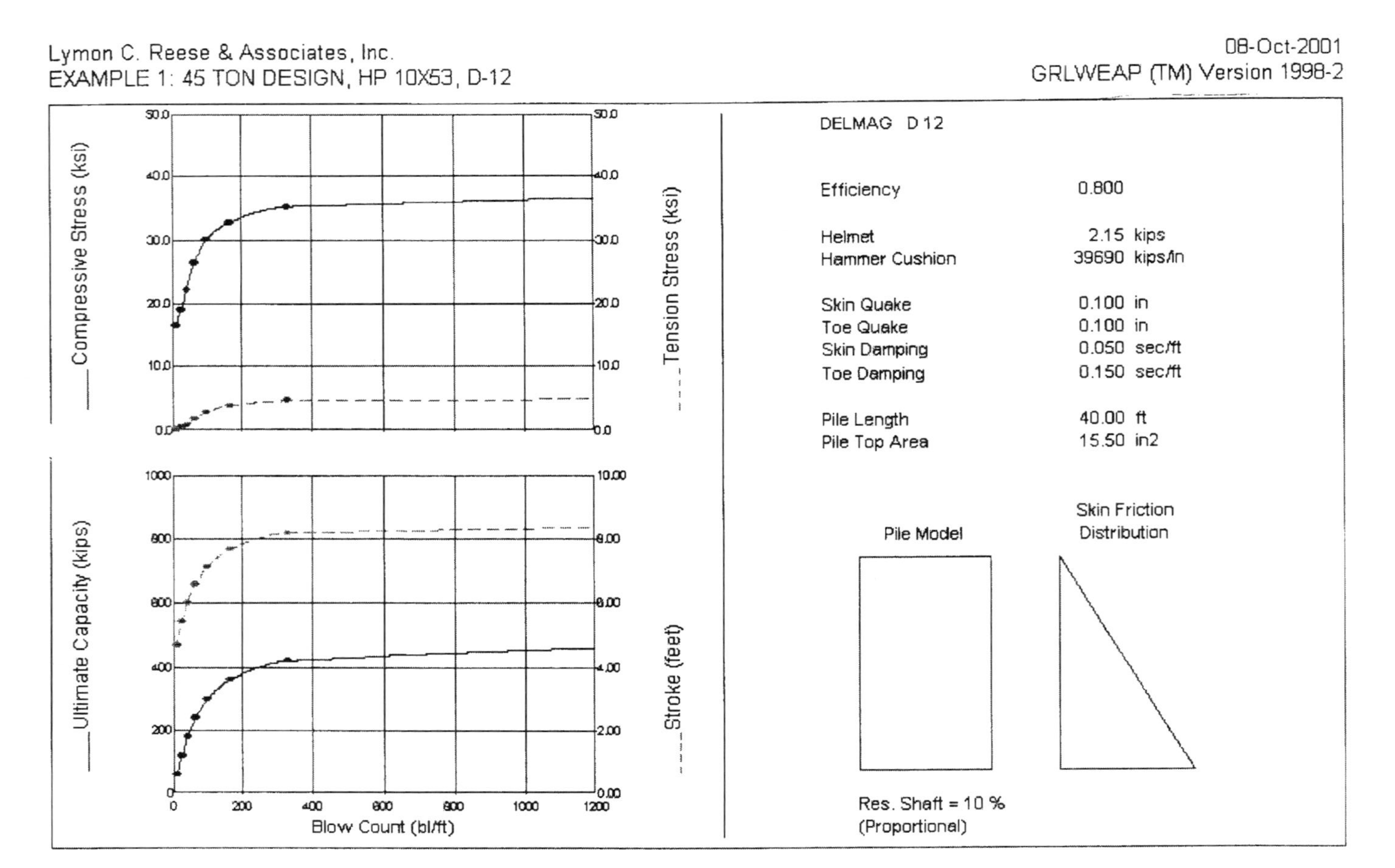

Figure 10.24 Relationship between pile capacity and driving resistance.

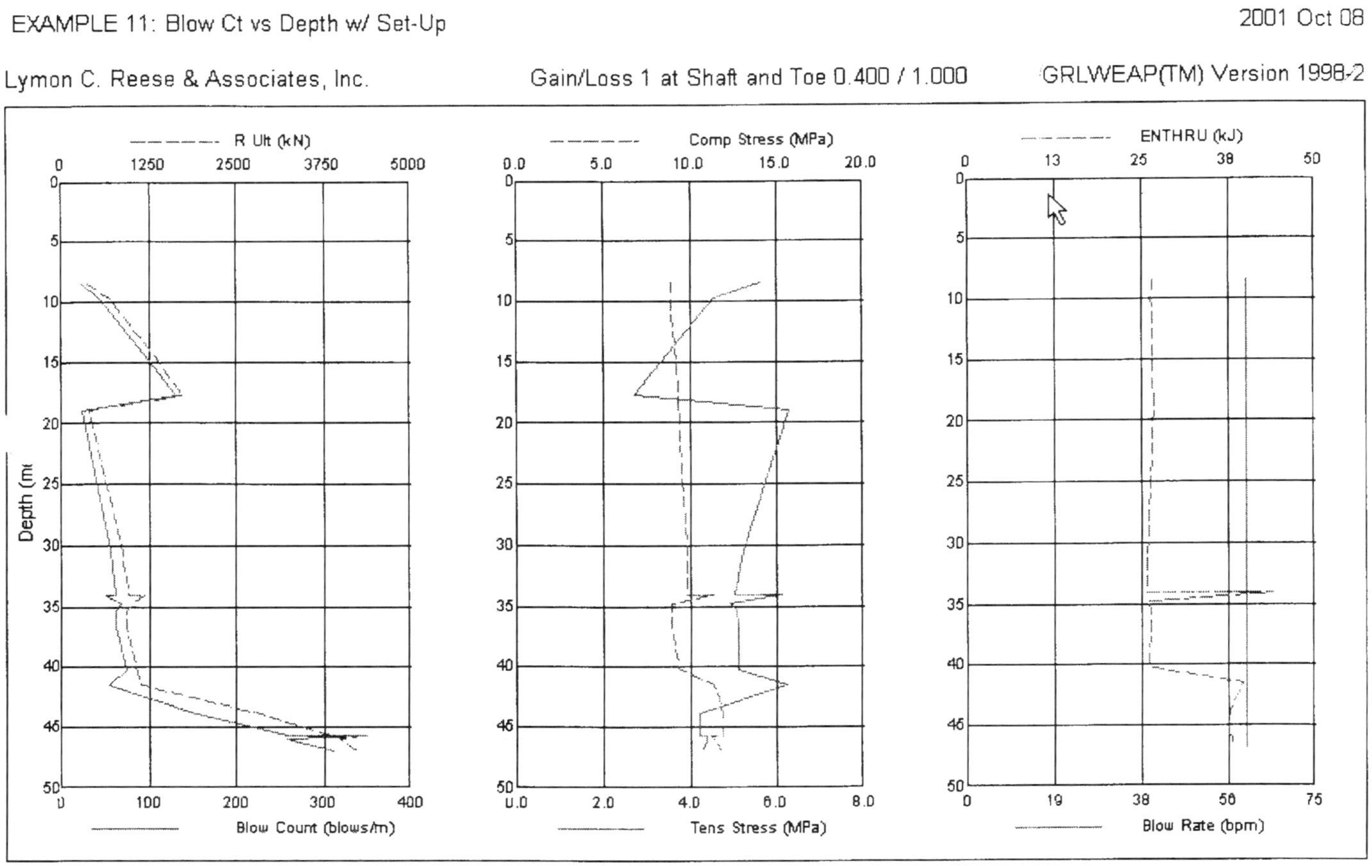

EXAMPLE 11: Blow Ct vs Depth w/ Set-Up 2001 Oct 08

Lymon C. Reese & Associates, Inc. Gain/Loss 1 at Shaft and Toe 0.400 / 1.000 GRLWEAP(TM) Version 1998-2

R Ult (kN) — Blow Count (blows/m) — Depth (m)

Comp Stress (MPa) — Tens Stress (MPa)

ENTHRU (kJ) — Blow Rate (bpm)

Figure 10.25 Relationship between driving stress and pile penetration.

between the pile and the soil or whether some soil moves down with the pile. In regard to downward movement of an upper stratum of soil, Tomlinson (1980) reported that driving a pile causes a significant amount of soil in an upper stratum to move into the stratum just below. At the completion of driving there will be a stress in the soil that is indeterminate, probably complex, and completely different from the in situ state of stress. The state of stress immediately after driving will reflect the residual stresses in the pile that exist on completion of driving.

10.7.6 Effects of Time After Pile Driving with No Load

Cohesive Soil Because of the time required for excess porewater stress to dissipate in soils with low permeability, there are significant effects over time on piles driven into saturated clays. An increase in axial capacity occurs over time for such piles. Engineers have long been aware that the piles "set up" over time and that this effect can be quite dramatic. There are reports of pile-driving crews stopping for lunch with a pile moving readily; when they began driving after lunch, the pile had set up so that is refused to drive farther. Dr. Karl Terzaghi was called to a location where it was impossible to drive piles. The piles were timber and were designed to carry a relatively light load. The caller reported that it was difficult to drive the piles to grade, that they floated up after each hammer blow, and that their bearing capacity was quite small. Terzaghi asked that the piles be driven to grade and deliberately delayed going to the site. After finally arriving at the site, he directed load tests to be performed. It was found that the piles were performing satisfactorily.

The state of stress of clay immediately after pile driving is such that there are excess porewater pressures that are undoubtedly at a maximum in the vicinity of the pile wall, but measurements have shown that these excess porewater pressures extend several diameters, and perhaps more, away from the pile wall. The excess porewater pressures decay relatively rapidly where gradients are high, with the decay being an exponential function of time. Along with the decay of excess porewater pressures is movement of water in the clay, with evidence indicating that its flow is predominantly horizontal. The horizontal movement of water leads to a decrease in water content at the pile wall; measurements have indicated that this change in water content is significant. Along with the decrease in water content, an increase in the shearing strength of the clay occurs. The process is complex and not well understood because it seems to indicate an exchange between equal volumes of solids and water. Part of the strength increase may be due to the thixotrophy of the clay.

Along with the movement of water due to the excess porewater pressures created by pile driving, there is consolidation of the clay due to its own weight. The remolding of the clay due to pile driving changes its consolidation characteristics so that the soil consolidates under its own weight, with the result that the ground surface tends to move down. If the ground surface moves down, negative skin friction tends to develop at the pile wall, with a consequent change in the residual stresses that existed immediately after pile driving.

Cohesionless Soil Driving a pile into sand or cohesionless soil, with the resulting vibration, causes the sand deposit to decrease in volume sufficiently to allow the pile to penetrate. The volume decrease is frequently such that the ground surface is lowered, particularly in the vicinity of the pile. While there are few experimental data, there is a strong indication that the lateral vibration of a pile driven into sand results in arching. A particle of sand will move outward and downward. Lateral vibration creates a space that is slightly larger than the pile, and arching reduces the lateral stress against the pile wall. Therefore, the soil stresses against the wall of a pile in sand are usually much lower than passive pressures and can be lower than the earth pressure at rest.

If the sand is so dense that volume change cannot take place, the driving resistance becomes extremely high and crushing of the sand grains will result. The above points apply principally to a sand composed of quartz or feldspar; a pile driven into calcareous sand will crush the sand, and lateral earth pressures against the pile wall can be quite low, especially if the calcareous sand is cemented. It is usually assumed that an open-ended-pipe pile driven into quartz sand will plug; this may not be the case for calcareous sand. As with clay, driving a pile into cohesionless soil will cause shearing deformation of a large magnitude at the pile wall; sand above a clay can be moved downward; there will be residual stresses in the driven pile; and there will be a complex state of stress in the soil surrounding the pile after the pile has been driven.

PROBLEMS

P10.1. A steel pile has an outer diameter (OD) of 1.0 m and an inner diameter (ID) of 0.86 m. The subsurface condition consists mainly of cohesionless soils, and the pile was driven to a depth of 33.3 m. The total unit weight of soils is 18 kN/m^3 and the internal friction angle is 32° at the ground surface and linearly increases to 38° at a depth of 40 m. The water table is 10 m below the ground surface. Compute the ultimate axial capacity for this pile.

P10.2. A contractor plans to install a 0.5-m-OD steel pile embedded in cohesive soils. The pile is an open-ended steel pipe, and a plug of soil may be forced up the inside of the pile. The soil resistance from the plug consists of the friction from the plug inside the pile and the bearing capacity from the steel metal area only. The soil resistance from the plug is compared to the resistance from the end bearing over the full area of the base, and the smaller of the two values is used for tip resistance. Use both API method and Lambda methods for computations.

Pile and Soil Data

Pile length: = 39 m
Pile diameter: OD = 0.50 m
ID = 0.48 m

Soil Information

Depth (m)	Soil Type	γ' kN/m^3	C kN/m^2
0–4	Clay	19	9.8
4–10	Clay	9	9.8
10–20	Clay	9	19.6
20–36	Clay	9	58.8
36–40	Clay	9	78.4

CHAPTER 11

GEOTECHNICAL DESIGN OF DRILLED SHAFTS UNDER AXIAL LOADING

11.1 INTRODUCTION

The methods used to analyze and design drilled shafts under axial loading have evolved since the 1960s, when drilled shafts came into wide use. The design methods recommended for use today reflect the evolution of construction practices developed since that time.

11.2 PRESENTATION OF THE FHWA DESIGN PROCEDURE

11.2.1 Introduction

This chapter presents methods for computation of the capacity of drilled shafts under axial loading. The methods for computation of axial capacity were developed by O'Neill and Reese (1999).

The methods of analysis assume that excellent construction procedures have been employed. It is further assume that the excavation remained stable, and was completed with the proper dimensions, that the rebar was placed properly, that a high-slump concrete was used, that the concrete was placed in a correct manner, that the concrete was placed within 4 hours of the time that the excavation was completed, and that any slurry that was used was properly conditioned before the concrete was placed. Much additional information on construction methods is given in O'Neill and Reese (1999). Another FHWA publication (LCPC, 1986), translated from French, also gives much useful information.

While the design methods presented here have proved to be useful, they are not perfect. Research continues on the performance of drilled shafts, and improved methods for design are expected in the future. An appropriate factor of safety must be employed to determine a safe working load. The engineer may elect to employ a factor of safety that will lead to a conservative assessment of capacity if the job is small. A load test to develop design parameters or to prove the design is strongly recommended for a job of any significance.

11.3 STRENGTH AND SERVICEABILITY REQUIREMENTS

11.3.1 General Requirements

Two methods are available for determining the factored moments, shears, and thrusts for designing structures by the strength design method: the single-load factor method and a method based on the American Concrete Institute Building Code (ACI 318).

In addition to satisfying strength and serviceability requirements, many structures must satisfy stability requirements under various loading and foundation conditions.

11.3.2 Stability Analysis

The stability analysis of structures such as retaining walls must be performed using unfactored loads. The unfactored loads and the resulting reactions are then used to determine the unfactored moments, shears, and thrusts at critical sections of the structure. The unfactored moments, shears, and thrusts are then multiplied by the appropriate load factors, and the hydraulic load factor when appropriate, to determine the required strengths used to establish the required section properties.

The single-load factor method must be used when the loads on the structural component being analyzed include reactions from a soil–structure interaction stability analysis, such as footings for walls. For simplicity and ease of application, this method should generally be used for all elements of such structures. The load factor method based on the ACI 318 Building Code may be used for some elements of the structure, but must be used with caution to ensure that the load combinations do not produce unconservative results.

11.3.3 Strength Requirements

Strength requirements are based on loads resulting from dead and live loads, hydraulic loading, and seismic loading.

11.4 DESIGN CRITERIA

11.4.1 Applicability and Deviations

The design criteria for drilled shafts generally follow the recommendations for structures founded on driven piles for loading conditions and design factors of safety.

11.4.2 Loading Conditions

Loading conditions are generally divided into cases of usual, unusual, and extreme conditions.

11.4.3 Allowable Stresses

No current design standard contains limitations on allowable stresses in concrete or steel used in drilled shafts. However, the recommendations of ACI 318-02, Section A.3 may be used as a guide for design. These recommendations are summarized in Table 11.1.

In cases where a drilled shaft is fully embedded in clays, silts, or sands, the structural capacity of the drilled shaft is usually limited by the allowable moment capacity. In the case of drilled shafts socketed into rock, the structural capacity of the drilled shaft may be limited by the allowable stress in the shaft. Maximum shear force usually occurs below the top of rock.

11.5 GENERAL COMPUTATIONS FOR AXIAL CAPACITY OF INDIVIDUAL DRILLED SHAFTS

The axial capacity of drilled shafts should be computed by engineers who are thoroughly familiar with the limitations of construction methods, any special requirements for design, and the soil conditions at the site.

TABLE 11.1 Permissible Service Load Stresses Recommended by ACI 318-95, Section A.3

Case	Stress Level*
Flexure—extreme fiber stress in compression	$0.45\ f'_c$
Shear—shear carried by concrete	$1.1\sqrt{f'_c}$
Shear—maximum shear carried by concrete plus shear reinforcement	$v_c + 4.4\sqrt{f'_c}$
Tensile stress—Grade 40 or 50 reinforcement	20,000 psi
Tensile stress—Grade 60 reinforcement	24,000 psi

*v_c = permissible shear stress carried by concrete, psi, f'_c = compressive strength of concrete, psi.

The axial capacity of a drilled shaft may be computed by the following formulas:

$$Q_{\text{ult}} = Q_s + Q_b \tag{11.1}$$

$$Q_s = f_s A_s \tag{11.2}$$

$$Q_b = q_{\max} A_b \tag{11.3}$$

where

Q_{ult} = axial capacity of the drilled shaft,
Q_s = axial capacity in skin friction,
Q_b = axial capacity in end bearing,
f_s = average unit side resistance,
A_s = surface area of the shaft in contact with the soil along the side of the shaft,
$q_{\max}$ = unit end-bearing capacity, and
A_b = area of the base of the shaft in contact with the soil.

11.6 DESIGN EQUATIONS FOR AXIAL CAPACITY IN COMPRESSION AND IN UPLIFT

11.6.1 Description of Soil and Rock for Axial Capacity Computations

The following six subsections present the design equations for axial capacity in compression and in uplift. The first five subsections present the design equations in side resistance and end bearing for clay, sand, clay shales, gravels, and rock. The last section presents a discussion of a statistical study of the performance of the design equations in clays and sands.

O'Neill and Reese (1999) introduced the descriptive terms used in this book to describe soil and rock for axial capacity computations. Collectively, all types of soil and rock are described as *geomaterials*. The basic distinction between soil and rock types is the characteristic of cohesiveness. All soil and rock types include the descriptive terms *cohesive* or *cohesionless*.

11.6.2 Design for Axial Capacity in Cohesive Soils

Side Resistance in Cohesive Soils The basic method used for computing the unit load transfer in side resistance (i.e., in skin friction) for drilled shafts in cohesive soils is the *alpha* (α) *method*. The profile of undrained shear strength c_u of the clay versus depth is found from unconsolidated-undrained triaxial tests, and the following equation is employed to compute the ultimate value of unit load transfer at a depth z below the ground surface:

$$f_s = \alpha \, c_u \tag{11.4}$$

where

f_s = ultimate load transfer in side resistance at depth z,
c_u = undrained shear strength at depth z, and
α = empirical factor that can vary with the magnitude of undrained shear strength, which varies with depth z.

The value of α includes the effects of sampling disturbance on the soil's shear strength, migration of water from the fluid concrete into the soil, and other factors that might affect axial capacity in side resistance.

The total load Q_s in side resistance is computed by multiplying the unit side resistance by the peripheral area of the shaft. This quantity is expressed by

$$Q_s = \int_{z_{top}}^{z_{bot}} f_{sz} dA \tag{11.5}$$

where

dA = differential area of perimeter along sides of drilled shaft over penetration depth,
L = penetration of drilled shaft below ground surface,
z_{top} = depth to top of zone considered for side resistance, and
z_{bot} = depth to bottom of zone considered for side resistance.

The peripheral areas over which side resistance in clay is computed are shown in Figure 11.1. The upper portion of the shaft is excluded for both compression and uplift to account for soil shrinkage in the zone of seasonal moisture change. In areas where the depth of seasonal moisture change is greater than 5 ft (1.5 m) or when substantial groundline deflection results from lateral loading, the upper exclusion zone should be extended to deeper depths. The lower portion of the shaft is excluded when the shaft is loaded in compression because downward movement of the base will generate tensile stresses in the soil that will be relieved by cracking of soil, and porewater suction will be relieved by inward movement of groundwater. If a shaft is loaded in uplift, the exclusion of the lower zone for straight-sided shafts should not be used because these effects do not occur during uplift loading.

The value of α is the same for loading in both compression and uplift.

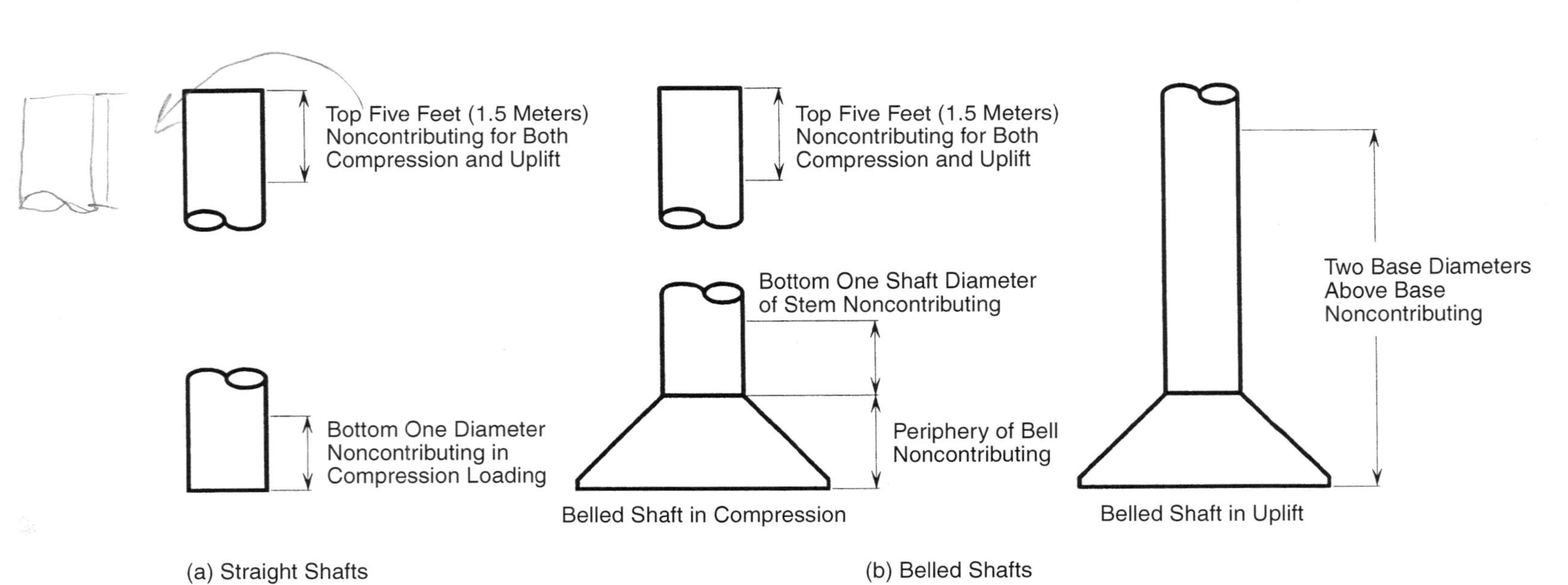

Figure 11.1 Portions of drilled shaft not considered in computing side resistance in clay.

If the shaft is constructed with an enlarged base (also called an *underream* or a *bell*), the exclusion zones for side resistance at the bottom of the shaft differ for loading in compression and uplift, as shown in Figure 11.1b. If the shaft is loaded in compression, the lower exclusion zone includes the upper surface of the bell and the lower one diameter of the shaft above the bell. When a belled shaft is loaded in uplift, the lower exclusion zone for side resistance extends two base diameters above the base. If the lower exclusion zone overlaps the upper exclusion zone, then no side resistance is considered in the computations of axial capacity in uplift.

Equation 11.4 indicates that the unit load transfer in side resistance at depth z is a product of α and undrained shear strength at depth z. Research on the results of load tests of instrumented drilled shafts has found that α is not a constant and that it varies with the magnitude of the undrained shear strength of cohesive soils. O'Neill and Reese (1999) recommend using

$$\alpha = 0.55 \quad \text{for} \quad \frac{c_u}{p_a} \leq 1.5 \tag{11.6a}$$

and

$$\alpha = 0.55 - 0.1\left(\frac{c_u}{p_a} - 1.5\right) \quad \text{for} \quad 1.5 \leq \frac{c_u}{p_a} \leq 2.5 \tag{11.6b}$$

where p_a = atmospheric pressure = 101.3 kPa = 2116 psf. (Note that 1.5 p_a = 152 kPa or 3,170 psf and 2.5 p_a = 253 kPa or 5290 psf.)

For cases where c_u/p_a exceeds 2.5, side resistance should be computed using the methods for cohesive intermediate geomaterials, discussed later in this chapter.

Some experimental measurements of α obtained from load tests are shown in Figure 11.2. The relationship for α as a function of c_u/p_a defined by Eq. 11.6 is also shown in this figure. For values of c_u/p_a greater than 2.5, side resistance should be computed using the recommendations for cohesive intermediate geomaterials, discussed later in this chapter.

When an excavation is open prior to the placement of concrete, the lateral effective stresses at the sides of the drilled hole are zero if the excavation is drilled in the dry or small if there is drilling fluid in the excavation. Lateral stresses will then be imposed on the sides of the excavation because of the fluid pressure of the fresh concrete. At the ground surface, the lateral stresses from the concrete will be zero or close to zero. It can be expected that the lateral stress from the concrete will increase almost linearly with depth, assuming that the concrete has a relatively high slump. Experiments conducted by Bernal and Reese (1983) found that the assumption of a linear increase in the lateral stress from fluid concrete for depths of concrete of 3.0 m (10 ft) or more is correct. For depths greater than 3.0 m, the lateral stress is strongly

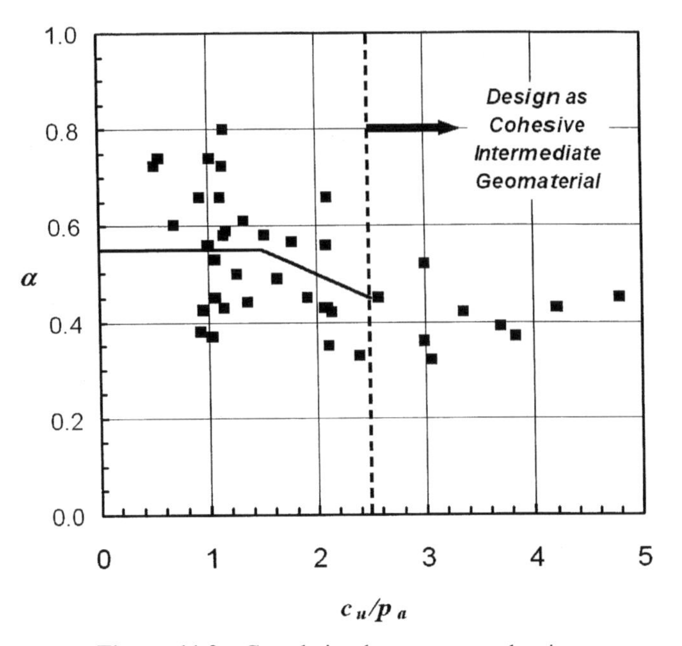

Figure 11.2 Correlation between α and c_u/p_a.

dependent on the character of the fresh concrete. From available experimental evidence, it follows that a rational recommendation for α is that α should vary perhaps linearly with depth, starting at zero at the groundline, to its ultimate value at some critical depth below the groundline. However, insufficient data are available for making such a detailed recommendation. The recommendations for design in Eq. 11.6 generally lead to a reasonable correlation between experimental measurements and computed results.

With regard to the depth over which α is assumed to be zero, consideration must be given to those cases where there are seasonal changes in the moisture content of the soil. It is conceivable, perhaps likely, that clay near the ground surface will shrink away from the drilled shaft so that the load transfer is reduced to zero in dry weather over the full depth of the seasonal moisture change. There may also be other instances where the engineer may wish to deviate from the recommendations of Eq. 11.6 because of special circumstances at a particular site. A drilled shaft subjected to a large lateral load is an example of such a circumstance; if the lateral deflection at the groundline is enough to open a gap between the shaft and soil, the portion of the drilled shaft above the first point of zero deflection should be excluded from side resistance.

End Bearing in Cohesive Soils The computation of load transfer in end bearing for deep foundations in clays is subject to much less uncertainty than is the computation of load transfer in side resistance. Skempton (1951) and other investigators have developed consistent formulas for the computation of end bearing. In addition, the accuracy of Skempton's work has been confirmed by results from instrumented drilled shafts where general base failure was observed. Equation 11.7 is employed for computing the ultimate unit end bearing q_{max} for drilled shafts in saturated clay:

$$q_{max} = N_c^* c_u \tag{11.7}$$

where c_u is an average undrained shear strength of the clay computed over a depth of two diameters below the base, but judgment must be used if the shear strength varies strongly with depth.

The bearing capacity factor N_c^* is computed using

$$N_c^* = 1.33(\ln|I_r| + 1) \tag{11.8}$$

where I_r is the rigidity index of the soil, which for a saturated, undrained material ($\phi = 0$) soil is expressed by

$$I_r = \frac{E_s}{3c_u} \tag{11.9}$$

where E_s is Young's modulus of the soil undrained loading. E_s should be measured in laboratory triaxial tests or in-situ by pressuremeter tests to apply the above equations. For cases in which measurements of E_s are not available, I_r can be estimated by interpolation using Table 11.2.

When L/B is less than 3.0, q_{max} should be reduced to account for the effect of the presence of the ground surface by using

$$q_{max} = 0.667\left[1 + 0.1667\frac{L}{B}\right] N_c^* c_u \tag{11.10}$$

where:

TABLE 11.2 I_r and N_c^* Values for Cohesive Soil

c_u	I_r	N_c^*
24 kPa (500 psf)	50	6.55
48 kPa (1000 psf)	150	8.01
96 kPa (2000 psf)	250	8.69
192 kPa (4000 psf)	300	8.94

L = length of the shaft, and
B = diameter of the base of the shaft.

Uplift Resistance of Straight-Sided Shafts in Cohesive Soil Base resistance for uplift loading on straight-sided shafts should be assumed to be zero unless confirmed by load testing.

Uplift Resistance of Belled Shafts in Cohesive Soil Unit base resistance for belled shafts can be computed from

$$q_{\max}\ (\text{uplift}) = s_u N_u \qquad (11.11)$$

where

N_u = bearing capacity factor for uplift, and
s_u = average undrained shear strength between the base of the bell and 2 B_b above the base.

N_u can be computed from

$$N_u = 3.5 \frac{D_b}{B_b} \leq 9 \quad \text{for unfissured clay} \qquad (11.12)$$

or from

$$N_u = 0.7 \frac{D_b}{B_b} \leq 9 \quad \text{for fissured clay} \qquad (11.13)$$

In the expressions above, D_b is the depth of the base of the bell below the top of the soil stratum that contains the bell, but not counting any depth within the zone of seasonal moisture change.

The unit uplift resistance should be applied over the projected area of the bell, A_u. The projected area is computed from

$$A_u = \frac{\pi}{4}(B_b^2 - B^2) \qquad (11.14)$$

Example Problem 1—Shaft in Cohesive Soil This is an example of a shaft drilled into clay. It is based on a case history, referred to as "Pile A," reported by Whitaker and Cooke (1966).

Soil Profile. The soil profile is shown in Figure 11.3. The clay is overconsolidated. The depth to the water table was not given and is not needed in making capacity calculations. However, the range of depth of the water table

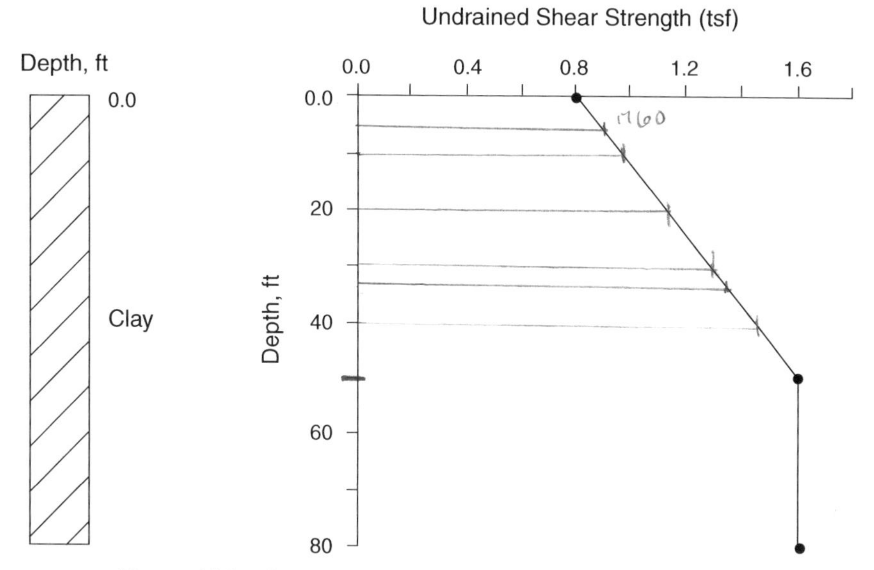

Figure 11.3 General soil description of Example Problem 1.

should be determined and always reported in the construction documents for construction considerations.

Soil Properties. Values of undrained shear strength obtained from laboratory tests are included in Figure 11.3.

Construction. High-quality construction, good specifications, and excellent inspection are assumed.

Loadings. The working axial load is 230 tons. No downdrag acting on the shaft is expected, and vertical movement of the soil due to expansive clay is not a problem. Effects due to lateral loading are also thought to be negligible. The depth to the zone of seasonal moisture change is judged to be about 10 ft.

Factor of Safety. It is assumed that a load test has been performed in the area, that the design parameters have been proven, and that the soil conditions across the site are relatively uniform; therefore, an overall factor of safety of 2 was selected.

Ultimate Load. Using a factor of safety of 2, the ultimate axial load was computed to be 460 tons.

Geometry of the Drilled Shaft. An underreamed shaft was designed to penetrate a total of 40 ft into the clay. The height of the bell is 4 ft, making the length of the straight-sided portion 36 ft. The diameter of the straight-sided portion of the shaft is 2.58 ft, and the diameter of the bell is 5.5 ft.

Computations

SIDE RESISTANCE. For ease of hand computations, a constant value of α_z equal to 0.55 and an average c_u of 2280 psf are assumed. However, the computer program interpolates linearly the top and bottom values of c_u with depth. The hand computations are as follows:

Depth Interval, ft	ΔA, ft^2	Avg. Effective Stress, tsf	α_Z	ΔQ_s, tons
0–5			0	0
5–33.4	230.4	1.14	0.55	144.4
33.4–40			0	0
				Q_s = 144.4

BASE RESISTANCE. The average undrained shear strength over two base diameters below the base is 1.48 tsf, and the area of the base is 23.76 ft^2.

By interpolation between the values of N_c^* shown in Table 11.2, N_c^* = 8.81.

$$q_{max} = N_c^* c_u = (8.81)\ (1.48\ \text{tsf}) = 13.04\ \text{tsf}$$
$$A_b = 23.76\ \text{ft}^2$$
$$Q_b = (12.92\ \text{tsf})\ (23.76\ \text{ft}^2) = 309.8\ \text{tons}$$

TOTAL RESISTANCE

$$Q_T = 144.3 + 307 = 454\ \text{tons}$$

11.6.3 Design for Axial Capacity in Cohesionless Soils

Side Resistance in Cohesionless Soils The shear strength of sands and other cohesionless soils is characterized by an angle of internal friction that ranges from about 30° up, depending on the kinds of grains and their packing. The cohesion of such soils is assumed to be zero. The friction angle at the interface between the concrete and the soil may be different from that of the soil itself. The unit side resistance, as the drilled shaft is pushed downward, is equal to the normal effective stress at the interface times the tangent of the interface friction angle.

Unsupported shaft excavations are prone to collapse, so excavations in cohesionless soil are made using drilling slurry to stabilize the borehole, and the normal stress at the face of the completed drilled shaft depends on the construction method. The fluid stress from the fresh concrete will impose a normal stress that is dependent on the characteristics of the concrete. Experiments have found that concrete with a moderate slump (up to 6 in., 150 mm) acts hydrostatically over a depth of 10 to 15 ft (3 to 4.6 m) and that there is a leveling off in the lateral stress at greater depths, probably due to arching (Bernal and Reese, 1983). Concrete with a high slump (about 9 in., 230 mm) acts hydrostatically to a depth of 32 ft (10 m) or more. Thus, the construction procedures and the nature of the concrete will have a strong influence on the magnitude of the lateral stress at the concrete–soil interface. Furthermore, the angle of internal friction of the soil near the interface will be affected by the details of construction.

In view of the above discussion, the method of computing the unit load transfer in side resistance must depend on the results from field experiments as well as on theory. The following equations are recommended for design. The form of the equations is based on theory, but the values of the parameters that are suggested for design are based principally on the results of field experiments.

$$f_{sz} = K\sigma'_z \tan \phi_c \tag{11.15}$$

$$Q_s = \int_0^L K\sigma'_z \tan \phi_c \, dA \tag{11.16}$$

where

f_{sz} = ultimate unit side resistance in sand at depth z,
K = a parameter that combines the lateral pressure coefficient and a correlation factor,
σ'_z = vertical effective stress in soil at depth z,
ϕ_c = friction angle at the interface of concrete and soil,
L = depth of embedment of the drilled shaft, and
dA = differential area of the perimeter along sides of drilled shaft over the penetration depth.

Equations 11.15 and 11.16 can be used in the computations of side resistance in sand, but simpler expressions can be developed if the terms for K and $\tan \phi_c$ are combined. The resulting expressions are shown in Eqs. 11.17 through 11.20.

$$f_{sz} = \beta \, \sigma'_z \leq 2.1 \text{ tsf} \quad (200 \text{ kPa}) \tag{11.17}$$

$$Q_s = \int_0^L \beta\sigma'_z \, dA \tag{11.18}$$

$$\beta = 1.5 - 0.135\sqrt{z\ (\text{ft})} \quad \text{or}$$
$$\beta = 1.5 - 0.245\sqrt{z\ (\text{m})}; \quad 0.25 \leq \beta \leq 1.20 \tag{11.19a}$$

where z = depth below the ground surface, in feet or meters, as indicated.

When the uncorrected SPT resistance, N_{60}, is less than or equal to 15 blows/ft, β is computed using

$$\beta = \frac{N_{60}}{15}\left(1.5 - 0.135\sqrt{z\ (\text{ft})}\right) \quad \text{or}$$
$$\beta = \frac{N_{60}}{15}\left(1.5 - .245\sqrt{z\ (\text{m})}\right) \quad \text{for } N_{60} \leq 15 \tag{11.19b}$$

Note that for sands

$$\beta = 0.25 \quad \text{when} \quad z > 85.73 \text{ ft or } 26.14 \text{ m} \tag{11.20}$$

For very gravelly sands or gravels

$$\beta = 2.0 - 0.0615\ [z\ (\text{ft})]^{0.75} \quad \text{or}$$
$$\beta = 2.0 - 0.15\ [z\ (\text{m})]^{0.75}; \quad 0.25 \leq \beta \leq 1.8 \tag{11.21}$$

For gravelly sands or gravels

$$\beta = 0.25 \quad \text{when} \quad z > 86.61 \text{ ft} \quad \text{or} \quad 26.46 \text{ m} \tag{11.22}$$

The design equations for drilled shafts in sand use SPT N_{60}-values uncorrected for overburden stress. The majority of the load tests on which the design equations are based were performed in the Texas Gulf Coast region and the Los Angeles Basin in California. The N_{60}-values for these load tests were obtained using donut hammers with a rope-and-pulley hammer release system. If a designer has N-values that were measured with other systems or were corrected for level of overburden stress and rod energy, it will be necessary to adjust the corrected N-values to the uncorrected N_{60} form for donut hammers with rope-and-pulley hammer release systems before use in the design expressions of Eq. 11.19b and Table 11.4. Guidance for methods used to correct SPT penetration resistances is presented in Chapter 3 of EM 1110-1-1905.

The parameter β combines the influence of the coefficient of lateral earth pressure and the tangent of the friction angle. The parameter also takes into account the fact that the stress at the interface due to the fluid pressure of the concrete may be greater than that from the soil itself. In connection with the lateral stress at the interface of the soil and the concrete, the assumption implicit in Eq. 11.17 is that good construction procedures are employed. See Chapter 5 for further construction information. Among other factors, the slump of the concrete should be 6 in. or more and drilling slurry, if employed, should not cause a weak layer of bentonite to develop at the wall of the excavation. The reader is referred to O'Neill and Reese (1999) for further details on methods of construction.

The limiting value of side resistance shown in Eq. 11.17 is not a theoretical limit but is the largest value that has been measured (Owens and Reese, 1982). Use of higher values can be justified by results from a load test.

A comparison of β values computed from Eq. 11.19 and β values derived from loading tests in sand on fully instrumented drilled shafts is presented in Figure 11.4. As can be seen, the recommended expression for β yields values that are in reasonable agreement with experimental values.

Equation 11.17 has been employed in computations of f_{sz}, and the results are shown in Figure 11.5. Three values of β were selected; two of these are in the range of values of β for submerged sand, and the third is an approximate value of β for dry sand. The curves are cut off at a depth below 60 ft (18 m) because only a small amount of data has been gathered from instrumented drilled shafts in sand with deep penetrations. Field load tests are indicated if drilled shafts in sand are to be built with penetrations of over 70 ft (21 m).

It can be argued that Eqs. 11.17 and 11.18 are too elementary and that the angle of internal friction, for example, should be treated explicitly. However, the drilling process has an influence on in situ shearing properties, so the true friction angle at the interface cannot be determined from a field investigation that was conducted before the shaft was constructed. Furthermore, Eqs. 11.17 and 11.18 appear to yield an satisfactory correlation with results from full-scale load tests.

The comparisons of results from computations with those from experiments, using the above equations for sand, show that virtually every computed value is conservative (i.e., the computed capacity is less than the experimentally measured capacity). However, it is of interest that most of the tests in sand are at locations where the sand was somewhat cemented. Therefore, caution should be observed in using the design equations for sand if the sand is clean, loose, and uncemented.

Either Eq. 11.15 or Eq. 11.17 can be used to compute the side resistance in sand. The angle of internal friction of the sand is generally used in Eq. 11.15 in place of the friction angle interface at the interface of the concrete and soil if no information is available. In some cases, only SPT resistance data are available. In such cases, the engineer can convert the SPT penetration resistance to the equivalent internal friction angle by using Table 11.3 as a guide.

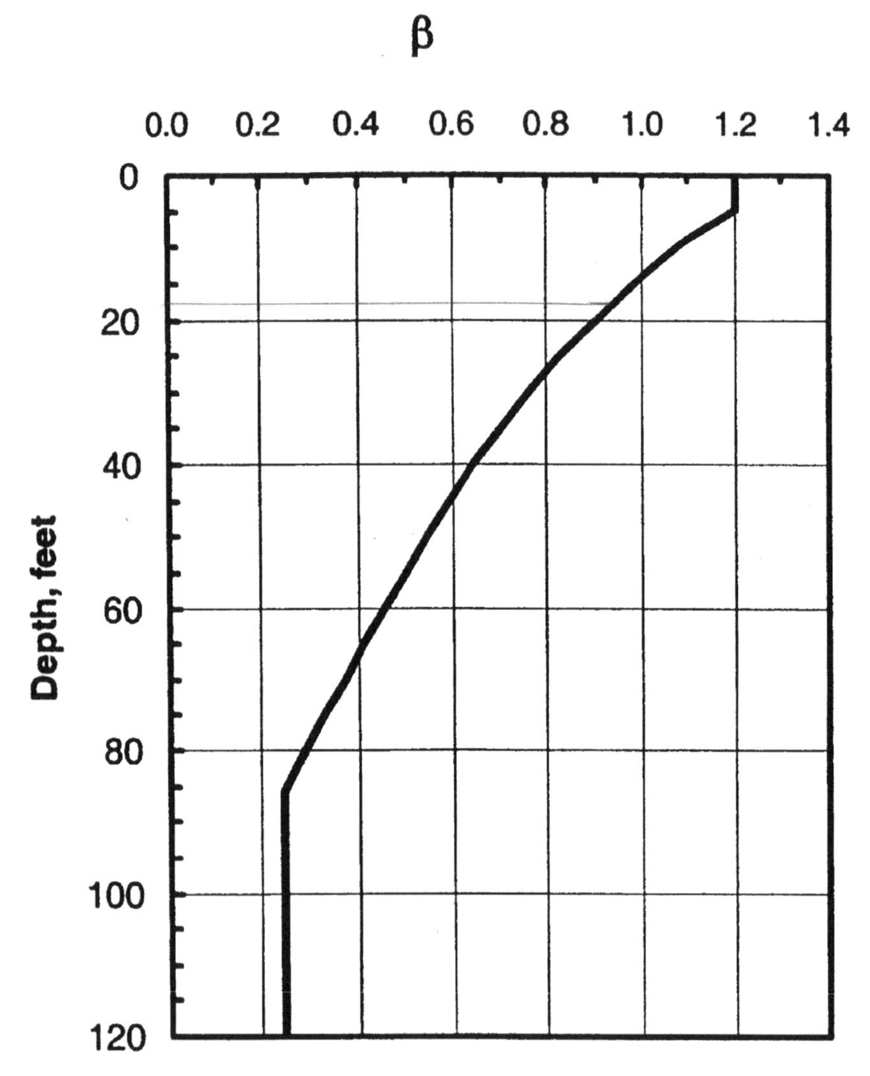

Figure 11.4 Plot of experimental values of β.

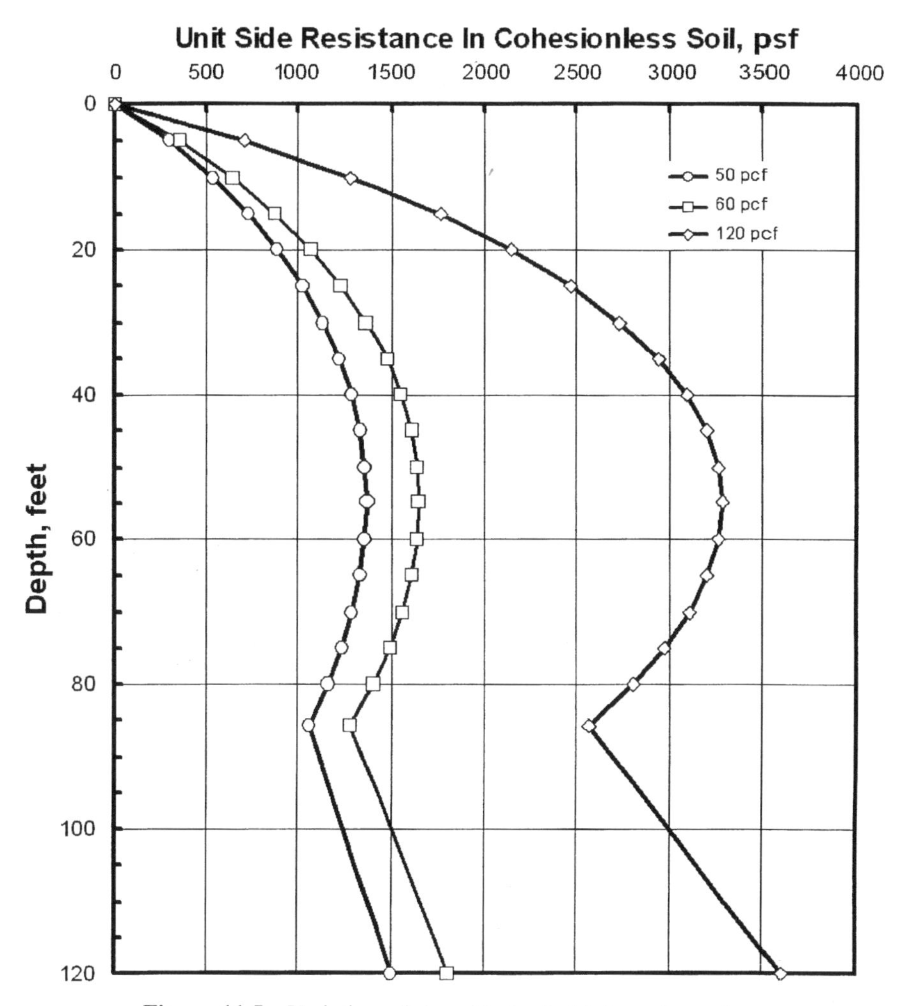

Figure 11.5 Variation of f_{sz} with depth (z) for values of γ.

Side Resistance in Cohesionless Soils—Uplift Loading For uplift loading in granular materials, the side resistance developed under drained, uplift conditions is reduced due to the Poisson effect, lowering the normal stress at the side of the shaft. Based on analytical modeling and centrifuge research reported by de Nicola (1996), side resistance in uplift may be computed using

$$f_{max}\ (\text{uplift}) = \Psi\ f_{max}\ (\text{compression}) \tag{11.23}$$

The factor Ψ is computed using

TABLE 11.3 Relationship Between N, ϕ, and D_r

N	ϕ, deg.	D_r, %	ϕ, deg.	D_r, %	ϕ, deg.	D_r, %
2	32	45				
4	34	55				
6	36	65	30	37		
10	38	75	32	46	31	40
15	42	90	34	57	32	48
20	45	100	36	65	34	55
25			37	72	35	60
30			39	77	36	65
35			40	82	36	67
40			41	86	37	72
45			42	90	38	75
50			44	95	39	77
55			45	100	39	80
60					40	83
65					41	86
70					42	90
75					42	92
80					43	95
85					44	97
90					44	99

Source: After Gibbs and Holtz (1957).

$$\eta = \upsilon_p \tan \delta \left(\frac{L}{B}\right)\left(\frac{G_{\text{avg}}}{E_p}\right) \tag{11.24}$$

$$\Psi = \left\{1 - 0.2 \log_{10}\left|\frac{100}{(L/B)}\right|\right\}(1 - 8\eta + 25\eta^2) \tag{11.25}$$

where

E_p = Young's modulus of the shaft, and
G_{avg} = average shear modulus of the soil along the length of the shaft, estimated as the average Young's modulus of soil divided by 2.6.

O'Neill and Reese (1999) examined Eq. 11.25 and concluded that typical values of Ψ fall into the range 0.74 to 0.85 for L/B ratios of 5 to 20. They noted that Eq. 11.25 appears to overestimate Ψ for L/B ratios larger than 20. The value Ψ can be taken conservatively to be 0.75 for design purposes.

End Bearing in Cohesionless Soils Because of the relief of stress when an excavation is drilled into sand, the sand tends to loosen slightly at the bottom of the excavation. Also, there appears to be some densification of the sand beneath the base of a drilled shaft as settlement occurs. A similar phenomenon has also been observed in model tests on spread footings founded in sand by Vesić (1973). The load-settlement curves for the base of drilled shafts that have been obtained from load tests on instrumented test shafts are consistent with these concepts. In many load tests, the base load continued to increase to a settlement of more that 15% of the diameter of the base. Such a large settlement cannot be tolerated for most structures; therefore, it was decided to formulate the design equations for end bearing to limit values of end bearing for drilled shafts in granular soil to those that are expected to occur at a downward movement of 5% of the diameter of the base (O'Neill and Reese, 1999).

Values of q_b are tabulated as a function of N_{60} (standardized for hammer energy but uncorrected for overburden stress) in Table 11.4. The computation of tip capacity is based directly on the penetration resistance from the SPT near the tip of the drilled shaft.

The values in Table 11.4 can be expressed in equation form as follows:

$$\text{If } L \geq 10 \text{ m:} \quad q_b = 57.5\, N_{\text{SPT}} \leq 2.9 \text{ MPa} \tag{11.26}$$

$$\text{If } L < 10 \text{ m:} \quad q_b = \frac{L}{10 \text{ m}} 57.5\, N_{SPT} \leq \frac{L}{10 \text{ m}} 2.9 \text{ MPa} \tag{11.27}$$

or in U.S. Customary Units

$$\text{If } L \geq 32.8 \text{ ft:} \quad q_b = 0.60\, N_{SPT} \leq 30 \text{ tsf} \tag{11.28}$$

$$\text{If } L < 32.8 \text{ ft:} \quad q_b = \frac{L}{32.8 \text{ ft}} 0.60\, N_{SPT} \leq \frac{L}{32.8 \text{ ft}} 30 \text{ tsf} \tag{11.29}$$

where L is the shaft length in meters or feet as required.

These values are similar to those recommended by Quiros and Reese (1977). These authors recommended no unit end bearing for loose sand

TABLE 11.4 Recommended Values of Unit End Bearing for Shafts in Cohesionless Soils with Settlements Less Than 5% of the Base Diameter

Range of Value on N_{60}	Value of q_b, tons/ft^2	Value of q_b, MPa
0 to 50 blows/ft	0.60 N_{60}	0.0575 N_{60}
Upper limit	30 tsf	2.9 MPa

Note: For shafts with penetrations of less than 10 base diameters, it is recommended that q_b be varied linearly from zero at the groundline to the value computed at 10 diameters using Table 11.4.

($\phi \leq 30°$), a value of 16 tons/ft^2 (1.53 MPa) for medium-dense sand ($30° < \phi \leq 36°$), and a value of 40 tons/ft^2 (3.83 MPa) for very dense sand ($36° < \phi \leq 41°$).

Neither of the sets of recommendations involves the stress in the soil outside the tip of the drilled shaft. This concept is consistent with the work of Meyerhof (1976) and others. Furthermore, the values in Table 11.4 are based predominantly on experimental results for shaft settlements of less than 5% of the base diameter where the drilled shafts had various penetrations. However, implicit in the values of q_b that are given is that the penetration of the drilled shaft must be at least 10 diameters below the ground surface. For penetration of less than 10 diameters, it is recommended that q_b be varied linearly from zero at the groundline to the value computed at 10 diameters using Table 11.4.

Table 11.4 sets the limiting value of load transfer in end bearing at 30 tsf (2.9 MPa) at a settlement equal to 5% of the base diameter. However, higher unit end-bearing values are routinely used when validated by load tests. For example, a value of 58 tsf (5.6 MPa) was measured at a settlement of 4% of the diameter of the base at a site in Florida (Owens and Reese, 1982).

Example Problem 2—Shaft in Cohesionless Soil This is an example of a shaft drilled into sand. The example has been studied in the referenced literature of Reese and O'Neill (1988). It is also modeled after load tests in sand (Owens and Reese, 1982).

Soil Profile. The soil profile is shown in Figure 11.6. The water table is at a depth of 4 ft below the ground surface.

Soil Properties. N_{60}-values (blow counts per foot) from the SPT are included in Figure 11.6.

Construction. High-quality construction is assumed. The contractor will have all the required equipment in good order, and experienced personnel will be on the job.

Loadings. The working axial load is 170 tons, the lateral load is negligible, and no downdrag is expected.

Factor of Safety. It is assumed that a load test has been performed nearby, but considering the possible variation in the soil properties over the site and other factors, an overall factor of safety of 2.5 is selected. The diameter will be sufficiently small so that reduced end bearing will not be required. Consequently, the global factor of safety can be applied to both components of resistance.

Ultimate Load. The ultimate axial load is thus established as 2.5×170 tons $= 425$ tons, since a global factor of safety (of 2.5) is used.

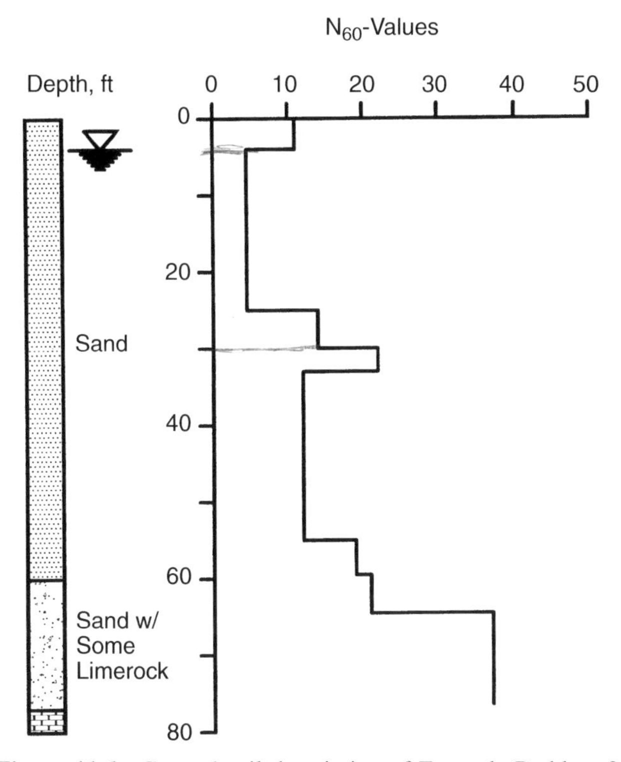

Figure 11.6 General soil description of Example Problem 2.

Geometry of the Drilled Shaft. A straight-sided drilled shaft is selected with a diameter of 3 ft and a penetration of 60 ft.

Computations

SIDE RESISTANCE. Computations are performed assuming a total unit weight of sand equal to 115 pcf. The hand computations are as follows:

Depths, ft	ΔA, ft^2	Avg. Effective Stress, tsf	β	ΔQ_s, tons
0–4	37.7	0.115	1.200	5.2
4–30	245.0	0.572	0.943	132.1
30–60	282.7	1.308	0.594	219.7
				Q_s = 357.0 tons

BASE RESISTANCE. Computations for base resistance are performed using the soil at the base of the shaft. At the 60-ft location, $N_{SPT} = 21$.

$$q_b = 0.60 N_{SPT} \leq 30 \text{ tsf}$$
$$q_b = (0.6)(21) = 12.6 \text{ tsf}$$

$$A_b = 7.07 \text{ ft}^2$$
$$Q_b = (7.07 \text{ ft}^2)\ (12.6 \text{ tsf}) = 89.1 \text{ tons}$$

TOTAL RESISTANCE

$$Q_T = 357 + 89.1 = 446.1 \text{ tons}$$

Example Problem 3—Shaft in Mixed Profile This is an example of a shaft drilled into a soil of a mixed profile with layers of clay and sand. It is modeled after load tests performed and reported by Touma and Reese (1972) at the G1 site.

Soil Profile. The soil profile is shown in Figure 11.7. The water table is at a depth of 17 ft below the ground surface.

Soil Properties. Values of undrained shear strength obtained from laboratory tests and N_{60}-values (blow counts per foot) from the SPT are included in Figure 11.7.

Construction. High-quality construction, good specifications, and excellent inspection are assumed.

Loadings. The working axial load is 150 tons, no downdrag is expected, and lateral loading is negligible. The depth to the zone of seasonal moisture change is judged to be about 10 ft.

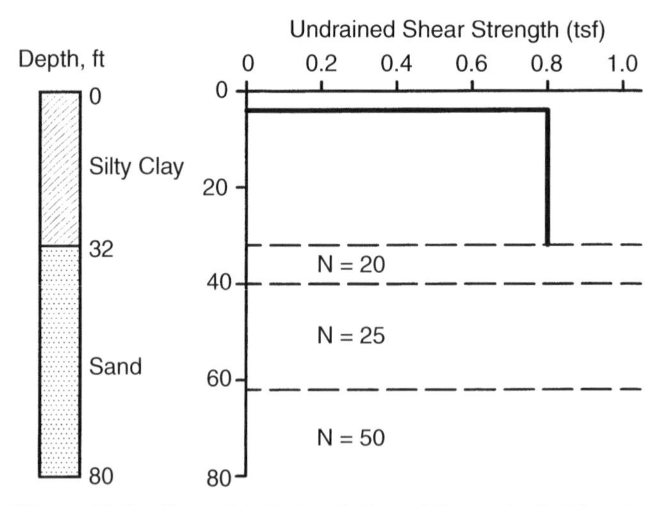

Figure 11.7 General soil description of Example Problem 3.

Factor of Safety. Soil conditions across the site are variable, and the foundation is for a major and complex structure. An overall factor of safety of 3 was selected.

Ultimate Load. Using the factor of safety of 3, the ultimate axial load is computed to be 450 tons.

Geometry of the Drilled Shaft. A straight-sided shaft is selected, with a diameter of 3 ft and a penetration of 59 ft.

Computations

SIDE RESISTANCE. Computations are performed assuming a total unit weight of clay equal to 125 pcf and a total unit weight of sand equal to 115 pcf. For ease of hand computations, an average value of β was selected for the sand layer. The computations are as follows:

Soil Type	Depth Interval, ft	ΔA, ft^2	Avg. c_u or Effective Stress, tsf	α_Z or β	ΔQ_s, tons
Clay	0–5	—	(Cased)	0	0
Clay	5–32	254.5	0.81	0.55	113.4
Sand	32–59	254.5	1.887	0.589	282.9
					Q_s = 396.3 tons

BASE RESISTANCE. Computations for base resistance are performed using the soil at the base of the shaft. At the 59-ft location, $N_{SPT} = 25$.

$$q_b = (0.6)\,(25 \text{ tsf}) = 15.0 \text{ tsf}$$
$$A_b = 7.07 \text{ ft}^2$$
$$Q_b = (7.07 \text{ ft}^2)\,(15.0 \text{ tsf}) = 106.0 \text{ tons}$$

TOTAL RESISTANCE

$$Q_T = 396.3 + 106.0 = 502.3 \text{ tons}$$

11.6.4 Design for Axial Capacity in Cohesive Intermediate Geomaterials and Jointed Rock

Side Resistance in Cohesive Intermediate Geomaterials *Weak rock* is the term for materials that some authors call *intermediate geomaterials.* In general, the soil resistances and settlements computed by these criteria are considered appropriate for weak rock with compressive strength in the range of 0.5 to 5.0 MPa (73 to 725 psi). The following intermediate geomaterials usually fall within this category: argillaceous geomaterials (such as heavily overconsolidated clay, hard-cohesive soil, and weak rock such as claystones) or calcareous rock (limestone and limerock, within the specified values of compressive strength).

Drilled shafts are attractive as a reliable foundation system for use in intermediate geomaterials. These geomaterials are not difficult to excavate, and provide good stability and excellent capacity.

Two procedures for computation of side resistance in cohesive intermediate geomaterials are presented by O'Neill and Reese (1999). One procedure is a simplified version of a more detailed procedure developed by O'Neill et al. (1996). Both procedures are presented in the following sections.

Simplified Procedure for Side Resistance in Cohesive Intermediate Geomaterial. The first decision to be made by the designer is whether to classify the borehole as smooth or rough. A borehole can be classified as rough only if artificial means are used to roughen its sides and to remove any smeared material from its sides. If conditions are otherwise, the borehole must be classified as smooth for purposes of design.

The term *smooth* refers to a condition in which the borehole is cut naturally with the drilling tool without leaving smeared material on the sides of the borehole wall. To be classified as rough, a borehole must have keys cut into its wall that are at least 76 mm (3 in.) high, 51 mm (2 in.) deep, and spaced vertically every 0.46 m (1.5 ft) along the depth of shafts that are at least 0.61 m (24 in.) in diameter.

The peak side resistance for a smooth borehole in Layer i is computed using

$$f_{\max,i} = \alpha \varphi q_{u,i} \tag{11.30}$$

where α (not equal in value to the α for cohesive soils) is obtained from Figure 11.8. The terms in the figure are defined as follows. E_m is Young's modulus of the rock (i.e., the rock mass modulus), q_u is the unconfined strength of the intact material, and w_t is the settlement at the top of the rock socket at which α is developed. The rock mass modulus can be estimated from measurements of Young's modulus of intact rock cores using Table 11.6. The curves in Figure 11.8 are based on the assumption that the interface friction angle between the rock and concrete is 30°. If the interface friction is different from 30°, it should be modified using

$$\alpha = \alpha_{\phi_{rc}=30^\circ} \frac{\tan\phi_{rc}}{\tan 30^\circ} \tag{11.31}$$

or

$$\alpha = 1.73\alpha_{\phi_{rc}=30^\circ} \tan \phi_{rc}$$

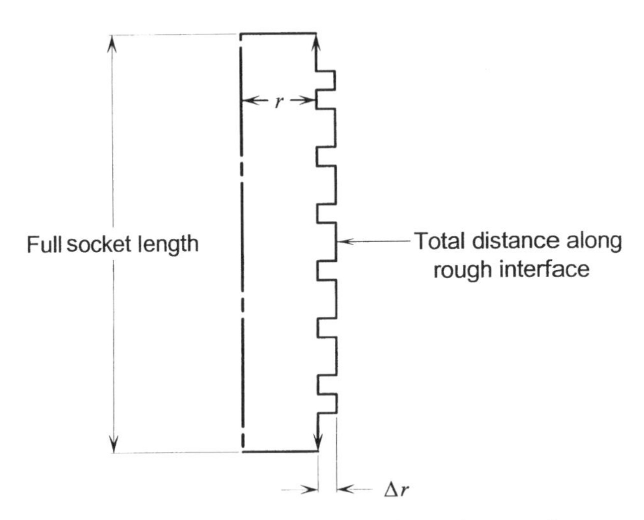

Figure 11.8 Definition of terms for surface roughness.

To use Figure 11.8, the designer must estimate the horizontal pressure of the fluid concrete acting at the middle of Layer i, σ_n. If the concrete has a slump of 175 mm (7 in.) or more and is placed at a rate of 12 m (40 ft) per hour, then σ_n at a depth z_i^* below the cutoff elevation up to 12 m (40 ft) can be estimated from

$$\sigma_n = 0.65\gamma_c z_i^* \tag{11.32}$$

φ is a joint effect factor that accounts for the presence of open joints that are voided or filled with soft gouge material. The joint effect factor can be estimated from Table 11.5.

$q_{u,i}$ is the design value for q_u in Layer i. This is usually taken as the mean value from intact cores larger than 50 mm (2 in.) in diameter. The possibility of the presence of weaker material between the intact geomaterial that could be sampled is considered through the joint effect factor, φ.

For a smooth rock socket in cohesive intermediate geomaterial, the side resistance is computed using

TABLE 11.5 Estimation of E_m/E_i Based on RQD

RQD (%)	Closed Joints	Open Joints
100	1.00	0.60
70	0.70	0.10
50	0.15	0.10
20	0.05	0.05

$$f_{\max,i} = 0.65\varphi\, p_a \sqrt{\frac{q_{u,i}}{p_a}} \tag{11.33}$$

If the rock socket has been roughened, the side resistance for a rough rock socket in cohesive intermediate geomaterial is

$$f_{\max,i} = 0.8\varphi \left[\frac{\Delta r}{r} \left(\frac{L'}{L} \right) \right]^{0.45} q_{u,i} \tag{11.34}$$

where

Δr = depth of shear keys,
r = radius of borehole,
L' = vertical spacing between the shear keys, and
L = depth covered by the shear keys, as defined in Fig. 11.9.

Detailed Procedure for Side Resistance in Cohesive Intermediate Geomaterial. O'Neill et al. (1996) recommend methods for estimating side and base resistances as well as settlement of drilled shafts under axial loads in this type of geomaterial. Their primary method, called *direct load-settlement sim-*

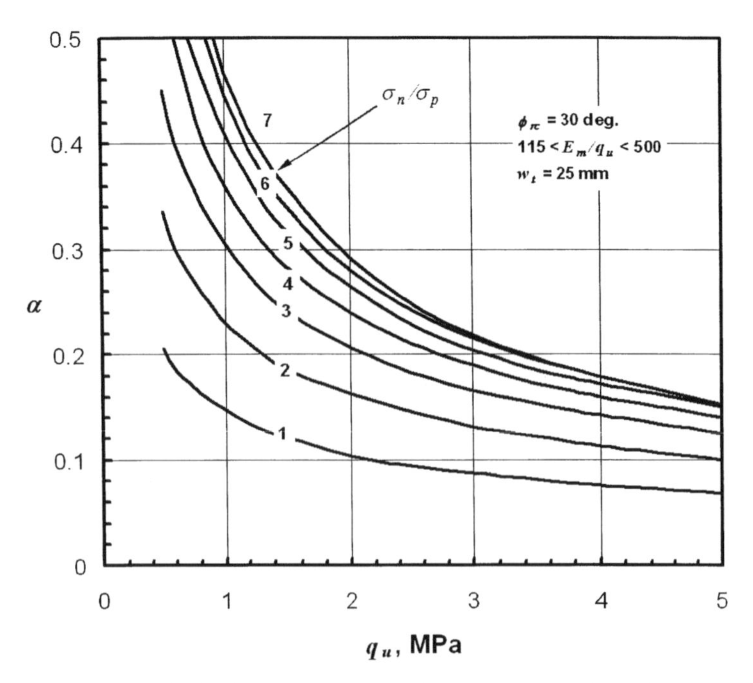

Figure 11.9 Factor α for smooth category 1 or 2 IGMs.

ulation, is used to compute the axial capacity of drilled shafts socketed into weak rock.

The direct simulation design model, based on an approximation of the broad range of FEM solutions, is as follows:

Decide whether the socket of weak rock in which the drilled shaft is placed requires subdivision into sublayers for analysis. If the weak rock is relatively uniform, the behavior of axially loaded drilled shafts can probably be simulated satisfactorily for design purposes using the simple procedure outlined below. If there is significant layering of the weak rock in the depth range of the socket, a load transfer function analysis should be modeled by a special FEM, as recommended by O'Neill et al. (1996). Significant layering in this respect will exist if the weak rock at the base of the shaft is considerably stronger and stiffer than that surrounding the sides and/or if changes in the stiffness and strength of the weak rock occur along the sides of the shaft. Load transfer function analyses should also be conducted if sockets exceed about 7.6 m (25 ft) in length.

Obtain representative values of the compressive strength q_c of the weak rock. It is recognized in practice that q_u is often used to represent compressive strength. Accordingly, q_u will be used to symbolize q_c in this criteria. Whenever possible, the weak rock cores should be consolidated to the mean effective stress in the ground and then subjected to undrained loading to establish the value of q_u. This solution is valid for soft rocks with $0.5 < q_u < 5.0$ MPa ($73 < q_u < 725$ psi). The method also assumes that high-quality samples, such as those obtained using triple-walled core barrels, have been recovered.

Determine the percentage of core recovery. If core recovery using high-quality sampling techniques is less than 50%, this method does not apply, and field loading tests are recommended to establish the design parameters.

Determine or estimate the mass modulus of elasticity of the weak rock, E_m. If Young's modulus of the material in the softer seams within the harder weak rock, E_s, can be estimated, and if Young's modulus of the recovered, intact core material, E_i, is measured or estimated, then the following expression, can be used:

$$\frac{E_m}{E_i} = \frac{L_c}{\dfrac{e_i}{E_s} \sum t_{\text{seams}} + \sum t_{\text{intact core segments}}} \tag{11.35}$$

In Eq. 11.32, L_c is the length of the core and Σt_{seams} is the summation of the thickness of all of the seams in the core, which can be assumed to be $(1 - r_c)\, L_c$, where r_c is the core recovery ratio (percent recovery/100%) and can be assumed equal to $r_c\, L_c$. If the weak rock is uniform and without significant soft seams or voids, it is usually conservative to take $E_m = 115\, q_u$. If the core recovery is less than 100%, it is recommended that appropriate in situ tests be conducted to determine E_m. If the core recovery is at least

50%, the recovered weak rock is generally uniform and the seams are filled with soft geomaterial, such as clay, but moduli of the seam material cannot be determined. Table 11.6 can be used, with linear interpolation if necessary, to estimate very approximately E_m/E_i. Use of this table is not recommended unless it is impossible to secure better data.

The designer must decide whether the walls in the socket can be classified as rough. If experience indicates that the excavation will produce a borehole that is rough according to the following definition, then the drilled shaft may be designed according to the method for the rough borehole. If not, or if the designer cannot predict the roughness, the drilled shaft should be designed according to the method for the smooth borehole.

A borehole can be considered rough if the roughness factor R_f will reliably exceed 0.15. The roughness factor is defined by

$$R_f = \left[\frac{\Delta r}{r}\left(\frac{L_t}{L_s}\right)\right] \tag{11.36}$$

where the terms in this equation are defined in Figure 11.9.

Estimate whether the soft rock is likely to smear when drilled with the construction equipment that is expected on the job site. *Smear* in this sense refers to the softening of the wall of the borehole due to drilling disturbance and/or exposure of the borehole to free water. If the thickness of the smear zone is expected to exceed about 0.1 times the mean asperity height, the drilled shaft should be designed as if it were smooth.

The effects of roughness and smear on both resistance and settlement are very significant, as will be demonstrated in the design examples. As part of the site exploration process for major projects, full-sized drilled shaft excavations should be made so that the engineer can quantify these factors, either by entering the borehole or by using appropriate down-hole testing tools such as calipers and sidewall probes. Rough borehole conditions can be assured if the sides of the borehole are artificially roughened by cutting devices on the drilling tools immediately prior to placing concrete such that $R_f > 0.15$ is attained.

Estimate f_a, the apparent maximum average unit side shear at infinite displacement. Note that f_a is not equal to f_{max}, which is defined as a displacement defined by the user of this method.

TABLE 11.6 Estimation of E_m/E_i Based on RQD

RQD (%)	Closed Joints	Open Joints
100	1.00	0.60
70	0.70	0.10
50	0.15	0.10
20	0.05	0.05

For weak rock with a rough borehole, use

$$f_a = \frac{q_u}{2} \tag{11.37}$$

For weak rock with a smooth borehole, use

$$f_a = \alpha\, q_u \quad \text{where} \quad \alpha \leq 0.5 \tag{11.38}$$

where α is a constant of proportionality that is determined from Figure 11.9 based on the finite element simulations. The factor σ_p in Figure 11.9 is the value of atmospheric pressure in the units employed by the designer. The maximum value of α that is permitted is 0.5. The parameter ϕ_{rc} in Figure 11.9 represents the angle of internal friction at the interface of the weak rock and concrete. The curves of Figure 11.9 are based on the use of $\phi_{rc} = 30°$, a value measured at a test site in clay-shale that is believed to be typical of clay-shales and mudstones in the United States. If evidence indicates that ϕ_{rc} differs from 30°, then α should be adjusted using Eq. 11.39:

$$\alpha = \alpha_{\text{Figure 11.8}} \frac{\tan \phi_{rc}}{\tan 30°} = 1.73\, (\alpha_{\text{Figure 11.8}}) \tan \phi_{rc} \tag{11.39}$$

If $E_m/E_i < 1$, adjust f_a for the presence of soft geomaterial within the soft rock matrix using Table 11.7. Define the adjusted value of f_a as f_{aa}. E_m can be estimated from the E_m/E_i ratios based on RQD of the cores. In cases where RQD is less than 50%, it is advisable to make direct measurements of E_m in situ using plate loading tests, borehole jacks, large-scale pressuremeter test, or by back-calculating E_m from field load tests of drilled shafts. The correlations shown in Table 11.7 become less accurate with decreasing values of RQD.

Estimate σ_n, the normal stress between the concrete and borehole wall at the time of loading. This stress is evaluated when the concrete is fluid. If no other information is available, general guidance on the selection of σ_n can be

TABLE 11.7 Adjustment of f_a for the Presence of Soft Seams

E_m/E_i	f_{aa}/f_a
1	1.0
0.5	0.8
0.3	0.7
0.1	0.55
0.05	0.45
0.02	0.3

obtained from Eq. 11.40, which is based on measurements made by Bernal and Reese (1983):

$$\sigma_n = M\ \gamma_c z_c \tag{11.40}$$

where

γ_c = unit weight of the concrete,

z_c = distance from the top of the completed column of concrete to the point in the borehole at which σ_n is desired, usually the middle of the socket, and

M = an empirical factor which depends upon the fluidity of the concrete, as indexed by the concrete slump (obtained from Figure 11.10).

The values shown in Fig. 11.10 represent the distance from the top of the completed column of concrete to the point in the borehole at which σ_n is desired. Figure 11.10 may be assumed valid if the rate of placement of concrete in the borehole exceeds 12 m/hr and if the ratio of the maximum coarse aggregate size to the borehole diameter is less than 0.02. Note that σ_n for slump outside the range of 125 to 225 mm (5 to 9 in.) is not evaluated. Unless there is information to support larger values of σ_n, the maximum value of z_c should be taken as 12 m (40 ft) in these calculations. This statement is predicated on the assumption that arching and partial setting will become significant after the concrete has been placed in the borehole for more than 1 hour.

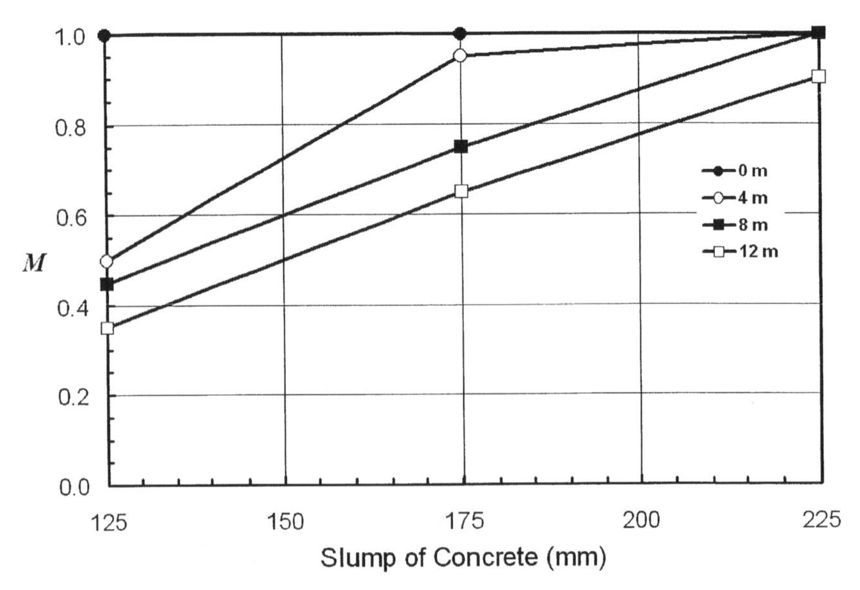

Figure 11.10 Factor M versus slump of concrete.

Note that E_m increases with increasing q_u, and the Poisson effect in the shaft causes an increase in the lateral normal interface stresses as E_m increases, producing higher values of side load transfer at the frictional interface.

Determine the *characteristic parameter n*, which is a fitting factor for the load-settlement syntheses produced by the finite element analyses. If the weak rock socket is rough:

$$n = \frac{\sigma_n}{q_u} \tag{11.41}$$

If the weak rock socket is smooth, estimate n from Figure 11.11. Note that n was determined in Figure 11.11 for $\phi_{rc} = 30°$. It is not sensitive to the value of ϕ_{rc}. However, α is sensitive to ϕ_{rc}, as indicated in Eq. 11.31.

If the socket has the following conditions—relatively uniform, and the soft rock beneath the base of the socket has a consistency equivalent to that of the soft rock along the sides of the shaft, $2 < L/D < 20$, $D > 0.5$ m, and $10 < E_c/E_m < 500$—then compute the load-settlement relation for the weak rock socket as follows. Under the same general conditions, if the socket is highly stratified and/or if the geomaterial beneath the base of the socket has a consistency considerably different from that along the sides of the socket, use the unit load transfer function version of this method described later.

Compute Q_t (load still in the shaft at the top of the socket) versus w_t (settlement at the top of the socket) from Eq. 11.42 or Eq. 11.43, depending on the value of n. These equations apply to both rough and smooth sockets.

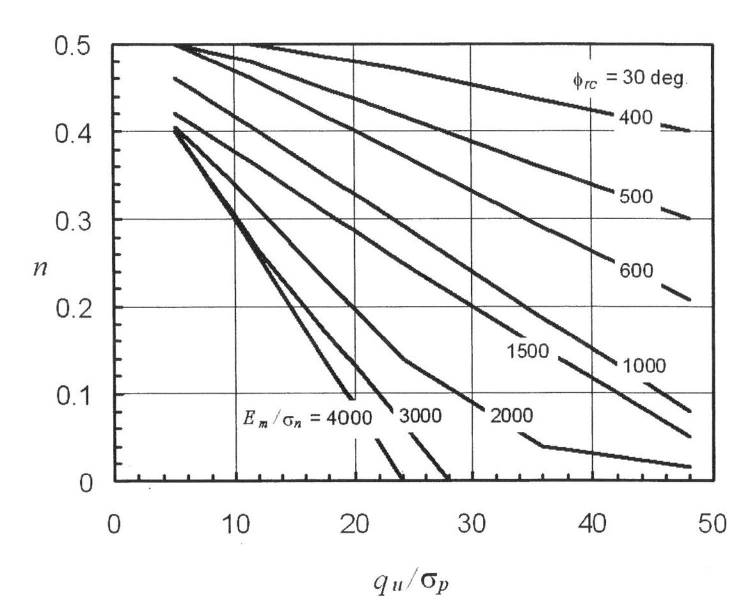

Figure 11.11 Factor n for smooth sockets for various combinations of parameters.

$$\text{If } H_f \leq n: \quad Q_t = \pi D L H_f f_{aa} = \frac{\pi D^2}{4} q_b \tag{11.42}$$

$$\text{If } H_f > n: \quad Q_t = \pi D L K_f f_{aa} = \frac{\pi D^2}{4} q_b \tag{11.43}$$

Equation 11.42 applies in the elastic range before any slippage has occurred at the shaft–weak rock interface, and an elastic base response, as represented by the last expression on the right-hand side of the equation, also occurs. Equation 11.43 applies during interface slippage (nonlinear response). To evaluate Q_t, a value of w_t is selected, and H_f, which is a function of w_t, is evaluated before deciding which equation to use. If $H_f > n$, evaluate K_f and use Eq. 11.43; otherwise, use Eq. 11.42. Equations 11.44 and 11.45 are used to evaluate H_f and K_f, respectively.

$$H_f = \frac{E_m \Omega}{\pi L \Gamma f_{aa}} w_t \tag{11.44}$$

$$K_f = n + \frac{(H_f - n)(1 - n)}{H_f - 2n + 1} \leq 1 \tag{11.45}$$

where

$$\Omega = 1.14 \left(\frac{L}{D}\right)^{0.5} - 0.05 \left[\left(\frac{L}{D}\right)^{0.5} - 1\right] \log_{10} \left|\frac{E_c}{E_m}\right| - 0.44 \tag{11.46}$$

with $D < 1.53$ m and

$$\Gamma = 0.37 \left(\frac{L}{D}\right)^{0.5} - 0.15 \left[\left(\frac{L}{D}\right)^{0.5} - 1\right] \log_{10} \left|\frac{E_c}{E_m}\right| + 0.13 \tag{11.47}$$

Finally,

$$q_b = \Lambda \, w_t^{0.67} \tag{11.48}$$

where

$$\Lambda = 0.0134 E_m \left(\frac{L/D}{L/D + 1}\right) \left[\frac{200(\sqrt{L/D} - \Omega)(1 + L/D)}{\pi L \Gamma}\right] \tag{11.49}$$

Check the values computed for q_b. If core recovery in the weak rock surrounding the base is 100%, q_b should not exceed $q_{max} = 2.5\, q_u$. At working loads, q_b should not exceed $0.4\, q_{max}$.

Graph the load-settlement curve resulting from the computations. Select ultimate and service limit resistances based on settlements. For example, the ultimate resistance might be selected as the load Q_t corresponding to a settlement w_t of 25 mm (1 in.), while the service limit resistance might be selected as the load Q_t corresponding to a value of $w_t < 25$ mm ($w_t < 1$ in.).

Example Problem 4—Shaft in Cohesive Intermediate Geomaterial
This is an example of a drilled shaft in weak rock.

Description of the Problem—Rough Socket. Consider the shaft and soil profile shown in Figure 11.12. The user is asked to compute the load-settlement relation for the socket and to estimate the ultimate resistance at a settlement of $w_t = 25$ mm. The socket is assumed to be rough. The RQD for the sample is 100%.

Computations

1. Since the core recovery and RQD are high, assume that $E_m = 115\ q_u$. Note that $E_c/E_m = 100\%$.
2. $f_{aa} = f_a = 2.4/2 = 1.2$ MPa, or 1200 kPa.
3. $z_c = 6.1$ m (depth from the top of the concrete to the middle of the socket). Considering concrete placement specifications:

$$\sigma_n = 0.92\ \gamma_c\ z_c \text{ from Figure 11.12 or}$$

$$\sigma_n = 0.92\ (20.4)\ (6.1) = 114.5\text{ kPa} = 1.13\ \sigma_p.$$

4. $n = 115\text{ kPa}/2400\text{ kPa} = 0.0477.$
5. $L/D = 6.1/0.61 = 10.$

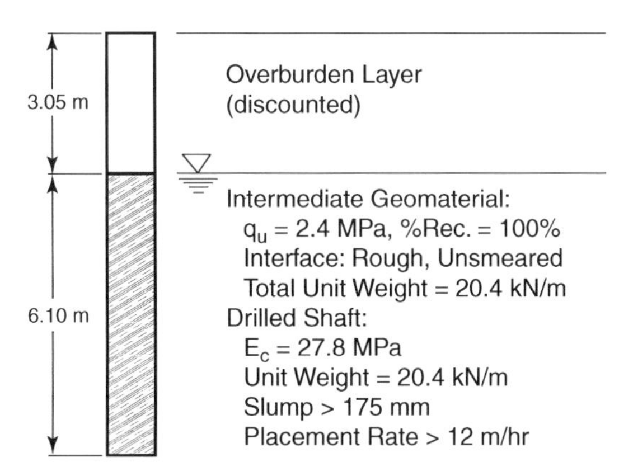

Figure 11.12 General soil description of Example Problem 4.

6. $\Omega = 1.14\sqrt{10} - 0.05\lfloor\sqrt{10} - 1\rfloor\log|100| - 0.44 = 2.949.$
7. $\Gamma = 0.37\sqrt{10} - 0.15\lfloor\sqrt{10} - 1\rfloor\log|100| + 0.13$ or $\Gamma = 0.651.$
8. $H_f = \{[115\ (2400)\ 2.94)]/[3.14\ (6100\ \text{mm})(0.651)\ 1200]\}w_i$

   $H_f = 0.0541\ w_t$
9. $K_f = 0.0477 + (0.0541w_t - 0.0477)\ (1 - 0.0477)/(0.0541w_t - 0.096 + 1)$

   $K_f = 0.0477 + (0.0541w_t - 0.0477)\ (0.952)/(0.0541w_t + 0.904)$
10. At the end of the elastic stage, $\Theta_f = n$ (implied by Eq. 3.44). Therefore,

$$w_{te} = n/\Theta = 0.0477/0.0541 = 0.882 \text{ mm}$$

   where w_{te} signifies w_t at the end of the elastic stage. (Note that the elastic response occurs only up to a very small settlement in this example.)
11. $q_b = \Lambda w_t^{0.67}$ (Eqs. 3.50 and 3.51)

   $q_b = \{[(115)\ (2400)\ (10/11)\}\ \{[200\ (10^{0.5} - 2.949)\ (11)]/[3.14\ (6100)\ 0.651]\}^{0.67}\ w_t^{0.67}$

   $q_b = 373.4\ w_t\ (\text{mm})^{0.67}$ (kPa). Note that $\Lambda = 373.4$.
12. $\pi DL = 11.69\ \text{m}^2$; $\pi D^2/4 = 0.2922\ \text{m}^2$.
13. Compute Q_t corresponding to w_{te}, signified by Q_{te}:

$$Q_{te} = 11.69\ (0.0541)(0.882)(1200) + (0.2922)(373.4)(0.881)^{0.67}$$

$$Q_{te} = 669 + 100 = 769 \text{ kN (Eq. 3.44).}$$

   Note that at this point, 670 kN is transferred to the weak rock in side resistance and 103 kN is transferred in base resistance. (Q_{te}, w_{te}) is a point on the load-settlement curve, and a straight line can be drawn from ($Q_t = 0$, $w_t = 0$) to this point.
14. Compute the values of Q_t for selected values of w_t on the nonlinear portion of the load-settlement curve. Numerical evaluations are made in the following table.

Rough Socket

w_t, mm	H_f	K_f	Q_s, kN	q_b, kPa	Q_b, kN	Q_t, kN
1	0.0541	0.0540	758.2	373.4	109.1	867.3
2	0.1082	0.1046	1466.9	594.2	173.6	1640.5
3	0.1623	0.1500	2103.7	779.7	227.9	2331.6
4	0.2164	0.1910	2679.1	945.4	276.3	2955.4

5	0.2705	0.2282	3201.5	1097.8	320.8	3522.3
6	0.3245	0.2622	3677.9	1240.5	362.5	4040.5
7	0.3786	0.2933	4114.2	1375.5	402.0	4516.2
8	0.4327	0.3219	4515.2	1504.2	439.6	4954.8
9	0.4868	0.3482	4885.0	1627.7	475.7	5360.7
10	0.5409	0.3726	5227.1	1746.8	510.5	5737.6
11	0.5950	0.3953	5544.5	1861.9	544.1	6088.7
12	0.6491	0.4163	5839.9	1973.7	576.8	6416.7
13	0.7032	0.4359	6115.3	2082.4	608.6	6723.9
14	0.7573	0.4543	6372.9	2188.5	639.6	7012.5
15	0.8114	0.4715	6614.2	2292.0	669.8	7284.0
16	0.8654	0.4877	6840.7	2393.3	699.4	7540.2
17	0.9195	0.5028	7053.9	2492.5	728.4	7782.3
18	0.9736	0.5172	7254.7	2589.8	756.9	8011.6
19	1.0277	0.5307	7444.3	2685.3	784.8	8229.1
20	1.0818	0.5435	7623.6	2779.2	812.2	8435.8
21	1.1359	0.5556	7793.3	2871.6	938.2	8632.5
22	1.1900	0.5670	7954.3	2962.5	865.8	8820.1
23	1.2441	0.5779	8107.2	3052.0	891.9	8999.2
24	1.2982	0.5883	8252.6	3140.3	917.7	9170.4
25	1.3523	0.5982	8391.0	3227.4	943.2	9334.2

Note that

$$q_b \text{ (at } w_t = 25 \text{ mm)} = 3.23 \text{ MPa} = 1.34\, q_u < q_{\max} = 2.5\, q_u$$

which is acceptable for the definition of ultimate resistance. Based on base resistance, the working load should be limited to $q_b = q_u$ or w_t should be limited to about 12 mm at the working load. Note also that the compressive stress in the shaft at $w_t = 15$ mm is 24,900 kPa (7284 kN/cross-sectional area), which may be approaching the structural failure load in the drilled shaft.

15. The numerical values from Steps 13 and 14 are graphed in Figure 11.13. Also shown in the figure is the case with a smooth socket for the same problem. Hand computations for the case of a smooth socket are included in the next section.

    The physical significance of the parameters Q_f and K_f is evident from the numerical solution. Q_f is a constant of proportionality for elastic resistance for side shear, and K_f is a proportionality parameter for actual side shear, including elastic, plastic, and interface slip effects.

Description of the Problem—Smooth Socket. Consider the same example as before (shaft and soil profile shown in Figure 11.12). The rock socket is now assumed to be smooth. Estimate that $\phi_{rc} = 30°$. The user is asked to

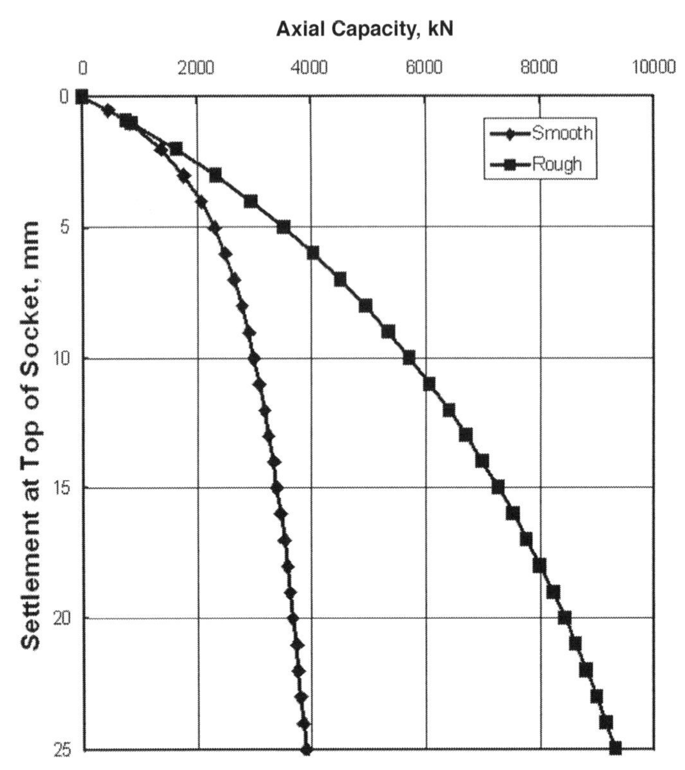

Figure 11.13 Computed axial load versus settlement for Example Problem 4.

compute the load-settlement relation for the socket and to estimate the ultimate resistance at a settlement of w_t = 25 mm.

Computations

1. $f_a = f_{aa} = \alpha q_u$.
2. Referring to Figure 11.13, for σ_n/σ_p = 1.13 and q_u = 2.4 MPa, we have α = 0.12.
3. $f_a = f_{aa}$ = 0.12(2400) = 288 kPa.
4. q_u/σ_p = 2400/101.3 = 23.7 and E_m/σ_n = 115 (2.4) (1000)/114.5 = 2411.
5. From Figure 11.13, n = 0.11.
6. Ω = 2.949 (unchanged); Γ = 0.651 (unchanged).
7. H_f = {[115(2400) 2.949)]/[3.14 (6100 mm)(0.651) 288]}w_t

   = 0.226 w_t.
8. K_f = 0.11 + [(0.226 w_t − 0.11)(1 − 0.11)]/[0.226 w_t − 2(0.11) + 1]

   = 0.11 + (0.226 w_t − 0.11)(0.89)/(0.226 w_t + 0.78).

9. At $H_f = n$, $w_{te} = 0.11/0.226 = 0.486$ mm.
10. $q_b = 373.4\ (w_t)^{0.67}$, where w_t is in millimeters and q_b is in kPa.
11. $Q_{te} = 11.69\ (0.226)(0.486)(288) + (0.2922)(373.4)(0.486)^{0.67}\ Q_{te} = 370 + 67.3 = 437$ kN (Eq. 3.44).
12. $Q_{te} = 437$ kN, $w_{te} = 0.486$ mm is the point at the end of the linear portion of the load settlement curve.
13. Compute the values of Q_t for selected values of w_t on the nonlinear portion of the load-settlement curve. Numerical evaluations are made below.

Smooth Socket

w_t mm	H_f	$K_f =$	q_b kPa	Q_b kN	Q_s kN	Q_t kN
1	0.23	0.213	373	109.1	717	826
2	0.45	0.357	594	173.6	1203	1377
3	0.68	0.457	780	227.9	1539	1767
4	0.91	0.530	945	276.3	1785	2061
5	1.13	0.586	1098	320.8	1972	2293
6	1.36	0.630	1240	362.5	2120	2482
7	1.58	0.665	1375	402.0	2239	2641
8	1.81	0.694	1504	439.6	2337	2777
9	2.04	0.719	1628	475.7	2420	2896
10	2.26	0.740	1747	510.5	2491	3001
11	2.49	0.758	1862	544.1	2551	3095
12	2.72	0.773	1974	576.8	2604	3181
13	2.94	0.787	2082	608.6	2650	3259
14	3.17	0.799	2188	639.6	2691	3331
15	3.40	0.810	2292	669.8	2728	3398
16	3.62	0.820	2393	699.4	2761	3460
17	3.85	0.829	2492	728.4	2791	3519
18	4.08	0.837	2590	756.9	2817	3574
19	4.30	0.844	2685	784.8	2842	3627
20	4.53	0.851	2779	812.2	2864	3676
21	4.75	0.857	2872	839.2	2885	3724
22	4.98	0.862	2692	865.8	2904	3770
23	5.21	0.868	3052	891.9	2921	3813
24	5.43	0.873	3140	917.7	2937	3855
25	5.66	0.877	3227	943.2	2953	3896

14. The numerical values for a smooth socket are graphed in Figure 11.13 in comparison with the values from a rough socket to illustrate the effect of borehole roughness in this problem. Note again that $q_b < 2.5\ q_u$.

Side Resistance in Cohesive Intermediate Geomaterials—Uplift Loading Side resistance in uplift loading for cohesive intermediate geomaterials is identical to that developed in compressive loading, provided that the shaft borehole is rough. If the shaft borehole is smooth, Poisson's effect reduces shaft resistance because the normal stress at the side of the borehole is reduced.

When the borehole is smooth, f_{max} should be reduced from the value computed for compressive loading by a factor Ψ:

$$f_{max}\text{ (uplift)} = \Psi\, f_{max}\text{ (compression)} \tag{11.50}$$

The value of Ψ is taken to be 1.0 when $(E_c/E_m)\,(B/D)^2 \geq 4$, or 0.7 when $(E_c/E_m)\,(B/D)^2 < 4$, unless field loading tests are performed. E_c and E_m are the composite Young's modulus of the shaft's cross section and rock mass, respectively, B is the socket diameter, and D is the socket length.

End Bearing in Cohesive Intermediate Geomaterials Several procedures are available for estimating the undrained unit end bearing in jointed rock. The procedures of Carter and Kulhawy (1988) and of the *Canadian Foundation Engineering Manual* (Canadian Geotechnical Society, 1992) are presented.

The method developed by Carter and Kulhawy (1988) can be used to estimate a lower bound for end bearing on randomly jointed rock. The same solution is used either for a shaft base bearing on the surface of the rock or for a base socketed into rock. It is assumed that

- the joints are drained,
- the rock between the joints is undrained,
- the shearing stresses in the rock mass are nonlinearly dependent on the normal stresses at failure,
- the joints are not necessarily oriented in a preferential direction,
- the joints may be open or closed, and
- the joints may be filled with weathered geomaterial (gouge).

End-bearing resistance is computed by

$$q_{max} = \left(\sqrt{s} + \sqrt{m\sqrt{s} + s}\right) q_u \tag{11.51}$$

The parameters s and m for the cohesive intermediate geomaterial are roughly equivalent to c' and ϕ' for soil. The term in parentheses is analogous to the parameter N_c for clay soils. Values of s and m for cohesive intermediate geomaterial are obtained from Tables 11.8 and 11.9.

The second method for computing resistance in end bearing for drilled shafts in jointed, sedimentary rock, where the joints are primarily horizontal,

TABLE 11.8 Descriptions of Rock Types for Use in Table 11.9

Rock Type	Rock Description
A	Carbonate rocks with well-developed crystal cleavage (dolostone, limestone, marble)
B	Lithified argillaceous rocks (mudstone, siltstone, shale, slate)
C	Arenaceous rocks (sandstone, quartz)
D	Fine-grained igneous rocks (andesite, dolerite, diabase, rhyolite)
E	Coarse-grained igneous and metamorphic crystalline rocks (amphibolite, gabbro, gneiss, granite, norite, quartzdirorite)

is the method of the *Canadian Foundation Engineering Manual* (Canadian Geotechnical Society, 1992):

$$q_{\max} = 3K_{sp}\, d\, q_u \tag{11.52}$$

$$d = 1 + 0.4\frac{D_s}{B} \leq 3.4 \tag{11.53}$$

$$K_{sp} = \frac{3 + c_s/B_b}{10\sqrt{1 + 300\,\delta/c_s}} \tag{11.54}$$

where

TABLE 11.9 Values of *s* and *m* Based on Rock Classification

Rock Quality	Joint Description and Spacing		Value of *m* as a Function of Rock Type (A–E) from Table 11.8				
		s	*A*	*B*	*C*	*D*	*E*
Excellent	Intact (closed); spacing <3 m	1	7	10	15	17	25
Very good	Interlocking spacing of 1 to 3 m	0.1	3.5	5	7.5	8.5	12.5
Good	Slightly weathered; spacing of 1 to 3 m	0.04	0.7	1	1.5	1.7	2.5
Fair	Moderately weathered; spacing of 0.3 to 1 m	10^{-4}	0.14	0.2	0.3	0.34	0.5
Poor	Weathered with gouge (soft material); spacing of 30 to 300 mm	10^{-5}	0.04	0.05	0.08	0.09	0.13
Very poor	Heavily weathered; spacing of less than 50 mm	0	0.007	0.01	0.015	0.017	0.025

q_{max} = unit end-bearing capacity,
K_{sp} = empirical coefficient that depends on the spacing of discontinuities,
q_u = average unconfined compressive strength of the rock cores,
c_s = spacing of discontinuities,
δ = thickness of individual discontinuities,
D_s = depth of the socket measured from the top of rock (not the ground surface), and
B_b = diameter of the socket.

The above equations are valid for a rock mass with spacing of discontinuities greater than 12 in. (0.3 m) and thickness of discontinuities less than 0.2 in. (5 mm) (or less than 1 in. [25 mm] if filled with soil or rock debris) and for a foundation with a width greater than 12 in. (305 mm). For sedimentary or foliated rocks, the strata must be level or nearly so. Note that Eq. 11.52 is different in form from that presented in the *Canadian Foundation Engineering Manual* in that a factor of 3 has been added to remove the implicit factor of safety of 3 contained in the original form. Further, note that the form of Eq. 11.53 differs from that in the first edition of the *Manual*.

11.6.5 Design for Axial Capacity in Cohesionless Intermediate Geomaterials

Cohesionless intermediate geomaterials are residual or transported materials that exhibit N_{60}-values of more than 50 blows per foot. It is common practice to treat these materials as undrained because they may contain enough fine-grained material to significantly lower permeability.

The following design equations are based on the original work of Mayne and Harris (1993) and modifications by O'Neill et al. (1996). The theory was proposed for gravelly soils, either transported or residual, with penetration resistances (from the SPT) between 50 and 100. The method has been used by Mayne and Harris to predict and verify the behavior of full-scale drilled shafts in residual micaceous sands from the Piedmont province in the eastern United States. Further verification tests were reported by O'Neill et al. for granular glacial till in the northeastern United States.

Side Resistance in Cohesionless Intermediate Geomaterial The maximum load transfer in side resistance in Layer i can be estimated from

$$f_{max,i} = \sigma'_{vi} K_{0i} \tan \phi'_i \qquad (11.55)$$

where σ'_{vi} is the vertical effective stress at the middle of Layer i. The earth pressure coefficient K_{0i} and effective angle of internal friction of the gravel ϕ' can be estimated from field or laboratory testing or from

$$K_{0i} = (1 - \sin \phi_i')\left(\frac{0.2N_{60,i}}{\sigma_{vi}'/p_a}\right)^{\sin\phi_i'} \tag{11.56}$$

$$\phi_i' = \tan^{-1}\left\{\left[\frac{N_{60,i}}{12.2 + 20.3(\sigma_v'/p_a)}\right]^{0.34}\right\} \tag{11.57}$$

where $N_{60,i}$ is the SPT penetration resistance, in blows per foot (or blows per 300 mm), for the condition in which the energy transferred to the top of the drive string is 60% of the drop energy of the SPT hammer, uncorrected for the effects of overburden stress; and p_a is the atmospheric pressure in the selected system of units (usually 1 atmosphere, which converts to 101.4 kPa or 14.7 psi).

Side Resistance in Cohesionless Intermediate Geomaterial—Uplift Loading The equations for side resistance in uplift for cohesionless soils can be used for cohesionless intermediate geomaterials. The pertinent equations are Eqs. 11.23, 11.24, and 11.25.

End Bearing in Cohesionless Geomaterial Mayne and Harris (1993) developed the following expression for unit end-bearing capacity for cohesionless intermediate geomaterials:

$$q_{max} = 0.59\left[N_{60}\left(\frac{p_a}{\sigma_{vb}'}\right)\right]^{0.8} \sigma_{vb}' \tag{11.58}$$

where N_{60} is the blow count immediately below the base of the shaft.

The value of q_{max} should be reduced when the diameter of the shaft B_b is more than 1.27 m (50 in.). If the diameter of the shaft is between 1.27 and 1.9 m (50 to 75 in.), $q_{max,r}$ is computed using

$$q_{max,r} = \frac{1.27}{B_b\ (\text{m})} q_{max} \tag{11.59}$$

and if the shaft diameter is more than 1.9 m:

$$q_{max,r} = q_{max}\left[\frac{2.5}{aB_b\ (\text{m}) + 2.5b}\right] \tag{11.60}$$

where

$$a = 0.28B_b \text{ (m)} + 0.083 \frac{L}{B_b} \tag{11.61}$$

$$b = 0.065\sqrt{s_{ub}} \text{ (kPa)} \tag{11.62}$$

where

L = depth of base below the ground surface or the top of the bearing layer if the bearing layer is significantly stronger than the overlying soils, and

s_{ub} = average undrained shear strength of the soil or rock between the elevation of the base and $2B_b$ below the base. If the bearing layer is rock, s_{ub} can be taken as $q_u/2$.

The above equations are based on load tests of large-diameter underreamed drilled shafts in very stiff clay and soft clay-shale. These equations were developed assuming $q_{\max,r}$ to be the net bearing stress at a base settlement of 2.5 in. (64 mm) (O'Neill and Sheikh, 1985; Sheikh et al., 1985). When one-half or more of the design load is carried in end bearing and a global factor of safety is applied, the global factor of safety should not be less than 2.5, even if soil conditions are well defined, unless one or more site-specific load tests are performed.

Commentary on the Direct Load-Settlement Method This method is intended for use with relatively ductile weak rock, in which deformations occur in asperities prior to shear. If the weak rock is friable or unusually brittle, the method may be unconservative, and appropriate loading tests should be conducted to ascertain the behavior of the drilled shaft for design purposes. The method is also intended for use with drilled shafts in weak rocks produced in the dry. If it is necessary to produce the shaft using water, or with mineral or synthetic drilling slurries, the shaft should be treated as smooth for design purposes unless it can be proved that rough conditions apply. The method also assumes that the bearing surface at the base of the socket is clean, such that the shaft concrete is in contact with undisturbed weak rock. If base cleanliness cannot be verified during construction, base resistance (q_b) should be assumed to be zero.

The design examples did not consider the effect of a phreatic surface (water table) above the base of the socket. This effect can be handled by computing σ_n, assuming that the unit weight of the concrete below the phreatic surface is its buoyant unit weight:

$$\sigma_n = M[\gamma_c z_w + \gamma'_c (z_c - z_w)] \tag{11.63}$$

where

γ'_c = buoyant unit weight of concrete,
M is obtained from Figure 11.7, and
z_w = depth from top of concrete to elevation of water table.

11.6.6 Design for Axial Capacity in Massive Rock

Computation Procedures for Rock A broad view of the classification of intact rock can be obtained by referring to Figure 11.14 (Deere, 1968, and

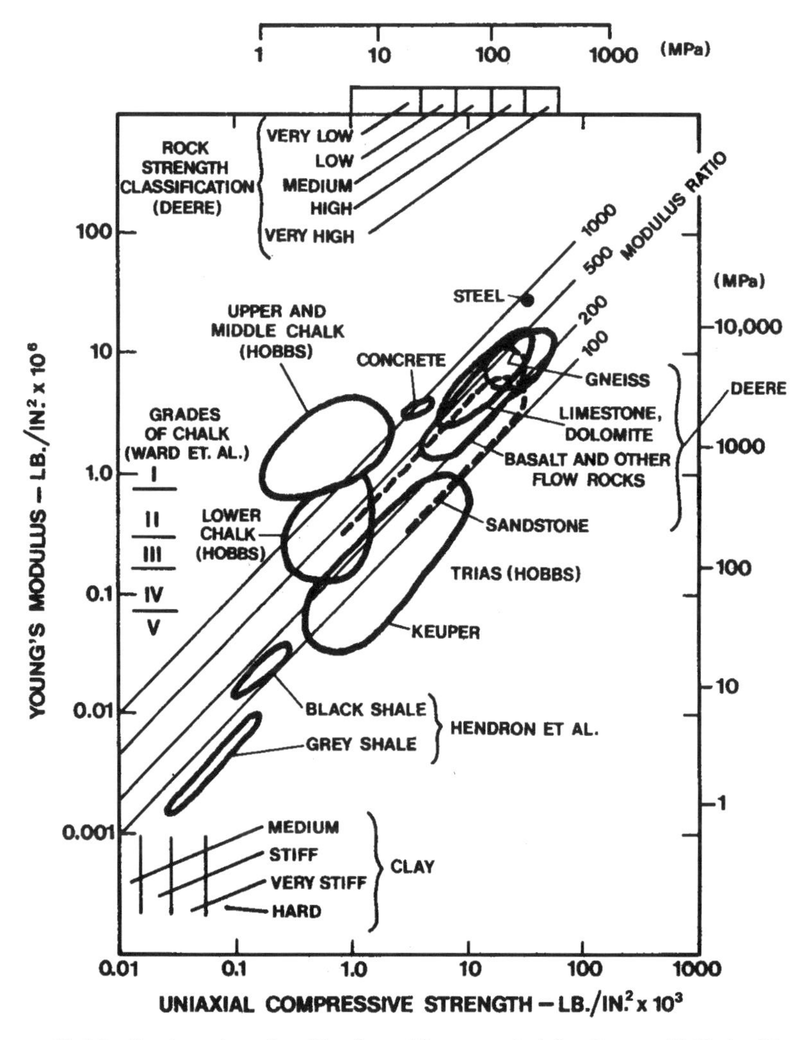

Figure 11.14 Engineering classification of intact rock (after Deere, 1968, by Horvath and Kenney, 1979).

Peck, 1976, as presented by Horvath and Kenney, 1979). The figure shows medium clay at the low range and gneiss at the high range. Concrete and steel are also shown for reference. Several of the rock categories have compressive strengths that are in the range of that for concrete or higher. As can be expected, many of the design procedures for drilled shafts in rock are directed at weak rock because strong rock could well be as strong as or stronger than the concrete in the drilled shaft. In this situation, the drilled shaft would fail structurally before any bearing capacity failure could occur.

Except for instances where drilled shafts were installed in weak rocks such as shales or mudstones, there are virtually no occasions where loading has resulted in failure of the drilled shaft foundation. An example of a field test where failure of the drilled shaft was impossible is shown in Figures 11.15 and 11.16. The rock at the site was a vuggy limestone that was difficult to core without fracture. Only after considerable trouble was it possible to obtain the strength of the rock. Two compression tests were performed in the laboratory, and in situ grout-plug tests were performed under the direction of Schmertmann (1977).

The following procedure was used for the in situ grout-plug tests. A hole was drilled into the limestone, followed by placement of a high-strength steel bar into the excavation, casting of a grout plug over the lower end of the bar, and pulling of the bar after the grout had cured. Five such tests were per-

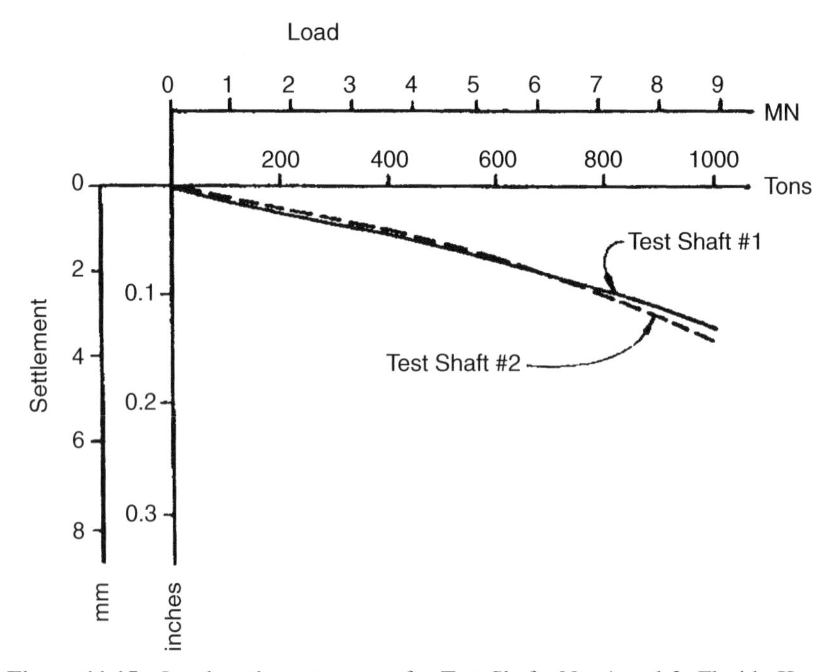

Figure 11.15 Load-settlement curves for Test Shafts No. 1 and 2, Florida Keys.

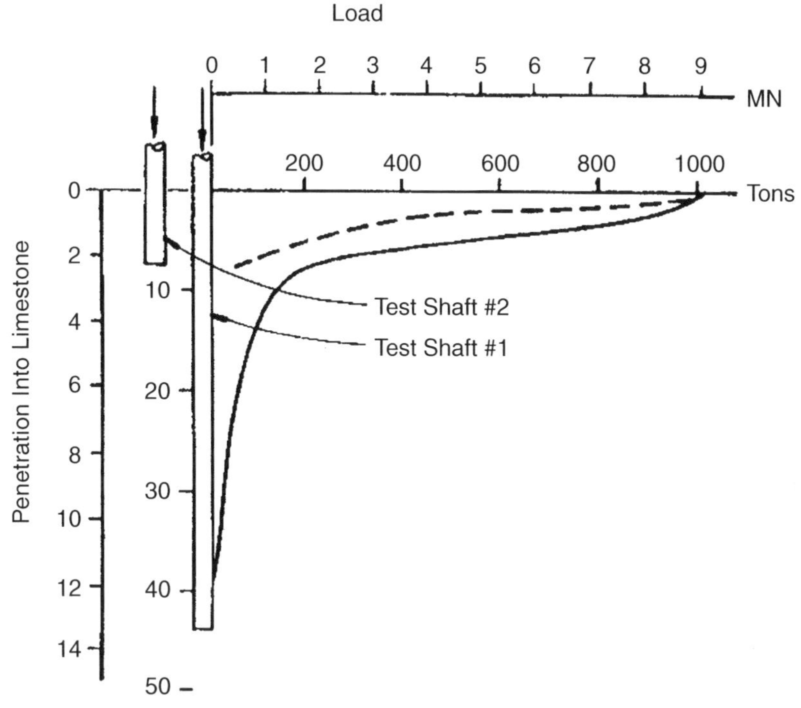

Figure 11.16 Load-distribution curves for Test Shafts Nos. 1 and 2, Florida Keys.

formed over the top 10 ft of the rock. Side resistance ranged from 12.0 to 23.8 tons/ft^2 (1.15 to 2.28 MPa), with an average of approximately 18.0 tons/ft^2 (1.72 MPa). The compressive strength of the rock was approximately 500 psi (3.45 MPa), putting the vuggy limestone in the lower ranges of the strength of the chalk shown in Figure 11.14.

Two axial load tests were performed at the site on cylindrical drilled shafts that were 36 in. (914 mm) in diameter (Reese and Nyman, 1978). Test Shaft No. 1 penetrated 43.7 ft (13.3 m) into the limestone, and Test Shaft No. 2 penetrated 7.6 ft (2.32 m). Test Shaft No. 1 was loaded first, with the results shown in Figures 11.15 and 11.16, and it was then decided to shorten the penetration and construct Test Shaft No. 2. As may be seen in Figure 11.15, the load-settlement curves for the two shafts are almost identical, with Test Shaft No. 2 showing slightly more settlement at the 1000-ton (8.9-MN) load (the limit of the loading system). The settlement of the two shafts under the maximum load is quite small, and most of the settlement (about 0.10 in., 2.5 mm) was due to elastic shortening of the drilled shafts.

The distribution of load with depth, determined from internal instrumentation in the drilled shafts, for the maximum load is shown in Figure 11.16. As may be seen, no load reached the base of Test Shaft No. 1, and only about

60 tons (530 kN) reached the base of Test Shaft No. 2. The data allowed a design for the foundations to be made at the site with confidence; however, as indicated, it was impossible to find the ultimate values of load transfer in side resistance and in end bearing because of the limitations of the loading equipment in relation to the strength of the rock. The results are typical for drilled shafts that are founded in rock that cannot develop the ultimate values of load transfer.

A special program of subsurface exploration is frequently necessary to obtain the in situ properties of the rock. Not only is it important to obtain the compressive strength and stiffness of the sound rock, but it is necessary to obtain detailed information on the nature and spacing of joints and cracks so that the stiffness of the rock mass can be determined. The properties of the rock mass will normally determine the amount of load that can be imposed on a rock-socketed drilled shaft. The pressuremeter has been used to investigate the character of in situ rock, and design methods have been proposed based on such results.

An example of the kind of detailed study that can be made concerns the mudstone of Melbourne, Australia. The Geomechanics Group of Monash University in Melbourne has written an excellent set of papers on drilled shafts that give recommendations in detail for subsurface investigations, determination of properties, design, and construction (Donald et al., 1980; Johnston et al., 1980a, 1980b; Williams, 1980; Williams et al., 1980a, 1980b; Williams and Erwin, 1980). These papers imply that the development of rational methods for the design of drilled shafts in a particular weak rock will require an extensive study and, even so, some questions may remain unanswered. It is clear, however, that a substantial expenditure for the development of design methods for a specific site could be warranted if there is to be a significant amount of construction at the site.

Williams et al. (1980b) discussed their design concept and stated: "A satisfactory design cannot be arrived at without consideration of pile load tests, field and laboratory parameter determinations and theoretical analyses; initially elastic, but later hopefully also elastoplastic. With the present state of the art, and the major influence of field factors, particularly failure mechanisms and rock defects, a design method must be based primarily on the assessment of field tests."

Other reports on drilled shafts in rock confirm the above statements about a computation method; therefore, the method presented here must be considered to be approximate. Detailed studies, including field tests, are often needed to confirm a design.

The procedure recommended by Kulhawy (1983) presents a logical approach. The basic steps are as follows.

1. The penetration of the drilled shaft into the rock for the given axial load is obtained by using an appropriate value of side resistance (see the later recommendation).

2. If the full load is taken by the base of the drilled shaft, the settlement of the drilled shaft in the rock is computed by adding the elastic shortening to the settlement required to develop end bearing. The stiffness of the rock mass is needed for this computation.
3. If the computed settlement is less than about 0.4 in. (10 mm), the side resistance will dominate and little load can be expected to reach the base of the foundation.
4. If the computed settlement is more than about 0.4 in. (10 mm), the bond in the socket may be broken and the tip resistance will be more important.

Kulhawy (1983) presents curves that will give the approximate distribution of the load for Steps 3 and 4; however, the procedure adopted here is to assume that the load is carried entirely in side resistance or in end bearing, depending on whether or not the computed settlement is less or more than 0.4 in.

The recommendations that follow are based on the concept that side resistance and end bearing will not develop simultaneously. The concept is conservative, of course, but it is supported by the fact that the maximum load transfer in side resistance in the rock will occur at the top of the rock, where the relative settlement between the drilled shaft and the rock is greatest. If the rock is brittle, which is a possibility, the bond at the top of the rock could fail, with the result that additional stress is transferred downward. There could then be a progressive failure in side resistance.

Note that the settlement will be small if the load is carried only in side resistance. The settlement in end bearing could be considerable and must be checked as an integral part of the analysis.

The following specific recommendations are made to implement the above general procedure:

1. Horvath and Kenney (1979) did an extensive study of the load transfer in side resistance for rock-socketed drilled shafts. The following equation is in reasonable agreement with the best-fit curve that was obtained where no unusual attempt was made to roughen the walls:

$$f_s \text{ (psi)} = 2.5\sqrt{q_u \text{ (psi)}} \tag{11.64}$$

where

f_s = ultimate side resistance in units of lb/in.^{2}, and

q_u = uniaxial compressive strength of the rock or concrete, whichever is less in units of lb/in^2.

(Note: Equation 11.61 is nonhomogeneous, and the value of q_u must be converted to English units, the equation solved for f_s in English units, and f_s

then converted to SI units before performing further computations with SI units.)

Note that there was a large amount of scatter in the data gathered by Horvath and Kenney (1979), but Eq. 11.64 can be used to compute the necessary length of the socket. If the drilled shaft is installed in clay-shale, the ultimate side resistance may be predicted more accurately by the procedures described in the previous section for clay-shale rather than by using Eq. 11.64.

2. The shortening ρ_c of the drilled shaft can be computed by elementary mechanics, employing the dimensions of the shaft and the stiffness of the concrete:

$$\rho_c = \frac{Q_{ST}L}{AE_c} \tag{11.65}$$

where

L = penetration of the socket,
Q_{ST} = load at the top of the socket,
A = cross-sectional area of the socket, and
E_c = equivalent Young's modulus of the concrete in the socket, considering the stiffening effects of any steel reinforcement.

3. The settlement of the base of the shaft can be obtained by assuming that the rock will behave elastically. The following equation will give an acceptable result:

$$w = \frac{Q_{ST}I_\rho}{B_bE_m} \tag{11.66}$$

where

w = settlement of the base of the drilled shaft,
I_ρ = influence coefficient,
B_b = diameter of drilled shaft, and
E_m = modulus of the in situ rock, taking the joints and their spacing into account.

4. The value of I_ρ can be found by using Figure 11.17. The symbol E_c in the figure refers to Young's modulus of the concrete in the drilled shaft. The value of Young's modulus of the intact rock E_L can be obtained by test or by selecting an appropriate value from Figure 11.14. The value of the modulus of the in situ rock can be found by test, or the intact modulus can be modified in an approximate way. Figure 11.18 allows modification of the modulus of

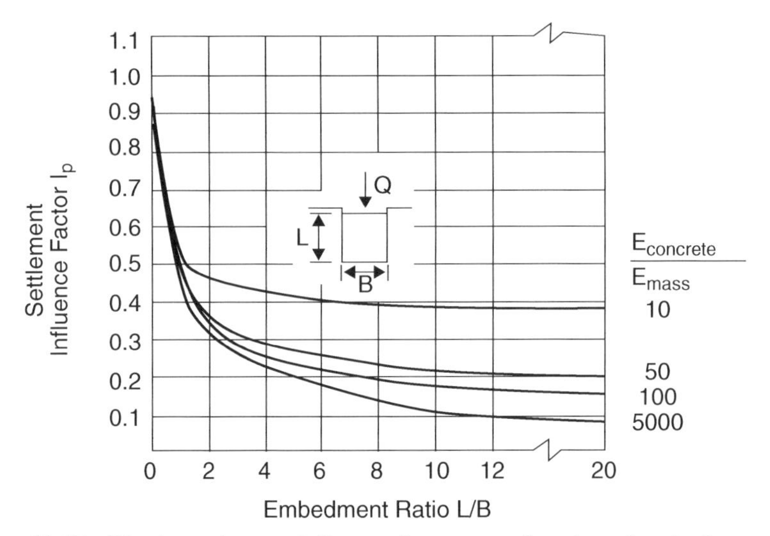

Figure 11.17 Elastic settlement influence factor as a function of embedment ratio and modular ratio (after Donald et al., 1980).

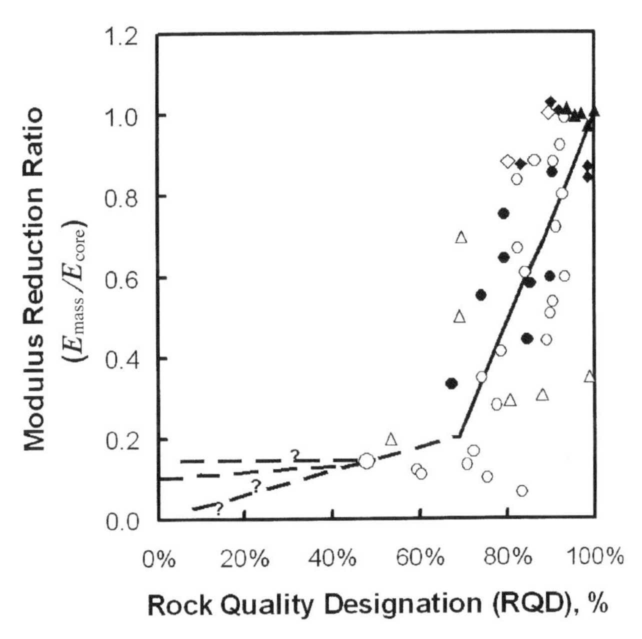

Figure 11.18 Modulus reduction ratio as a function of RQD (from Bieniawski, 1984).

the intact rock by using the RQD. As may be seen, the scatter in the data is great but the trend is unmistakable.

5. The bearing capacity of the rock can be computed by a method proposed by the Canadian Geotechnical Society (1978):

$$q_a = K_{sp}\, q_u \tag{11.67}$$

$$K_{sp} = \frac{3 + c_s/B_b}{10\sqrt{1 + 300\, \delta/c_s}} \tag{11.68}$$

where

q_a = allowable bearing pressure,
K_{sp} = empirical coefficient that depends on the spacing of discontinuities and includes a factor of safety of 3,
q_u = average unconfined compressive strength of the rock cores,
c_s = spacing of discontinuities,
δ = thickness of individual discontinuities, and
B_b = diameter of socket.

6. Equation 11.65 is valid for a rock mass with spacing of discontinuities greater than 12 in. (305 mm) and thickness of discontinuities less than 0.2 in. (5 mm) (or less than 1 in. [25 mm] if filled with soil or rock debris), and for a foundation with a width greater than 12 in. (305 mm). For sedimentary or foliated rocks, the strata must be level or nearly so (Canadian Geotechnical Society, 1978). Again, if the drilled shaft is seated on clay-shale, the procedures described in the previous section should provide a better prediction.

7. If the rock is weak (compressive strength of less than 1000 psi), the design should depend on load transfer in side resistance. The settlement should be checked to see that it does not exceed 0.4 in.

8. If the rock is strong, the design should be based on end bearing. The settlement under working load should be computed to see that it does not exceed the allowable value as dictated by the superstructure.

For the equations for the design of drilled shafts in rock to be valid, the construction must be carried out properly. Because the load-transfer values are higher for rock, the construction requires perhaps more attention than does construction in other materials. For example, for the load transfer in side resistance to attain the allowable values, there must be a good bond between the concrete and the natural rock. An excellent practice is to roughen the sides of the excavation if this appears necessary. There may be occasions when the drilling machine is underpowered and water is placed in the excavation to facilitate drilling. In such a case, the sides of the excavation may be "gun barrel" slick, with a layer of weak material. Roughening of the sides of the excavation is imperative.

Any loose material at the bottom of the excavation should be removed even though the design is based on side resistance.

Another matter of concern with regard to construction in rock is whether or not the rock will react to the presence of water or drilling fluids. Some shales will lose strength rapidly in the presence of water.

Example Problem 5—Shaft in Rock This is an example of a drilled shaft into strong rock.

Soil Profile. The soil profile is shown in Figure 11.19. Only a small amount of water was encountered at the site during the geotechnical investigation.

Soil Properties. The dolomite rock found at the site had a compressive strength of 8000 psi, and the RQD was 100%. Young's modulus of the intact rock was estimated as 2.0×10^6 psi, and the modulus of the rock mass was identical to this value. Assume that the spacing of discontinuities is about 7 ft and that the thickness of the discontinuities is negligible.

Construction. The excavation can be made dry. A socket can be drilled into the strong rock and inspected carefully before concrete is poured.

Loading. The lateral load is negligible. The working axial load is 300 tons. No downdrag or uplift is expected.

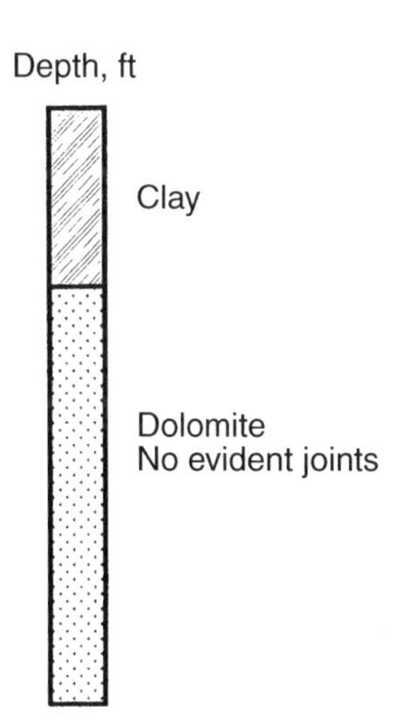

Figure 11.19 General soil and rock description of Example Problem 5.

Factor of Safety. An overall factor of safety of 3 is selected.

Geometry of the Drilled Shaft. A diameter of 3.5 ft is selected, and a socket of 3.5 ft into the dolomite is specified.

Hand Computations. Assuming that all load is transferred in end bearing and using the method proposed by the Canadian Geotechnical Society (1978):

$$q_a = K_{sp}\, q_u$$
$$q_a = (0.5)(8000) = 4000 \text{ psi}$$
$$Q_B = (4000)(\pi/4)(42)^2 = 5.54 \times 10^6 \text{ lb} = 2771 \text{ tons}$$

Note: The value of end bearing includes a factor of safety of 3.

11.6.7 Addition of Side Resistance and End Bearing in Rock

The decision to add or not to add side resistance and end bearing in rock must be made on a case-by-case basis using engineering judgment. A short discussion of several commonly encountered cases follows.

If only compression loading is applied and massive rock exists, it may only be necessary to penetrate the massive rock a short distance, large enough to expose the sound rock. In this situation, only end bearing is counted on because the distance of penetration (usually just a few inches) is too short for any significant side resistance to be developed.

In cases where a rock socket is formed to provide uplift resistance, the decision on whether to add side and base resistance for establishing the ultimate bearing capacity in compression is a matter of engineering judgment in which the nature of the rock must be considered. There are two possible cases. If the rock is brittle, it is possible that the majority of side resistance may be lost as settlement increases to the amount required to fully mobilize end bearing. It would not be reasonable to add side and base resistances together in this situation because they would not occur at large amounts of settlement. If the rock is ductile in shear and deflection softening does not occur, then side resistance and end bearing resistance can definitely be added together.

In cases where end bearing is questionable because of poor rock quality, the length of shafts is often extended in an effort to found the tip of the shaft in good-quality rock. In such situations, the rock socket may become long enough to develop substantial axial capacity in side resistance. In such situations, it is permissible to add side and base resistance together to obtain the total axial capacity of the shaft. Often, questions about the nature of the rock are best answered by conducting a well-planned load test on a full-size test shaft using an Osterberg load cell. The results of such a load test can establish the actual axial load capacity. If the actual axial capacity is sufficiently high, the results of the load test may be used to reduce the size of the foundation shafts to obtain a more economically efficient design.

11.6.8 Commentary on Design for Axial Capacity in Karst

Design of drilled shaft foundations in karst presents the foundation designer with dangerous situations not found elsewhere. In karst geology, the depth to rock is uncertain (i.e., pinnacled rock), solution cavities may be present, numerous seams of softer materials may be present in the rock, and rock ledges and floaters may be encountered above the bearing strata.

These situations make it difficult for the designer to obtain accurate subsurface information from the field exploration program. The uncertain conditions caused by karst make it difficult for the designing engineer to prepare foundation plans and specifications. In such situations, it is mandatory that the engineer recognize the following:

- The field exploration must be more extensive than usual.
- Any change in structure location, or increased loading, will require existing field data to be reviewed and additional field exploration to be conducted to obtain the necessary information.
- Foundation plans and specifications should be written to anticipate the likely changed conditions to be encountered.
- It may be necessary to provide a means in the specifications to set the maximum length of drilled shafts based on side resistance alone when poor end-bearing conditions are encountered during construction.
- It is mandatory to prepare the plans and specifications in a manner consistent with existing purchasing regulations and in sufficient detail that each bid proposal will cover an identical scope of work.

The above conditions make it difficult to estimate the cost of construction accurately. In conditions with pinnacled rock, it may be necessary to drill additional exploratory borings at the location of every drilled shaft to obtain realistic estimates of foundation depths prior to the finalization of bid documents. In such situations, it is mandatory that the additional borings extend to depths greater than the planned depths of the foundations to ensure that no solution cavities exist below the foundations.

The issues discussed in the preceding paragraphs can only be resolved by a thorough field investigation. In some situations with karst, it will be necessary to conduct two or more phases of field boring programs to obtain the necessary information on which to base bid documents.

In large projects with highly variable conditions, use of *technique* shafts may be warranted prior to submission of bids for foundation construction. Technique shafts are full-sized demonstration shafts that are drilled on site. In this practice, all interested contractors are required to be on site to observe the drilling of the technique shafts before they submit their bids for shaft construction. Observation of the technique shafts allows the interested contractors to observe the local conditions to determine the tools and equipment required to complete the job successfully. Use of technique shafts has reduced claims for changed conditions on many projects.

11.6.9 Comparison of Results from Theory and Experiment

Several studies have compared axial capacities computed using the static design methods to experimentally measured axial capacities. The results from one study by Isenhower and Long (1997) are presented in Figure 11.20. The data in this figure represent a database of axial load tests on drilled shafts that was compiled independently of the database used by Reese and O'Neill when developing the FHWA axial capacity methods.

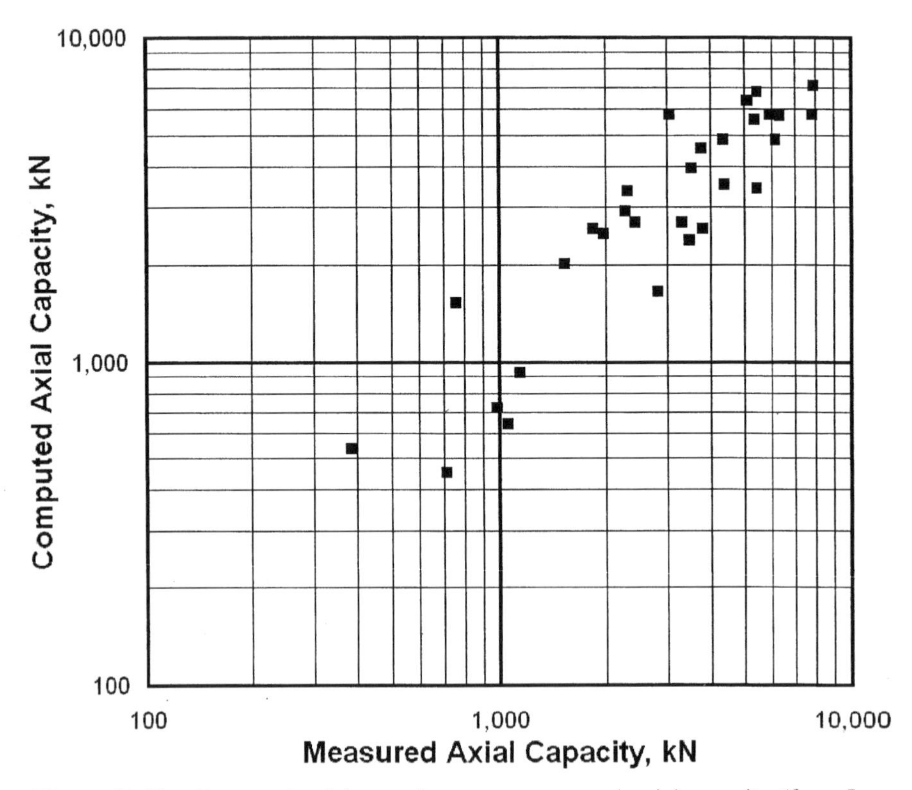

Figure 11.20 Computed axial capacity versus measured axial capacity (from Isenhower and Long, 1997).

PROBLEMS

P11.1. Compute the axial capacity of drilled sand in sand. The shaft diameter is 3 ft and the length is 50 ft.

The soil profile is:

0–20 ft, average N_{60} = 11 blows/ft, γ = 115 pcf
20–50 ft, average N_{60} = 32 blows/ft, γ = 120 pcf
50–80 ft, average N_{60} = 35 blows/ft, γ = 123 pcf

The depth of the water table is 10 ft.

P11.2. What is the allowable axial capacity of the foundation in P11.1 if a factor of safety of 3 is used?

P11.3. Develop a bearing graph for the shaft of P11.1. Plot axial capacity on the horizontal axis and shaft length on the vertical axis. Include shaft lengths from 10 ft to 50 ft, at 5 ft intervals.

P11.4. Develop a bearing graph for shafts in the soil profile of P11.1 for shaft diameters of 2, 3, 4, 5, 6, 7, and 8 ft. Include shaft lengths from 10 ft to 50 ft, at 5 ft intervals.

P11.5. Using the bearing graph of P11.4, estimate the cost of the shaft dimensions that would have allowable capacities of 100 tons, using a factor of safety of 3 and unit cost figures provided by your instructor.

P11.6. Consider a deep layer of sand extending from the ground surface to a depth of 40 meters. For this sand, $\gamma = 18\ \text{kN/m}^3$ and the water table is at the ground surface. Calculate the axial capacity in side resistance for a shaft with a 1 m diameter and a length of 32 m. Perform the computations using a single layer, two equally thick layers, and four equally thick layers. How does the thickness of the layer affect the computed results?

P11.7. Compute the axial capacity of a drilled shaft in a cohesive soil profile. The shaft diameter is 4 ft and the length is 65 ft.

The soil profile is:

0–15 ft, c = 2200 psf, γ = 120 pcf
15–30 ft, c = 2700 psf, γ = 123 pcf
30–55 ft, c = 3250 psf, γ = 125 pcf
55–100 ft, c linearly varying with depth from 3250 to 4400 psf, γ = 130 pcf

The depth of the water table was 15 ft.

P11.8. What is the axial uplift capacity of the shaft of P11.7.

P11.9. Compute the axial capacity of the drilled shaft of P11.7 with a 3-m, 45 deg. Bell.

P11.10. What are the exclusion zones from side resistance for compression loading for the shaft of P11.7?

P11.11. What are the exclusion zones from side resistance for uplift loading for the shaft of P11.7?

P11.12. What are the exclusion zones from side resistance for compression loading for the shaft of P11.9?

P11.13. What are the exclusion zones from side resistance for uplift loading for the shaft of P11.9?

P11.14. Compute the axial capacity of a rock socket. The rock joints are horizontal and closed. The uniaxial compressive strength is 12 MPa, and the RQD = 75%. The diameter of the socket is 1 m and the depth is 3.5 m. Show the computations for side resistance and base resistance.

P11.15. Discuss the situations when axial side resistance and end bearing can be combined for rock sockets and when they cannot be combined.

CHAPTER 12

FUNDAMENTAL CONCEPTS REGARDING DEEP FOUNDATIONS UNDER LATERAL LOADING

12.1 INTRODUCTION

12.1.1 Description of the Problem

A pile subjected to lateral loading is one of a class of problems that involve the interaction of soils and structures. Soil–structure interaction is encountered in every problem in foundation engineering, but in some cases the structure is so stiff that a solution can be developed assuming nonlinear behavior for the soil and no change in shape for the structural unit. But for a pile under lateral loading, a solution cannot be obtained without accounting for the deformation of both the pile and the soil. The deflection of the pile and the lateral resistance of the soil are interdependent; therefore, because of the nonlinearity of the soil, and sometimes of the pile, iterative techniques are almost always necessary to achieve a solution for a particular case of loading on the pile.

This chapter is aimed at developing an understanding of soil–structure interaction as related to the single pile. Many structures are supported by groups of piles, and methods of analyzing such groups will be presented in Chapter 15. However, the methods of analysis of pile groups under lateral loading begin with the analysis of single piles; soil response are then modified according to pile spacing. While this chapter describes the nature of the problem of the laterally loaded pile and presents a useful method of analysis, Chapter 15 presents comprehensive methods of analysis of single piles and utilizes them in the analysis of groups.

12.1.2 Occurrence of Piles Under Lateral Loading

As noted below, a principal use of piles is to support offshore platforms. A system of design and construction is employed to minimize the length of time that an expensive derrick barge with special equipment and personnel is needed at the site. A jacket or template is constructed onshore, floated to the site or transported on a barge, and placed on the ocean floor by the derrick barge. One or more soil borings are made from a special rig prior to construction. Prefabricated piles of open-ended steel pipe are stabbed into slots in the template and driven into position with an impact hammer. The piles are designed to achieve the desired axial load with the largest wall thickness in the area of computed maximum moment, usually near the mudline, and problems arise if the piles cannot be driven to the required depth. Sometimes a smaller-diameter pile is placed to the needed penetration by drilling and grouting, with a sizable increase in cost. Other problems arise if a pile is damaged during the stabbing operation and cannot be driven to the necessary depth. The system, however, has been used successfully at hundreds of offshore locations around the world.

A view of an offshore drilling platform in the Gulf of Mexico from a helicopter is presented in Figure 12.1. The helipad is shown. Helicopters are used for much of the transport of personnel at the offshore location. An interesting feature in the design of fixed offshore platforms is the distance the deck is raised above the water surface. A distance designed to be above the crest of the largest wave is expected. The projected area of the members below the deck is relatively austere, with the purpose of minimizing the forces generated by waves during a hurricane.

Another important use of piles under lateral loading is to support overhead structures sustaining a variety of facilities. Examples are wind farms, trans-

Figure 12.1 Helicopter view of an offshore drilling platform in the Gulf of Mexico.

mission lines, microwave towers, highways, signs, and a variety of units in industrial plants. A view of wind turbines that form a wind farm is presented in Figure 12.2. The height of the tower can range from 22 to 80 m, with a rotor diameter of up to 76 m. The largest turbine can generate power of up to 2000 kW (DNV/Risø, 2001). The environmental characteristics of a wind farm may be understood by viewing the grazing cow in the right foreground of the photograph.

Many pile-supported bridges are subjected to wind and forces from storm water. A special problem is impact from floating vessels. Some high-rise buildings are subjected to large lateral forces from wind, earthquakes, and sometimes from earth pressure.

12.1.3 Historical Comment

Pile-supported structures in the past were frequently subjected to lateral loading, but the piles were designed by using judgment or by referring to building codes, which usually allowed a modest load on each pile if the soil met certain conditions. Designs became more critical when platforms were built offshore in the Gulf of Mexico by the oil industry or off the East Coast of the United States by the Federal government to support early-warning radar equipment.

Engineers understood early that a pile under lateral load would act as a beam. Hetenyi (1946) wrote a book giving the solution of the differential equation for a beam on a foundation, with a linear relationship between pile deflection and soil response. In the early 1950s, Shell Oil Company was planning to install an offshore platform at Block 42 in 25 m (82 ft) of water near the Louisiana coast, where the soil was predominantly soft clay. A pro-

Figure 12.2 View of a wind farm at an onshore location.

cedure was available for computing lateral forces on the platform during a hurricane. Two groups of engineers were asked to work independently and compute the deflection and bending stresses for the pipe piles supporting the platform. One team used the work of Hetenyi, and the other team used limit analysis to obtain the reaction from the soft clay at the site and estimated the pile response. The answers varied widely, and a joint meeting revealed that research was needed on predicting *p-y* curves.

A comprehensive research program was initiated by Shell, and joined later by other companies, for two parallel developments: (1) methods for predicting *p-y* curves and (2) techniques for solving the relevant differential equation. Much of the Shell research was released in time and has been followed by contributions from many investigators. Research is continuing, but given here is practical information for understanding the nature of the problem and for obtaining solutions for some particular cases.

12.2 DERIVATION OF THE DIFFERENTIAL EQUATION

In most instances, the axial load on a laterally loaded pile has relatively little influence on bending moment. However, there are occasions when it is desirable to find the buckling load for a pile; thus, the axial load is needed in the derivation. The derivation for the differential equation for the beam column on a foundation was given by Hetenyi (1946).

The assumption is made that a bar on an elastic foundation is subjected to horizontal loading and to a pair of compressive forces P_x acting in the center of gravity of the end cross sections of the bar. If an infinitely small unloaded element, bounded by two horizontals a distance dx apart, is cut out of this bar (see Figure 2.1), the equilibrium of moments (ignoring second-order terms) leads to the equation

$$(M + dM) - M + P_x dy - V_v dx = 0 \tag{12.1}$$

or

$$\frac{dM}{dx} + P_x \frac{dy}{dx} - V_v = 0 \tag{12.2}$$

Differentiating Eq. 12.2 with respect to x, the following equation is obtained:

$$\frac{d^2M}{dx^2} + P_x \frac{d^2y}{dx^2} - \frac{dV_v}{dx} = 0. \tag{12.3}$$

The following identities are noted:

$$\frac{d^2M}{dx^2} = E_p I_p \frac{d^4y}{dx^4}$$

$$\frac{dV_v}{dx} = p$$

and

$$p = E_{py} y$$

And making the indicated substitutions, Eq. 12.3 becomes

$$E_p I_p \frac{d^4y}{dx^4} + P_x \frac{d^2y}{dx^2} + E_{py} y = 0 \tag{12.4}$$

The direction of the shearing force V_v is shown in Figure 12.3. The shearing force in the plane normal to the deflection line can be obtained as follows:

$$V_n = V_v \cos S - P_x \sin S \tag{12.5}$$

Because S is usually small, $\cos S = 1$ and $\sin S = \tan S = dy/dx$. Thus, Eq. 12.6 is obtained:

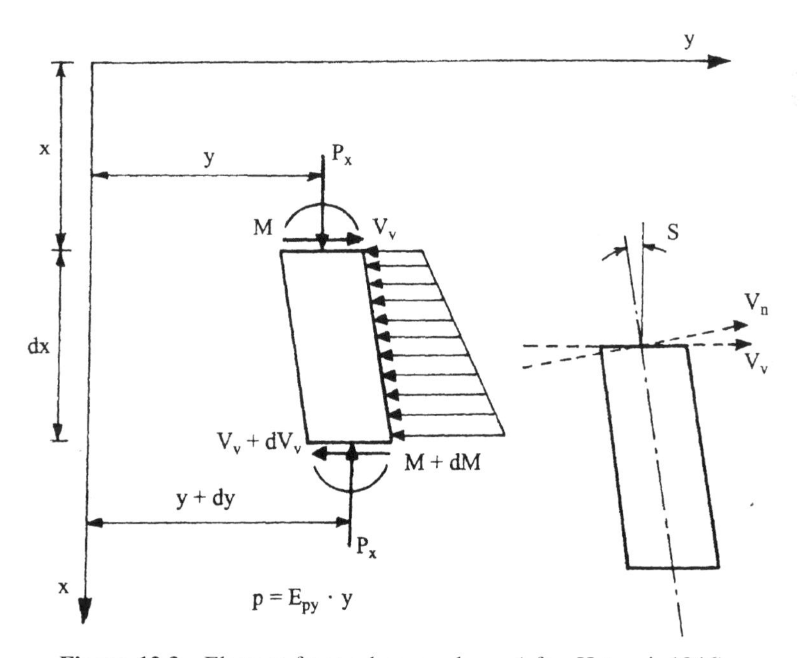

Figure 12.3 Element from a beam column (after Hetenyi, 1946).

$$V_n = V_v - P_x \frac{dy}{dx} \tag{12.6}$$

V_n will be used mostly in computations, but V_v can be computed from Eq. 12.6, where dy/dx is equal to the rotation S.

The ability to allow a distributed force W per unit of length along the upper portion of a pile is convenient in solving a number of practical problems. The differential equation is then given by Eq. 12.7:

$$E_p I_p \frac{d^4}{dx^4} + P_x \frac{d^2y}{dx^2} - p + W = 0 \tag{12.7}$$

where

P_x = axial load on the pile,
y = lateral deflection of the pile at a point x along the length of the pile,
p = soil reaction per unit length,
$E_p I_p$ = bending stiffness, and
W = distributed load along the length of the pile.

Other beam formulas that are needed in analyzing piles under lateral loads are:

$$E_p I_p \frac{d^3y}{dx^3} + P_x \frac{dy}{dx} = V \tag{12.8}$$

$$E_p I_p \frac{d^2y}{dx^2} = M \tag{12.9}$$

$$\frac{dy}{dx} = S \tag{12.10}$$

where

V = shear in the pile,
M = bending moment of the pile, and
S = slope of the elastic curve defined by the axis of the pile.

Except for the axial load P_x, the sign conventions are the same as those usually employed in the mechanics for beams, with the axis for the pile rotated 90° clockwise from the axis for the beam. The axial load P_x does not normally appear in the equations for beams. The sign conventions are presented graphically in Figure 12.4. A solution of the differential equation yields a set of curves such as those shown in Figure 12.5. The mathematical relationships for the various curves that give the response of the pile are shown in the figure for the case where no axial load is applied.

The assumptions that are made in deriving the differential equation are as follows:

1. The pile is straight and has a uniform cross section.
2. The pile has a longitudinal plane of symmetry; loads and reactions lie in that plane.

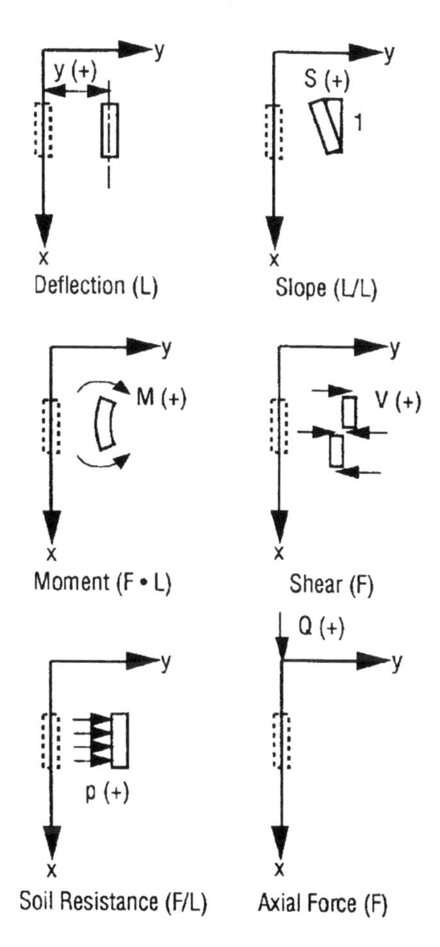

Figure 12.4 Sign conventions for a pile under lateral loading.

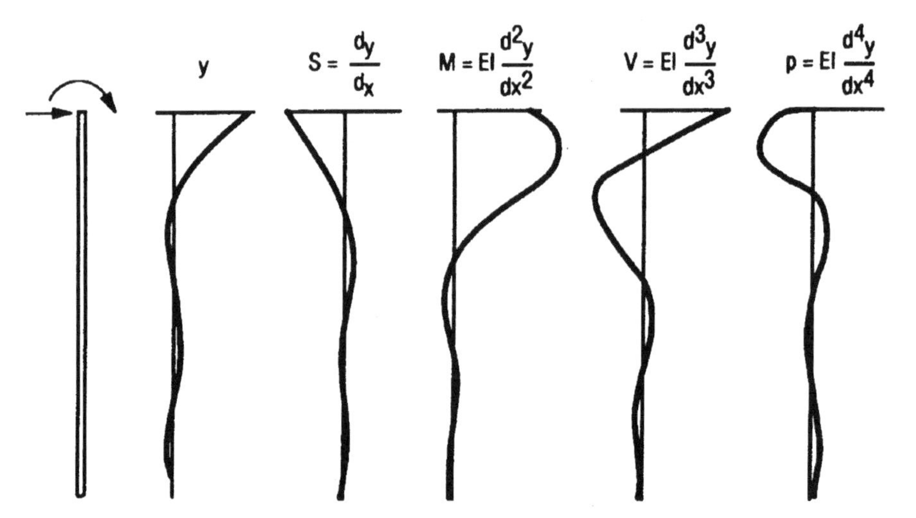

Figure 12.5 Form of the results obtained from a complete solution.

3. The pile material is homogeneous and isotropic.
4. The proportional limit of the pile material is not exceeded.
5. The modulus of elasticity of the pile material is the same in tension and compression.
6. Transverse deflections of the pile are small.
7. The pile is not subjected to dynamic loading.
8. Deflections due to shearing stresses are small.

Assumption 8 can be addressed by including more terms in the differential equation, but errors associated with omission of these terms are usually small. The numerical method presented later can deal with the behavior of a pile made of materials with nonlinear stress-strain properties.

12.2.1 Solution of the Reduced Form of the Differential Equation

A simpler form of the differential equation results from Eq. 12.4 if the assumptions are made that no axial load is applied, that the bending stiffness E_pI_p is constant with depth, and that the soil reaction E_{py} is a constant and equal to α. The first two assumptions can be satisfied in many practical cases; however, the third assumption is seldom satisfied in practice.

The solution shown in this section is presented for two important reasons: (1) the resulting equations demonstrate several factors that are common to any solution; thus, the nature of the problem is revealed; and (2) the closed-form solution allows for a check of the accuracy of the numerical solutions given later in this chapter.

If the assumptions shown above and the identity shown in Eq. 12.11 are employed, a reduced form of the differential equation is shown in Eq. 12.12:

$$\beta^4 = \frac{\alpha}{4E_pI_p} \tag{12.11}$$

$$\frac{d^4y}{dx^4} + 4\beta^4y = 0 \tag{12.12}$$

The solution to Eq. 12.12 may be directly written as

$$y = e^{\beta x}(\chi_1 \cos \beta x + \chi_2 \sin \beta x) + e^{-\beta x}(\chi_3 \cos \beta x - \chi_4 \sin \beta x) \tag{12.13}$$

The coefficients χ_1, χ_2, χ_3, and χ_4 must be evaluated for the various boundary conditions that are desired. If one considers a long pile, a simple set of equations can be derived. An examination of Eq. 12.13 shows that χ_1 and χ_2 must approach zero for a long pile because the term $e^{\beta x}$ will be large with large values of x.

The sketches in Figure 12.6 show the boundary conditions for the top of the pile employed in the reduced form of the differential equations. A more complete discussion of boundary conditions is presented in the next section. The boundary conditions at the top of the long pile that are selected for the first case are illustrated in Figure 12.6a and in equation form are

$$\text{At } x = 0: \quad \frac{d^2y}{dx^2} = \frac{M_t}{E_pI_p} \tag{12.14}$$

$$\frac{d^3y}{dx^3} = \frac{P_t}{E_pI_p} \tag{12.15}$$

From Eq. 12.13 and substitution of Eq. 12.14, one obtains for a long pile

$$\chi_4 = \frac{-M_t}{2E_pI_p\beta^2} \tag{12.16}$$

The substitutions indicated by Eq. 12.15 yield the following:

$$\chi_3 + \chi_4 = \frac{P_t}{2E_pI_p\beta^3} \tag{12.17}$$

Equations 12.16 and 12.17 are used, and expressions for deflection y, slope S, bending moment M, shear V, and soil resistance p for the long pile can be written in Eqs. 12.18 through 12.22:

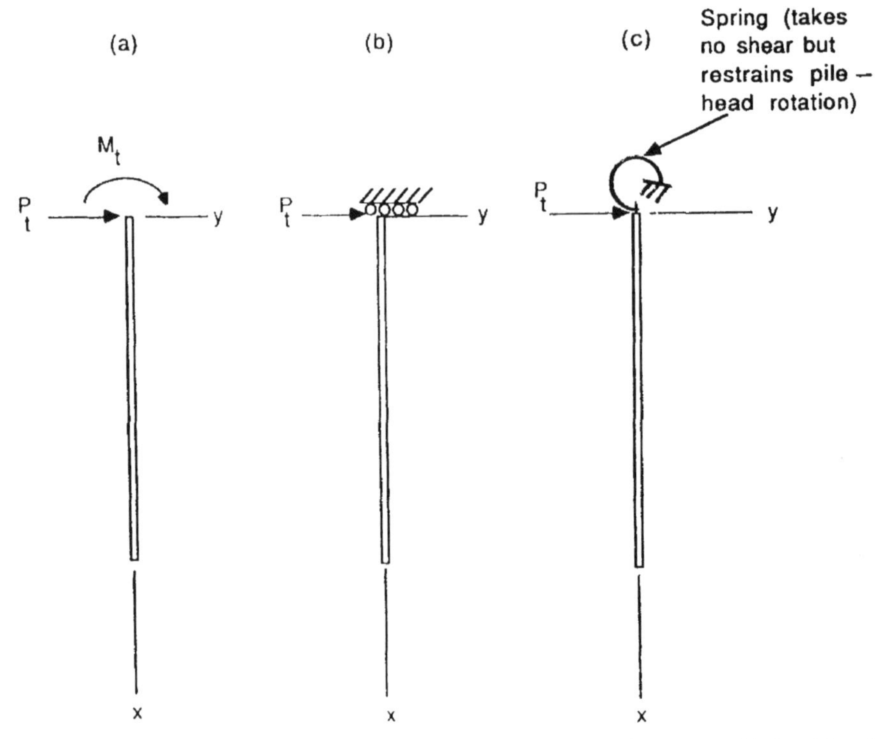

Figure 12.6 Boundary conditions at the top of a pile: (a) free-head. (b) fixed-head. (c) partially restrained.

$$y = \frac{2\beta^2 e^{-\beta x}}{\alpha}\left[\frac{P_t}{\beta}\cos\beta x + M_t(\cos\beta x - \sin\beta x)\right] \tag{12.18}$$

$$S = e^{-\beta x}\left[\frac{2P_t\beta^2}{\alpha}(\sin\beta x + \cos\beta x) + \frac{M_t}{E_p I_p \beta}\cos\beta x\right] \tag{12.19}$$

$$M = e^{-\beta x}\left[\frac{P_t}{\beta}\sin\beta x + M_t(\sin\beta x + \cos\beta x)\right] \tag{12.20}$$

$$V = e^{-\beta x}[P_t(\cos\beta x - \sin\beta x) - 2M_t\beta\sin\beta x] \tag{12.21}$$

$$p = -\beta^2 e^{-\beta x}\left[\frac{P_t}{\beta}\cos\beta x + M_t(\cos\beta x - \sin\beta x)\right] \tag{12.22}$$

It is convenient to define some functions for simplifying the written form of the above equations:

$$A_1 = e^{-\beta x}(\cos \beta x + \sin \beta x) \tag{12.23}$$

$$B_1 = e^{-\beta x}(\cos \beta x - \sin \beta x) \tag{12.24}$$

$$C_1 = e^{-\beta x} \cos \beta x \tag{12.25}$$

$$D_1 = e^{-\beta x} \sin \beta x \tag{12.26}$$

Using these functions, Eqs. 12.18 through 12.22 become

$$y = \frac{2P_t\beta}{\alpha} C_1 + \frac{M_t}{2E_pI_p\beta^2} B_1 \tag{12.27}$$

$$S = \frac{2P_t\beta^2}{\alpha} A_1 - \frac{M_t}{E_pI_p\beta} C_1 \tag{12.28}$$

$$M = \frac{P_t}{\beta} D_1 + M_tA_1, \tag{12.29}$$

$$V = P_tB_1 - 2M_t\beta D_1, \text{ and} \tag{12.30}$$

$$p = -2P_t\beta C_1 - 2M_t\beta^2B_1. \tag{12.31}$$

Values for A_1, B_1, C_1, and D_1, are shown in Table 12.1 as a function of the nondimensional distance βx along the long pile.

For a long pile whose head is fixed against rotation, as shown in Figure 12.6b, the solution may be obtained by employing the boundary conditions given in Eqs. 12.32 and 12.33:

$$\text{At } x = 0: \quad \frac{dy}{dx} = 0 \tag{12.32}$$

$$\frac{d^3y}{dx^3} = \frac{P_t}{E_pI_p} \tag{12.33}$$

Using the same procedures as for the first set of boundary conditions, the results are as follows:

$$\chi_3 = \chi_4 = \frac{P_t}{4E_pI_p\beta^3} \tag{12.34}$$

The solution for long piles, finally, is given in Eqs. 12.35 through 12.39:

TABLE 12.1 Table of Functions for a Pile of Infinite Length

βx	A_1	B_1	C_1	D_1
0	1.0000	1.0000	1.0000	0.0000
0.1	0.9907	0.8100	0.9003	0.0903
0.2	0.9651	0.6398	0.8024	0.1627
0.3	0.9267	0.4888	0.7077	0.2189
0.4	0.8784	0.3564	0.6174	0.2610
0.5	0.8231	0.2415	0.5323	0.2908
0.6	0.7628	0.1431	0.4530	0.3099
0.7	0.6997	0.599	0.3798	0.3199
0.8	0.6354	−0.0093	0.3131	0.3223
0.9	0.5712	−0.0657	0.2527	0.3185
1.0	0.5083	−0.1108	0.1988	0.3096
1.1	0.4476	−0.1457	0.1510	0.2967
1.2	0.3899	−0.1716	0.1091	0.2807
1.3	0.3355	−0.1897	0.0729	0.2626
1.4	0.2849	−0.2011	0.0419	0.2430
1.5	0.2384	−0.2068	0.0158	0.2226
1.6	0.1959	−0.2077	−0.0059	0.2018
1.7	0.1576	−0.2047	−0.0235	0.1812
1.8	0.1234	−0.1985	−0.0376	0.1610
1.9	0.0932	−0.1899	−0.0484	0.1415
2.0	0.0667	−0.1794	−0.0563	0.1230
2.2	0.0244	−0.1548	−0.0652	0.0895
2.4	−0.0056	−0.1282	−0.0669	0.0613
2.6	−0.0254	−0.1019	−0.0636	0.0383
2.8	−0.0369	−0.0777	−0.0573	0.0204
3.2	−0.0431	−0.0383	−0.0407	−0.0024
3.6	−0.0366	−0.0124	−0.0245	−0.0121
4.0	−0.0258	0.0019	−0.0120	−0.0139
4.4	−0.0155	0.0079	−0.0038	−0.0117
4.8	−0.0075	0.0089	0.0007	−0.0082
5.2	−0.0023	0.0075	0.0026	−0.0049
5.6	0.0005	0.0052	0.0029	−0.0023
6.0	0.0017	0.0031	0.0024	−0.0007
6.4	0.0018	0.0015	0.0017	0.0003
6.8	0.0015	0.0004	0.0010	0.0006
7.2	0.0011	−0.00014	0.00045	0.00060
7.6	0.00061	−0.00036	0.00012	0.00049
8.0	0.00028	−0.00038	−0.0005	0.00033
8.4	0.00007	−0.00031	−0.00012	0.00019
8.8	−0.00003	−0.00021	−0.00012	0.00009
9.2	−0.00008	−0.00012	−0.00010	0.00002
9.6	−0.00008	−0.00005	−0.00007	−0.00001
10.0	−0.00006	−0.00001	−0.00004	−0.00002

$$y = \frac{P_t \beta}{\alpha} A_1 \tag{12.35}$$

$$S = \frac{P_t}{2E_p I_p \beta^2} D_1 \tag{12.36}$$

$$M = -\frac{P_t}{2\beta} B_1 \tag{12.37}$$

$$V = P_t C_1 \tag{12.38}$$

$$p = -P_t \beta A_1 \tag{12.39}$$

It is sometimes convenient to have a solution for a third set of boundary conditions, as shown in Figure 12.6c. These boundary conditions are given in Eqs. 12.40 and 12.41:

$$\text{At } x = 0: \quad \frac{E_p I_p \dfrac{d^2y}{dx^2}}{\dfrac{dy}{dx}} = \frac{M_t}{S_t} \tag{12.40}$$

$$\frac{d^3y}{dx^3} = \frac{P_t}{E_p I_p} \tag{12.41}$$

Employing these boundary conditions, for the long pile the coefficients χ_3 and χ_4 were evaluated and the results are shown in Eqs. 12.42 and 12.43. For convenience in writing, the rotational restraint M_t/S_t is given the symbol k_θ.

$$\chi_3 = \frac{P_t(2E_p I_p + k_\theta)}{E_p I_p(\alpha + 4\beta^3 k_\theta)} \tag{12.42}$$

$$\chi_4 = \frac{k_\theta P_t}{E_p I_p(\alpha + 4\beta^3 k_\theta)} \tag{12.43}$$

These expressions can be substituted into Eq. 12.13 and differentiation performed as appropriate. Substitution of Eqs. 12.23 through 12.26 will yield a set of expressions for the long pile similar to those in Eqs. 12.27 through 12.31 and 12.35 through 12.39.

Timoshenko (1941) stated that the solution for the long pile is satisfactory where $\beta L \geq 4$; however, there are occasions when the solution of the reduced differential equation is desired for piles that have a nondimensional length less than 4. The solution at any pile length L can be obtained by using the following boundary conditions at the tip of the pile:

$$\text{At } x = L: \quad \frac{d^2y}{dx^2} = 0 \quad (M \text{ is zero at pile tip}) \tag{12.44}$$

$$\text{At } x = L: \quad \frac{d^3y}{dx^3} = 0 \quad (V \text{ is zero at pile tip}) \tag{12.45}$$

When the above boundary conditions are fulfilled, along with a set for the top of the pile, the four coefficients χ_1, χ_2, χ_3, and χ_4 can be evaluated.

The influence of the length of a pile on the groundline deflection is illustrated in Figure 12.7. A pile with loads applied is shown in Figure 12.7a; an axial load is shown, but the assumption is made that this load is small, so the lateral load P_t and the moment M_t will control the length of the pile. Computations are made with constant loading, constant pile cross section, and an initial length in the long-pile range. The computations proceed with the length being reduced in increments; the groundline deflection is plotted as a function of the selected length, as shown in Figure 12.7b. The figure shows that the groundline deflection is unaffected until the critical length is approached. At

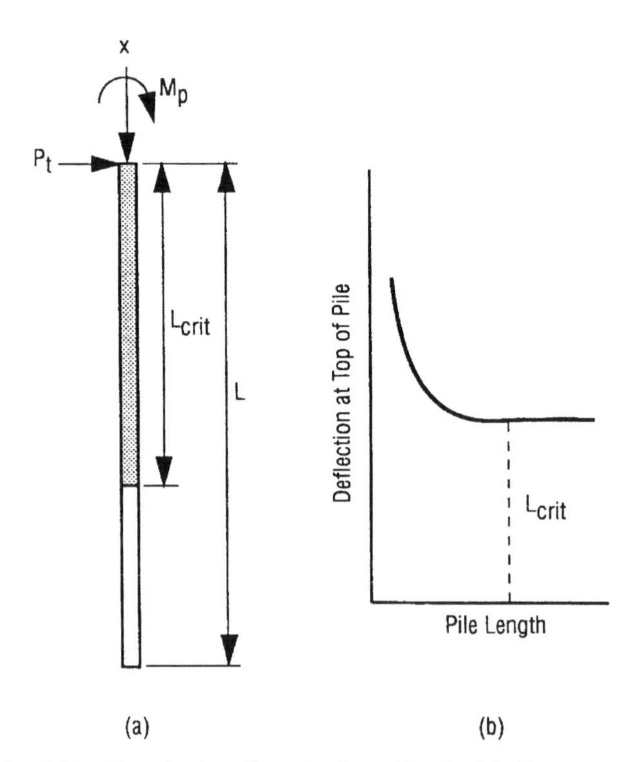

Figure 12.7 Critical length of a pile under lateral loads: (a) pile geometry and loads. (b) pile-head deflection vs. pile length.

this length, only one point of zero deflection will occur in the computations. There will be a significant increase in the groundline deflection as the length in the solution is made less than critical. The engineer can select a length that will give an appropriate factor of safety against excessive groundline deflection. The accuracy of the solution will depend, of course, on how well the soil-response curves reflect the actual situation in the field.

The reduced form of the differential equation will not normally be used for the solution of problems encountered in design. However, the influence of pile length, pile stiffness, and other parameters is illustrated with clarity.

12.3 RESPONSE OF SOIL TO LATERAL LOADING

The above discussion suggests that the key to determining the behavior of a pile under lateral loading is the ability to develop a family of curves (*p-y* curves) that give soil reaction as a function of the lateral deflection of a pile. The nature of *p-y* curves may be understood by referring to Figure 12.8, where a slice of soil is examined (Reese, 1983). Figure 12.6b shows a uniform distribution of unit stresses with units of F/L^2 around a cylindrical pile at some depth z_1 below the ground surface, assuming that the pile was installed without bending. If the pile is assumed to be moved laterally through a distance y_1 (exaggerated in the figure for ease of presentation), the distribution of stresses after the deflection is symmetrical but no longer uniform (Figure 12.8c). The stresses on the side in the direction of movement (front side) have increased, and those on the back side have decreased. Integration of the unit stresses in Figure 12.8c yields the value of p_1 that acts in a direction opposite the deflection y and is called the *soil resistance* or *soil reaction,* with units for force per unit of pile length (F/L). The soil resistance p is similar in concept to the distributed loading encountered in designing a beam supporting a floor slab. Reflection shows that, for the particular value of z below the ground surface, the value of p will vary with the deflection y of the pile and with the kind of soil into which the pile is installed, p_{ult}.

A pile subjected to an axial load, a lateral load, and a moment is shown in Figure 12.9a. A possible shape of the deflected pile is shown in Figure 12.9b, along with a set of nonlinear mechanisms that serve to resist the deflection. A set of *p-y* curves is shown in Figure 12.9c, representing the response of the soil as simulated by the mechanisms. Procedures for predicting *p-y* curves in a variety of soils and rocks are given in Chapter 14; however, the information here demonstrates the method and leads to practical solutions for a class of problems.

The shape of a *p-y* curve conceptually is shown in Figure 12.10. An initial straight-line portion and a final straight-line portion, p_{ult}, define the curve. The two lines are connected by a curve. By referring to Figure 12.8, one can reason that at small deflections, the initial slope of the *p-y* curve should be

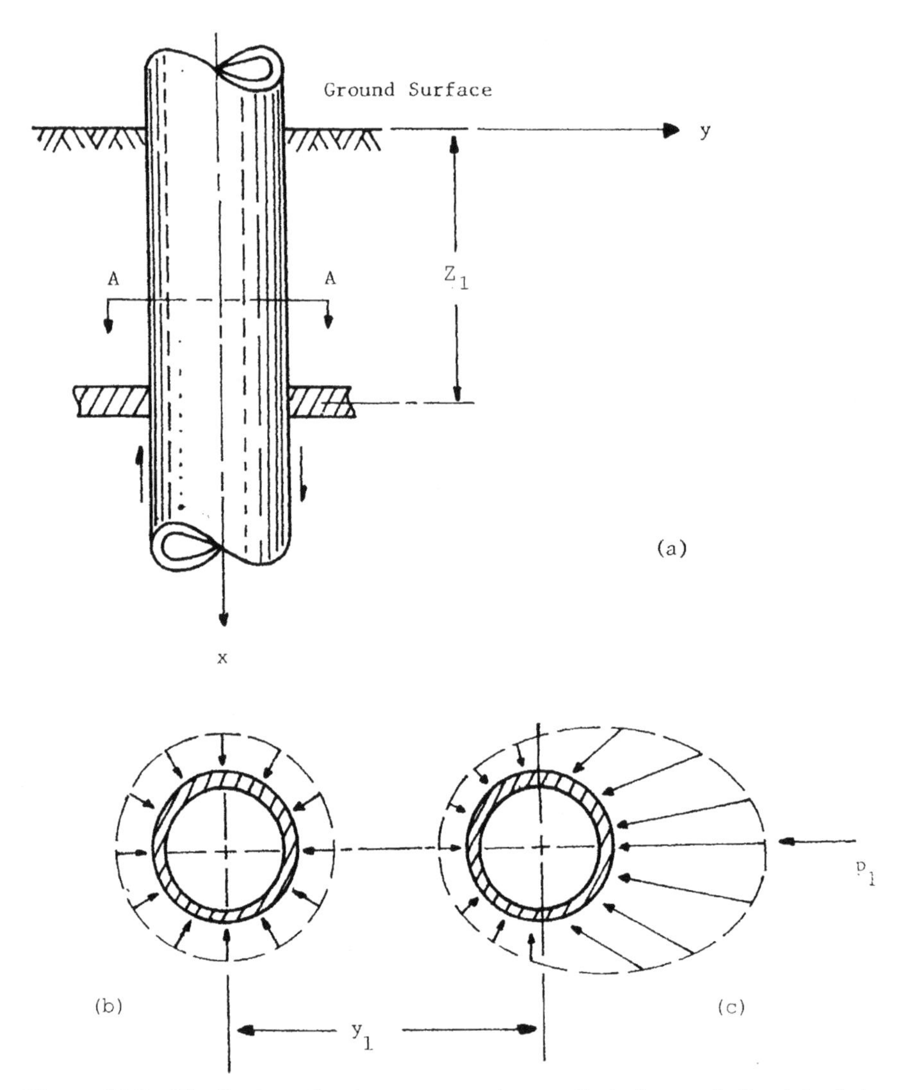

Figure 12.8 Distribution of unit stresses against a pile before and after lateral deflection: (a) elevation view of section of pile. (b) earth pressure prior to lateral loading. (c) earth pressure after lateral loading.

related in some direct way to the initial modulus of the stress-strain curve for the soil. Further, the value of p_{ult} should be related in some direct way to the failure of the soil in bearing capacity.

Figure 12.10 shows a dashed line labeled E_{py}, a secant to the p-y curve that will be implemented in the analyses that follow. As shown in the figure, E_{py} will have the greatest value with small values of y and will decrease with

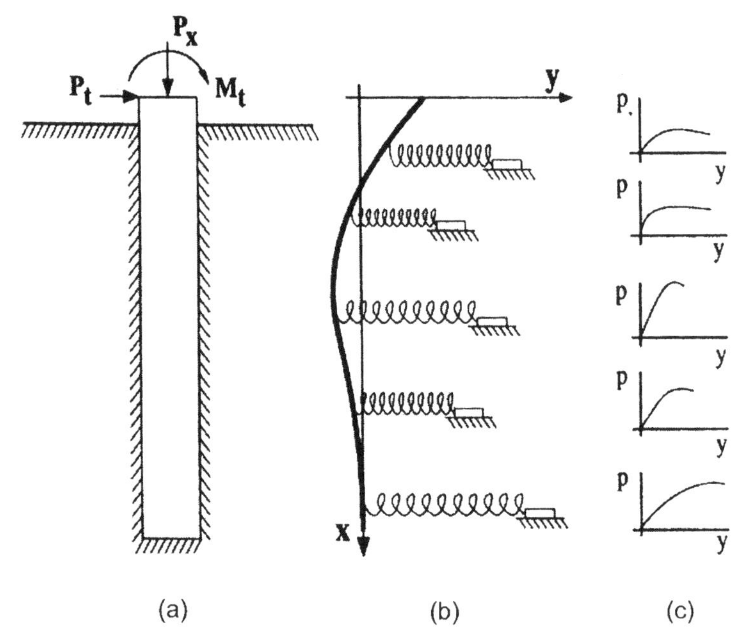

Figure 12.9 Model of a pile subjected to lateral loading with a set of *p-y* curves: (a) pile under lateral loading. (b) soil spring model. (c) *p-y* curves.

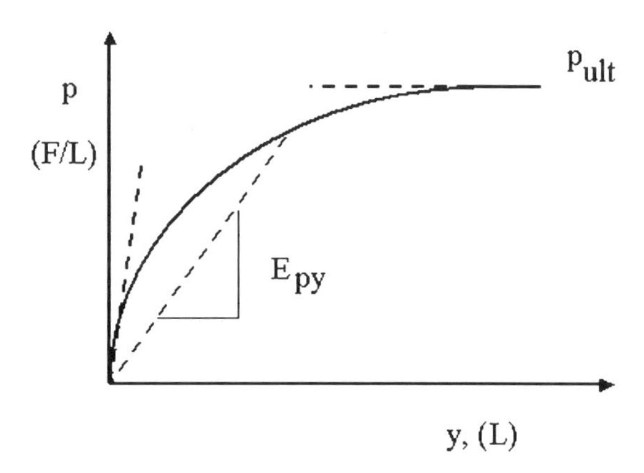

Figure 12.10 Conceptual shape of a *p-y* curve.

increasing pile deflection. Because of the nonlinearity of the response of the soil to deflection, a solution must proceed for a given pile and a given set of loadings by selecting trial values of E_{py}, solving a set of equations for deflection of the pile, obtaining improved values of E_{py}, and continuing until convergence is achieved. The computer solves such a problem with ease, but hand solutions are also available, as shown later in this chapter.

12.4 EFFECT OF THE NATURE OF LOADING ON THE RESPONSE OF SOIL

Four types of loading can be defined as affecting the response of soil around a pile when loaded laterally: short-term static, sustained, repeated, and dynamic. Dynamic loading includes loads from machines and from seismic events. Short-term static loading will lead to *p-y* curves that conceptually can be developed from properties of the supporting soil but do not occur often in practice. But *p-y* curves for short-term static loading are useful for cases where deflections are small and for sustained loading if the supporting soil is granular or overconsolidated clay. If sustained loading is present and the supporting soil is soft to medium clay, the engineer can expect that consolidation will occur. Lateral earth pressure can be computed, and the equations of consolidation can be applied to the extent possible. Installing a pile in the clay and imposing lateral loading, while observing deflection with time, is a useful procedure (Reuss et al., 1992).

Repeated and cyclic loading that will occur against an offshore structure is another matter. If deflection is small where the supporting soil is acting elastically, then the *p-y* curves for static loading may be used. For larger deflection, a significant loss of resistance will occur for overconsolidated clay and some for granular soil. If the mudline deflection of a pile is large enough that a gap will occur and if water exists at the site, erosion of overconsolidated clay will occur as water alternately falls into the gap and is squeezed out. The prediction of *p-y* curves for such a case does not readily yield to analysis but depends strongly on results from appropriate experiments. Figure 12.11a shows experimental *p-y* curves for cases of short-term static loading, and Figure 12.11b shows curves for cyclic loading for a pile in overconsolidated clay (Reese et al., 1975). The influence of cyclic loading on the *p-y* curves is apparent.

The *p-y* curves for short-term static loading can usually be used without modification for dynamic loading if deflections are very small. The prediction of *p-y* curves affected by earthquakes is a complicated problem. Some of the factors affecting the problem can be identified: the precise definition of the earthquake, the distance of the earthquake from the site, and the nature of the soils in the area of the site. The lateral and vertical accelerations at the site must be predicted utilizing the concept of microzonation. An earthquake in Mexico caused extensive damage in the Valley of Mexico, consisting of a deep deposit of clay, but the damage in the hills outside the Valley was

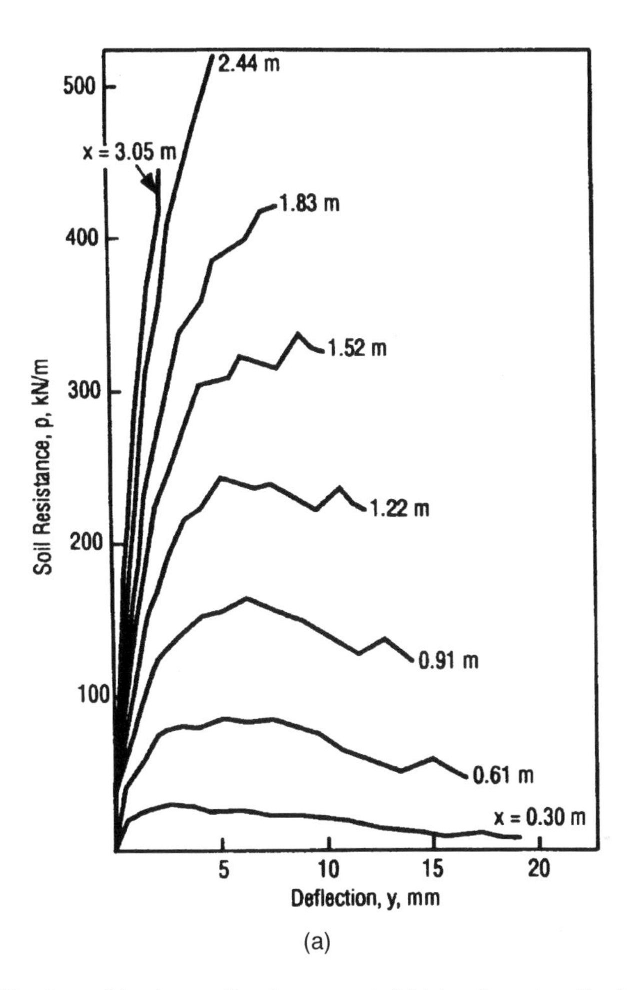

Figure 12.11 Lateral load test of an instrumented 24-in.-diameter pile: (a) *p-y* curves developed from a static load. (*Continued on page 398.*)

minimal. Along with accelerations of the ground surface, movements of the soil will occur as a function of depth. The *p-y* curves presented here do not include effects of inertia, nor are movements of the soil during earthquakes with respect to depth taken into account; therefore, the curves cannot be applied to an analysis of a pile-supported structure subjected to a seismic event without special consideration.

12.5 METHOD OF ANALYSIS FOR INTRODUCTORY SOLUTIONS FOR A SINGLE PILE

Engineers understood many years ago that the physical nature of some soils led to the argument that E_{py} should be zero at the mudline and increase

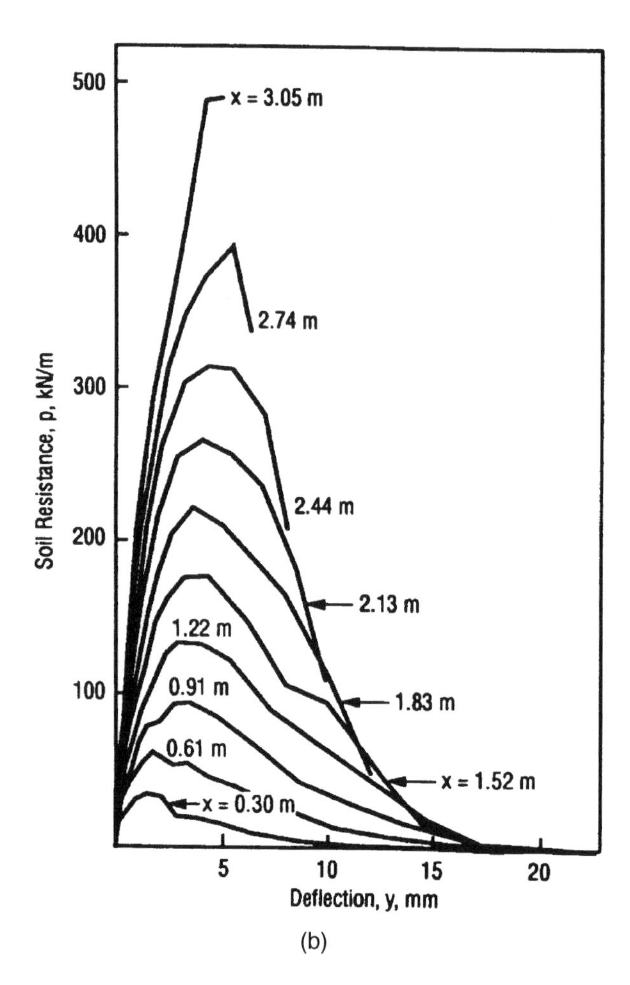

Figure 12.11 (*Continued*) (b) *p*-*y* curves developed from a cyclic load.

linearly with depth or that E_{py} is equal to $k_{py}x$. Granular soil and normally consolidated clays follow such a pattern and, even for overconsolidated clays, the value of the p_{ult} will be low near the ground surface. Therefore, crude but useful numerical solutions were made using hand-operated calculators, with the assumption that $E_{py} = k_{py}x$. Examination of the analytical parameters in the numerical solutions led to the proposal of a formal analytical procedure for $E_{py} = k_{py}x$ (Reese and Matlock, 1956) and later to the use of nondimensional methods to develop a wide range of solutions for a pattern of variations of E_{py} with depth (Matlock and Reese, 1962). While the availability of cheap and powerful personal computers has made rapid numerical solutions to Eq. 12.4 practical and inexpensive (see Chapter 14), reasons exist to present solutions for $E_{py} = k_{py}x$ where no axial load is applied ($p_x = 0$). The method presents effectively the nature of the solution, yields the relationship between relevant parameters, and provides a useful method of solution for a range of practical problems.

The following equations can be developed by numerical analysis (Matlock and Reese, 1962) for the case where $E_{py} = k_{py}x$. A lateral load and a moment may be imposed at the pile head, and the length of the pile can be considered.

$$y = A_y \frac{P_t T^3}{E_p I_p} + B_y \frac{M_t T^2}{E_p I_p} \tag{12.46}$$

$$S = A_s \frac{P_t T^2}{E_p I_p} + B_s \frac{M_t T}{E_p I_p} \tag{12.47}$$

$$M = A_m P_t T + B_m M_t \tag{12.48}$$

$$V = A_v P_t + B_v \frac{M_t}{T} \tag{12.49}$$

$$T = \sqrt[5]{\frac{E_p I_p}{k_{py}}} \tag{12.50}$$

$$Z_{\max} = \frac{L}{T} \tag{12.51}$$

where

y, S, M, and V = deflection, slope, moment, and shear, respectively, along the length of the pile,

P_t and M_t = lateral load (shear) and applied moment at the pile head,

T = termed the *relative stiffness factor*, and

A_y, B_y, A_s, B_s, A_m, B_m, A_v, and B_v = nondimensional parameters shown in Figures 12.12 through 12.19.

The curves are entered with Z, where Z is equal to x/T. Because the soil is extending to the top of the pile, the distance below the ground surface z and the coordinate giving the distance along the pile from its head x are identical for this nondimensional method.

Several curves are shown in each of the figures where the nondimensional length of the pile is given. As may be seen in Figure 12.12, a pile with a nondimensional length of 2 exhibits only one point of zero deflection, with a

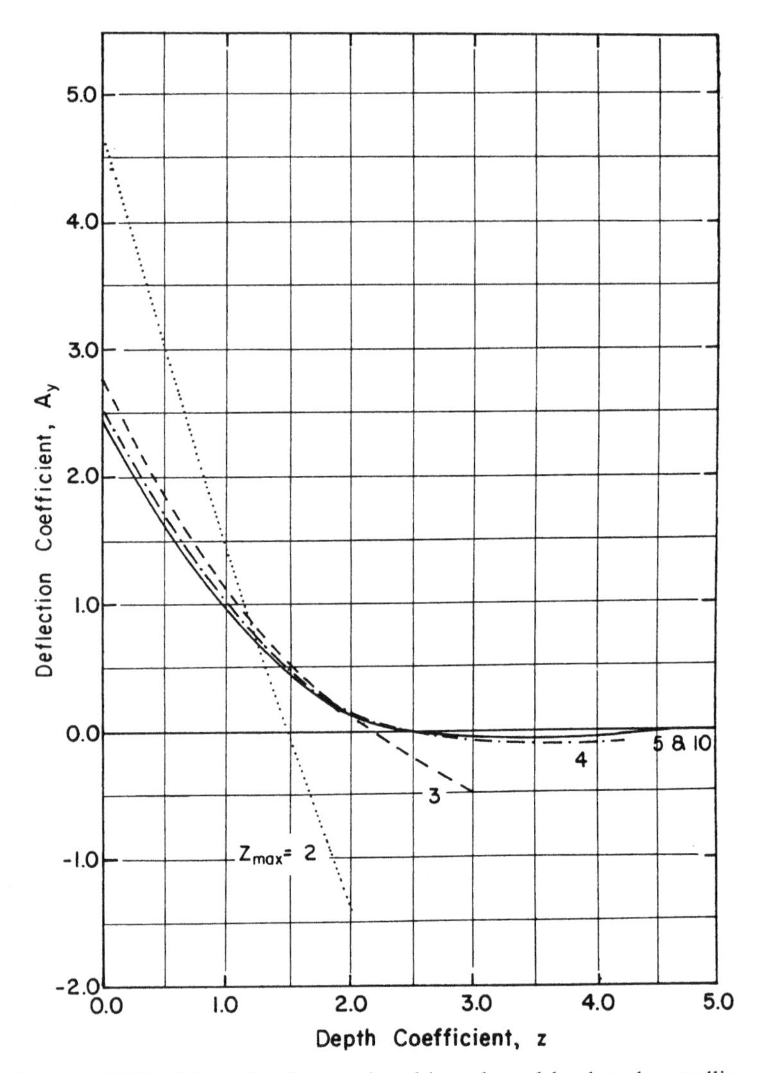

Figure 12.12 Pile deflection produced by a lateral load at the mudline.

substantial deflection of the bottom of the pile. In a number of practical cases, the embedded length of a pile is determined from the equations for axial resistance, and multiple points of zero deflection will appear during the analysis for lateral loading. If the length of the pile is dependent only on the response to lateral loading, piles are usually designed to have an embedded length to indicate two or three points of zero deflection. A "short" pile can behave like a fence post, typical of the case for a nondimensional length of 2.

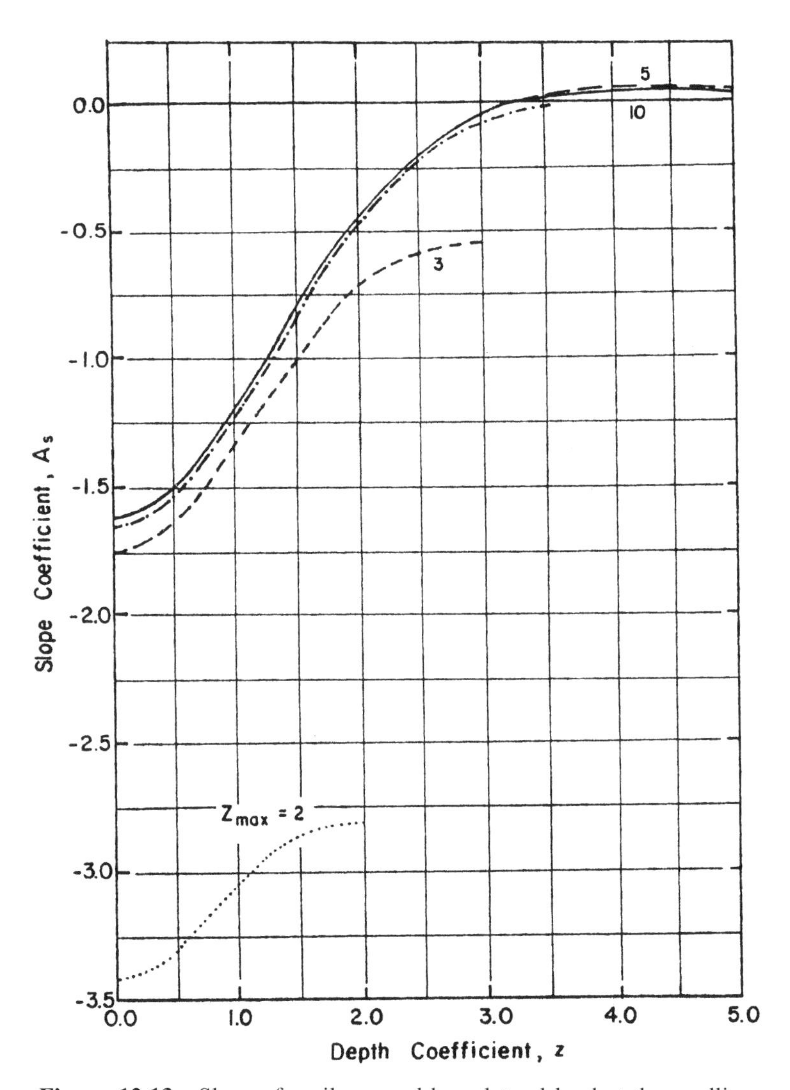

Figure 12.13 Slope of a pile caused by a lateral load at the mudline.

12.6 EXAMPLE SOLUTION USING NONDIMENSIONAL CHARTS FOR ANALYSIS OF A SINGLE PILE

A steel-pipe pile has been driven into the ground to serve as an anchor. A lateral load of 100 kN is to be applied to the top of the pile that is free to rotate. The load is increased to 300 kN to achieve an appropriate level of safety against failure of the pile where failure will occur with excessive bending. The magnitude of the deflection of the pile at the groundline is not judged to be critical.

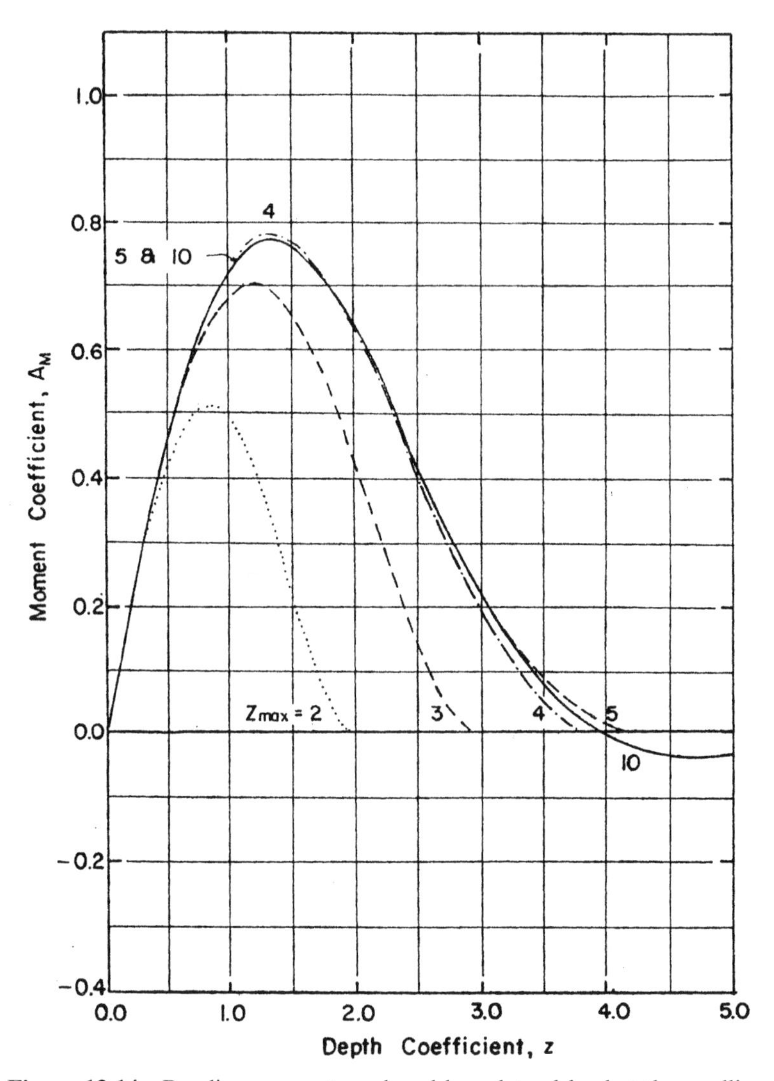

Figure 12.14 Bending moment produced by a lateral load at the mudline.

For purposes of the example, a steel pipe was selected with an outside diameter of 380 mm and an inside diameter of 330 mm. The moment of inertia I was computed to be 4.414×10^{-4} m^4, and the modulus of elasticity E was taken as 2.0×10^8 kPa, yielding the E_pI_p value of 88,280 kN-m^2. The yield strength of the steel was taken as 250,000 kPa, the stress at which a plastic hinge would occur. The length of the pipe was selected as 20 m.

A family of p-y curves was developed to represent the resistance of the soil to lateral loading, as shown in Figure 12.20. The curves are for an over-

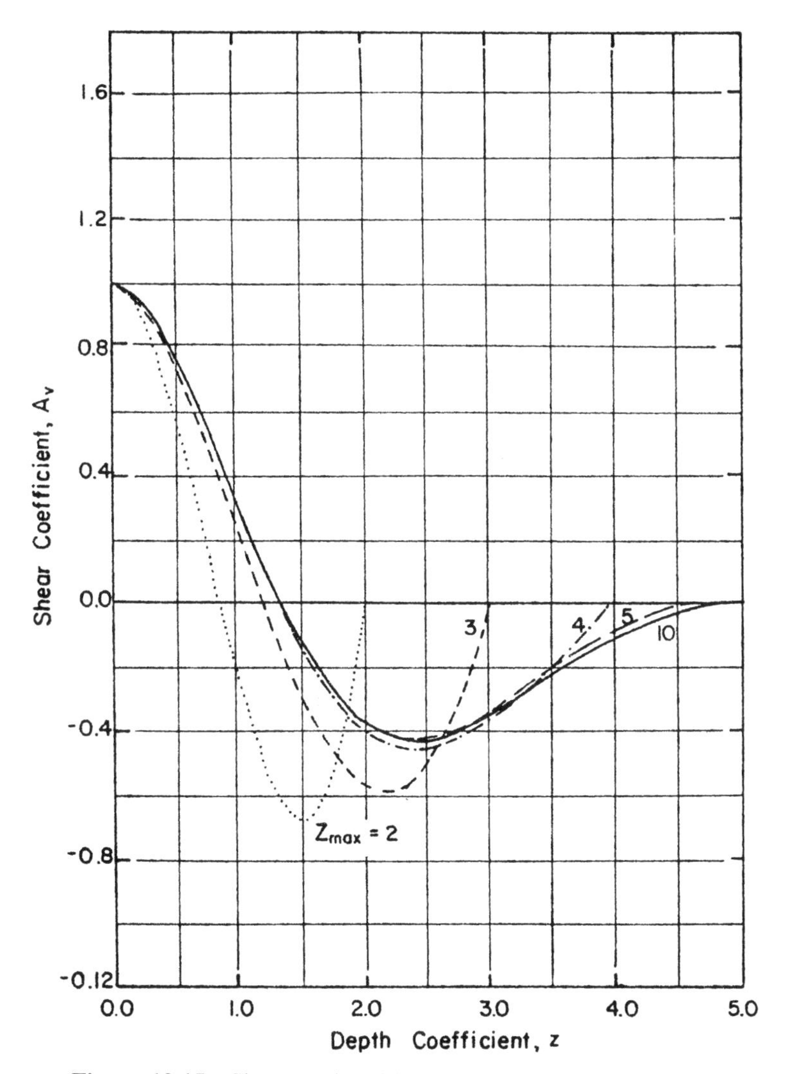

Figure 12.15 Shear produced by a lateral load at the mudline.

consolidated clay, assumed to be the soil at the site, for a pile with an outside diameter of 380 mm, and for the loading expected on the pile. Some aspects of the curves are instructive. For example, the initial portions of the curves form straight lines, representing a linear relationship between stress and strain. The initial portions of the curves show increased stiffness with depth, which is to be expected. The final portions of the curves are also straight lines, reflecting the fact that the deflection of the pile was sufficient to mobilize the full shear strength of the soil at the various depths. The *p-y* curve for the

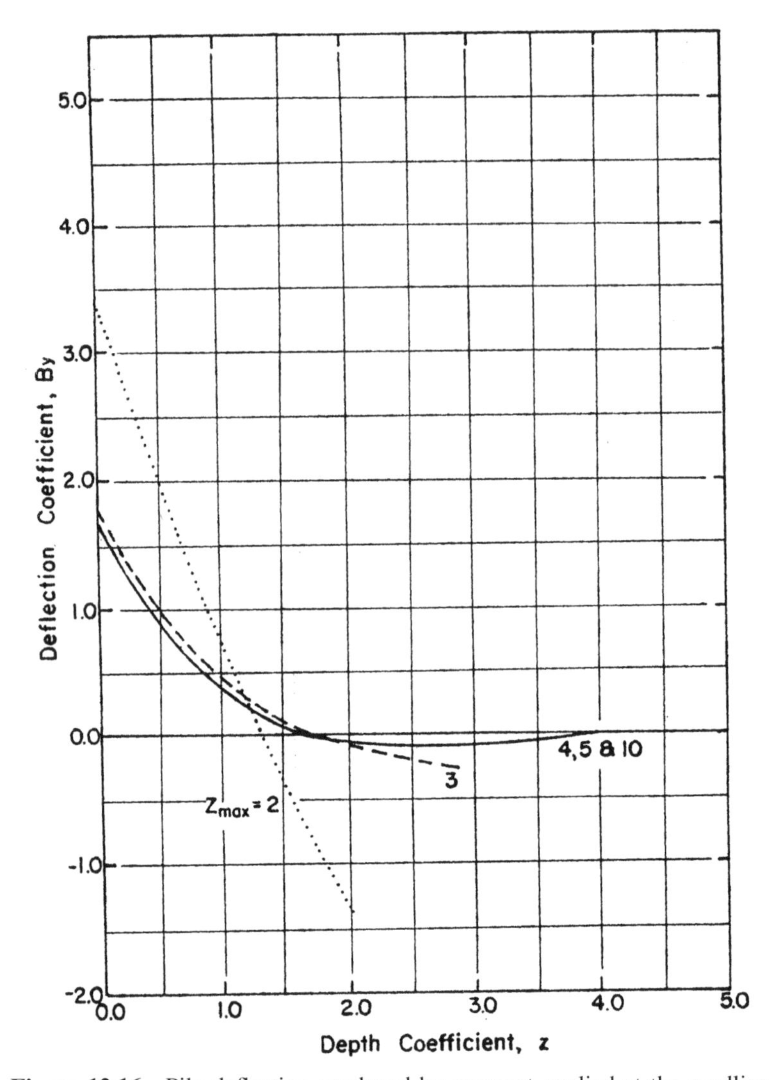

Figure 12.16 Pile deflection produced by moment applied at the mudline.

ground surface has some resistance; therefore, the assumption of $E_{py} = k_{py}x$ is not ideal for the example problem being solved. As noted earlier, the assumption that $E_{py} = k_{py}x$ should apply better to uniform sand or to normally consolidated clay.

Even though the nondimensional method is not ideally applied to the family of curves shown in Figure 12.20, the exercise that follows shows that useful results can be obtained when less than ideal *p-y* curves are used.

The procedure for a solution with the nondimensional method starts with (1) selection of the relative stiffness factor T (T_{tried}); (2) computation of Z_{max}

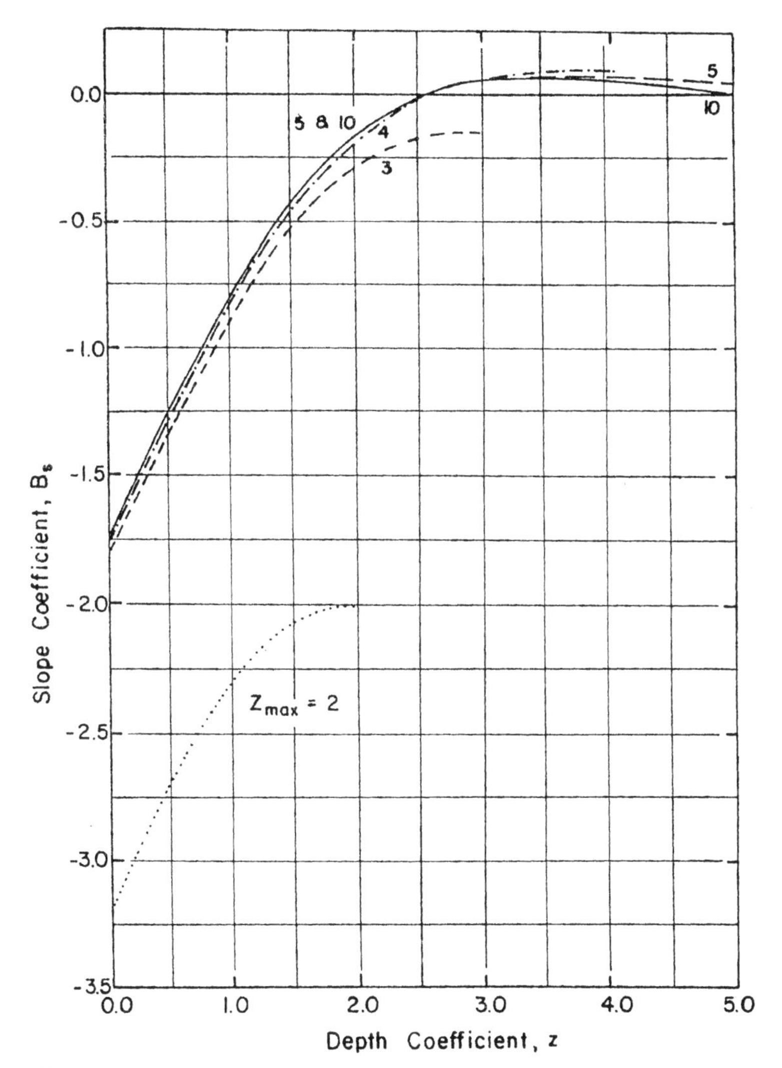

Figure 12.17 Slope of pile caused by moment applied at the mudline.

to select the curve to use in the nondimensional plots (Figures 12.12 through 12.19); (3) computation of a trial deflection of the pile using Eq. 12.46, selection of a corresponding value of p from Figure 12.20, computing E_{py} by dividing values of p by the corresponding values of y, plotting E_{py} as a function of x, fitting the best straight line from the origin through the plotted points to obtain the best value of k_{py}, where k_{py} is equal to E_{py} divided by x; and, finally, (4) computing a new value of T (T_{obt}) by using Eq. 12.50. If T_{obt} does not equal T_{tried}, a new value of T_{tried} is selected for a second trial. Two

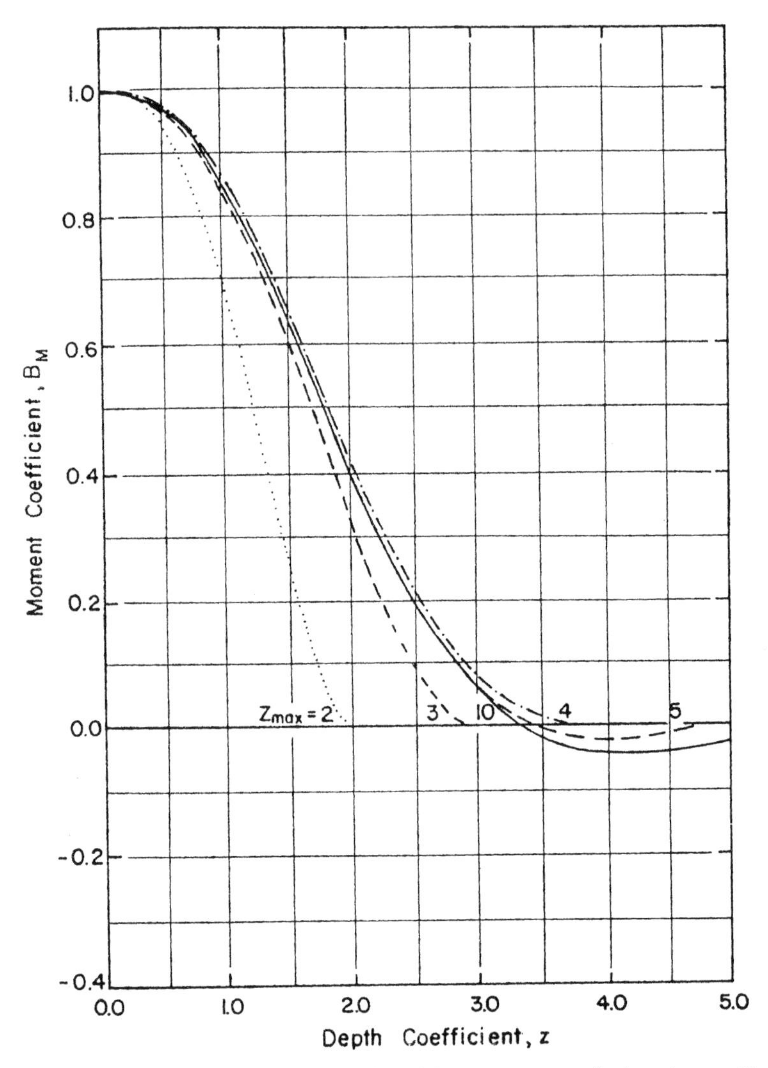

Figure 12.18 Bending moment produced by moment applied at the mudline.

trials are frequently sufficient to obtain convergence, where successive trials are necessary to accommodate the nonlinearity of the problem.

Try T = 2 m, $Z_{\max} = \dfrac{L}{T} = 10$; use the appropriate curve for A_y in Figure 12.12.

$$y_A = A_y \frac{P_t T^3}{E_p I_p} = A_y \frac{(300)(2^3)}{88{,}280} = A_y(0.027186)\ \text{m} = A_y(27.186)\ \text{mm}$$

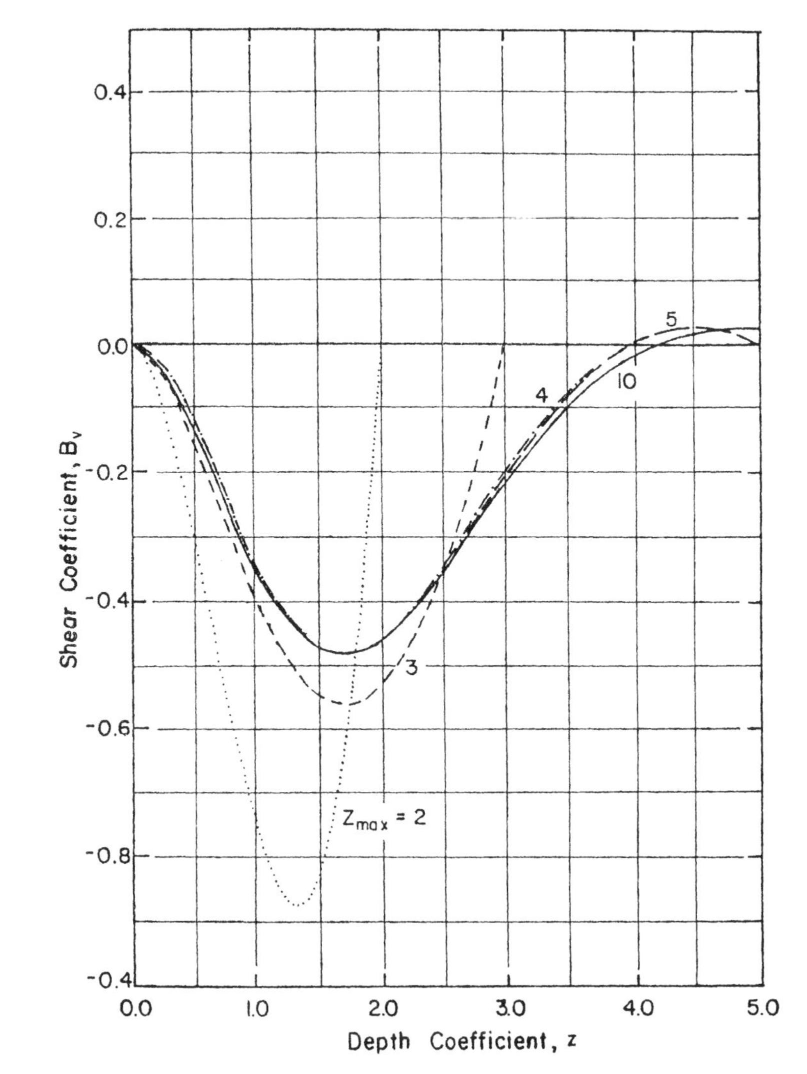

Figure 12.19 Shear produced by moment applied at the mudline.

x, m	Z	A_y	y_A, mm	p, kN/m	E_{py}, kPa
0	0	2.435	66.2	110	1,662
0.25	0.125	2.2	59.8	107	1,789
0.51	0.255	2.0	54.4	115	2,114
1.02	0.51	1.6	43.5	133	3,057
1.78	0.89	1.1	29.9	150	5,017
2.54	1.27	0.7	19.0	162	8,526
3.81	1.91	0.15	4.1	140	34,146

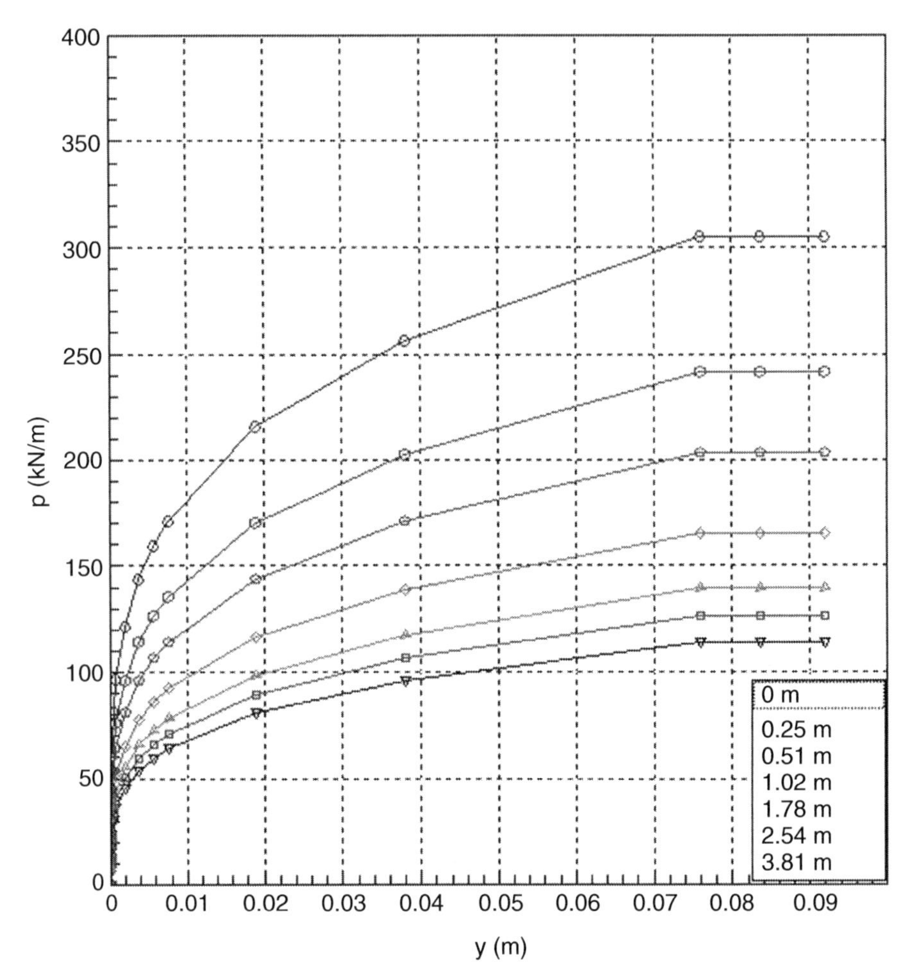

Figure 12.20 Family of *p-y* curves for an example problem.

The derived values of E_{py} versus x for a value of T of 2.0 are plotted in Figure 12.21. A straight line is plotted that passes through the origin. As can be seen, the first three points are on the right side of the line, the next two are on the left side, and the final point is on the right side. Even though the line seems to be a poor fit of the data points, the results are useful. The computations show that the slope of the line was employed to find an obtained value of T, a value of 1.91 m.

Because the value of T_{obt} was lower than that of T_{tried}, the second trial was made with a T_{tried} value of 1.5 in order to limit the number of trials to achieve convergence. The computations proceeded as before. The new value of T led

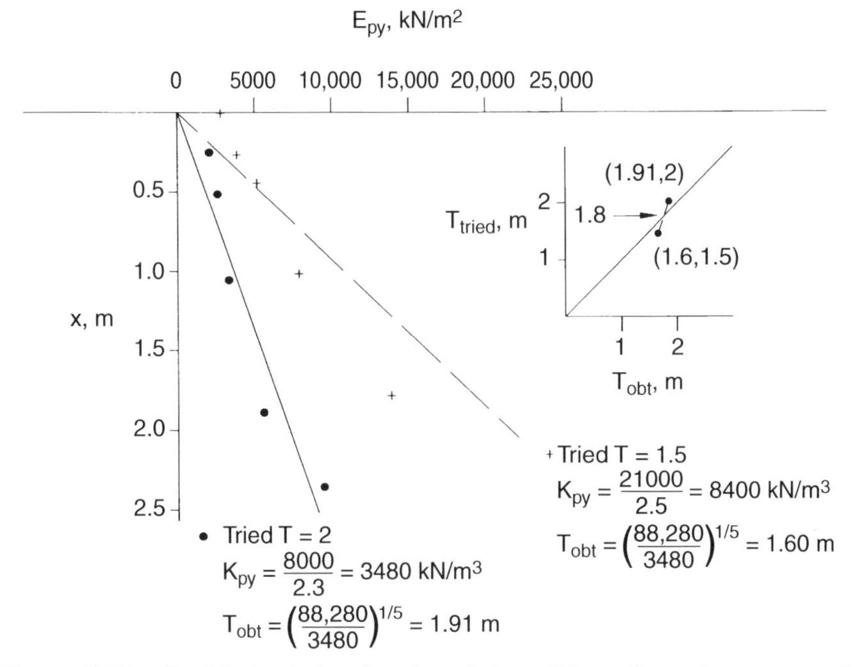

Figure 12.21 Graphical solution for the relative stiffness factor for an example problem.

to new values of Z and A_y, and hence to new values of p and E_{py}. The plotted results are shown in Figure 12.21, where the value of T_{tried} of 1.5 m led to a value of T_{obt} of 1.60 m. The values of the relative stiffness factor tried and obtained are plotted in the sketch in Figure 12.21. As shown in the plot, convergence occurred at a value of T of 1.80 m.

The response of the pile under the lateral load of 300 kN can now be computed using the first terms in Eqs. 12.46 and 12.48. After computing $Z_{\max}$, the curves to use in Figs. 12.12 and 12.14 may be selected.

$$Z_{\max} = 20/1.8 = 11.1$$

so use the curves for $Z_{\max} = 10$.

$$y = A_y P_t T^3 / E_p I_p = A_y \frac{(300)(1.8^3)}{88{,}280} = A_y\ (0.01982)\ \text{m} = A_y\ (19.82)\ \text{mm}$$

$$M = A_m P_t T = A_m\ (540)\ \text{kN-m}$$

Z	x, m	A_y	y, mm	A_m	M, kN-m
0	0	2.4	47.6	0	0
0.5	0.9	1.6	31.7	0.46	248
1.0	1.8	0.9	17.8	0.72	389
1.5	2.7	0.4	7.9	0.76	410
2.0	3.6	0.2	4.0	0.64	346
2.5	4.5	0	0	0.41	221
3.0	5.4	−0.05	−1.0	0.22	119
4.0	7.2	−0.05	−1.0	0	0
5.0	9.0	0	0	−0.04	−22

$$f_{\max} = \frac{(410)\left(\frac{0.38}{2}\right)}{4.414 - 10^{-4}} = 176{,}500 \text{ kPa} < 250{,}000$$

As the above computation shows, the load factor of 3 led to a computed bending stress less than that required to cause a plastic hinge. The computed values of deflection and bending moment are plotted in Figure 12.22. The digital computer was used to solve the problem with the method presented in

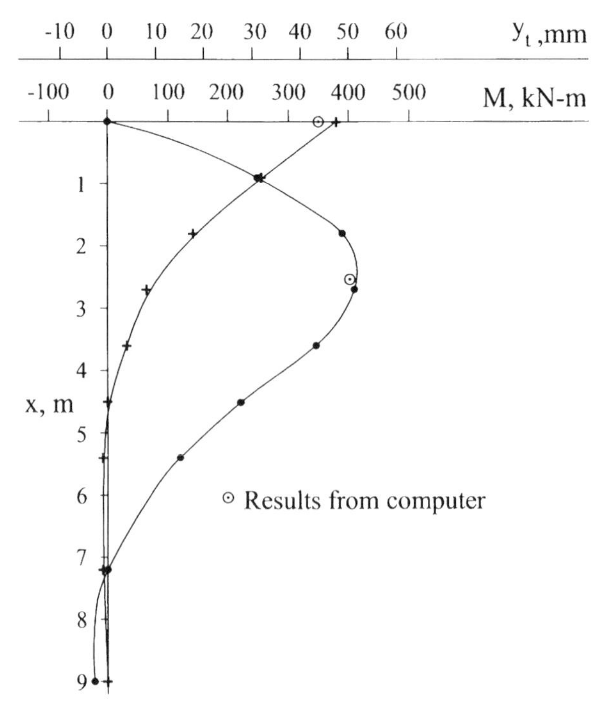

Figure 12.22 Computed values of deflection and maximum bending moment for an example problem.

Chapter 14, and the results for maximum bending moment and groundline deflection are plotted in Figure 12.22. While the computer gave a value of groundline deflection about 10% lower than the value from the nondimensional method, the position and magnitude of the maximum bending moment of the computer solution agreed closely with the results of the nondimensional method.

The deflection at the groundline of 47.6 mm may be considered excessive even though the lateral load had been factored up. At one time, excessive deflection at the groundline caused some investigations to speculate that there had been a "soil failure," but the soil does not actually fail; the deflection becomes excessive only for some applications. For the case analyzed, the designer might wish to make a computation without factoring up the load to obtain an idea of the deflection under the working load.

Several comments are pertinent concerning the solution obtained with the nondimensional method with the assumption that $E_{py} = k_{py}x$.

1. The values of deflection for the top of the pile obtained with the nondimensional method and the computer are in reasonable agreement. The values of maximum bending moment achieved agree almost exactly, both in regard to value and to location. One could reason that the good agreement between the results is only accidental, but close agreement has also been found when other problems are solved with both methods.
2. Compared with the computer solution, the nondimensional method is time-consuming and tedious because values must be estimated from curves and the nonlinearity of the problem requires iteration.
3. However, the nondimensional method is valuable because a computer solution can be checked, which may be required in some cases, and because the hand solution gives the beginning user an excellent introduction to a fairly complex problem.
4. The nondimensional method is limited in that no axial load may be imposed, the pile stiffness must be constant with depth and independent of the bending moment, and the method is not applicable to layered soils. All of these limitations are overcome by the methods shown in Chapter 14.

PROBLEMS

P12.1. List examples of piles under lateral loading in the community where you live and sketch one example, showing rough dimensions and the source of lateral loading, along with a note showing the condition of the soil at the ground surface.

P12.2. For the solved example in the section on nondimensional solutions, compute the deflection at the top of an unloaded extension of the pile to a height of 5 m.

P12.3. **a.** For the pile and *p-y* curves used in the example in the nondimensional method of solution, work out the case where the pile is loaded with a horizontal load of 65 kN at 8 m above the groundline, yielding a value of P_t of 65 kN and a value of M_t of 520 kN-m. Solve for the value of T_{obt} with values of T_{tried} assigned by the instructor.

b. Bring your values of T_{tried} and T_{obt} to class to make a plot to find the value of T where the solution converges.

c. Using the value of T found in (b), make a plot of pile deflection y and bending moment M as a function of length along the pile. Compute the value of the maximum bending stress and compare it with the stress that will cause a plastic hinge.

CHAPTER 13

ANALYSIS OF INDIVIDUAL DEEP FOUNDATIONS UNDER AXIAL LOADING USING THE *t*-*z* MODEL

13.1 SHORT-TERM SETTLEMENT AND UPLIFT

13.1.1 Settlement and Uplift Movements

Two analytical methods are used to compute the load-settlement curve of an axially loaded pile. One method uses the theory of elasticity. The theories discussed by D'Appolonia and Romualdi (1963), Thurman and D'Appolonia (1965), Poulos and Davis (1968), Poulos and Mattes (1969), Mattes and Poulos (1969), and Poulos and Davis (1980) use methods based on the theory of elasticity. These methods use Mindlin's (1936) equations for stress and deformations at any point in the interior of semi-infinite, elastic, and isotropic solids resulting from a force applied at another point in the solids.

The displacement of the pile is calculated by superimposing the influences of the load transfer in side resistance along the length of the pile and in end bearing at the pile tip. The compatibility of those forces and the displacement of the pile are obtained by solving a set of simultaneous equations. This method takes the stress distribution within the soil into consideration; therefore, the elasticity method presents the possibility of solving for the behavior of a group of closely spaced piles under axial loadings (Poulos, 1968; Poulos and Davis, 1980).

The drawback of the elasticity method lies in the basic assumptions that must be made. The actual ground conditions rarely if ever satisfy the basic assumption of uniform and isotropic material. In spite of the highly nonlinear stress-strain characteristics of soils, the only soil properties considered in the elasticity method are Young's modulus E and the Poisson's ratio υ. The use of only two constants, E and υ, to represent soil characteristics is too much of an oversimplification to allow the elasticity-based methods to work in conditions involving stratified soils with differing strengths and compressibilities.

The other method used to calculate the load-settlement curve for an axially loaded pile may be called the *load-transfer method* (commonly referred to as the t-z method). Finite-difference equations are employed to achieve compatibility between pile displacement and the load transfer along a pile and between displacement and resistance at the tip of the pile. This method was first used by Seed and Reese (1957); other studies have been reported by Coyle and Reese (1966), Coyle and Sulaiman (1967), and Kraft et al. (1981). The t-z difference method assumes the Winkler concept; that is, the load transfer at a certain pile section and the pile tip resistance are independent of the pile displacement elsewhere. The close agreement between the prediction and the loading test results in clays (Coyle and Reese, 1966) and the scattering of prediction values for the loading test in sands (Coyle and Sulaiman, 1967) may possibly be explained by the relative sensitivity of a soil to changes in patterns of stress. Admitting the deficiency in the displacement-shear force criteria of sand, the finite-difference method is still practical and potential because it can deal with any complex composition of soil layers with any nonlinear relationship of displacement versus shear force. Furthermore, the method can accommodate improvements in soil criteria with no modifications of the basic theory.

In the following sections, the derivation of the finite-difference equations is shown. The fundamental technique employed here is the same as that used in previous investigations. The difference lies in the computation procedure. The method shown here gives a solution first for the pile displacement at all stations. Then the pile force at each station is calculated. Convergence of the iterative computation is quite fast even near the ultimate load.

13.1.2 Basic Equations

The mechanical model for an axially loaded pile is presented in Figure 13.1. The pile head is subjected to an axial force P_t and undergoes a displacement z_t. The pile-tip displacement is z_{tip}, and the pile displacement at the depth x is z. Displacement z is positive downward, and the compressive force P is positive.

In an element dx (Figure 13.1a), the strain in the element due to the axial force P is calculated by neglecting the second-order term dP.

$$\frac{dz}{dx} = -\frac{P}{EA} \tag{13.1a}$$

or

$$P = -EA\frac{dz}{dx} \tag{13.1b}$$

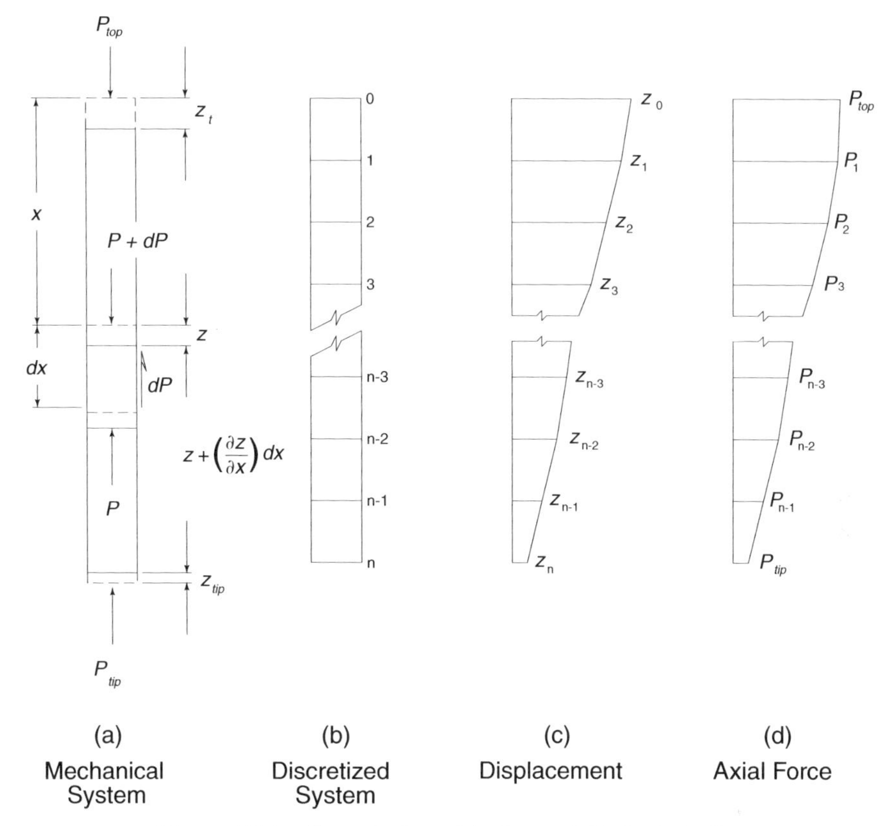

Figure 13.1 Numerical model of an axially loaded pile.

where

P = axial force in the pile in pounds (downward positive),
E = Young's modulus of pile material in psi, and
A = cross-sectional area of the pile in square inches.

The total load transfer through an element dx is expressed by using the modulus μ in the load transfer curve (Figure 13.2a).

$$dP = -\mu \, z \, \ell \, dx \tag{13.2}$$

or

$$\frac{dP}{dx} = -\mu \, z \, \ell \tag{13.3}$$

where

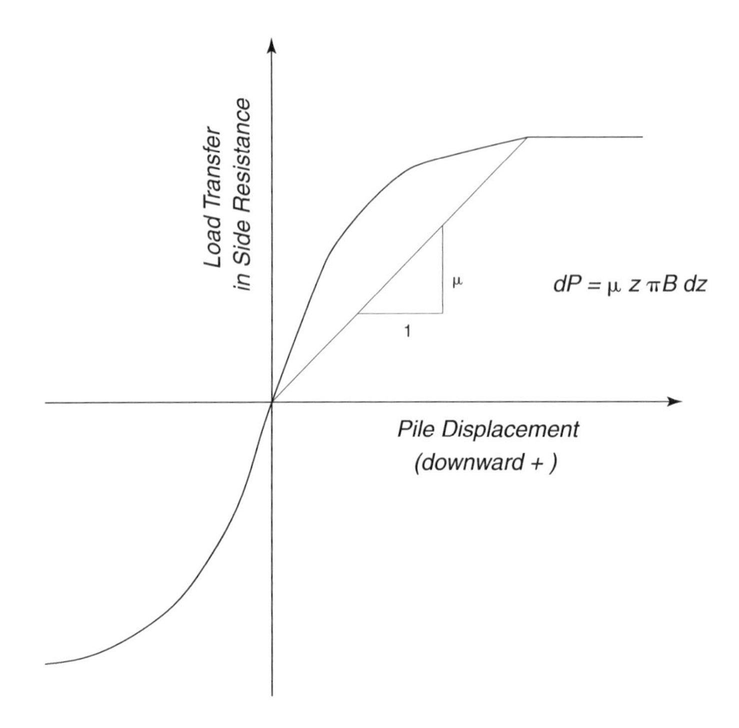

(a) Load Transfer Curve for Side Resistance

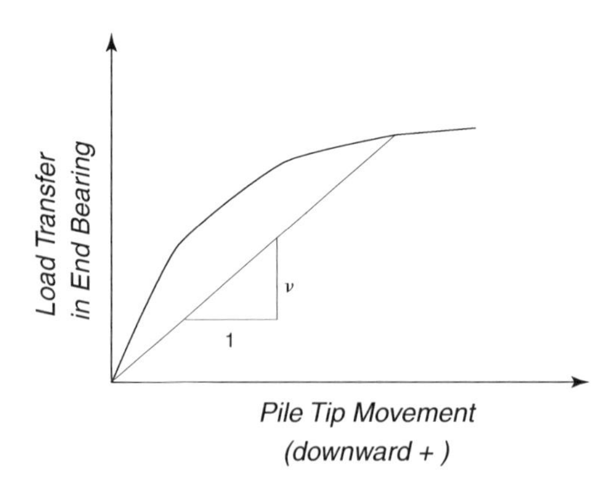

(b) Load Transfer Curve in End Bearing

Figure 13.2 Load-transfer curves for side and tip resistance.

ℓ = circumference of a cylindrical pile or the perimeter encompassing an H-pile, and
μ = modulus in the load transfer curve in Figure 13.2a.

Equation 13.1 is differentiated with respect to x and equated with Eq. 13.3 to obtain Eq. 13.4:

$$\frac{d}{dx} EA \frac{dz}{dx} = \mu \, z \, \ell \tag{13.4}$$

The pile-tip resistance is the product of a secant modulus v and the pile-tip movement z_{tip}. (See the curves of pile-tip movement versus resistance, Figure 13.2b).

$$P_{tip} = v \, z_{tip} \tag{13.5}$$

Equation 13.4 is the basic differential equation that must be solved. Boundary conditions at the tip and at the top of the pile must be established. The boundary condition at the tip of the pile is given by Eq. 13.5. At the top of the pile, the boundary condition may be either a force or a displacement. Treatment of these two cases is presented later.

13.1.3 Finite Difference Equations

Equation 13.6 gives in difference-equation form the differential equation (Eq. 13.4) for solving the axial pile displacement at discrete stations.

$$a_i z_{i+1} + b_i z_i + c_i z_{i-1} = 0 \tag{13.6}$$

for

$$a_i = \tfrac{1}{4} EA_{i+1} + EA_i - \tfrac{1}{4} EA_{i-1} \tag{13.7}$$

$$b_i = -\mu l h^2 - 2EA_i \tag{13.8}$$

$$c_i = -\tfrac{1}{4} EA_{i+1} + EA_i + \tfrac{1}{4} EA_{i-1} \tag{13.9}$$

and where h = increment length or dx.

13.1.4 Load-Transfer Curves

The acquisition of load-transfer curves from a load test requires that the pile be instrumental internally for the measurement of axial load with depth. The number of such experiments is relatively small, and in some cases the data are barely adequate; therefore, the amount of information that can be used to develop analytical expressions is limited.

Undoubtedly, additional studies will be reported in technical literature from time to time. Any improvements made in load-transfer curves can be readily incorporated into the analyses.

13.1.5 Load-Transfer Curves for Side Resistance in Cohesive Soil

Coyle and Reese (1966) examined the results from three instrumented field tests and the results from rod tests in the laboratory and developed a recommendation for a load-transfer curve. The curve was tested by using results of full-scale experiments with uninstrumented piles. The comparisons of computed load-settlement curves with those from experiments showed agreement that was excellent to fair. Table 13.1 presents the fundamental curve developed by Coyle and Reese.

This table shows that the movement to develop full load transfer is quite small. Furthermore, the curve is independent of soil properties and pile diameter.

Reese and O'Neill (1988) studied of the results of several field-load tests of instrumented bored piles and developed the curves shown in Figure 13.3. This figure shows that the maximum load transfer occurred at approximately 0.6% of the diameter of a pile. Because the piles tested had diameters of 24 to 36 in., the movement at full load transfer would be on the order of 0.2 in., a much larger value than was obtained by Coyle and Reese.

Kraft et al. (1981) studied the theory of load transfer in side resistance and noted that pile diameter, axial pile stiffness, pile length, and distribution of soil strength and stiffness along the pile are all factors that influence load-transfer curves. Equations for computing the curves were presented. Vijayvergiya (1977) also presented a method for obtaining load-transfer curves.

TABLE 13.1 Load Transfer versus Pile Movement for Cohesive Soil

Ratio of Load Transfer to Maximum Load Transfer	Pile Movement, in.
0	0
0.18	0.01
0.38	0.02
0.79	0.04
0.97	0.06
1.00	0.08
0.97	0.12
0.93	0.16
0.93	0.20

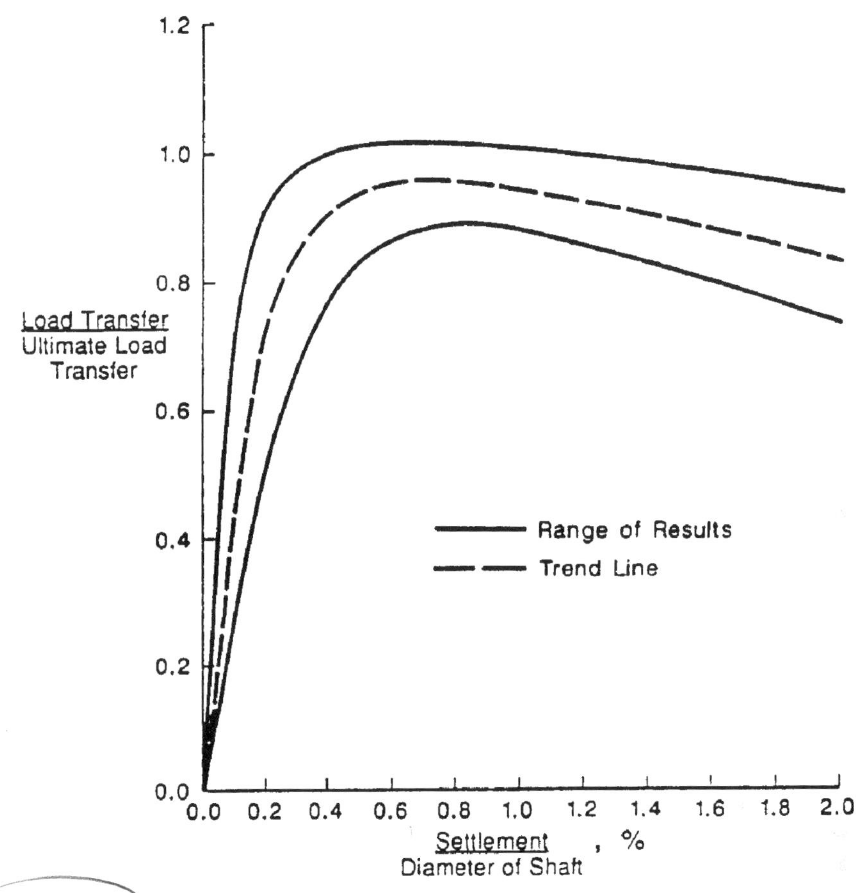

Figure 13.3 Normalized curves showing load transfer in side resistance versus settlement for drilled shafts in clay (after Reese and O'Neill, 1988).

The details of the methods of Kraft et al. and Vijayvergiya are not presented here.

13.1.6 Load-Transfer Curves for End Bearing in Cohesive Soil

The work of Skempton (1951) (also cited in the discussion of lateral loading) was employed, and a method was developed for predicting the load in end bearing of a pile in clay as a function of the movement of the pile tip. The laboratory stress-strain curve for the clay at the base of the pile was obtained by testing, or was estimated from values given by Skempton for laboratory strain, ε_{50}, at one-half of the ultimate compressive strength of the clay. Skempton reported that ε_{50} ranged from 0.005 to 0.02. He also used the theory of elasticity to develop approximate equations for the settlement of a footing (base of a pile) using laboratory stress-strain curves. Skempton's equations are as follows:

$$q_b = N_c \frac{\sigma_f}{2} \tag{13.10}$$

$$\frac{w}{B} = 2\varepsilon \tag{13.11}$$

where

q_b = failure stress in bearing at the base of the footing,
σ_f = failure compressive stress in the laboratory unconfined-compression or quick triaxial test,
N_c = bearing capacity factor (Skempton recommended 9.0)
B = diameter of the footing or equivalent length of a side for a square or rectangular shape,
ε = strain measures from the unconfined-compression or quick-triaxial test, and
w_b = settlement of the footing or base of the pile.

Stress-strain curves from a number of laboratory tests have been found to have a slope of approximately 0.5 when plotted on logarithmic scales. Therefore, in the absence of a laboratory stress-strain curve, a parabola can be used to yield a stress-strain curve up to the failure stress. The value of ε_{50} selected will depend on whether the clay is brittle or plastic.

The following is an example of the Skempton relationships. The basic data are:

$$B = 30 \text{ in. } (2.5 \text{ ft})$$

$$c = 780 \text{ psf}$$

$$\varepsilon_{50} = 0.01$$

The basic equation for load versus settlement at the tip of the pile is

$$Q_b = K_b \,(w_b)^n \tag{13.12}$$

where K_b is a fitting factor. If n is selected as 0.5, K_b can be evaluated as follows:

$$\frac{Q_b}{2} = \frac{N_c c A_{\text{tip}}}{2} = \frac{(9)(780)\pi(1.25)^2}{2} = 17{,}230 \text{ lb}$$

$$w_b = 2\ B\ \varepsilon_{50} = (2)(30 \text{ in.})(0.01) = 0.60 \text{ in.}$$

$$K_b = \frac{Q_b}{w_b^{0.5}} = 22{,}240$$

The curve shown in Table 13.2 can then be computed using Equation 13.12. In this example, it is assumed that the load will not drop as the tip of the pile penetrates the clay.

Reese and O'Neill (1988) studied the results of several tests of bored piles in clay where measurements yielded load in end bearing versus settlement. Figure 13.4 resulted from their studies. A study of the mean curve for the example given above shows that a value of about 1.2 in. is obtained at the ultimate bearing stress rather than the value of 2.4 in. found by the method adapted from Skempton's work. However, the range of values shown by Wright (1977) will easily encompass the earlier result.

Note that the movement of a pile causing the full load transfer in end bearing is several times than that which is necessary to develop full load transfer in skin friction. The above concept is easily demonstrated by considering the soil elements that are strained in end bearing compared to the soil elements that are strained in skin friction.

13.1.7 Load-Transfer Curves for Side Resistance in Cohesionless Soil

Coyle and Sulaiman (1967) studied the load transfer in skin friction of steel piles driven into sand and obtained the curves shown in Figure 13.5. The piles

TABLE 13.2 Load Transfer in Side Resistance for Cohesive Soil

Tip load Q_b, lb	Tip Movement w_b, in.
0	0
2,200	0.01
4,450	0.04
7,000	0.10
12,100	0.30
17,230	0.60
24,500	1.20
34,460	2.40
34,460	10.00

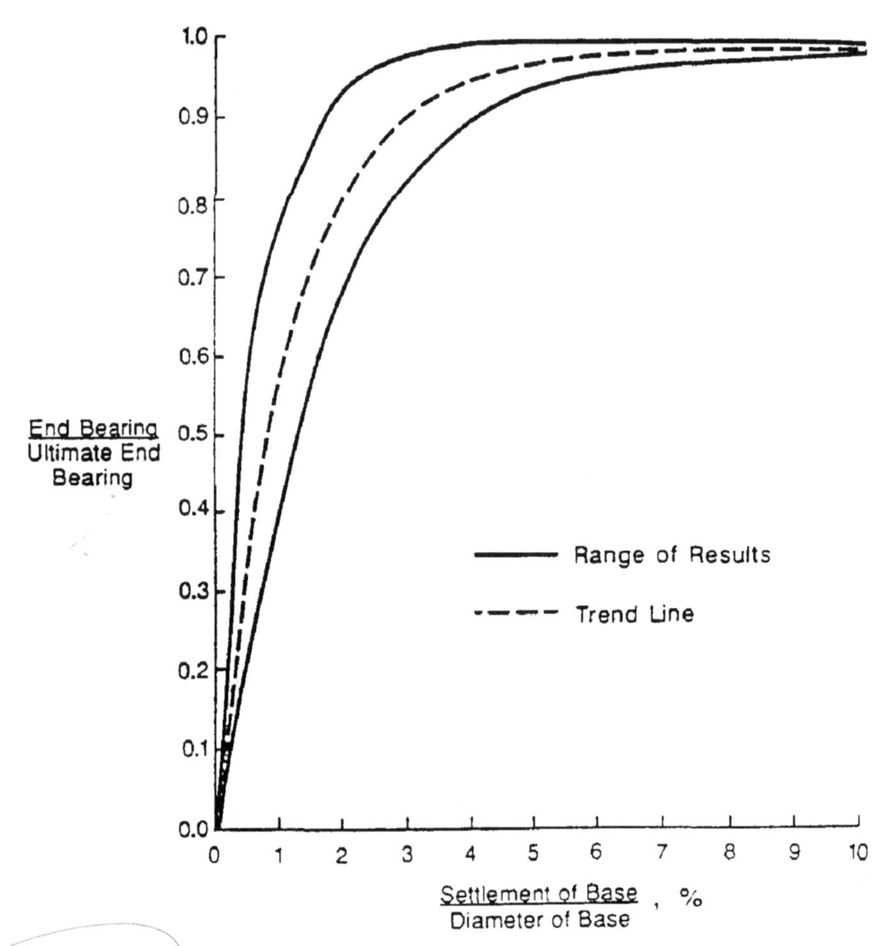

Figure 13.4 Normalized curves showing load transfer in end bearing versus settlement for drilled shafts in clay (after Reese and O'Neill, 1988).

had diameters ranging from 13 to 16 in. and a penetration of about 50 ft. An examination of the shape of the curves showed that they could be fitted with the following equation:

$$f = K_s\left(\frac{w}{B}\right)^{0.15}, \quad \left(\frac{w}{B}\right) \leq 0.07 \tag{13.13}$$

where

f = load transfer in side resistance,
w = movement of pile,

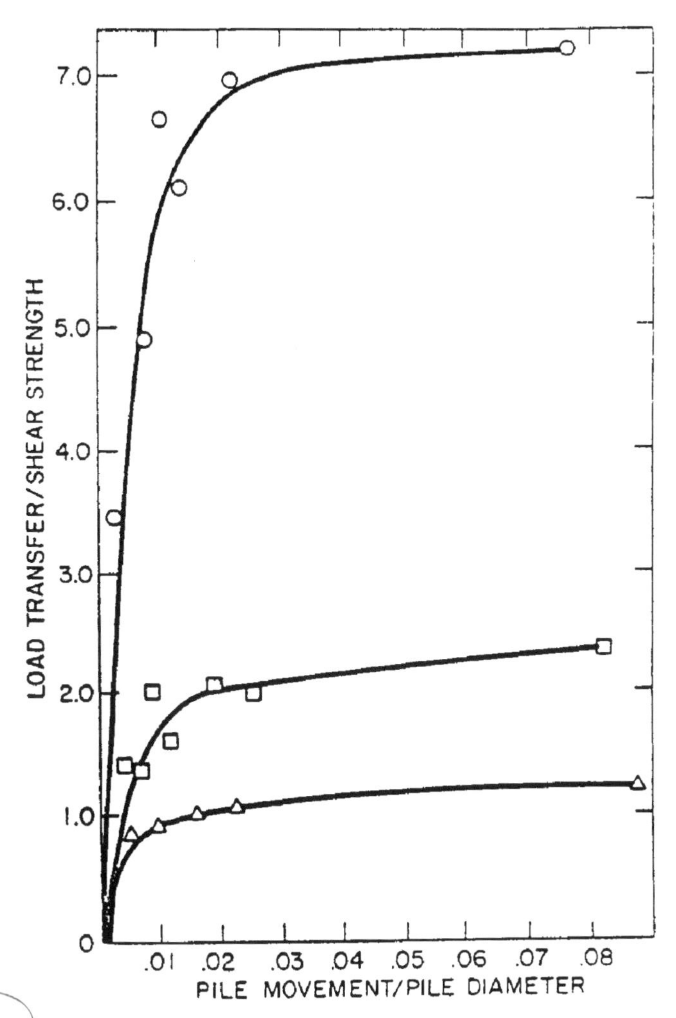

Figure 13.5 Experimental load-transfer curves in side resistance for sand (after Coyle and Sulaiman, 1967).

B = pile diameter, and
K_s = fitting factor.

Equation 13.13 can be used to obtain the approximate load transfer in skin friction for a pile in sand. The value of the maximum unit load transfer f_{max} would be obtained by use of appropriate equations. The value of the fitting factor K_s can be found by using the following equation:

$$f_{max} = K_s(0.07)^{0.15} \tag{13.14}$$

The curves are assumed to be flat for movements larger than 0.07*B*. While Eq. 13.14 and Figure 13.5 show that the load transfer continues to increase to a movement of 0.07*B*, the maximum load transfer essentially occurs at a movement of about 0.03*B*.

O'Neill and Reese (1999) examined the results of load tests of a number of full-sized bored piles that were instrumented for the measurement of axial load with respect to depth. The results of this study showed that the curves for cohesionless soils were similar to those for cohesive soils and that Figure 13.6 can be used for cohesionless soils.

Mosher (1984) studied the transfer of load in skin friction (side resistance) of axially loaded piles driven into sand. He developed the following equation and Table 13.3 for obtaining *t-z* (load transfer curves):

$$f = \frac{w}{\frac{1}{E_i} + \frac{1}{f_{\max}} w} \tag{13.15}$$

where

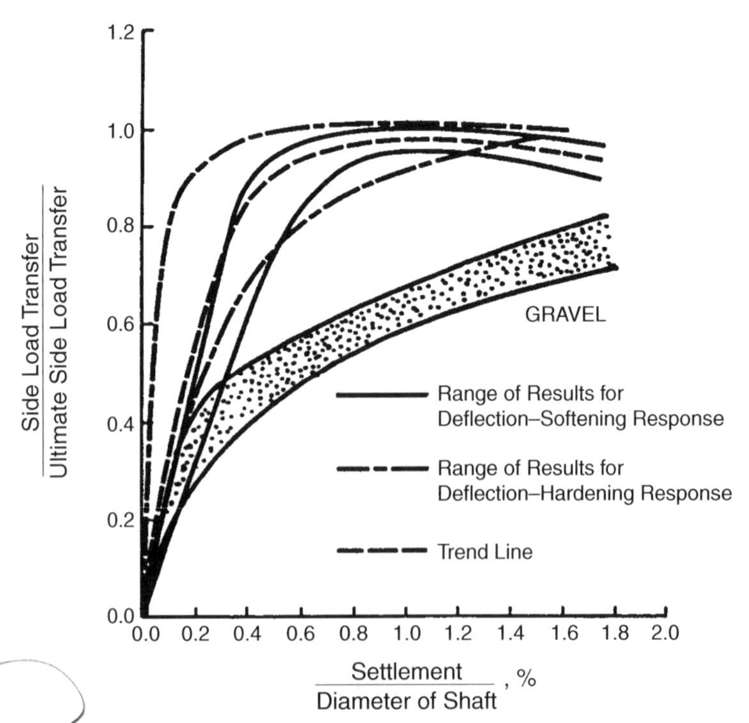

Figure 13.6 Normalized curves showing load transfer in side resistance versus settlement for drilled shafts in cohesionless soil (from O'Neill and Reese, 1999).

TABLE 13.3 Mosher's Values of E_s for Sand for Use in f-w Curves

Relative Density	ϕ, degrees	E_s, kPa/m	E, lb/ft²/in.
Loose	28–31	6.3×10^{-2}–10.5×10^{-2}	(6,000–10,000)
Medium	32–34	10.5×10^{-2}–14.7×10^{-2}	(10,000–14,000)
Dense	35–38	14.7×10^{-2}–18.9×10^{-2}	(14,000–18,000)

f = unit load transfer, lb/ft²
w = pile movement, in.
f_{max} = maximum unit load transfer, lb/ft², and
E_s = soil modulus, lb/ft²/in.

The mixed dimensions for E_s should be noted and followed when using Eq. 13.15 in design computation.

13.1.8 Load-Transfer Curves for End Bearing in Cohesionless Soil

Vesić (1970) proposed the following equation for computing the load versus tip settlement for piles in sand after studying the literature and performing several careful experiments:

$$w = \frac{C_w Q_p}{(1 + D_r^2)\, B\, q_b} \tag{13.16}$$

where

w = settlement,
Q_p = tip load,
D_r = relative density, decimal,
B = diameter of tip,
q_b = ultimate base resistance, and
C_w = settlement coefficient (the author found the following values: 0.0372 for driven piles, 0.0465 for jacked piles, and 0.167 for buried, or bored, piles).

(Note: Equation 13.16 is not dimensionally homogeneous, so values of C_w were recomputed and are dependent on the system of units being used.)

Reese and O'Neill (1988) studied the results of experiments with drilled shafts and developed Figure 13.7. The information in this figure was developed from a relatively small amount of data and this method, like other methods presented in this chapter, should be used with appropriate discretion.

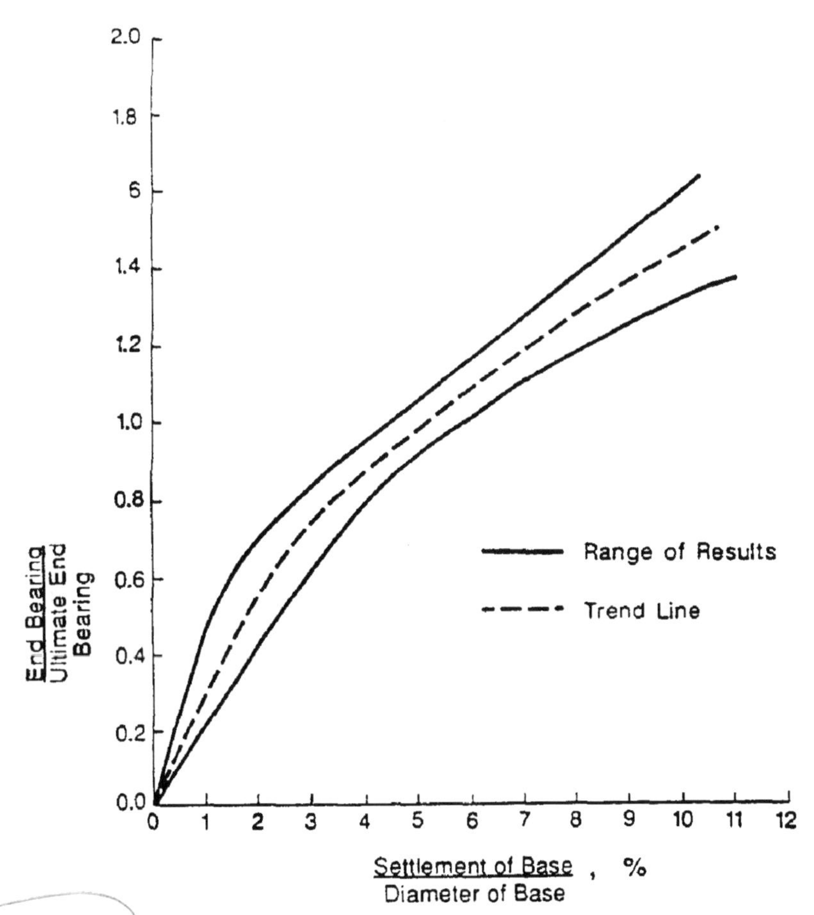

Figure 13.7 Normalized curves showing load transfer in end bearing versus settlement for drilled shafts in cohesionless soil (after Reese and O'Neill, 1988).

13.1.9 Load-Transfer Curves for Cohesionless Intermediate Geomaterials

Load-deformation behavior of drilled shaft sockets in cohensionless intermediate geomaterials can be computed using methods similar to those described for drilled shafts in soft rock. A total load-settlement method, as originally developed by Randolph and Wroth (1978), is recommended.

In the following, only the load-settlement behavior of the socket is described. Elastic shortening in the overburden (generally 0.25–2.0 mm, depending on load and socket depth) will need to be added to the computed settlement to obtain the settlement at the shaft head.

The load-settlement relation for the rock socket is the three-branched curve shown in Figure 13.8. For a given load Q_t at the top of the socket, the cor-

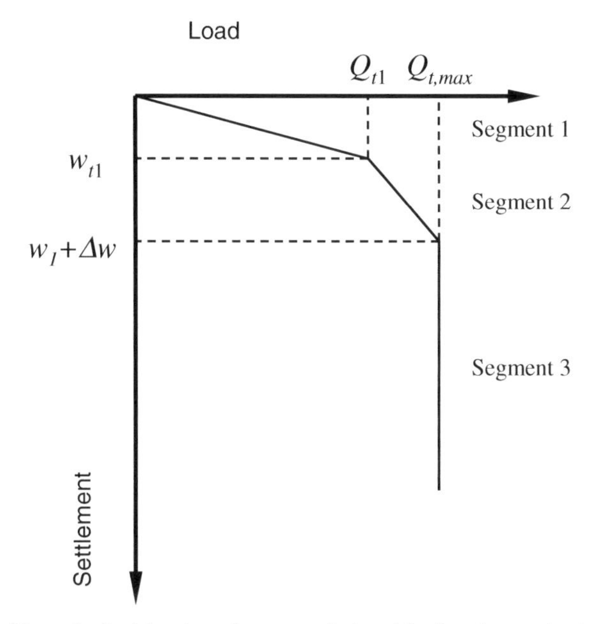

Figure 13.8 Hypothetical load-settlement relationship for the method of Mayne and Harris.

responding elastic settlement along Segment 1, w_t, is computed from Eq. 13.17:

$$w_t = \frac{Q_t I}{E_{sL} B} \tag{13.17}$$

Here E_{sL} is Young's modulus of the granular geomaterial along the sides of the socket at the base level (as distinguished from the geomaterial below the base.)

Based on correlations between energy-corrected SPT tests and Young's moduli (E_s) determined from dilatometer testing in Piedmont residuum, Mayne and Harris (1993) suggest

$$E_s = 22 P_a N_{60}^{0.82} \tag{13.18}$$

in which N_{60} is the SPT blow count for this condition in which the energy transferred to the top of the drivestring is 60% of the dwp energy of the SPT hammer, P_a, is the atmospheric pressure (101.3 kPa, or 14.7 psi). More accurate estimates of E_s (and K_0) might be possible if field test results obtained from pressuremeter, flat plate dilatometer, or seismic data are available for the project site.

Mayne and Harris provided a closed-form solution for I for straight-sided shafts from the original solution of Randolph and Wroth, given in Eq. 13.19:

$$I = 4(1 + v) \frac{1 + \dfrac{8 \tanh(\mu L)L}{\pi\lambda(1 - v)\xi(\mu L)D}}{\left[\dfrac{4}{(1 - v)\xi}\right] + \left[\dfrac{4\pi \dfrac{E_{sm}}{E_{sL}} \tanh(\mu L)L}{\zeta(\mu L)D}\right]} \tag{13.19}$$

where

v = Poisson's ratio of the geomaterial, which can be taken as approximately 0.3 for gravel unless evidence indicates otherwise.
L = socket length, and
μL = a lateral extent influence factor for elastic settlement, which can be taken to be

$$\mu L = 2\sqrt{\frac{2}{\zeta\lambda}}\left(\frac{L}{D}\right) \tag{13.20}$$

where

$$\zeta = \ln\left|\left\{0.25 + \left[2.5 \frac{E_{sm}}{E_{sL}}(1 - v) - 0.25\right]\xi\right\}\frac{2L}{D}\right| \tag{13.21}$$

$$\lambda = 2(1 + v)\frac{E_c}{E_{sL}} \tag{13.22}$$

E_c = Young's modulus of the composite (steel and concrete) cross section of the drilled shaft, and
E_{sm} = Young's modulus of soil at the mid-depth of the socket.
E_{sL} = Young's modulus of soil at the base of the socket.

Where the decomposed rock becomes stronger with depth (N increases with depth along the socket), E_{sm}/E_{sL} can ordinarily be taken to be 0.5.

$$\xi = \frac{E_{sL}}{E_b}$$

where E_b = Young's modulus of the granular geomaterial beneath the base of the drilled shaft, which can be different from E_{sL}.

In modeling drilled shaft load tests in Piedmont residuum in the Atlanta, Georgia, area, E_b must be taken to be about 0.4 E_{sL} to obtain an optimum match with the measured load-settlement relations. That is, $\xi = 2.5$.

A schematic of the variation of soil moduli for this method is shown in Figure 13.9.

Equation 13.17 is used to model load versus settlement only until the maximum side resistance, $Q_{s\ max}$, has been reached (segment 1, Figure 13.8).

$$Q_{s,\max} = f_{\max}\ \pi DL \tag{13.23}$$

and

$$Q_t \text{ (end of segment 1)} = Q_{t1}$$

$$Q_{t1} = \frac{Q_{s,\max}}{1 - \left\{ \dfrac{I}{[\xi \cosh(\mu L)][(1 - v)(1 + v)]} \right\}} \tag{13.24}$$

Equation 13.24 is valid approximately for $\xi < 20$. The settlement at the top of the socket at the end of segment 1, w_{t1}, can be determined by letting $Q_t = Q_{t1}$, in Eq. 13.17.

Equations 13.17 and 13.24 define the end of linear segment 1 and the beginning of linear segment 2. At this point, the load on the base at the end of segment 1 is

$$Q_{b1} = Q_{t1} - Q_{s\ \max}.$$

The load at the end of segment 2 is the maximum total resistance of the shaft in the given gravel:

$$Q_{t,\max} = Q_{s,\ \max} + Q_{b,\ \max}.$$

If the side resistance is perfectly plastic (no load softening or hardening after a movement of w_{t1}), then

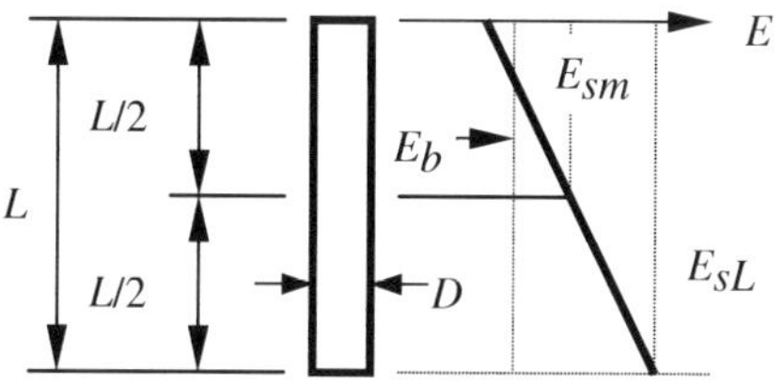

Figure 13.9 Potential soil modulus for computing settlement in granular decomposed rock.

$$Q_{t,\max} = f_{\max}(\pi DL) + q_{\max}\left(\frac{\pi D^2}{4}\right) \tag{13.25}$$

The corresponding settlement at the end of segment 2 is approximately w_{t1} plus the base settlement, Δw_b, due to the increment of base load $Q_{t,\max} - Q_{t1}$, which is given by

$$\Delta w_b = (Q_{t,\max} - Q_{t1})\frac{(1 - \upsilon)(1 + \mu)}{E_b D} \tag{13.26}$$

Finally, the end of segment 2 is defined by $Q_{t,\max}$ and $(w_{t1} + \Delta w_b)$. Segment 3 is a line defining continued settlement at no increasing load, which is probably conservative for most decomposed rock.

13.1.10 Example Problem

The *t-z* method is illustrated by the solution of the problem shown in Figure 13.10. The *t-z* curves and the *q-w* curve are hypothetical but include nonlinear behavior for small movements of the pile and constant values of load transfer after a given amount of movement of the pile.

The example was solved by use of the computer. The first step was to assume a number of tip movements, starting with a small value and with increasing values, as shown in Table 13.4. The final tip movement was large enough to ensure the computation of the ultimate value of pile-head load.

The pile is subdivided into a number of increments, depending on its length and the variability of the soil. With the smallest tip movement, the end-bearing curve is assessed to obtain a tip load. The bottom increment of the pile is then selected. The computer interpolates a *t-z* curve for the mid-height of the bottom increment, employs the tip movement, and computes the load transfer in side resistance along the bottom increment. The axial load at the top of the bottom increment can now be computed. The loads at the bottom and top of the bottom increment, the elastic shortening can now be computed for the bottom increment.

The assumption is made that the movement is linear over the bottom increment and a new midpoint movement can be obtained. The new mid-point movement is now used to obtain a new value of load transfer from the interpolated *t-z* curve. The procedure is continued until convergence is achieved for the bottom increment, where the load-transfer from the present solution and the just previous one is within the tolerance set by the user. The next increment above the bottom increment is then selected and solutions are made, increment by increment, up the length of the pile. The next value of tip load is selected and the procedure is repeated.

The output of the program may be tabulated or presented graphically. For the example problem, curves showing distribution of load with depth are given

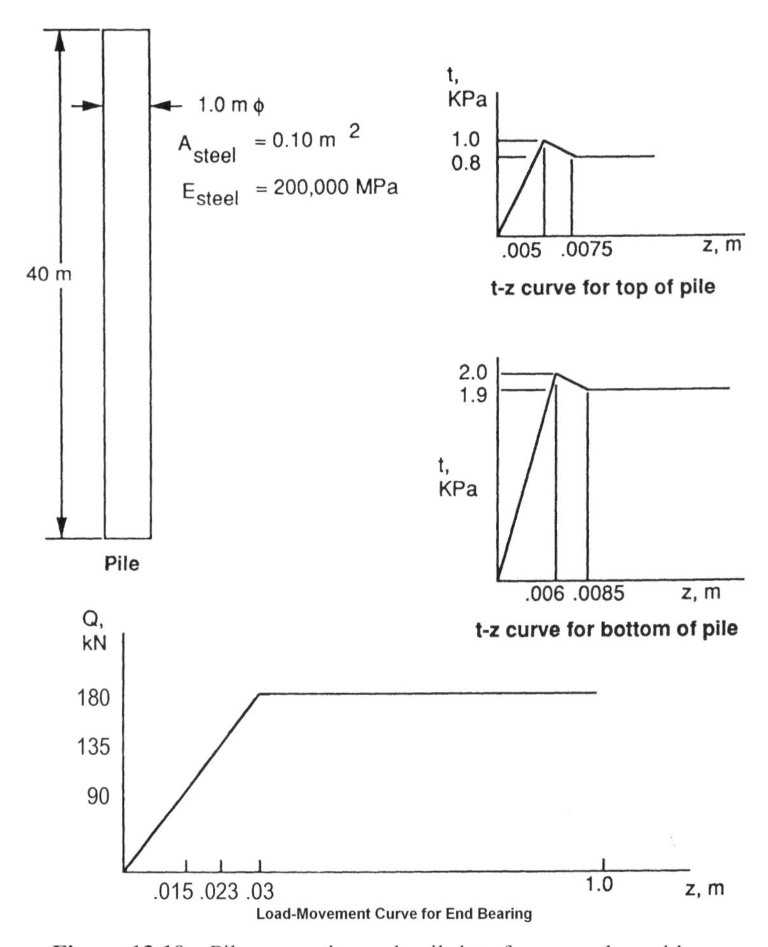

Figure 13.10 Pile properties and soil data for example problem.

TABLE 13.4 Summary of Load vs. Settlement Data at Pile Top and Pile Tip

Top Load KN.	Top Movement M.	Tip Movement M.	Tip Load KN.
0.4005E+02	0.1049E−02	0.1000E−02	0.6000E+01
0.8010E+02	0.2097E−02	0.2000E−02	0.1200E+02
0.1202E+03	0.3146E−02	0.3000E−02	0.1800E+02
0.1602E+03	0.4194E−02	0.4000E−02	0.2400E+02
0.2187E+03	0.8290E−02	0.8000E−02	0.4800E+02
0.2596E+03	0.1537E−01	0.1500E−01	0.9000E+02
0.3496E+03	0.3055E−01	0.3000E−01	0.1800E+03

in Figure 13.11 and a curve of the load versus settlement of the pile head is shown in Fig. 13.12. The numerical tabulation of output for points along the length of the pile is not shown but the engineer may consult the graphical and tabulated results to learn the stress and pile movement at any point along its length.

Partial Solution of Example Problem by Hand Computations A solution of the example was made by hand with an assumed tip movement of 0.0150 m. Only two increments of 20 m each were used along the pile. The curve in Figure 13.10 was consulted to find a tip load Q_b of 90 kN. With the assumption that the movement at the mid-height of the bottom increment was 0.015 m, Figure 13.13 was used to find the unit load transfer in side resistance of 1.625 kPa. The load in side resistance was then computed for the bottom increment.

$$Q_{s1} = (1.625)(\pi)(1)(20) = 102 \text{ kN}$$

The load at the top of the increment was 90 + 102 = 192 kN. The elastic shortening over the bottom increment may now be computed.

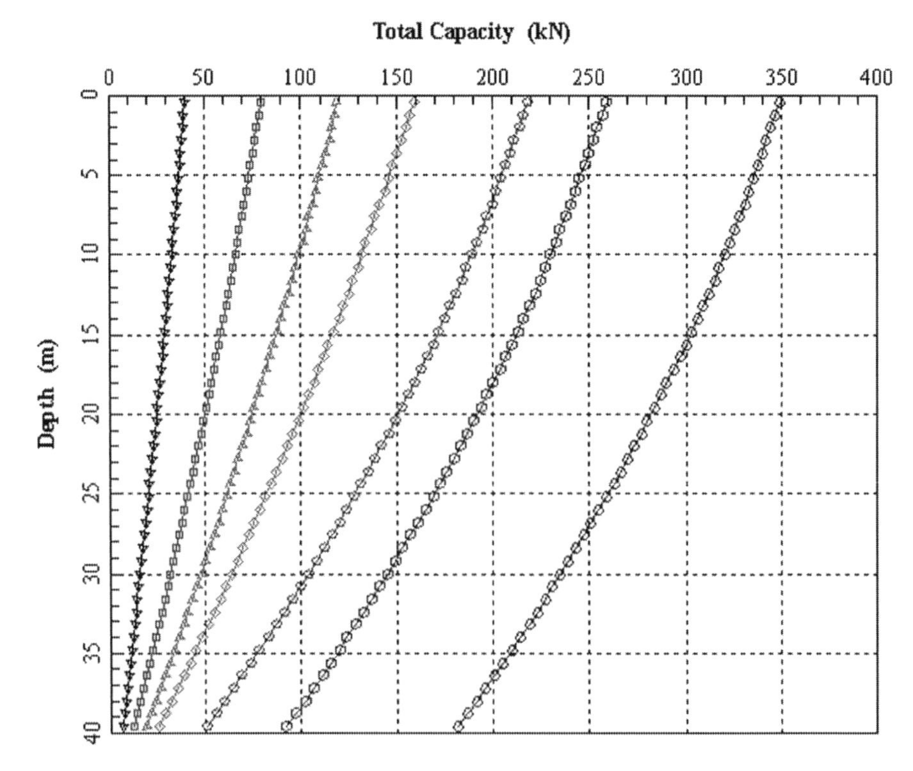

Figure 13.11 Load distribution curve for example problem.

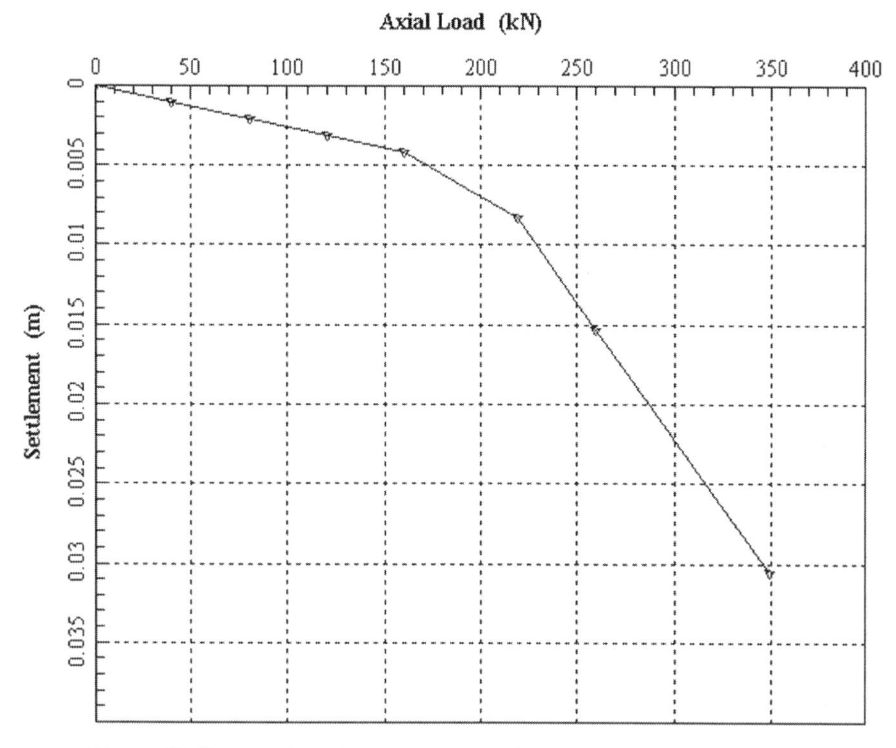

Figure 13.12 Load-settlement curve at pile head for example problem.

$$\delta_1 = \left(\frac{90 + 192}{2}\right)\left(\frac{20}{2x10^7}\right) = 0.0001410m$$

The movement at the top of the bottom increment may now be computed.

$$z_{\text{top}_1} = 0.0150 + 0.0001410 = 0.0151410m$$

Linear movement along the length of the bottom increment and the movement at the midpoint of the bottom increment are found.

$$z_{\text{mid}_1} = \left(\frac{0.0150 + 0.0151410}{2}\right) = 0.0150705m$$

The t-z curve for the bottom increment may now be consulted and the value of the load transfer is unchanged and no adjustment is needed in the computation.

The second increment is now considered and the movement at the top of the first increment may be used to obtain an estimate of the movement of the mid-height of the second increment, yielding a value of 1.075 kPa from Figure

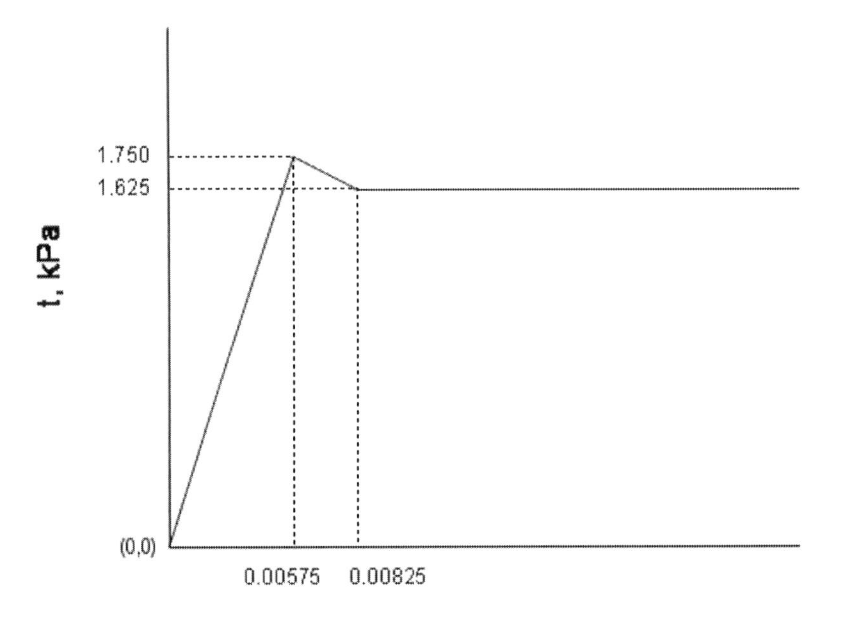

Figure 13.13 Interpolated t-z curve for 10 m above bottom of pile.

13.14. The load in side resistance of the second increment may now be computed.

$$Q_{s2} = (1.075)(\pi)(1)(20) = 67.5kN$$

The load at the top of the pile is 192 + 67.5 = 259.5. The elastic shortening of the top increment may now be computed.

$$\delta_2 = \left(\frac{192 + 259.5}{2}\right)\left(\frac{20}{2x10^7}\right) = 0.000226m$$

The movement at the top of the pile may now be computed.

$$z_{top2} = 0.0151410 + 0.000226 = 0.015367m$$

The midpoint movement of the top increment may now be computed.

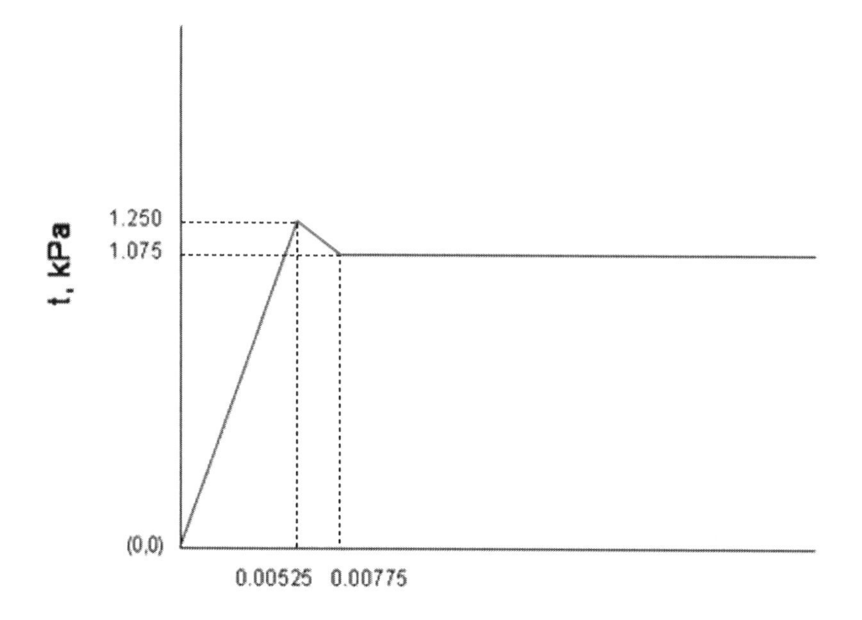

Figure 13.14 Interpolated t-z curve for 30 m above bottom of pile.

$$z_{\text{mid}_2} = \left(\frac{0.0151410 + 0.015367}{2}\right) = 0.0152540m$$

The *t-z* curve for the top increment may now be consulted and the value of load transfer is unchanged and no adjustment is needed in the computation.

The results for the hand solution and for the computer solution may now be shown.

	Load at top of pile, kN	Movement at top of pile, m
Hand	259.5	0.015367
Computer	259.6	0.01537

The hand solution and the computer solution yielded almost exact results because the pile was relatively stiff and because the computed mid-point movements did not change the load transfer values in side resistance from the values initially assumed. However, the hand computations shown above serve to illustrate the method of computation used in the *t-z* method of analysis.

13.1.11 Experimental Techniques for Obtaining Load-Transfer Versus Movement Curves

With the completion of a field load test on an instrumented pile, curves such as shown in Figs. 13.15a and 13.15b should be available. Figure 13.15a shows a load-settlement curve for the top of the pile. This curve may be obtained

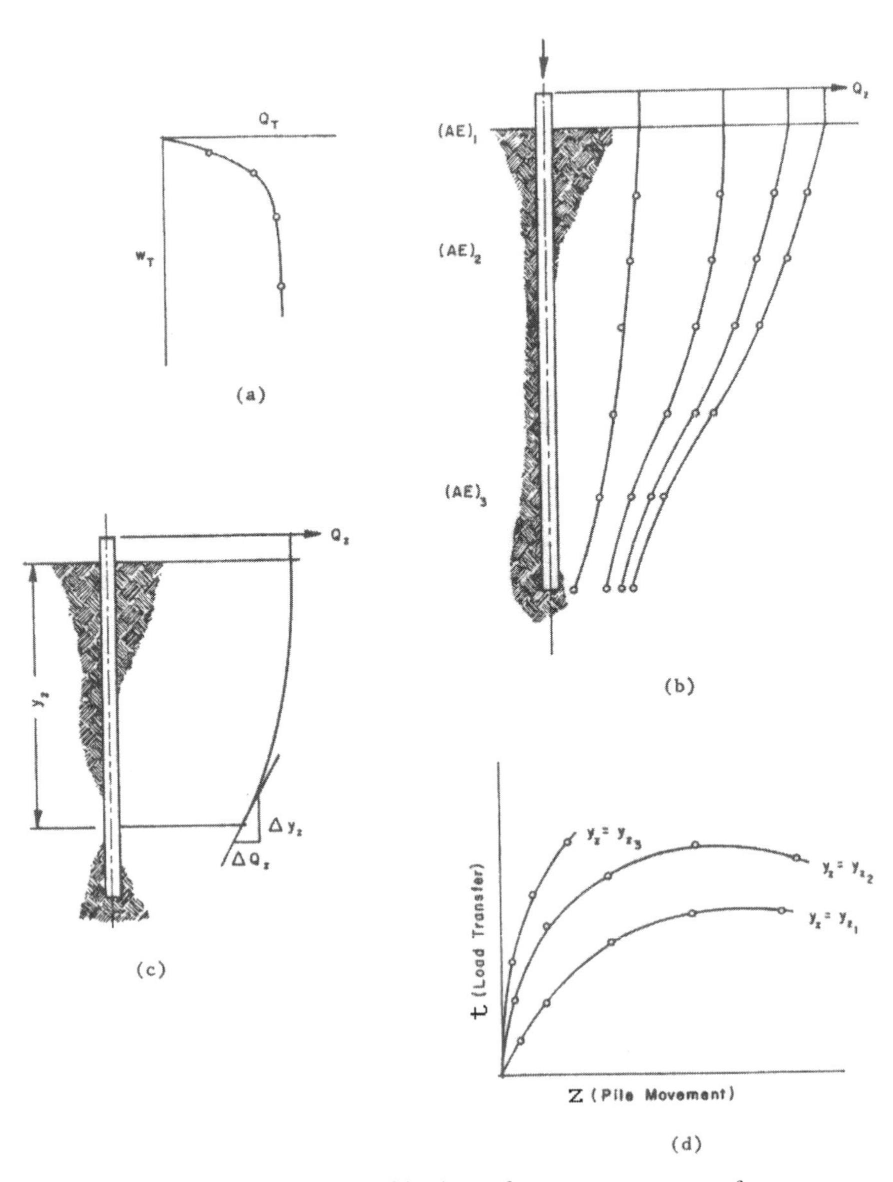

Figure 13.15 Development of load transfer versus movement of curves.

by measuring the load with a load cell and the downward movement of the top of the shaft with dial gauges. Figure 13.15b shows a set of curves which gives load in the pile at various points along its length for each of the applied loads. These data are obtained from instrumentation for measuring internal load in the pile at points along its length. Figures 13.15a and 13.15b indicate that four loads were applied to the pile; however, in the general case, several more loads would have been applied.

From the data in Figs. 13.15a and 13.15b, it is desired to produce a set of load-transfer curves such as are shown in Fig. 13.15d. Such curves can be produced for any desired depth. Figure 13.15c illustrates the procedure for obtaining a point on one of the curves. In this instance, the procedure is illustrated for obtaining load transfer at a depth y_z below the ground surface. For a particular load-distribution curve, corresponding to a particular load Q_z, the slope of the load-distribution curve is obtained at point y_z. In Fig. 13.15c this slope is indicated as the quantity $\Delta Q_z / \Delta y_z$. To obtain the load transfer t, the quantity is then divided by the pile circumference at the point y_z. Thus, the load transfer t will have the units of force per square of length.

The downward movement of the pile corresponding to the computed load transfer may be obtained as follows: (1) the settlement corresponding to the particular load in question is obtained from the curve in Fig. 13.15a; (2) the shortening of the pile is computed by dividing the cross-hatched area, shown in Fig. 13.15b, by the pile cross-sectional area times an effective modulus of elasticity; and (3) the downward movement of the pile at Point y_z is then computed by subtracting the shortening of the pile from the observed settlement.

The procedure enables one point on one load-transfer curve to be obtained for a particular depth. Other points may be obtained for the given depth by using the same procedures for other load-distribution curves. Other depths may be selected and a family of load-transfer curves may be obtained as shown in Fig. 13.15d. The settlement for a load-transfer curve for end bearing at the pile tip may be obtained as described above. The unit end bearing may be obtained by dividing the tip load by the area of the tip of the pile. As can be readily understood, accurate load-settlement and load-distribution data are required.

13.2 DESIGN FOR VERTICAL GROUND MOVEMENTS DUE TO DOWNDRAG OR EXPANSIVE UPLIFT

Any vertical movement of soil will affect the performance of a deep foundation. The problem of downdrag occurs when the direction of the soil movement is down relative to the foundation. The problem of uplift occurs when the direction of soil movement is up.

Peck et al. (1974) state: "Several examples of unexpected settlement of large magnitude have been attributed to neglect of negative skin friction."

Other failures have occurred when structures were founded on drilled shafts in expansive clay that swelled significantly after the shafts were constructed.

Tomlinson (1980) addressed the difficulty facing design engineers in designing for downdrag, stating: "The calculation of the total negative skin friction or downdrag force on a pile is a matter of great complexity, and the time factor is of importance." This statement applies equally well to the case of drilled shafts in expansive soil. To solve these problems, the design engineer must know the axial load–transfer relationship, as well as the distribution of these movements versus depth and how these movements may vary with time.

13.2.1 Downward Movement Due to Downdrag

Loading by downdrag occurs when a shaft penetrates through a compressible layer that settles after the shaft is installed. The net effect of the downdrag is to increase the magnitude of axial loading in the shaft in the zone moving downward around the shaft. The severity of this problem depends on the interaction of the skin friction in the settling soil, the type of soil layering, and the characteristics of the bearing strata. Each case should be examined individually.

An example of loading by downdrag is shown in Figure 13.16. In this case, the soil profile consists of two layers of compressible soft clay, with an upper layer of sand between the layers of soft clay and a lower layer of sand providing end-bearing support for the shaft. The presence of two compressible layers of clay permits the upper sand layer to move downward around the pile. This situation may not be recognized because downdrag is mistakenly thought to occur only in layers of clay. One should recognize that downdrag could occur in any type of soil that moves downward around the shaft.

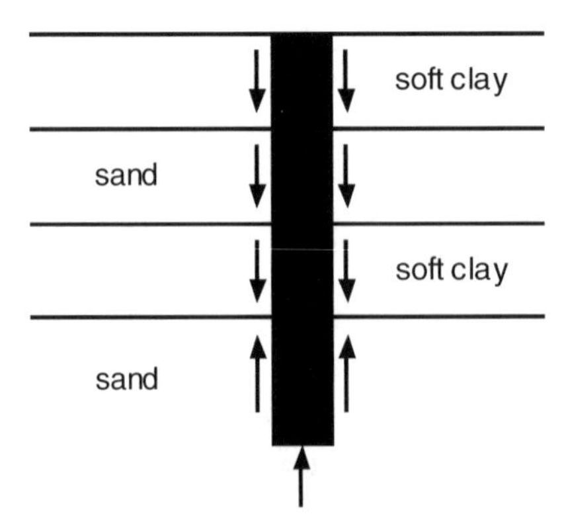

Figure 13.16 Soil profile susceptible to loading by downdrag.

The design of drilled shafts that are subjected to downdrag requires an analysis of the vertical settlement of soil versus depth. The methods of (EM) 1110-1-1904 Settlement Analysis may be used to estimate the vertical soil movements.

A simplified illustration of the mechanics of downdrag is shown in Figure 13.17. The relative movement of the soil to the shaft in shown in Parts (a) and (b) of this figure. The effect of the downward-moving soil is shown in Part (c). This effect increases the axial load in the drilled shaft above that applied at the top to the maximum value, Q_{max}, at an intermediate point along the depth of the shaft.

13.2.2 Upward Movement Due to Expansive Uplift

Swelling soils are found throughout the United States. The severity of problems due to swelling soils depends on the variation of the moisture content of the soil from dry to wet conditions. In general, the worst problems are found in areas with highly expansive clays where the amount of rainfall varies widely from year to year. In arid areas, problems are minimized because access to water is limited and the soils usually remain dry. In areas with high rainfall, problems are minimized because the wet conditions allow the soils to remain fully swollen.

A good indicator of the shrink-swell potential of the soil is the plasticity index. Plasticity index values above 35 are usually associated with clays with high swell potential. However, some highly plastic soils may be nonexpansive. Soils with plasticity index values less than 25 are generally considered to

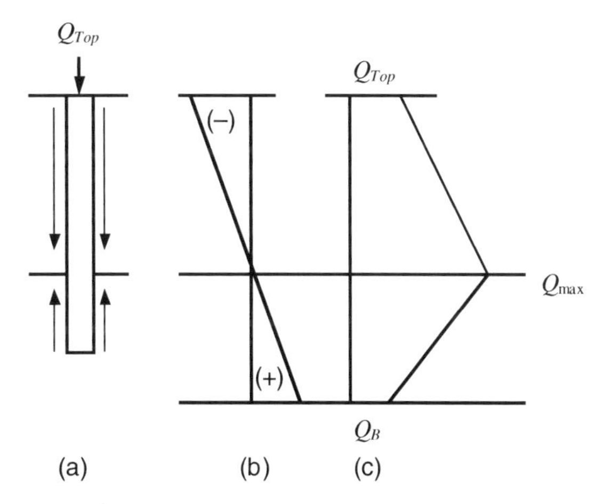

Figure 13.17 Mechanics of downdrag: (a) example problem; (b) relative movement of a drilled shaft relative to soil; (c) distribution of axial load along a drilled shaft.

have low swell potential, and soils with values between 25 and 35 are considered to have marginal swell potential.

PROBLEMS

P13.1. The computer program that solves a problem by use of the *t-z* method makes use of difference equations, where the difference between successive values of axial displacement of a pile is employed in the computer solution. What two sources of computational error can occur and what steps can the engineer take to minimize the errors?

P13.2. Your design is heavily dependent on the accuracy of the load-transfer curves shown in Figs. 13.3 and 13.4. What options do you have if you wish to validate the values shown in the curves?

P13.3. Present your ideas of why the curve in Fig. 13.7 for bored piles in cohesionless soil shows increased values of end bearing with increased settlement.

P13.4. (a) Re-work the hand computations in the example by keeping all of the input the same except that the thickness of the wall of the pile is 20 mm. (b) Plot curves showing distribution of load along the length of the pile using results from the computer and from the hand computations. Explain the differences in the curves. (c) Make a numerical comparison of the elastic shortening of the pile for the three solutions. (d) How would the solution differ if the pile had a nonlinear *AE* curve along its length?

P13.5. The calcareous soil at the Northwest Shelf in the ocean off the coast of Australia is strain-softening to a severe extent. What technique could be used to ensure that long piles supporting offshore platforms do not plunge under axial loading?

CHAPTER 14

ANALYSIS AND DESIGN BY COMPUTER OF PILES SUBJECTED TO LATERAL LOADING

14.1 NATURE OF THE COMPREHENSIVE PROBLEM

Chapter 12 demonstrated that the analysis and design of a pile under lateral loading requires the solution of a nonlinear problem in soil–structure interaction. The nonlinearity was reflected in the reduced value of a secant E_{py} to a *p-y* curve with pile deflection where y represents the lateral deflection and p represents the resistance from the soil in force per unit length.

A second nonlinearity must be addressed. The value of the bending stiffness E_pI_p of a reinforced concrete pile, in particular, will be reduced as the bending moment along the pile is increased. For a pile of reinforced concrete, explicit expressions must be developed, based on the geometry of the cross section of the pile and the properties of the steel and concrete, yielding the value of E_pI_p as a function of the applied bending moment. The method can be applied to piles with cross sections of other materials as well. Then a code must be written to determine the bending moment during a computation so that the value of E_pI_p can be modified as iterations proceed.

The differential equation for a pile under lateral loading given in Chapter 12 must be modified to account for the effect of axial loading on bending moment, sometimes called the $P - \Delta$ *effect.* The modified differential equation will allow the buckling of the pile under axial loading to be investigated, as well as account for the increased bending moment. The solution of the modified differential equation proceeds in a relatively straightforward way, using the power of the computer. Numerical techniques involving double-precision arithmetic are used, and the engineer must understand the nature of the numerical solutions to ensure that no computational errors occur due to improper selection of word length and computational tolerance during iterations.

The concept of *p-y* curves was presented in Chapter 12, and recommendations will be presented here on soils and rock under both static and repeated loading conditions. The recommendations are based on mechanics as much as possible but, more importantly, are based strongly on full-scale experiments in the field. Sustained loading will not cause much change in soil resistance for granular soil and overconsolidated clay. Field tests are recommended for piles in normally consolidated clay under sustained loading. In addition to specific recommendations for the formulation of *p-y* curves for soils and rock, several applications are presented for the solution of problems encountered in practice.

14.2 DIFFERENTIAL EQUATION FOR A COMPREHENSIVE SOLUTION

The differential equation required for a comprehensive solution is an expansion of Eq. 12.12 in Chapter 12. The two added features are the ability to account for an axial load and a distributed load. In many problems, the additional bending caused by an axial load is moderate but an important capability is added: computation of the load to cause buckling, particularly for piles that extend some distance above the groundline. The ability to consider a distributed force is essential in analyzing a pile extending through flowing water or subjected to pressures from moving earth. The differential equation for a comprehensive solution is

$$E_p I_p \frac{d^4 y}{dx^4} + P_x \frac{d^2 y}{dx^2} - p + W = 0 \tag{14.1}$$

where

P_x = axial load on the pile (positive to cause compression on the pile), and

W = distributed load in force per unit of length along the pile (positive in the positive direction of y).

The solution of Eq. 14.1 by numerical techniques is shown later in this chapter.

14.3 RECOMMENDATIONS FOR *p-y* CURVES FOR SOIL AND ROCK

14.3.1 Introduction

Basis for Useful Solutions The ability to make detailed analyses and a successful design of a pile to sustain lateral loading depends principally on the prediction of the response of the soil with appropriate accuracy. The computer gives the engineer the ability to solve complex nonlinear problems speedily, and with *p-y* curves for the soil and pile loadings, the differential equation can be solved with dispatch to investigate a number of parameters.

Prediction of the *p-y* curves for a particular solution must be given careful and detailed attention, starting with the acquisition of data on soils at the site, with particular reference to the role of water, use of soil mechanics to the extent possible, consideration of the nature of the loading at the site, evaluation of available data on performance of soils similar to those at the site, and development of multiple solutions to evaluate the importance of various parameters in the methods of prediction.

As shown in the following sections, the mechanics of soils have been used to gain information on the slope of the early portion of *p-y* curves and on the magnitude of the ultimate soil resistance p_u that develops with relatively large deflection of a pile. However, the performance of full-scale tests in the field with instrumented piles has been indispensable in the development of prediction methods.

Field Tests with Instrumented Piles The preferred method is to instrument a pile at close intervals along its length with electrical-resistance strain gauges for the measurement of bending moment as a series of measured loads are applied to the pile (Matlock and Ripperger, 1956). Integration of the bending moments from one of the loadings, using the boundary conditions, will yield values of y as a function of depth. Differentiation of the bending moments from the same loading, usually requiring special techniques, will yield values of p as a function of depth. Repeating the process for all of the loadings and cross-plotting values of p versus values of y will yield a family of *p-y* curves. The curves in Figure 12.3 were obtained by using the technique just described. There is considerable expense in the careful instrumentation of a pile, the installation at a site where data are needed, the performance of the test and the acquisition of data, and the analysis of the data. A limited number of such tests have been done.

A second experimental technique involves the steps noted above, except for installation of the internal instrumentation along the pile. Four boundary conditions are measured at the pile head for a series of loadings: deflection, slope, moment, and shear. Nondimensional techniques are employed, and val-

$_{py}$ as a function of depth are found that will fit the measured data (Reese and Cox, 1968). Values of y and p are computed as a function of depth. The procedure is repeated for each of the applied loads; cross-plotting then yields a family of p-y curves.

The p-y curves from experiments were used to compute pile-head deflection as a function of applied load to compare with results from tests where limited data were obtained. Comparisons were possible only where relevant data were available on pile, loadings, and soil. Where comparisons could be made, the results were useful in validating the quality of p-y curves from tests of instrumented piles.

Characteristics of p-y Curves A typical p-y curve is shown in Figure 14.1a, drawn only in the first quadrant for convenience, and is meant to represent the case where a short-term monotonic loading, or "static" loading, is applied to a pile. The three curves in Figure 14.1 show a straight-line relationship between p and y from the origin to point a. Assuming that the strain of soil is linearly related to stress for small strains, the assumption follows that p is linearly related to y for small deflections. Analytical methods for establishing such a relationship are discussed later in this chapter.

The portion of the curve in Figure 14.1a from point a to point b shows that the value of p is increasing at a decreasing rate with respect to y. This

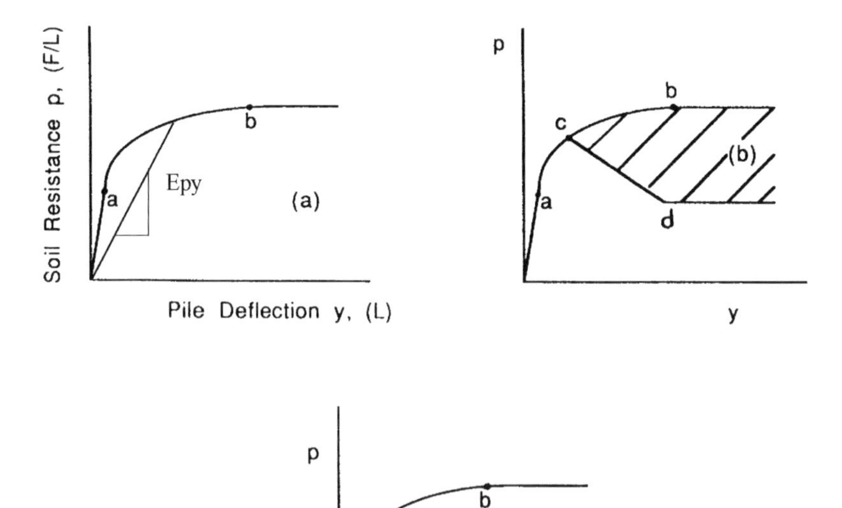

Figure 14.1 Conceptual p-y curves.

behavior reflects the nonlinear portion of the stress-strain curve for natural soil; no analytical method is currently available for computing the *a-b* portion of a *p-y* curve. The horizontal portion of the *p-y* curve in Figure 14.1a indicates that the soil is behaving plastically, with no loss of shear strength with increasing strain. Analytical models can be used to compute the ultimate resistance p_u as a function of pile dimensions and soil properties. These models will be demonstrated later in this chapter.

The shaded portion of the *p-y* curve shown in Figure 14.1b shows decreasing values of *p* from point *c* to point *d*. The decrease reflects the effect of cyclic loading. The curves in Figures 14.1a and 14.1b are identical up to point *c*, which implies that cyclic loading has little or no effect on a *p-y* curve for small deflections. The loss of resistance represented by the shaded area is, of course, related to the number of loading cycles. However, the number of cycles does not appear in the criteria shown later; rather, the assumption is made that the repeated loading is sufficient to cause the limiting value of soil resistance.

The possible effect of sustained loading is shown in Figure 14.1c, where there is an increase in *y* with a corresponding loss of *p*. For a pile in normally consolidated clay, lateral loading will cause an increase in porewater pressure and deflection will increase as the porewater pressure is dissipated. The decrease in the value of *p* suggests that resistance is shifted to other elements along the pile as deflection occurs at a particular point. A prediction of the effect of sustained loading for piles in soft or normally consolidated clays must be developed from field testing or estimated using the theory of consolidation. Sustained loading should have little effect on the behavior of piles in granular soils or in most overconsolidated clays.

Many soil profiles consist of a series of layers, with some, on occasion, being thin with respect to the diameter of the pile being designed. If relatively thin layers exist at a site, the engineer may wish to implement the method proposed by Georgiadis (1983).

Effects of Cyclic Loading Cyclic loading occurs in a number of designs; a notable example is an offshore platform. Therefore, a number of the field tests employing fully instrumented piles have used cyclic loading in the experimental procedures. Cyclic loading has invariably resulted in increased deflection and bending moment above the respective values obtained in short-term loading.

Cohesive Soils. A dramatic example of the loss of soil resistance due to cyclic loading may be seen by comparing the two sets of *p-y* curves in Figures 12.11a and 12.11b, where the pile was installed in overconsolidated clay with water being present above the ground surface. To gain information on the reasons for the dramatic loss of resistance shown in Figure 12.3b, Wang (1982) and Long (1984) did extensive studies of the influence of cyclic loading on *p-y* curves for clays. The following two reasons can be suggested for the reduction in soil resistance from cyclic loading: subjection of the clay to repeated strains of large magnitude, with a consequent loss of shearing strength due to remolding, and scour from the enforced flow of water in the vicinity of the pile. Long studied the first factor by performing some triaxial tests with repeated loading using specimens from sites where piles had been tested. The second factor is present when water is above the ground surface, and its influence can be severe.

Welch and Reese (1972) reported some experiments with a bored pile under repeated lateral loading in an overconsolidated clay with no free water present. During cyclic loading, the deflection of the pile at the groundline was on the order of 25 mm (1 in.). After a load was released, a gap was revealed at the face of the pile where the soil had been pushed back. Also, cracks a few millimeters wide radiated away from the front of the pile. Had water covered the ground surface, it is evident that water would have penetrated the gap and the cracks. With the application of a load, the gap would have closed and the water carrying soil particles would have been forced to the ground surface. This process was dramatically revealed during soil testing in overconsolidated clay at Manor (Reese et al., 1975) and at Houston (O'Neill and Dunnavant, 1984).

While the work of Long (1984), Wang (1982), and others produced considerable information on the factors that influence the loss of resistance in clays under free water due to cyclic loading, it did not produce a definitive method for predicting this loss of resistance. The analyst thus should use the numerical results presented here with caution in regard to the behavior of piles in clay under cyclic loading. Full-scale experiments with instrumented piles at a particular site are indicated for those cases where behavior under cyclic loading is a critical feature of the design.

Cohesionless Soils. The loss of resistance of granular soil due to the cyclic lateral loading of piles is not nearly as dramatic for cohesionless soil as it is for clay soils. Cyclic loading will usually cause a change in the void ratio of the soil, which may result in settlement of the ground surface. If the pile is cycled in the same direction with loads that cause deflection of more than a few millimeters, the granular soil with no cohesion will collapse behind the pile and cause deflection to be "locked in."

The relative density of granular soils should be determined as accurately as possible prior to construction of piles to be subjected to lateral loading. If the soil is dense, driving the piles may be impossible and an alternative procedure for installation may be selected. If the soil is loose, densification will occur during pile driving, and further densification can occur during cyclic loading. The engineer will want to use all details on the soil and the response of the pile, from full-scale experiments of piles under lateral loading in granular soils, to use most effectively the information show later in predicting *p-y* curves for a particular design.

Influence of Diameter The analytical developments presented to this point indicate that the term for the pile diameter appears to the first power in the expressions for *p-y* curves. This idea is reinforced when the concept of a stress *bulb* is employed for piles under static loading. Elementary analysis shows that the ratio of load per unit length and the pile deflection would be the same for any diameter. However, cyclic loading of piles in clay below the water surface involves phenomena that are probably not related to diameter.

Some experimental studies have been done that provide useful information. Reese et al. (1975) described tests of piles with diameters of 152 mm (6 in.) and 641 mm (24 in.) at the Manor site. The *p-y* formulations developed from the results with the larger piles were used to analyze the behavior of the smaller piles. The computation of bending moment led to good agreement between analytical and experimental results, but the computation of groundline deflection showed considerable disagreement, with the computed deflections being smaller than the measured ones.

O'Neill and Dunnavant (1984) and Dunnavant and O'Neill (1986) reported on tests performed at a site where the clay was overconsolidated and where lateral loading tests were performed on piles with diameters of 273 mm (10.75 in.), 1220 mm (48 in.), and 1830 mm (72 in.). They found that the site-specific response of the soil could best be characterized by a nonlinear function of the diameter.

In the recommendations that follow for the formulation of *p-y* curves, the diameter b appears to the first power, which does not seriously contradict any available experimental data. However, for piles of large diameter in overconsolidated clay below the water table, further experimental studies are recommended.

14.3.2 Recommendations for *p-y* Curves for Clays

Initial Portion of Curves The conceptual *p-y* curves in Figure 14.1 are characterized by an initial straight line from the origin to point a. A mass of soil with an assumed linear relationship between compressive stress and strain E_{smax} for small strains can be considered, where E_{smax} is the slope of the initial portion of the stress-strain curve in the laboratory. If a pile is made to deflect a small distance in such a soil, one can assume that the principles of mechanics can be used to find the initial slope K_{py} of the *p-y* curve. These are some difficulties in making the computations.

For one thing, the value of E_{smax} for soil is not easily determined. Stress-strain curves from unconfined compression tests were studied (Figure 14.2), and it was found that the initial modulus E_{smax} ranged from about 40 to about 200 times the undrained shear strength c (personal files). There is a considerable amount of scatter in the points, probably due to the heterogeneity of the soils at the two sites. The values of E_{smax}/c would probably have been higher had an attempt been made to get precise values for the early part of the curve. Stokoe (1989) reported that values of E_{smax} on the order of 2000 times c are found routinely in laboratory tests when soil specimens are subjected to very small strains. Johnson (1982) performed some tests with the self-boring pressuremeter, and computations with his results gave values of

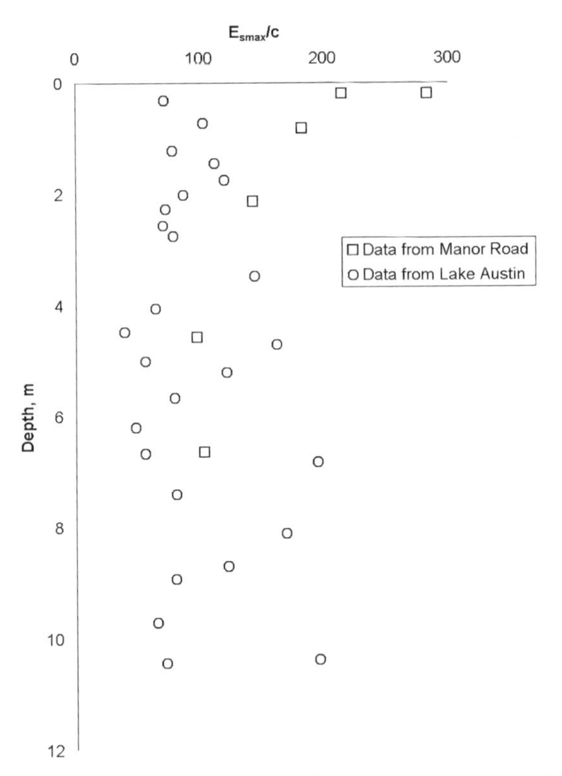

Figure 14.2 Plot of ratio of modulus to undrained shear strength from unconfined-compression test of specimens of clay.

E_{smax}/c that ranged from 1440 to 2840, with an average of 1990. The studies of the initial modulus from compressive stress-strain curves of clay seem to indicate that such curves are linear only over a very small range of strains.

The theory of elasticity (Vesić, 1961) and FEM (Yegian and Wright, 1973; Thompson, 1977; Kooijman, 1989; Brown et al., 1989) have been used to find values of K_{pymax}, the largest value of K_{pymax}, as a function of E_{smax} with limited success. With both methods, the presence of the ground surface presents analytical problems and no method of analysis can, at present, deal with the phenomena associated with cyclic loading, particularly if water is present at the ground surface. Considering the experimental difficulties in getting good values of E_{smax} and the analytical difficulties in obtaining values of K_{pymax}, the current best approach is to use values for the initial slope of p-y curves from experiments where careful controls were used. An example of such data is shown in Figure 14.3 (Reese et al., 1975). Table 14.1 was compiled after examining such data from experiments with a number of instrumented piles.

Ultimate Resistance p_u for Clays Two models can be used to gain some insight into the ultimate resistance p_u that will develop as a pile is deflected

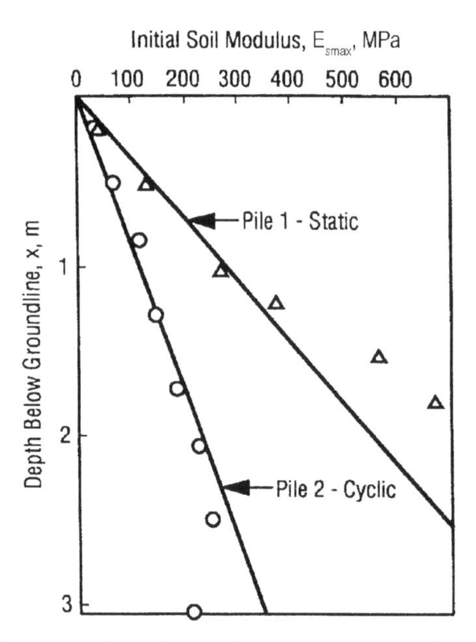

Figure 14.3 Initial soil modulus versus depth.

laterally. The first model assumes that the clay will move up and out at the ground surface; the second model takes over below the first and assumes horizontal movement of the clay. A wedge of soil for the first model is shown in Figure 14.4. Justification for the use of this model is presented in Figures 14.5a and 14.5b (Reese et al., 1975), where contours are shown of the rise of the ground surface at the front of a steel-pipe pile with a diameter of 641 mm (24 in.) in overconsolidated clay. As shown in Figure 14.5a, for a lateral load of 596 kN (134 kips), ground-surface movement occurred at a distance from the axis of the pile of about 4 m (13 ft). When the load was removed, the ground surface subsided somewhat, as shown in Figure 14.5b, suggesting that the soil behaved plastically.

TABLE 14.1 Representative Values of K_{py} for Clays

	Average Undrained Shear Strength*		
kPa	50–100	200–300	300–400
K_{py} (static) MN/m^3	135	270	540
(lb/in.^{3})	(500)	(1000)	(2000)
K_{py} (cyclic) MN/m^3	55	110	540
(lb/in.^{3})	(200)	(400)	(2000)

*The average shear strength should be computed to a depth of five pile diameters. It should be defined as half of the total maximum principal stress difference in an unconsolidated, undrained triaxial test.

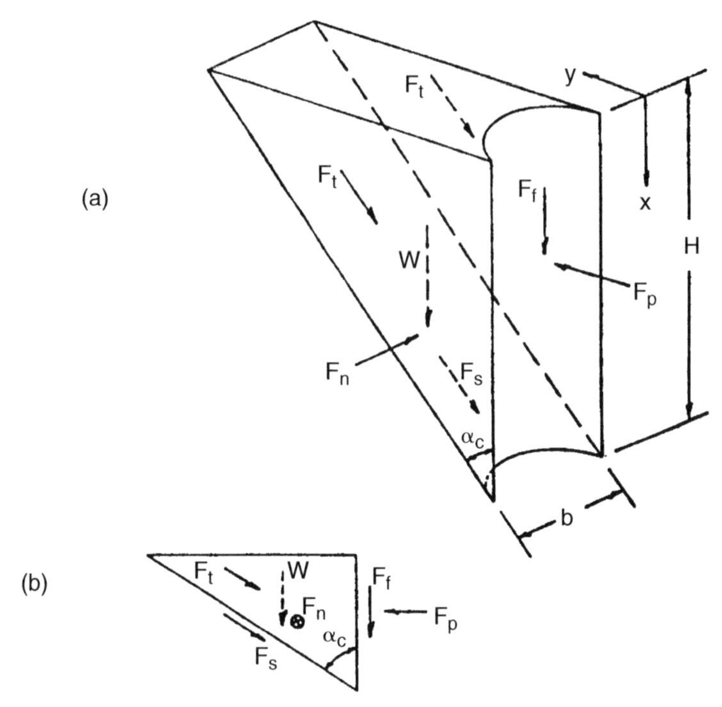

Figure 14.4 Assumed passive wedge-type failure for clay: (a) shape of the wedge; (b) forces acting on the wedge.

The use of plane sliding surfaces, shown in Figure 14.4, will obviously not model the movement indicated by the contours in Figure 14.5; however, a solution with the simplified model should give some insight into the variation of the ultimate lateral resistance p_u with depth. Taking the weight of the wedge into account, taking for the forces on the sliding surfaces into account, solving for F_p, and differentiating F_p with respect to H to solve for p_{u1} leads to Eq. 14.2:

$$p_{u1} = c_a b[\tan \alpha_c + (1 + \kappa)\cot \alpha_c] + \gamma bH + 2c_a H(\tan \alpha_c \sin \alpha_c + \cos \alpha_c) \tag{14.2}$$

where

p_{u1} = ultimate resistance near the ground surface per unit of length along the pile,
c_a = average undrained shear strength over the depth H,
α_c = angle of the inclined plane with the vertical,
γ = unit weight of soil, and
κ = reduction factor for shearing resistance along the face of the pile.

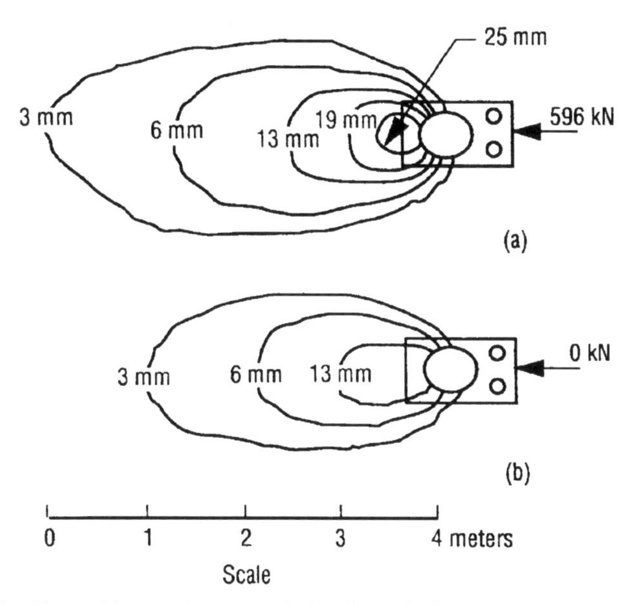

Figure 14.5 Ground heave due to static loading of pile 1: (a) heave at the maximum level; (b) residual heave.

If κ is set equal to zero, which is logical for the case of cyclic loading, and α_c is set at 45°, the following equation results:

$$p_{u1} = 2c_a b + \gamma bH + 2.83c_a H \tag{14.3}$$

The second model for computing the ultimate resistance p_u is shown in the plan view in Figure 14.6. At some point below the ground surface the maximum value of soil resistance will occur, with the soil moving horizontally. Movement in only one side of the pile is indicated, but movement, of course, will be around both sides of the pile. Again, planes are assumed for the sliding surfaces, with the acceptance of some approximation in the results.

A cylindrical pile is indicated in Figure 14.6, but for convenience in computation, a prismatic block of soil is assumed to be subjected to horizontal movement. Block 5 is moved laterally, as shown, and stress of sufficient magnitude is generated in that block to cause failure. Stress is transmitted to Block 4 and on around the pile to Block 1, with the assumed movements indicated by the dotted lines. Block 3 is assumed not to distort, but failure stresses develop on the sides of the block as it slides.

The Mohr-Coulomb diagram for undrained, saturated clay is shown in Figure 14.6b, and a free body of the pile is shown in Figure 14.6c. The ultimate soil resistance p_{c2} is independent of the value of σ_1 because the difference in the stress on the front σ_6 and back σ_1 of the pile is equal to $10c$. The shape of the cross section of a pile will have some influence on the magnitude of

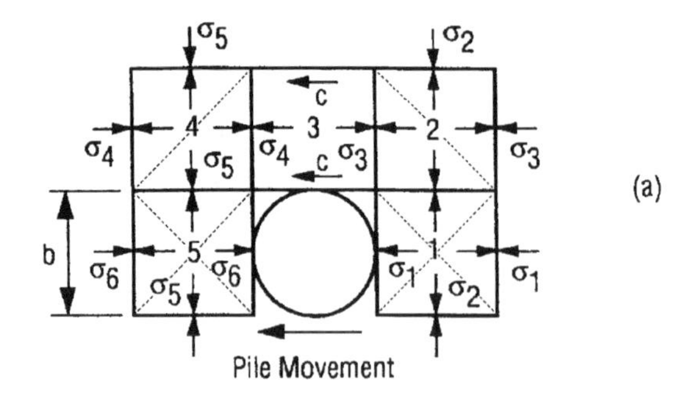

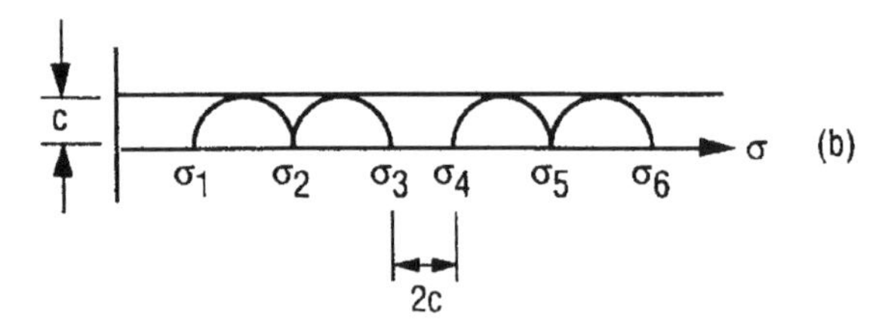

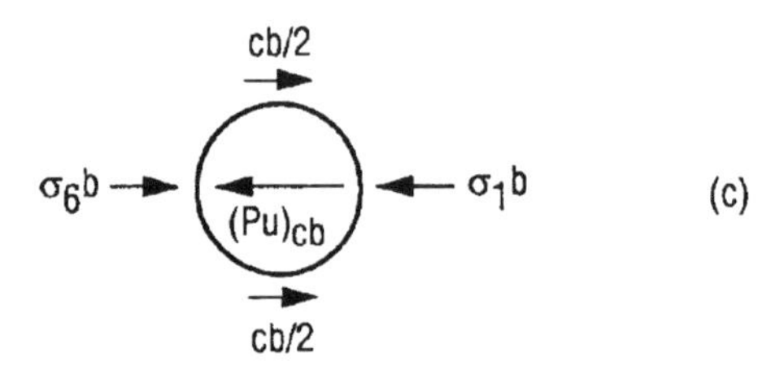

Figure 14.6 Assumed mode of soil failure by lateral flow around a pile in clay: (a) section through the pile; (b) Mohr-Coulomb diagram; (c) forces acting on a section of a pile.

p_{u2} for the circular cross section. It is assumed that the resistance developed on each side of the pile is equal to $cb/2$, and

$$p_{u2} = (\sigma_6 - \sigma_1 + c)\,b = 11cb \tag{14.4}$$

Equations 14.3 and 14.4 are similar to equations shown later in the recommendations for two of the sets of *p-y* curves. However, emphasis is placed directly on experimental results. The elementary analytical procedures shown here give considerable guidance to the recommendations that follow, but the results from experiments, similar in form to the above results, were favored. Note that many cases were analyzed where available data allowed the results of experiments to be compared with analytical results, confirming the value of the methods shown below.

Early Recommendations Designers used all available information in selecting the sizes of piles to sustain lateral loading in the period prior to the advent of instrumentation that allowed the development of *p-y* curves from experiments with instrumented piles. The methods yielded soil modulus values that were employed principally with closed-form solutions of the differential equation. The work of Skempton (1951) and the method proposed by Terzaghi (1955) were useful to early designers.

The method proposed by McClelland and Focht (1958), outlined below, appeared at the beginning of the period when large research projects were funded and is significant because those authors were the first to present the concept of *p-y* curves. Their paper is based on a full-scale experiment at an offshore site where a moderate amount of instrumentation was employed. Because the test was at an offshore site, rigorous control of the experiment was not possible. The paper shows conclusively that soil modulus is not a soil property but instead is a function of pile deflection and depth below the mudline, as well as of soil properties.

The paper recommends the performance of consolidated-undrained triaxial tests with the confining pressure equal to the overburden pressure. The full curve of compressive stress σ_1 and the corresponding strain ε is plotted. The following equation is recommended for obtaining the value of soil resistance p:

$$p = 5.5\sigma_1 b \tag{14.5}$$

To obtain values of the corresponding pile deflection y from stress-strain curves, the authors proposed the following equation:

$$y = 0.5\varepsilon b \tag{14.6}$$

The McClelland-Focht equations resemble the more comprehensive equations developed from tests of instrumented piles, shown below, except for the reduction in the value of p near the ground surface. The recommendations in the following three sections are currently used in many designs.

Response of Soft Clay in the Presence of Free Water Matlock (1970) performed lateral-load tests with a steel-pipe pile that was 324 mm (13 in.) in diameter and 12.8 m (42 ft) long. It was driven into clays near Lake Austin, Texas, that had a shear strength averaging about 38 kPa (800 lb/ft^2). The pile was recovered, taken to Sabine Pass, Texas, and driven into clay with a shear strength that averaged about 14.4 kPa (300 lb/ft^2) in the significant upper zone.

The initial loading was short-term (static), but the load remained on the pile long enough for readings of strain gauges to be taken by an extremely precise device; however, some creep occurred at the higher loads. The two sets of readings at each point along the pile were interpreted to find the assumed reading at a particular time, assuming that the change in moment due to creep was minor or had a constant rate. The accurate readings of bending moment allowed the soil resistance to be found by numerical differentiation.

The pile was extracted and redriven, and cyclic loading was applied. Readings of the strain gauges were taken under constant load after various numbers of loading cycles. The load was applied in two directions, with the load in the forward direction being more than twice as large as the load in the backward direction. After a significant number of cycles, the deflection at the top of the pile was changing not at all or only a small amount, and an equilibrium condition was assumed. Therefore, the *p-y* curves for cyclic loading are intended to represent a lower-bound condition.

The following procedure is for short-term static loading and is illustrated in Figure 14.7a. As noted earlier, the curves for static loading constitute the basis for indicating the influence of cyclic loading and would be used in design in special cases.

1. Obtain the best possible estimate of the variation of undrained shear strength c and submerged unit weight with depth. Also, obtain the value of ε_{50}, the strain corresponding to one-half of the maximum principal stress difference. If no stress-strain curves are available, typical values of ε_{50} are given in Table 14.2.

2. Compute the ultimate soil resistance per unit length of pile, using the smaller of the values given by the following equations:

$$p_u = \left[3 + \frac{\gamma'}{c} + \frac{J}{b}z\right]cb \tag{14.7}$$

$$p_u = 9cb \tag{14.8}$$

where

γ' = average effective unit weight from ground surface to *p-y* curve,
z = depth from ground surface to *p-y* curve,
c = undrained shear strength at depth z,
b = diameter (width) of pile, and
J = experimentally determined parameter.

Matlock (1970) stated that the value of J was determined to be 0.5 for a soft clay and about 0.25 for a medium clay. The value of 0.5 is frequently used for J. The value of p_u is computed at each depth where a *p-y* curve is desired based on shear strength at that depth.

3. Compute the deflection y_{50} at one-half of the ultimate soil resistance from the following equation:

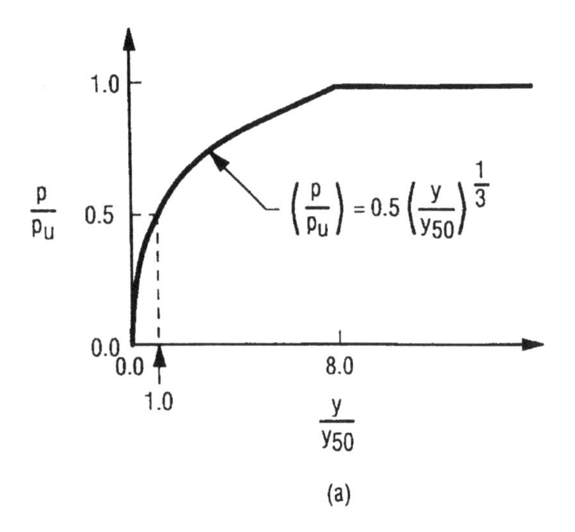

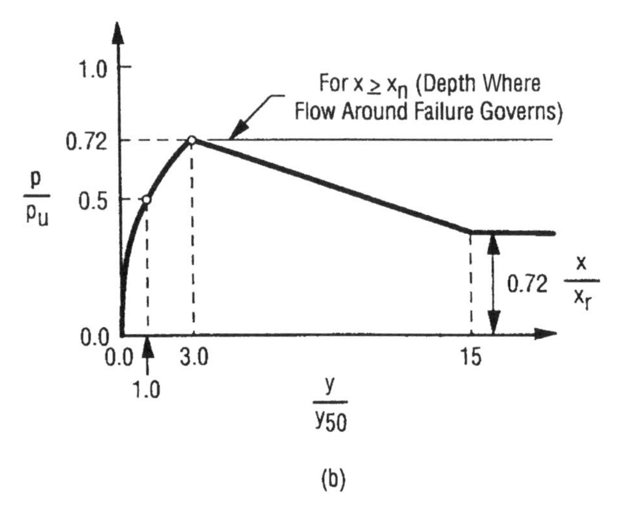

Figure 14.7 Characteristic shapes of p-y curves for soft clay in the presence of free water: (a) static loading; (b) cyclic loading (after Matlock, 1970).

TABLE 14.2 Representative Values of ε_{50}

Consistency of Clay, psf		Undrained Shear Strength, kPa	ε_{50}
Very soft	>250	>12	0.02
Soft	250–500	12–24	0.02
Medium	500–1000	24–48	0.01
Stiff	1000–2000	48–96	0.006
Very stiff	2000–4000	96–192	0.005
Hard	<2000	<192	0.004

$$y_{50} = 2.5\varepsilon_{50}b \tag{14.9}$$

4. Compute points describing the p-y curve from the following relationship:

$$\frac{p}{p_u} = 0.5\left(\frac{y}{y_{50}}\right)^{1/3} \tag{14.10}$$

The value of p remains constant beyond $y = 8y_{50}$. Equation 14.10 shows the slope of the p-y curve to be infinite at the origin, an anomalous result. The reasonable suggestion is made that the initial slope of the p-y curve be established by using K_{py} from Table 14.1.

With the modification shown above, the p-y curve for cyclic loading may be formulated by using the following procedure, illustrated in Figure 14.7b.

1. Construct the p-y curve in the same manner as for short-term static loading for values of p less than $0.72p_u$.
2. Solve Eqs. 14.7 and 14.8 simultaneously, taking the variation of c and γ' into account, to find the depth z_r where the transition occurs.
3. If the depth to the p-y curve is greater than or equal to z_r, select p as $0.72p_u$ for all values of y greater than $3y_{50}$.
4. If the depth of the p-y curve is less than z_r, note that the value of p decreases to $0.72p_u$ at $y = 3y_{50}$ and to the value given by the following expression at $y = 15y_{50}$:

$$p = 0.72p_u\left(\frac{z}{z_r}\right) \tag{14.11}$$

The value of p remains constant beyond $y = 15y_{50}$.

Response of Stiff Clay in the Presence of Free Water Reese et al. (1975) performed lateral-load tests with steel-pipe piles that were 641 mm (24 in.) in diameter and 15.2 m (50 ft) long. The piles were driven into stiff clay at a site near Manor, Texas. The clay had an undrained shear strength ranging from about 96 kPa (1 ton/ft^2) at the ground surface to about 290 kPa (3 ton/ft^2) at a depth of 3.7 m (12 ft).

The loading of the pile was carried out in a manner similar to that described for the tests performed by Matlock. A significant difference was that a data acquisition system was employed that allowed a full set of readings of the strain gauges to be taken in about 1 minute. Thus, the creep of the piles under sustained loading was small or negligible. The disadvantage of the system was that the accuracy of the curves of bending moment was such that curve fitting was necessary in doing the differentiations.

Also, as in the Matlock test recommendations for cyclic loading, the lower-bound case is presented. Cycling was continued until the deflection and bending moments appeared to stabilize. The number of loading cycles was on the order of 100, and 500 cycles were applied in a reloading test. During the experiment with repeated loading, a gap developed between the soil and the pile after deflection at the ground surface of perhaps 10 mm (0.4 in.), and scour of the soil at the face of the pile began at that time. There is reason to believe that scour would be initiated in overconsolidated clays after a given deflection at the mudline rather than at a given fraction of the pile diameter. However, the data available at present do not allow any change in the recommended procedures. But analysts could well recommend a field test at a particular site in recognition of some uncertainty regarding the influence of scour on *p-y* curves for overconsolidated clays.

The following procedure is for short-term static loading and is illustrated in Figure 14.8. As before, these curves form the basis for evaluating the effect of cyclic loading. They may also be used for sustained loading in some circumstances.

1. Obtain values of undrained shear strength c, soil submerged unit weight γ', and pile diameter b.
2. Compute the average undrained shear strength c_a over the depth z.
3. Compute the ultimate soil resistance per unit length of pile using the smaller of the values given by the following equations:

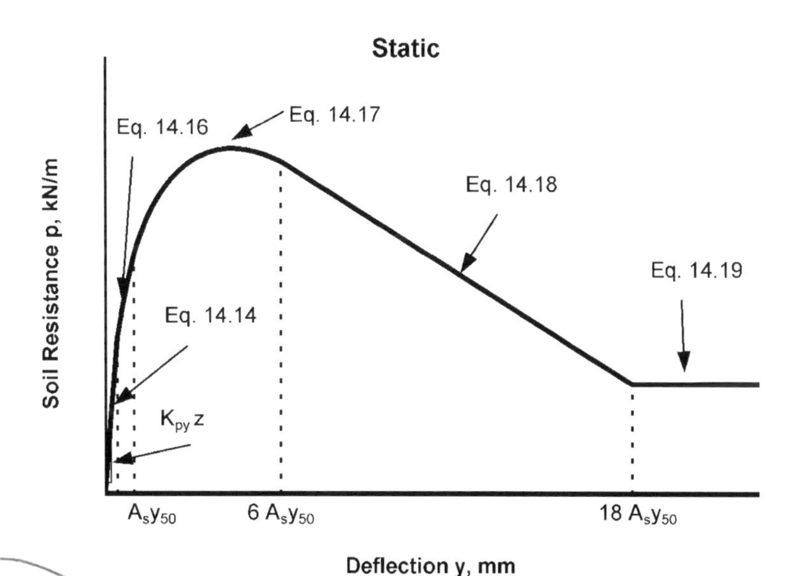

Figure 14.8 Characteristic shape of *p-y* curves for static loading in stiff clay in the presence of free water.

$$p_{ct} = 2c_a b + \gamma' bz + 2.83c_a z \tag{14.12}$$

$$p_{cd} = 11cb \tag{14.13}$$

4. Choose the appropriate value of A_s from Figure 14.9 for shaping the p-y curves.
5. Establish the initial straight-line portion of the p-y curve:

$$p = (K_{py}\, z)y \tag{14.14}$$

   Use the appropriate value of K_{py} from Table 14.1.
6. Compute the following:

$$y_{50} = \varepsilon_{50} b \tag{14.15}$$

   Use an appropriate value of ε_{50} from results of laboratory tests or, in the absence of laboratory tests, from Table 14.2.
7. Establish the first parabolic portion of the p-y curve, using the following equation and obtaining p_c from Eqs. 14.12 or 14.13:

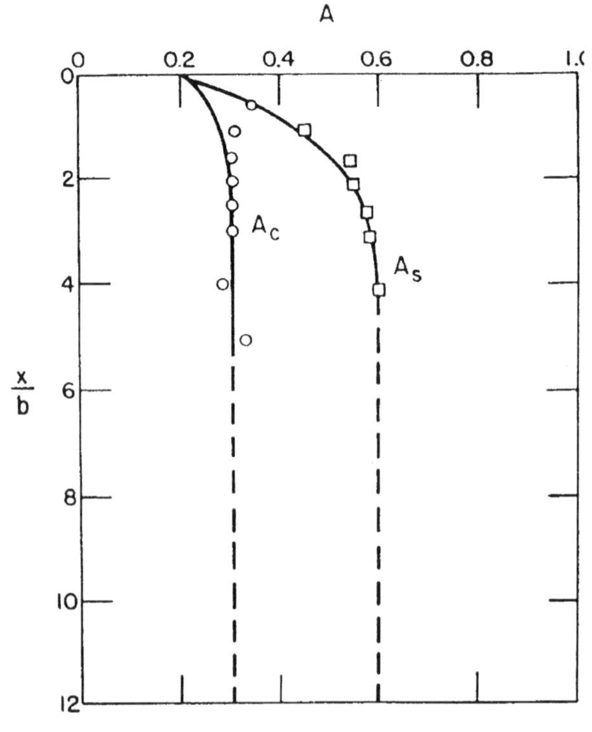

Figure 14.9 Values of constants A_s and A_c.

$$p = 0.5p_c\left(\frac{y}{y_{50}}\right)^{0.5} \tag{14.16}$$

Equation 14.16 should define the portion of the p-y curve from the point of the intersection with Eq. 14.14 to a point where y is equal to $A_s y_{50}$ (see the note in Step 10).

8. Establish the second parabolic portion of the p-y curve:

$$p = 0.5p_c\left(\frac{y}{y_{50}}\right)^{0.5} - 0.055p_c\left(\frac{y - A_s y_{50}}{A_s y_{50}}\right)^{1.25} \tag{14.17}$$

Equation 14.17 should define the portion of the p-y curve from the point where y is equal to $A_s y_{50}$ to a point where y is equal to $6A_s y_{50}$ (see note in Step 10).

9. Establish the next straight-line portion of the p-y curve:

$$p = 0.5p_c(6A_s)^{0.5} - 0.411p_c - \frac{0.0625}{y_{50}}p_c(y - 6A_s y_{50}) \tag{14.18}$$

Equation 14.18 should define the portion of the p-y curve from the point where y is equal to $6A_s y_{50}$ to a point where y is equal to $18A_s y_{50}$ (see the note in Step 10).

10. Establish the final straight-line portion of the p-y curve:

$$p = p_c[1.225(A_s)^{0.5} - 0.75A_s - 0.411] \tag{14.19}$$

Equation 14.19 should define the portion of the p-y curve from the point where y is equal to $18A_s y_{50}$. For all larger values of y, see the following note.

Note: The step-by-step procedure is outlined, and Figure 14.8 is drawn, as if there is an intersection between Eqs. 14.14 and 14.16. However, there may be no intersection of Eq. 14.14 with any of the other equations defining the p-y curve. Equation 14.14 defines the p-y curve until it intersects with one of the other equations; if no intersection occurs, Eq. 14.14 defines the complete p-y curve.

A second pile, identical to the pile used for static loading, was tested under cyclic loading. The following procedure is for cyclic loading and is illustrated in Figure 14.10. As may be seen from a study of the recommended p-y curves, the results for the Manor site showed a very large loss of soil resistance. The test data have been studied carefully, and the recommended p-y curves for cyclic loading reflect accurately the behavior of the soil at the site. Nevertheless, the loss of resistance due to cyclic loading at Manor is much greater than has been observed elsewhere. Therefore, use of the recommendations in this section for cyclic loading will yield conservative results for many clays.

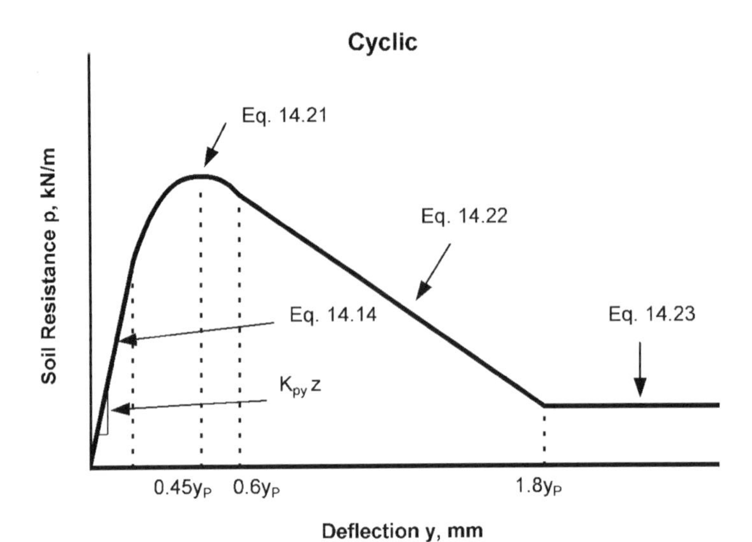

Figure 14.10 Characteristic shape of the *p-y* curve for cyclic loading in stiff clay in the presence of free water.

Long (1984) was unable to show precisely why the loss of resistance occurred during cyclic loading. One clue was that the clay from Manor was found to lose volume by slaking when a specimen was placed in fresh water. Thus, the clay was quite susceptible to erosion from the hydraulic action of the free water as the pile was pushed back and forth.

1. Steps 1, 2, 3, 5, and 6 are the same as for the static case.
4. Choose the appropriate value of A_c from Figure 14.9 for the particular nondimensional depth. Compute the following:

$$y_p = 4.1A_c y_{50} \tag{14.20}$$

7. Establish the parabolic portion of the *p-y* curve:

$$p = A_c p_c \left[1 - \left| \frac{y - 0.45y_p}{0.45y_p} \right|^{2.5} \right] \tag{14.21}$$

   Equation 14.21 should define the portion of the *p-y* curve from the point of the intersection with Eq. 14.14 to where y is equal to $0.6y_p$ (see the note in Step 9).
8. Establish the next straight-line portion of the *p-y* curve:

$$p = 0.936A_c p_c - \frac{0.085}{y_{50}} p_c(y - 0.6y_p) \tag{14.22}$$

Equation 14.22 should define the portion of the p-y curve from the point where y is equal to $0.6y_p$ to the point where y is equal to $1.8y_p$ (see the note in Step 9).

9. Establish the final straight-line portion of the p-y curve:

$$p = 0.936A_c p_c - \frac{0.102}{y_{50}} p_c y_p \qquad (14.23)$$

Equation 14.23 should define the portion of the p-y curve from the point where y is equal to $1.8y_p$ and for all larger values of y (see the following note).

Note: The step-by-step procedure is outlined, and Figure 14.10 is drawn, as if there is an intersection between Eq. 14.14 and Eq. 14.21. There may be no intersection of Eq. 14.14 with any of the other equations defining the p-y curve. If there is no intersection, the equation should be employed that gives the smallest value of p for any value of y.

Triaxial compression tests of the unconsolidated-undrained type with confining pressures conforming to in situ pressures are recommended for determining the shear strength of the soil. The value of ε_{50} should be taken as the strain during the test corresponding to the stress equal to one-half of the maximum total-principal-stress difference. The shear strength c should be interpreted as one-half of the maximum total-principal-stress difference. Values obtained from triaxial tests might be somewhat conservative but would represent more realistic strength values than values obtained from other tests. The unit weight of the soil must be determined.

Response of Stiff Clay with No Free Water A lateral-load test was performed at a site in Houston with a bored pile 915 mm (36 in.) in diameter. A 254-mm (10-in.)-diameter pipe, instrumented in intervals along its length with electrical-resistance-strain gauges, was positioned along the axis of the pile before concrete was placed. The embedded length of the pile was 12.8 m (42 ft). The average undrained shear strength of the clay in the upper 6 m (20 ft) was approximately 105 kPa (2200 lb/ft^2). The experiments and the interpretations are discussed by Welch and Reese (1972) and Reese and Welch (1975).

The same experiment was used to develop both the static and cyclic p-y curves, in contrast to the procedures employed for the two other experiments with piles in clays. The load was applied in only one direction, also in contrast to the other experiments.

A load was applied and maintained until the strain gauges were read with a high-speed data acquisition system. The same load was then cycled several times and held constant while the strain gauges were read at specific numbers of cycles. The load was then increased, and the procedure was repeated. The difference in the magnitude of successive loads was relatively large, and the

vas made that cycling at the previous load did not influence the Cycle 1 at the new load.

ırves at this site were relatively regular in shape, and yielded to an analysis that allowed the increase in deflection due to cyclic loading to be formulated in terms of the stress level and the number of loading cycles. Thus, the analyst can specify a number of loading cycles in doing the computations for a particular design.

The following procedure is for short-term static loading and is illustrated in Figure 14.11.

1. Obtain values for undrained shear strength c, soil unit weight γ, and pile diameter b. Also obtain the values of ε_{50} from stress-strain curves. If no stress-strain curves are available, use a value of ε_{50} as given in Table 14.2.
2. Compute the ultimate soil resistance p_u per unit length of pile using the smaller of the values given by Eqs. 14.7 and 14.8. (In using Eq. 14.7, the shear strength is taken as the average from the ground surface to the depth being considered and J is taken as 0.5. The unit weight of the soil should reflect the position of the water table.)
3. Compute the deflection y_{50} at one-half of the ultimate soil resistance from Eq. 14.9.
4. Compute points describing the p-y curve from the following relationship:

$$\frac{p}{p_u} = 0.5\left(\frac{y}{y_{50}}\right)^{0.25} \tag{14.24}$$

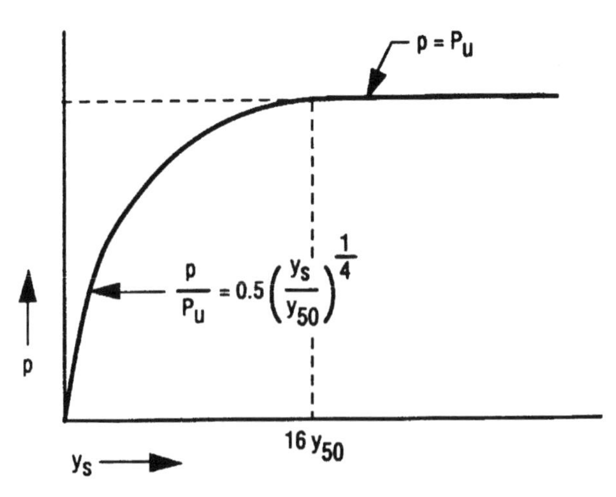

Figure 14.11 Characteristic shape of the p-y curve for static loading in stiff clay with no free water.

5. Beyond $y = 18y_{50}$, p is equal to p_u for all values of y.

The following procedure is for cyclic loading and is illustrated in Figure 14.12.

1. Determine the p-y curve for short-term static loading by the procedure previously given.
2. Determine the number of times the lateral load will be applied to the pile.
3. Obtain the value of C for several values of p/p_u, where C is the parameter describing the effect of repeated loading on deformation. The value of C is found from a relationship developed by laboratory tests (Welch and Reese, 1972) or, in the absence of tests, from the following equation:

$$C = 9.6\left(\frac{p}{p_u}\right)^4 \tag{14.25}$$

4. At the value of p corresponding to the values of p/p_u selected in Step 3, compute new values of y for cyclic loading from the following equation:

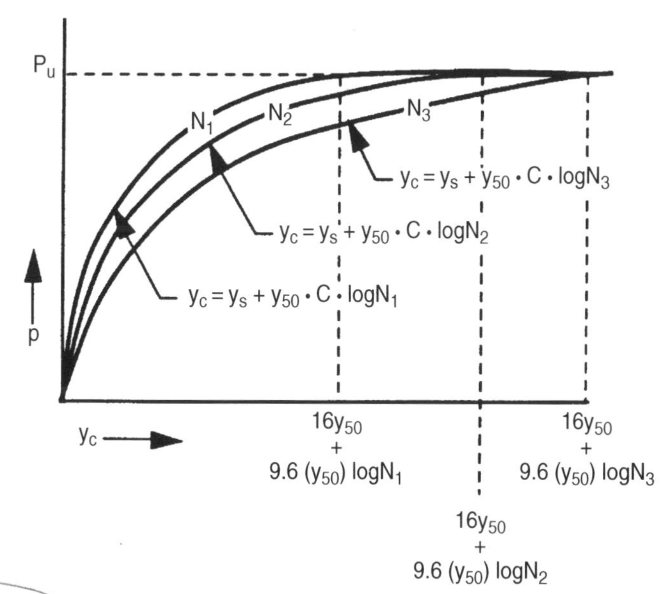

Figure 14.12 Characteristic shape of the p-y curve for cyclic loading in stiff clay with no free water.

$$y_c = y_s + y_{50} C \log N \tag{14.26}$$

ection under N cycles of load,

y_s = deflection under short-term static load,

y_{50} = deflection under short-term static load at one-half of the ultimate resistance, and

N = number of cycles of load application.

5. The *p-y* curve defines the soil response after N cycles of loading.

The properties of the clay at the site should be obtained by triaxial compression tests of the unconsolidated-undrained type with confining stresses equal to the overburden pressures at the elevations from which the samples were taken to determine the shear strength. The value of ε_{50} should be taken as the strain during the test corresponding to the stress equal to one-half of the maximum total-principal-stress difference. The undrained shear strength c should be defined as one-half of the maximum total-principal-stress difference. The unit weight of the soil must also be determined.

14.3.3 Recommendations for *p-y* Curves for Sands

Initial Portion of Curves The initial stiffness of stress-strain curves for sand is a function of the confining pressure; therefore, the use of mechanics for obtaining $E_{py\max}$ for sands is complicated. The *p-y* curve at the ground surface will be characterized by zero values of p for all values of y, and the initial slope of the curves and the ultimate resistance will increase approximately linearly with depth.

The recommendations for the initial portion of the *p-y* curves for sand were derived principally from the results of experiments. As noted later, Terzaghi (1955) presented values that were useful in form, that is, starting with zero at the groundline and increasing linearly with depth. However, his values were for the case where the computed value of p would be equal to one-half of the ultimate bearing stress. Experimental results led to the recommendations in Table 14.3.

Ultimate Resistance p_u for Sands Two models are used for computing the ultimate resistance for piles in sand, following a procedure similar to that used for clay. The first model for the soil resistance near the ground surface is shown in Figure 14.13. The total lateral force F_{pt} (Figure 14.13c) may be computed by subtracting the active force F_a, computed by using Rankine theory, from the passive force F_p, computed from the model by assuming that the Mohr-Coulomb failure condition is satisfied on planes, *ADE*, *BCF*, and

TABLE 14.3 Representative Values of K_{py} for Sands

Relative Density of Sand	Loose	Medium	Dense
	Submerged Sand		
MN/m^3	5.4	16.3	34.0
psi	20.0	60.0	125.0
	Sand Above Water Table		
MN/m^3	6.8	24.4	61.0
psi	25.0	90.0	225.0

AEFB (Figure 14.13a). The directions of the forces are shown in Figure 14.13b. Solutions other than the ones shown here have been developed by assuming a friction force on the surface *DEFC* (assumed to be zero in the analysis shown here) and by assuming the water table to be within the wedge (the unit weight is assumed to be constant in the analysis shown here).

The force F_{pt} may be computed by following a procedure similar to that used to solve the equation in the clay model (Figure 14.4). The resulting equation is as follows:

$$F_{pt} = \gamma H^2\left[\left(\frac{K_0 H \tan\phi \tan\beta}{3\tan(\beta-\phi)\cos\alpha_s}\right) + \left(\frac{\tan\beta}{\tan(\beta-\phi)}\left(\frac{b}{2}+\frac{H}{3}\tan\beta\tan\alpha_s\right)\right)\right]$$
$$+ \gamma H^2\left[\frac{K_0 H \tan\beta}{3}(\tan\varphi\sin\beta - \tan\alpha_s) - \frac{K_A b}{2}\right] \quad (14.27)$$

where

ϕ = friction angle,
K_0 = coefficient of earth pressure at rest, and
K_A = minimum coefficient of active earth pressure = $\tan^2(45 - \phi/2)$.

Sowers and Sowers (1970) have recommended K_0 values of 0.6 for loose sand and 0.4 for dense sand.

The ultimate soil resistance near the ground surface per unit length of the pile is obtained by differentiating Eq. 14.27:

$$(p_u)_{sa} = \gamma H\left[\frac{K_0 H \tan\phi\sin\beta}{\tan(\beta-\phi)\cos\alpha_s} + \frac{\tan\beta}{\tan(\beta-\phi)}(b + H\tan\beta\tan\alpha_s)\right.$$
$$\left. + K_0 H\tan\beta(\tan\phi\sin\beta - \tan\alpha_s) - K_a b\right] \quad (14.28)$$

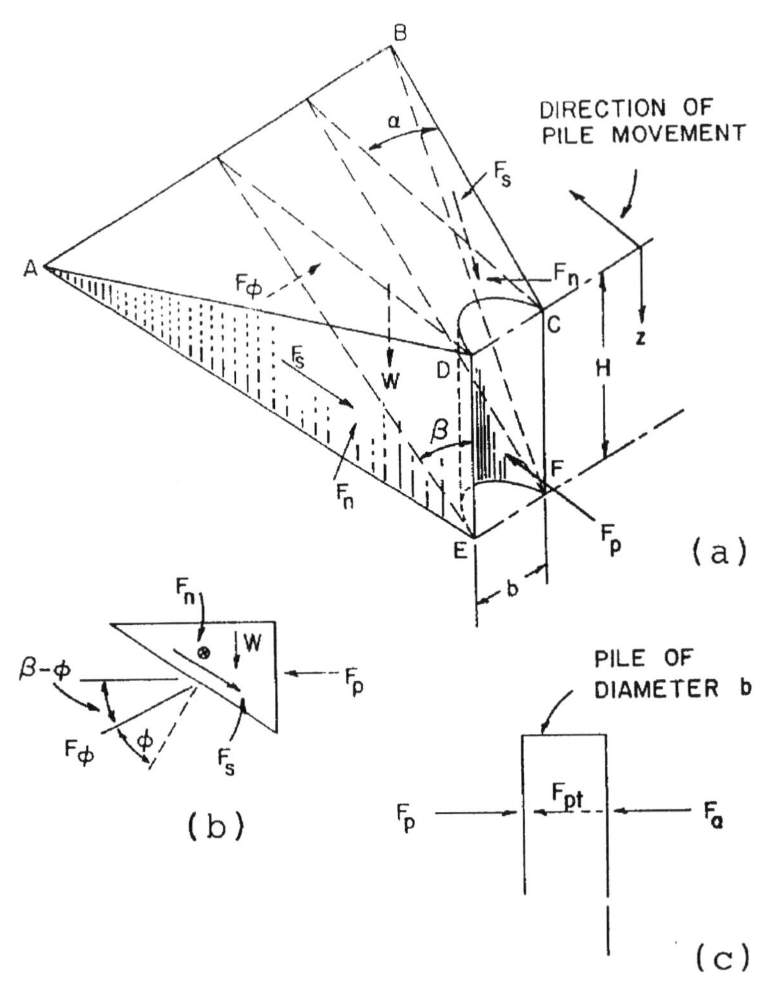

Figure 14.13 Assumed passive wedge-type failure of a pile in sand: (a) general shape of the wedge; (b) forces on the wedge, (c) forces on the pile.

Bowman (1958) performed laboratory experiments with careful measurements and suggested values of α_s ranging from $\phi/3$ to $\phi/2$ for loose sand and up to ϕ for dense sand. The value of β is approximated by the following equation:

$$\beta = 45 + \phi/2 \tag{14.29}$$

The model for computing the ultimate soil resistance at some distance below the ground surface is shown in Figure 14.14a. The stress σ_1 at the back of the pile must be equal to or larger than the minimum active earth pressure; if not, the soil could fail by slumping. The assumption is based on

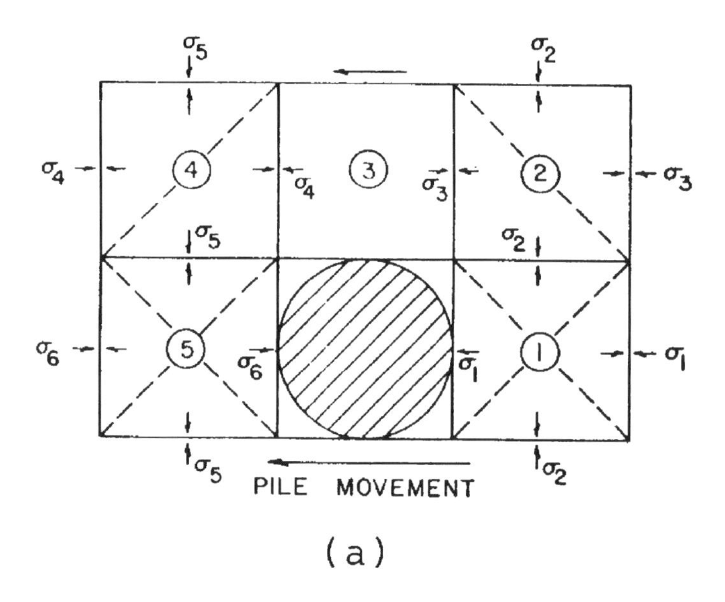

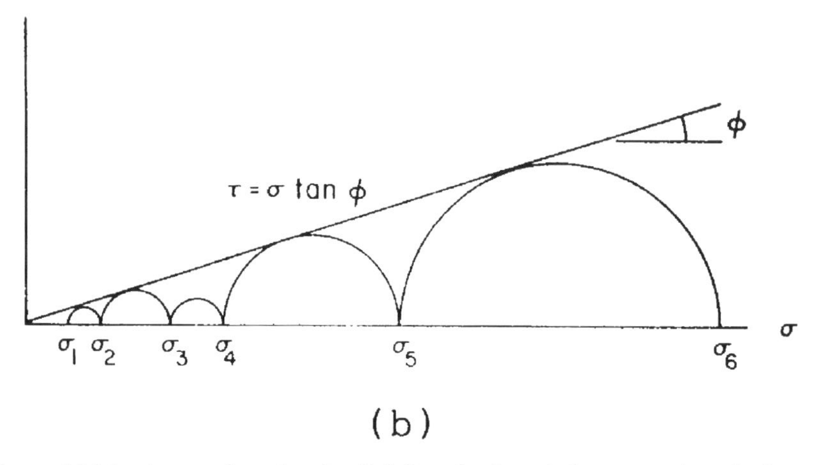

Figure 14.14 Assumed mode of soil failure by lateral flow around a pile in sand: (a) section through the pile; (b) Mohr-Coulomb diagram.

two-dimensional rather than three-dimensional behavior; therefore, some approximation is introduced. If the states of stress shown in Figure 14.14b are assumed, the ultimate soil resistance for horizontal movement of the soil is

$$(p_u)_{sb} = K_A b\gamma H(\tan^8 \beta - 1) + K_0 b\gamma H \tan \phi \tan^4 \beta \qquad (14.30)$$

The equations for $(p_u)_{sa}$ and $(p_u)_{sb}$ are approximate because of the elementary nature of the models used in the computations. However, the equations serve a useful purpose in indicating the form, if not the magnitude, of the ultimate soil resistance.

Early Recommendations The presentation of the recommendations of Terzaghi (1955) is of interest here, but it is recognized that his coefficients probably are meant to reflect the slope of secants to *p-y* curves rather than the initial moduli. As noted earlier, Terzaghi recommended the use of his coefficients up to the point where the computed soil resistance was equal to about one-half of the ultimate bearing stress. The numerical values given by Terzaghi for E_{py} started correctly with zero at the ground surface and increased linearly with depth, but they are not shown here because the values are for some undetermined points along the *p-y* curves. The recommendations provided a basis for computation, but his values could not be implemented very well until the digital computer became available.

Parker and Reese (1971) performed some small-scale experiments, examined unpublished data, and recommended procedures for predicting *p-y* curves for sand. The method apparently received little use because the method described below, based on a comprehensive testing program, soon became available.

Response of Sand Above and Below the Water Table The recommendations shown here are based principally on an extensive series of tests performed at a site on Mustang Island, near Corpus Christi, Texas (Cox et al., 1974). Two steel-pipe piles, 610 mm (24 in.) in diameter, were driven into sand in such a manner as to simulate the driving of an open-ended pipe and were subjected to lateral loading. The embedded length of the piles was 21 m (69 ft). One of the piles was subjected to short-term loading and the other to repeated loading.

The soil at the site was uniformly graded fine sand with an angle of internal friction of 39°. The submerged unit weight was 10.4 kN/m^3 (66 lb/ft^3). The water surface was maintained at 150 mm (6 in.) or so above the mudline throughout the test program.

The following procedure is for short-term static loading and for cyclic loading and is illustrated in Figure 14.15 (Reese et al., 1974).

1. Obtain values for the friction angle ϕ, the soil unit weight γ, and the pile diameter b (note: use buoyant unit weight for sand below the water table and total unit weight for sand above the water table).
2. Compute the ultimate soil resistance per unit length of pile using the smaller of the values given by the following equations:

$$p_{st} = \gamma z \left[\frac{K_0 \tan\phi \tan\beta}{\tan(\beta - \phi)\cos\alpha} + \frac{\tan\beta}{\tan(\beta - \phi)} (b + z\tan\beta\tan\alpha) \right.$$

$$\left. + K_0 z \tan\beta(\tan\phi\sin\beta - \tan\alpha) - K_A b \right] \tag{14.31}$$

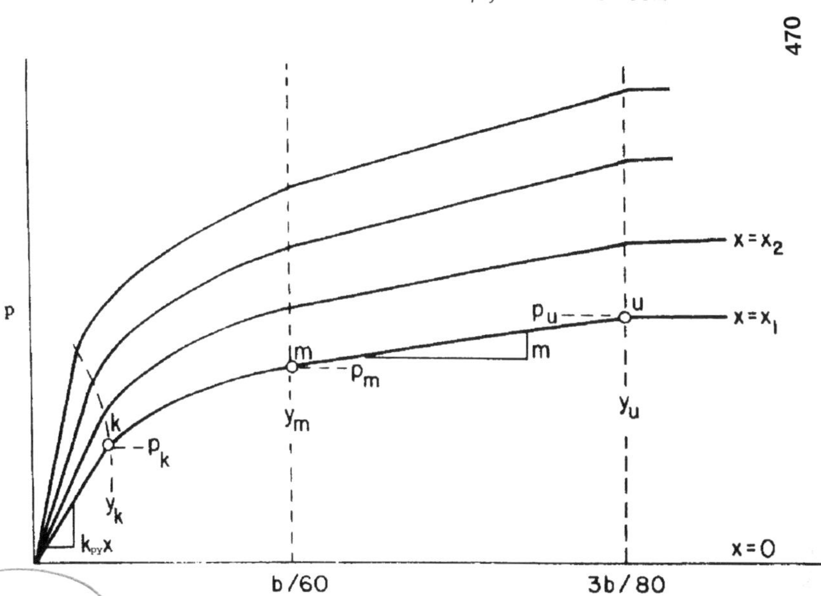

Figure 14.15 Characteristic shape of a family of *p-y* curves for static and cyclic loading in sand.

$$p_{st} = K_A b\gamma z(\tan^8 \beta - 1) + K_0 b\gamma z \tan \phi \tan^4 \beta \qquad (14.32)$$

where $\alpha = \phi/2$.

3. In making the computation in Step 2, find the depth z_t at which there is an intersection at Eqs. 14.31 and 14.32. Above z_t use Eq. 14.31. Below z_t use Eq. 14.32.
4. Select a depth at which a *p-y* curve is desired.
5. Establish y_u as $3b/80$. Compute p_u by the following equation:

$$p_u = \overline{A}_s p_s \quad \text{or} \quad p_u = \overline{A}_c p_s \qquad (14.33)$$

   Use the appropriate value of $\overline{A}_s$ or $\overline{A}_c$ from Figure 14.16 for the particular nondimensional depth and for either the static or cyclic case. Use the appropriate equation for p_s, Eq. 14.31 or Eq. 14.32, by referring to the computation in Step 3.
6. Establish y_m as $b/60$. Compute p_m by the following equation:

$$p_m = B_s p_s \quad \text{or} \quad p_m = B_c p_s \qquad (14.34)$$

   Use the appropriate value of B_s or B_c from Figure 14.17 for the particular nondimensional depth and for either the static or cyclic case. Use the appropriate equation for p_s. The two straight-line portions of

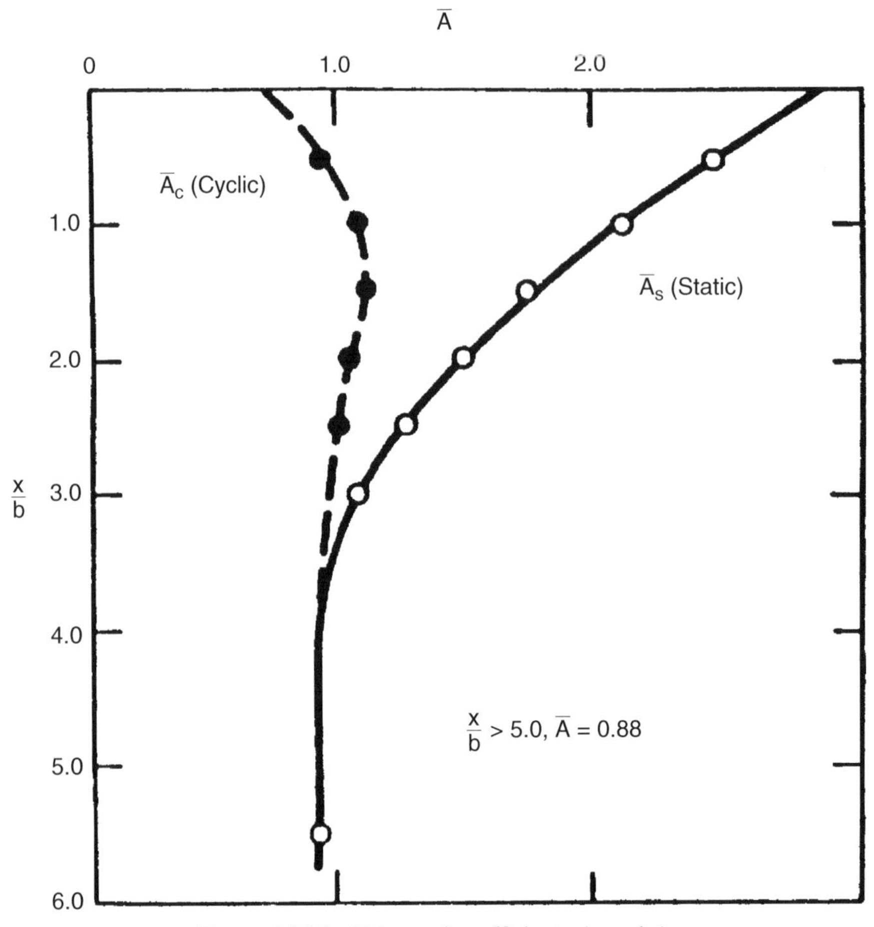

Figure 14.16 Values of coefficients A_c and A_s.

the *p-y* curve, beyond the point where y is equal to $b/60$, can now be established.

7. Establish the initial straight-line portion of the *p-y* curve:

$$p = K_{py}zy \tag{14.35}$$

Use the appropriate value of K_{py} from Table 14.3.

8. Establish the parabolic section of the *p-y* curve:

$$p = \overline{C}y^{1/n} \tag{14.36}$$

Fit the parabola between points k and m as follows:

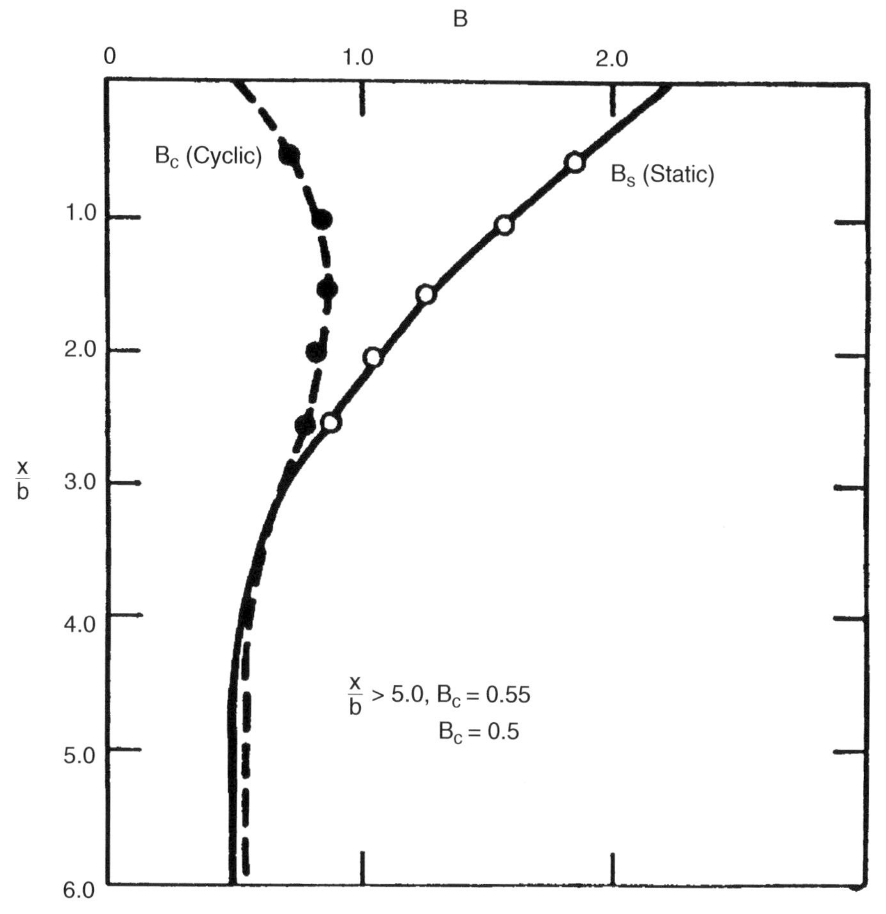

Figure 14.17 Values of coefficients B_c and B_s.

a. Find the slope of the line between points m and u using

$$m = \frac{p_u - p_m}{y_u - y_m} \tag{14.37}$$

b. Obtain the power of the parabolic section using

$$n = \frac{p_m}{m y_m} \tag{14.38}$$

c. Obtain the coefficient as follows:

$$\overline{C} = \frac{p_m}{y_m^{1/n}} \tag{14.39}$$

d. Determine point k as

$$y_k = \left(\frac{\overline{C}}{K_{py}\ z}\right) \tag{14.40}$$

e. Compute the appropriate number of points on the parabola using Eq. 14.36.

Note: The step-by-step procedure is outlined, and Figure 14.15 is drawn, as if there is an intersection between the initial straight-line portion of the p-y curve and the parabolic portion of the curve at point k. However, in some instances there may be no intersection with the parabola. Equation 14.35 defines the p-y curve until there is an intersection with another branch of the p-y curve; if no intersection occurs, Eq. 14.35 defines the complete p-y curve. This completes the development of the p-y curve for the desired depth. Any number of curves can be developed by repeating the above steps for each desired depth.

Triaxial compression tests are recommended for obtaining the friction angle of the sand. Confining pressures close to or equal to those at the depths being considered in the analysis should be used. Tests must be performed to determine the unit weight of the sand. However, it may be impossible to obtain undisturbed samples, and frequently the angle of internal friction is estimated from results of some type of in situ test. The procedure above can be used for sand above the water table if appropriate adjustments are made in the unit weight and angle of internal friction of the sand.

Another method for predicting the p-y curves for sand is presented by the API in its manual on recommended practice (RP2A). The two main differences between the recommendations given above and the API recommendations concern the initial slope of the p-y curves and the shape of the curves.

The following procedure is for short- term static loading and for cyclic loading as described in API RP2A (1987). The API recommendations were developed only for submerged sand, but the assumption is made that the method can be used for sand both above and below the water table, as was done for the recommendations above.

1. Obtain values for the friction angle ϕ, the soil unit weight γ, and the pile diameter b (note: use buoyant unit weight for sand below the water table and total unit weight for sand above the water table).
2. Compute the ultimate soil resistance at a selected depth z. The ultimate lateral bearing capacity (ultimate lateral resistance p_u) for sand has been found to vary from a value at shallow depths determined by Eq. 14.41 to a value at deep depths determined by Eq. 14.42. At a given depth, the equation giving the smallest value of p_u should be used as the ul-

timate bearing capacity.

$$p_{us} = (C_1 z + C_2 b)\gamma z \tag{14.41}$$

$$p_{ud} = C_3 b \gamma z \tag{14.42}$$

where C_1, C_2, and C_3 are coefficients determined from Figure 14.18.

3. Develop the *p-y* curve based on the ultimate soil resistance p_u, which is the smallest value of p_u calculated in Step 2 and using Eq. 14.43:

$$p = Ap_u \tanh\left(\frac{E_{py\ \max}\, z}{Ap_u}\, y\right) \tag{14.43}$$

where $A = 0.9$ for cyclic loading and $\left(3 - 0.8\,\frac{z}{b}\right) \geq 0.9$ for static loading, and K_{py} is found in Figure 14.19.

14.3.4 Modifications to *p-y* Curves for Sloping Ground

The recommendations for *p-y* curves presented to this point are developed for a horizontal ground surface. To allow designs to be made if a pile is installed on a slope, the *p-y* curves must be modified. The modifications

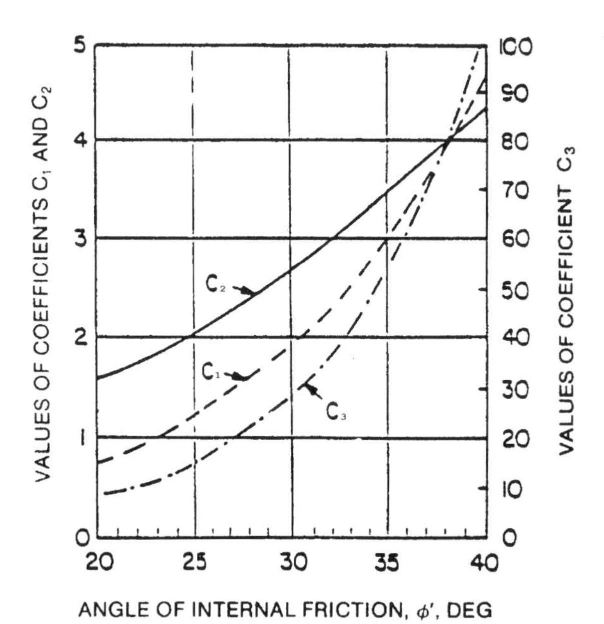

Figure 14.18 Coefficients as a function of Φ'.

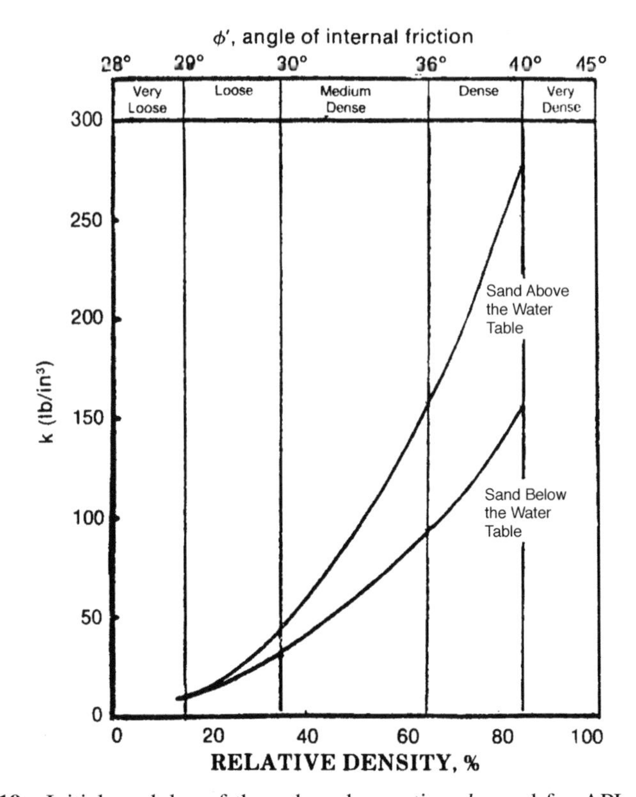

Figure 14.19 Initial modulus of the subgrade reaction, k, used for API sand criteria.

involve revisions in the manner in which the ultimate soil resistance is computed. In this regard, the assumption is made that the flow-around failure will not be influenced by sloping ground; therefore, only the equations for the wedge-type failure need modification.

The solutions presented here are entirely analytical and must be considered preliminary. Additional modifications may be indicated if it is possible to implement an extensive laboratory and field study.

Equations for Ultimate Resistance in Clay The ultimate soil resistance near the ground surface for saturated clay where the pile was installed in ground with a horizontal slope was derived by Reese (1958) and is shown in Eq. 14.44:

$$p_{\text{uca}} = 2c_a b + \gamma b H + 2.83 c_a H \tag{14.44}$$

If the ground surface has a slope angle θ, as shown in Figure 14.20, the ultimate soil resistance near the ground surface if the pile is pushed against the downhill slope is

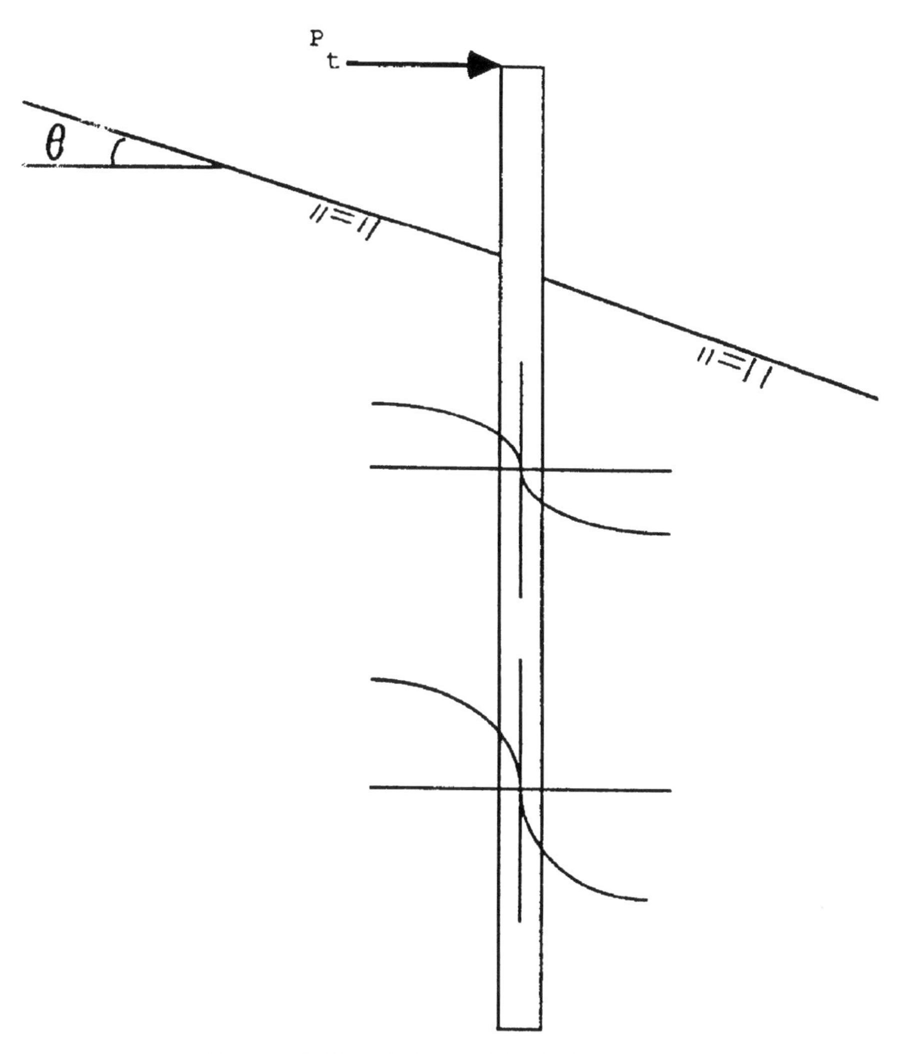

Figure 14.20 Pile installed in sloping ground.

$$p_{\text{uca}} = (2c_a b + \gamma bH + 2.83c_a H)\left(\frac{1}{1 + \tan\theta}\right) \tag{14.45}$$

The ultimate soil resistance near the ground surface if the pile is pushed against the uphill slope is

$$p_{\text{uca}} = (2c_a b + \gamma bH + 2.83c_a H)\left(\frac{\cos\theta}{\sqrt{2}\cos(45 + \theta)}\right) \tag{14.46}$$

where

c_a = average undrained shear strength in the vicinity of the ground surface, and

θ = angle of slope as measured from the horizontal.

A comparison of Eqs. 14.45 and 14.46 shows that the equations are identical except for the last terms in parentheses. If θ is equal to zero, the equations become equal to the original equation.

Equations for Ultimate Resistance in Sand The following equations apply to slopes in sand that are less steep than the friction angle of the sand. The ultimate soil resistance near the ground surface is shown in Eq. 14.31 and forms the basis for the equations shown below. If the pile is pushed against a downhill slope with an angle that is less than the friction angle ϕ, the ultimate soil resistance is

$$p_{usa} = \gamma H \left[\begin{array}{l} \dfrac{K_0 H \tan\phi \sin\beta}{\tan(\beta - \phi)\cos\alpha}(4D_1^3 - 3D_1^2 + 1) \\ + \dfrac{\tan\beta}{\tan(\beta - \phi)}(bD_2 + H \tan\beta \tan\alpha D_2^2) \\ + K_0 H \tan\beta(\tan\phi \sin\beta - \tan\alpha) \\ (4D_1^3 + 3D_1^2 + 1) - K_A b \end{array} \right] \tag{14.47}$$

If the pile is pushed against an uphill slope, the ultimate soil resistance is

$$p_{usa} = \gamma H \left[\begin{array}{l} \dfrac{K_0 H \tan\phi \sin\beta}{\tan(\beta - \phi)\cos\alpha}(4D_3^3 - 3D_3^2 + 1) \\ + \dfrac{\tan\beta}{\tan(\beta - \phi)}(bD_4 + H \tan\beta \tan\alpha D_4^2) \\ + K_0 H \tan\beta(\tan\phi \sin\beta - \tan\alpha) \\ (4D_3^3 + 3D_3^2 + 1) - K_A b \end{array} \right] \tag{14.48}$$

where

$$D_1 = \frac{\tan\beta \tan\theta}{\tan\beta \tan\theta + 1} \tag{14.49}$$

$$D_2 = 1 - D_1 \tag{14.50}$$

$$D_3 = \frac{\tan \beta \tan \theta}{1 - \tan \beta \tan \theta} \tag{14.51}$$

$$D_4 = 1 + D_3 \tag{14.52}$$

and

$$K_A = \cos \theta \frac{\cos \theta - (\cos^2 \theta - \cos^2 \phi)^{0.5}}{\cos \theta + (\cos^2 \theta - \cos^2 \phi)^{0.5}} \tag{14.53}$$

This completes the necessary derivations for modifying the equations for clay and sand to analyze a pile under lateral load in sloping ground.

14.3.5 Modifications for Raked (Battered) Piles

The effect of batter on the behavior of laterally loaded piles was investigated by the use of experiments. The lateral soil-resistance curves for a vertical pile in a horizontal ground surface were modified by a constant to account for the effect of the inclination of the pile. The values of the modifying constant as a function of the batter angle were deduced from the results in the test tank (Awoshika and Reese, 1971) and from the results from full-scale tests (Kubo, 1965). The modifier to be used is shown by the solid line in Figure 14.21.

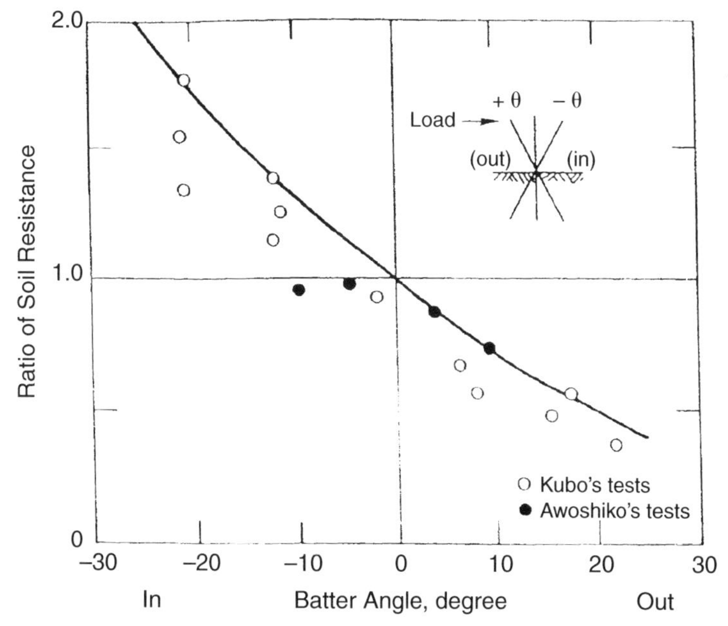

Figure 14.21 Modification of *p-y* curves for battered piles (after Kubo, 1964, and Awoshika and Reese, 1971).

This modifier is to be used to increase or decrease the value of p_u, which in turn will cause each of the p-values to be modified. While it is likely that the value of p_u for the deeper soils is not affected by batter, the behavior of a pile is affected only slightly by the resistance of the deeper soils; therefore, the use of the modifier for all depths is believed to be satisfactory.

As shown in Figure 14.21, the agreement between the empirical curve and the experiments for the out-batter piles (θ is positive) is somewhat better than that for the in-batter piles. The data indicate that the use of the modifier will yield somewhat questionable results; therefore, on an important project, the responsible engineer may wish to recommend full-scale testing.

14.3.6 Recommendations for *p-y* Curves for Rock

Introduction The use of deep foundations in rock is frequently required for support of transmission towers or other structures that sustain lateral loads of significant magnitude. Because the rock must be drilled in order to make the installation, drilled shafts (called *caissons* or *bored piles*) are frequently used. However, a steel pile could be grouted into the drilled hole. In any case, the designer must use appropriate mechanics, as shown elsewhere in this book, to compute the ultimate bending moment and the nonlinear bending stiffness *EI*. Experimental results show conclusively that *EI* must be reduced, as the bending moment increases, to achieve a correct result (Reese, 1997).

In some applications the axial load is negligible, so the penetration is controlled by the lateral load. Computations should be initiated by the designer with a relatively large penetration of the pile into the rock. After a suitable geometric section is found, the factored loads are employed and computer runs are made, with penetration being gradually reduced. The groundline deflection is plotted as a function of penetration, and a penetration is selected that provides adequate security against a sizable deflection of the bottom of the pile.

The following sections present concepts that form the basis for computing the response of piles in rock. The background for designing piles in rock is given and then two sets of criteria are presented, one for strong rock and the other for weak rock.

The secondary structure of rock is an overriding feature with respect to its response to lateral loading. Thus, an excellent subsurface investigation is assumed prior to making any design. The appropriate tools for investigating the rock are employed and the RQD should be obtained, along with the compressive strength of intact specimens. If possible, sufficient data should be taken to allow the computation of the rock mass rating (RMR). Sometimes, the RQD is so low that no specimens can be obtained for compressive tests. The performance of pressuremeter tests in such instances is indicated.

If investigation shows that there are soil-filled joints or cracks in the rock, the procedures suggested here should not be used but full-scale testing at the site is recommended. Furthermore, full-scale testing may be economical if a large number of piles are to be installed at a particular site. Such field testing will add to the data bank and lead to improvements in the recommendations shown below, which should be considered preliminary because of the meager amount of experimental data available.

In most designs, the deflection of the drilled shaft (or another kind of pile) will be so small that the ultimate strength p_{ur} of the rock is not developed. However, the ultimate resistance of the rock should be predicted to allow computation of the lateral loading that causes the failure of the pile. Contrary to the predictions of p-y curves for soil, where the unit weight is a significant parameter, the unit weight of rock is neglected in developing the prediction equations that follow. While a pile may move laterally only a small amount under the working loads, prediction of the early portion of the p-y curve is important because the small deflections may be critical in some designs.

Most specimens of intact rock are brittle and will develop shear planes under low amounts of shearing strain. This fact leads to an important concept about intact rock. *The rock is assumed to fracture and lose strength under small values of deflection of a pile.* If the RQD of a stratum of rock is zero or has a low value, the rock is assumed to have already fractured and, thus, will deflect without significant loss of strength. This concept leads to the recommendation of two sets of criteria for rock, one for strong rock and the other for weak rock. In the presentations here, strong rock is assumed to have a compressive strength of 6.9 MPa (1000 psi) or more.

The method for predicting the response of rock is based strongly on a limited number of experiments and on correlations that have been presented in technical literature. Some of the correlations are inexact; for example, if the engineer enters the figure for correlation between stiffness and strength with a value of stiffness from the pressuremeter, the resulting strength can vary by an order of magnitude, depending on the curve that is selected. The inexactness of the necessary correlations, plus the limited amount of data from controlled experiments, means that the methods for the analysis of piles in rock should be used with both judgment and caution.

Field Experiments An instrumented drilled shaft (bored pile) was installed in vuggy limestone in the Florida Keys (Reese and Nyman, 1978) and was tested under lateral loads. The test was performed to gain information for the design of foundations for highway bridges.

Two cores from the site were tested. The undrained shear strengths of the specimens were taken as one-half of the unconfined compressive strength and were 1.67 MPa (17.4 ton/ft^2) and 1.30 MPa (13.6 ton/ft^2). The rock at the site was also investigated by in situ grout-plug tests under the direction of Dr. John Schmertmann (1977). A 140-mm (5.5-in.)-diameter hole was drilled into the limestone, a high-strength steel bar was placed to the bottom of the hole, and a grout plug was cast over the lower end of the bar. The bar was pulled until failure occurred, and the grout was examined to see that failure occurred at the interface of the grout and limestone. The average of the eight tests was 1.56 MPa (18.3 ton/ft^2). However, the rock was stronger in the zone where the deflections of the drilled shaft were most significant, and a shear strength of 1.72 MPa (18.0 tons/ft^2) was selected for correlation.

The bored pile was 1220 mm (48 in.) in diameter and penetrated 13.3 m (43.7 ft) into the limestone. The overburden of fill was 4.3 m (14 ft) thick and was cased. The load was applied at 3.51 m (11.5 ft) above the limestone. A maximum horizontal load of 667 kN (75 tons) was applied to the pile. The maximum deflection at the point of load application was 18.0 mm (0.71 in.), and at the top of the rock (bottom of the casing) it was 0.54 mm (0.0213 in.). While the load versus deflection curve was nonlinear, there was no indication of rock failure.

The California Department of Transportation (Caltrans) performed lateral-load tests of two drilled shafts (bored piles) near San Francisco (Speer, 1992). The test results, while unpublished, have been provided with the courtesy of Caltrans. Two borings were made into the rock, and sampling was done with a NWD4 core barrel in a cased hole with a diameter of 102 mm (4 in.). The sandstone was medium to fine-grained, with grain sizes ranging from 0.1 to 0.5 mm (0.004 to 0.02 in.), well sorted, and thinly bedded, with a thickness of 25 to 75 mm (1 to 3 in.). Recovery was generally 100%. The reported RQD values ranged from 0 to 80, with an average of 45. Speer (1992) described the sandstone as very intensely to moderately fractured, with bedding joints, joints, and fracture zones.

Pressuremeter tests were performed, and the results were scattered. The values for the moduli of the rock were obtained, and a correlation was employed between the initial stiffness and the compressive strength. Three intervals of rock depth were identified with the following compressive strengths: 0 to 3.9 m, 1.86 MPa; 3.9 to 8.8 m, 6.45 MPa; and below 8.8 m, 16.0 MPa (0 to 12.8 ft, 270 psi; 12.8 to 28.9 ft, 936 psi; and below 28.9 ft, 2320 psi). The strength of the lowest interval is in the strong rock category; however, the first interval is mainly effective in providing the resistance to lateral loading.

Two drilled shafts, 2.25 m (7.38 ft) in diameter, with penetrations of 12.5 m (41 ft) and 13.8 m (45 ft), were tested simultaneously. Lateral loading was accomplished by hydraulic rams, acting on high-strength steel bars that were passed through tubes, transverse and perpendicular to the axes of the piles. Load was measured by load cells, and deflection was measured by transducers. The slope and deflection of the tops of the piles were obtained by readings from the slope indicators.

The load was applied in increments at 1.41 m (4.6 ft) above the ground line for Pile A and 1.24 m (4.1 ft) for Pile B. The pile-head deflection was measured at slightly different points above the rock, but the results were adjusted slightly to yield equivalent values for each of the piles. The load-deflection data for the two piles show that Pile A apparently had a structural weakness, so only Pile B was used in developing the recommendations for *p-y* curves. A groundline deflection of 17 mm was measured at a lateral load of 8000 kN, but the deflection increased to 50 mm at a lateral load of 8950 kN.

Interim Recommendations for Computing p-y Curves for Weak Rock
The *p-y* curve shown in Figure 14.22 is characteristic of the family representing the behavior of weak rock. The expression for the ultimate resistance p_{ur} for rock is derived from the mechanics for the ultimate resistance of a wedge of rock at its surface:

$$p_{ur} = \alpha_r q_{ur} b \left(1 + 1.4 \frac{z_r}{b}\right) \quad \text{for} \quad 0 \leq z_r \leq 3b \tag{14.54}$$

$$p_{ur} = 5.2 \alpha_r q_{ur} b \quad \text{for} \quad z_r > 3b \tag{14.55}$$

where

q_{ur} = compressive strength of the rock, usually lower-bound, as a function of depth,
α_r = strength reduction factor b = diameter of the pile, and
z_r = depth below the rock surface.

The assumption is made that failure will occur at the surface of the rock after small deflections of the pile. The compressive strength of intact specimens is reduced by multiplication by α_r to account for fractures in the rock. The value of α_r is assumed to be one-third for an RQD of 100 and to increase linearly to unity at an RQD of 0. The RQD values near the ground surface

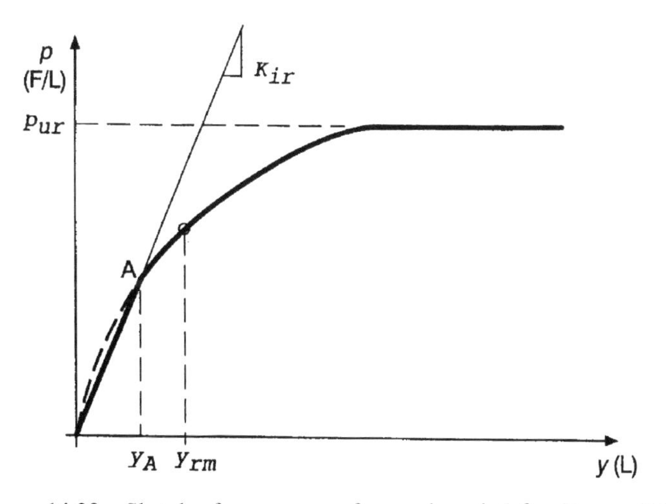

Figure 14.22 Sketch of a *p-y* curve for weak rock (after Reese, 1997).

should be given more weight, taking into account the random nature of the presence of fractures. If RQD is zero, the compressive strength may be obtained directly from a pressuremeter curve, or approximately from Figure 14.23, by entering the value of the pressuremeter modulus.

In considering a strip from a beam resting on an elastic, homogeneous, and isotropic solid, the initial modulus $(E_{py\,max})_r$ in Figure 14.22 may be shown to have the following value (using the symbols for rock):

$$(E_{py\ max})_r \cong k_{ir} E_{ir} \tag{14.56}$$

where

k_{ir} = dimensionless constant, and
E_{ir} = initial modulus of the rock.

Equations 14.57 and 14.58 for k_{ir} are derived from experimental data and reflect the assumption that the presence of the rock surface will have a similar effect on k_{ir} as was shown for p_{ur} for ultimate resistance:

$$k_{ir} = \left(100 + \frac{400 z_r}{3b}\right) \quad \text{for} \quad 0 \le z_r \le 3b \tag{14.57}$$

$$k_{ir} = 500 \quad \text{for} \quad z_r \ge 3b \tag{14.58}$$

With guidelines for computing p_{ur} and $(E_{py\,max})_r$ the equations for the three branches of the family of *p-y* curves for rock in Figure 14.22 can be computed with the equations that follow.

Branch 1:
$$p = (E_{py\ max})_r\, y \quad \text{for} \quad y \le y_A \tag{14.59}$$

Branch 2:
$$p = \frac{p_{ur}}{2}\left(\frac{y}{y_m}\right)^{0.25} \quad \text{for} \quad y \le y_A,\ p \le p_{ur} \tag{14.60}$$

and

$$y_{rm} = k_{rm} b \tag{14.61}$$

where k_{rm} = a constant ranging from 0.00005 to 0.0005. The value of y_A is computed by Eq. 14.62:

$$y_A = \left(\frac{p_{ur}}{2(y_m)^{0.25}\, k_{ir}}\right)^{1.333} \tag{14.62}$$

Branch 3:
$$p = p_u \tag{14.63}$$

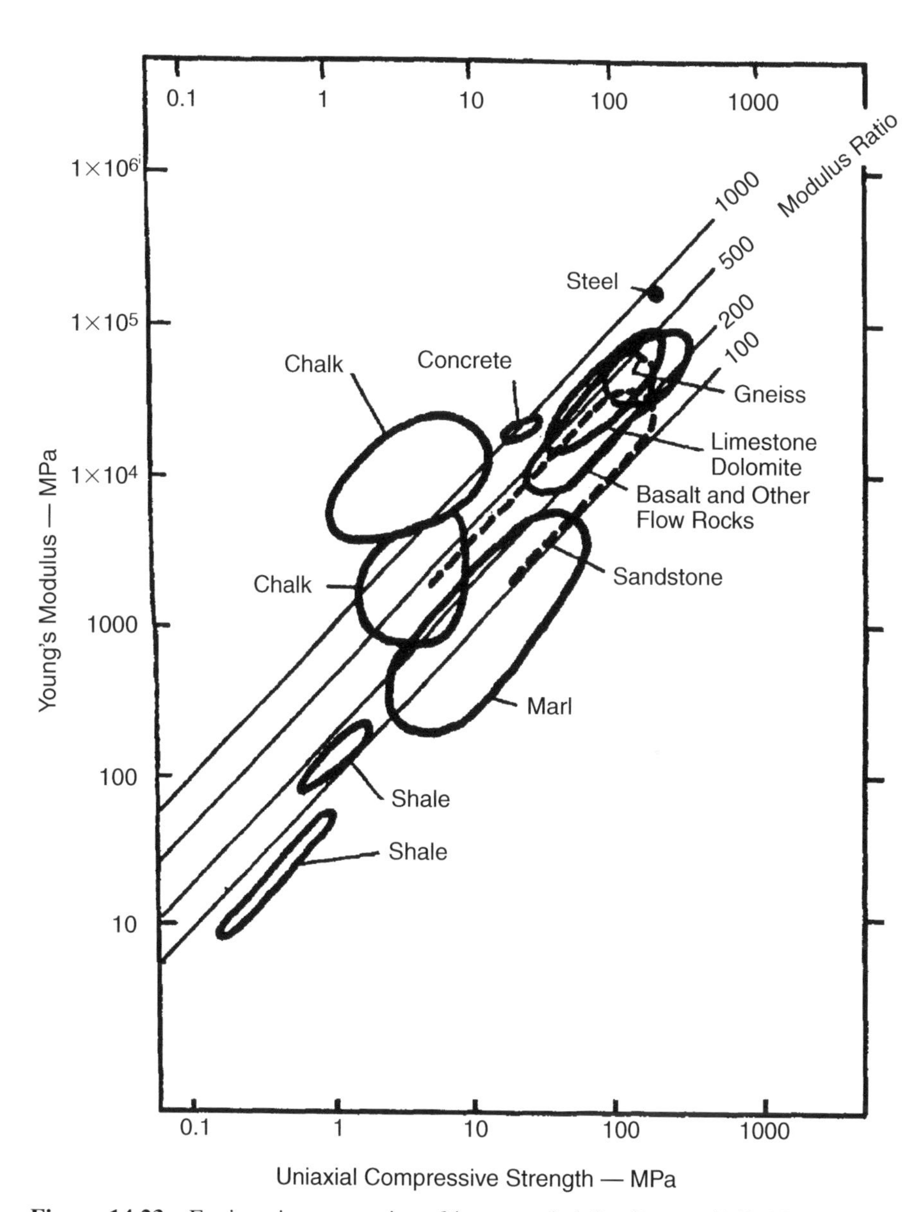

Figure 14.23 Engineering properties of intact rock (after Deere, 1968; Horvath and Kenney, 1979; Peck, 1976).

In selecting a value for k_{rm}, 0.0005 was used to obtain agreement with the experimental results for the test at Islamorada and 0.00005 was used to get agreement between the analytical and experimental results for the test at San Francisco (Reese, 1997). A review of the analyses of the experiments at Islamorada and San Francisco (Reese, 1997) shows that an important parameter is the nonlinear *EI* of the reinforced-concrete section of the drilled shaft. Therefore, computation of the nonlinear *EI* must be given careful attention, and parametric studies can be beneficial in selecting a value of k_{rm}, using a value from the given range to achieve conservatism.

Interim Recommendations for Computing p-y Curves for Strong Rock

The recommendations shown here for strong rock are based on elementary mechanics and on the assumption that there are few cracks and joints in the rock, especially near the rock surface. The *p-y* curve recommended for rock, with compressive strength of intact specimens q_{ur} larger than 6.9 MPa (1000 psi), is shown in Figure 14.24. If the rock increases in strength with depth, the strength at the top of the stratum will normally control. Cyclic loading is assumed to cause no loss of resistance.

As shown in Figure 14.24, load tests are recommended if deflection of the rock (and pile) is greater than 0.0004*b*, and brittle fracture is assumed if the lateral stress (force per unit length) against the rock becomes greater than one-half of the diameter times the compressive strength of the rock.

14.4 SOLUTION OF THE DIFFERENTIAL EQUATION BY COMPUTER

14.4.1 Introduction

The limitations imposed on the nondimensional method presented in Chapter 12 are eliminated by the use of finite differences. Some of the capabilities of the method are as follows: a unique *p-y* curve can be input at every increment along the pile; the nonlinear bending stiffness of a reinforced-concrete pile can be input; distributed loading from flowing water or moving soil can be taken into account; the effect of axial loading on bending can be considered explicitly; a variety of boundary conditions at the head and toe of the pile make possible the solution of a variety of practical problems; rapid solutions for a given set of inputs allow the investigation of various parameters and evaluation of various designs; and incrementing of the loading allows the investigation of pile failure by development of a plastic hinge or excessive deflection.

The problems with the finite-difference method are important. Writing the required computer program, using double-precision arithmetic, is complicated, and the available programs must be checked for correctness by the use of standard mechanics; in addition, the engineer must use care in entering the input for a particular problem. Some codes have built-in checks of certain

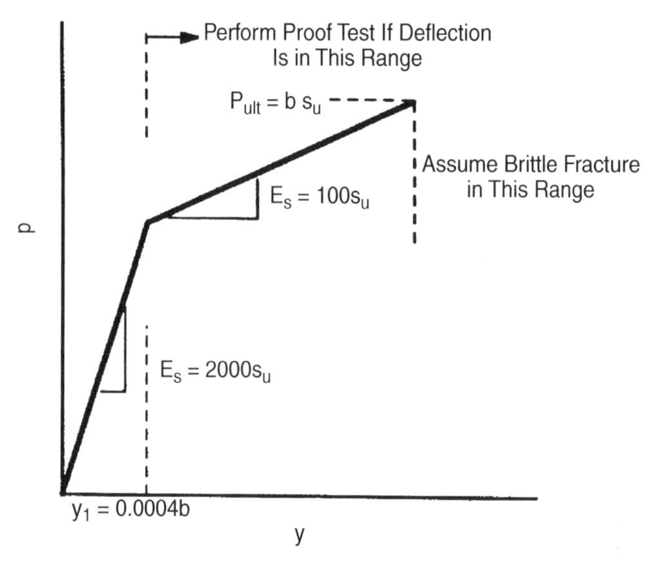

Figure 14.24 Recommended *p-y* curve for strong rock.

parameters to ensure correctness, but most codes require diligence on the part of the engineer.

A good example of necessary diligence involves selection of the number of increments into which the pile is subdivided (see Figure 14.25). Selection of too few increments means errors in solving the differential equation; se-

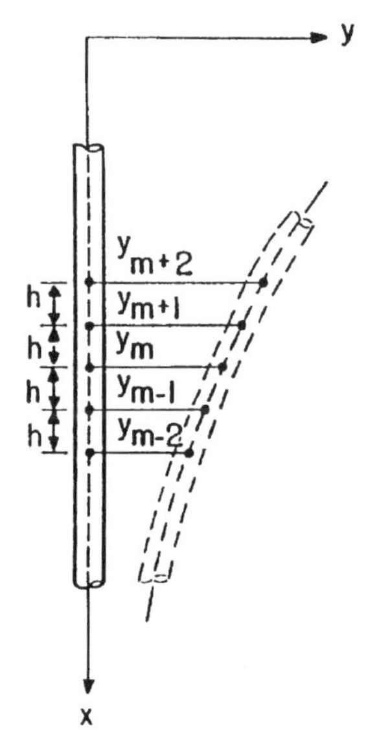

Figure 14.25 Computer representation of a deflected pile.

lection of too many increments leads to an erroneous solution because the difference between successive increments disappears. However, use of the computer program with understanding and care leads to good solutions of many practical problems.

14.4.2 Formulation of the Equation by Finite Differences

With the subdivision of the pile as shown in Figure 14.25, the derivatives can be written in difference form as follows:

$$\frac{dy}{dx} = \frac{y_{m-1} - y_{m+1}}{2h} \tag{14.64}$$

$$\frac{d^2y}{dx^2} = \frac{y_{m-1} - 2y_m + y_{m+1}}{h^2} \tag{14.65}$$

$$\frac{d^3y}{dx^3} = \frac{-y_{m-2} + 2y_{m-1} - 2y_{m+1} + y_{m+2}}{2h^3} \tag{14.66}$$

$$\frac{d^4y}{dx^4} = \frac{y_{m-2} - 4y_{m-1} + 6y_m - 4y_{m+1} + y_{m+2}}{h^4} \tag{14.67}$$

Using the expressions above, and by substitution into Equation 14.1, shown below,

$$E_pI_p \frac{d^4y}{dx^4} + P_x \frac{d^2y}{dx^2} - p + W = 0 \tag{14.1}$$

The differential equation for a pile under lateral loading in difference form is

$$\begin{aligned} & y_{m-2}R_{m-1} + y_{m-1}(-2R_{m-1} - 2R_m + P_xh^2) \\ & + y_m(R_{m-1} + 4R_m + R_{m+1} - 2P_xh^2 + E_{pym}h^4) \\ & + y_{m+1}(-2R_m - 2R_{m+1} + P_xh^2) + y_{m+2}R_{m-1}Wh^4 = 0 \end{aligned} \tag{14.68}$$

where $R_m = (E_pI_p)_m$, the bending stiffness of the pile at point m. If a pile is subdivided into n increments, $n + 1$ equations of the form of Eq. 14.68 can be written, leading to $n + 5$ unknowns. The following section shows the formulation of the four needed equations, two at the top of the pile and two at the bottom, to reflect appropriate boundary conditions.

14.4.3 Equations for Boundary Conditions for Useful Solutions

Bottom of the Pile Numbering of the increments along the pile assigns zero to the bottom point, giving the deflection at the bottom as y_0. Then the two imaginary points for use in implementing the boundary conditions are y_{-1} and y_{-2}. Assuming no eccentric axial load at the bottom of the pile, the first of the two boundary conditions sets the moment equal to zero:

$$y_{-1} - 2y_0 + y_1 = 0 \tag{14.69}$$

The second boundary condition needed is for the shear. In most solutions for piles under lateral loading, the pile will experience two or more points of zero deflection, and the deflection at the tip of the pile will be so small that any shear due to pile deflection will be negligible. Thus, the second equation needed is Eq. 14.70:

$$(y_{-2} - 2y_{-1} + 2y_1 - y_2) = 0 \tag{14.70}$$

There may be occasions when the engineer wishes to analyze the behavior of a "short" pile where there is only one point of zero deflection along the length of the pile. In that case, the deflection of the bottom of the pile could cause a shearing force to develop. Then Eq. 14.71 would be the second needed equation for the boundary conditions at the bottom of the pile:

$$\frac{R_0}{2h}(y_{-2} - 2y_{-1} + 2y_1 - y_2) = V_0 \tag{14.71}$$

If the engineer can formulate an equation for V_0 as a function of y_0, Eq. 14.71 can be revised to reflect the formulation. If there is a small deflection at the bottom of the pile, the lateral deflection at the top will be larger and could lead to excessive deflection. Designing a pile with only one point of zero deflection would probably be done only if the tip of the pile is founded in rock. The effect on shear of any axial deflection at the bottom of the pile is ignored in Eq. 14.71.

Top of the Pile The top of the pile can experience a number of boundary conditions, as shown in the following sections. Numbering of the increments along the pile assigned the letter t to the top point, giving the deflection at the top of the pile as y_t.

Shear and Moment. If a pile extends above the groundline and a lateral load is applied at its top, the boundary conditions at the groundline are a known moment and a shear. If M_t defines the moment at the top of the pile and P_t

defines the shear (or horizontal load), the two boundary equations for the top of the pile are

$$\frac{R_t}{h^2}(y_{t-1} - 2y_t + y_{t+1}) = M_t \tag{14.72}$$

$$\frac{R_t}{2h^3}(y_{t-2} - 2y_{t-1} + 2y_{t+1} - y_{t+2}) + \frac{P_x}{2h}(y_{t-1} - y_{t+1}) = P_t \tag{14.73}$$

An alternative way to solve the problem of a pile extending above the ground surface is to employ the known lateral load P_t and let M_t equal to zero. The p-y curves for the portion of the pile above the groundline can be input to shown zero values of p for all values of y.

Shear and Rotation. The top of a pile may be secured into a thick mat of concrete and essentially fixed against rotation. In other designs, the rotation of the top of the pile may be known or estimated. In both instances, the lateral load (shear) is assumed to be known. The following equation may be used with Eq. 14.73 for the boundary conditions at the top of the pile, where S_t is the rotation of the top of the pile:

$$\frac{y_{t-1} - y_{t+1}}{2h} = S_t \tag{14.74}$$

Shear and Rotational Restraint. There are occasions when the top of a pile is extended and becomes part of the superstructure. In the construction of an offshore platform, for example, a prefabricated template is set on the ocean floor and extends above the water surface. Piles are stabbed into open vertical pipes comprising the legs of the platform and then driven to a predetermined depth. Spacers or grout are used in the annular space between the template leg and the pile, allowing the rotational restraint at the pile head below the template to be computed by structural analysis. There may be other constructions where the rotational restraint at the pile head may be known or estimated. The following equation, along with Eq. 14.73, may be used for the boundary conditions at the top of the pile:

$$\frac{\frac{R_t}{h^2}(y_{t-1} - 2y_t + y_{t+1})}{\frac{y_{t-1} - y_{t+1}}{2h}} = \frac{M_t}{S_t} \tag{14.75}$$

Moment and Deflection. An example of the use of this boundary condition is when a pile supports a bridge abutment. The pile head is not restrained against rotation, leading to a zero moment at the pile head, but the lateral deflection Y_t at the pile head is known or can be estimated. In other instances, the pile-head moment may be known or estimated. The following equation, along with Eq. 14.72, can be used for the boundary conditions at the top of the pile:

$$y_t = Y_t \tag{14.76}$$

Deflection and Rotation. When a pile supports a superstructure, the structural engineer may compute a deflection and a rotation (slope) at the pile head. In that case, Eqs. 14.74 and 14.76 may be used for the boundary conditions.

14.5 IMPLEMENTATION OF COMPUTER CODE

A number of computer codes have been written both in the United States and abroad, and have been augmented from time to time, utilizing the equations and procedures presented here. The code used in the computations that follow is LPILE©, and a CD-ROM containing a Student Version of LPILE is included on the inside of the back cover of this book. Many of the computations shown below can be run with the Student Version of LPILE.

The form of the results from a solution with LPILE is shown in Figure 14.26. The deflection along the lower portion of the pile is exaggerated to illustrate the fact that the deflection oscillates, with decreasing values, from positive to negative over the full length of the pile consistent with input. The curves are related mathematically. For example, the values of p are negative when y is positive, and p and y are zero at the same points along the pile.

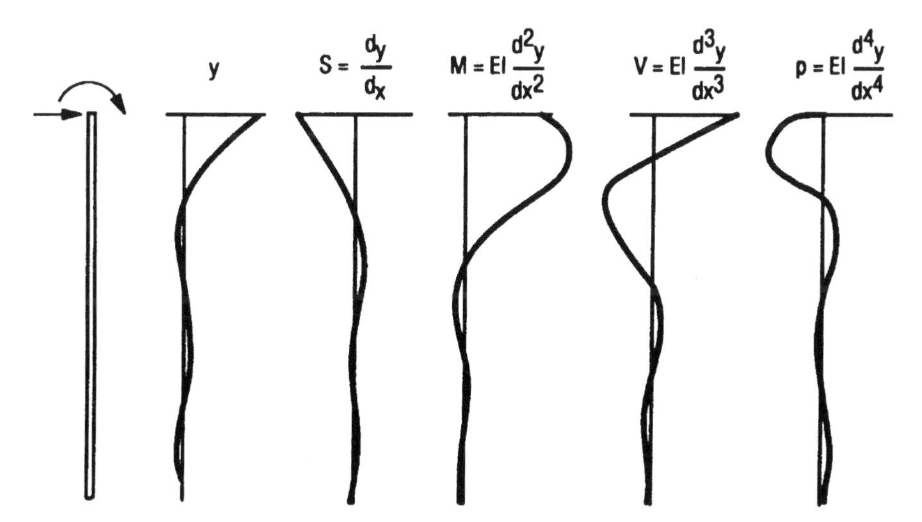

Figure 14.26 Form of the results obtained from a complete solution with LPILE.

The solution of the difference equations for behavior of a particular pile under lateral loading must be solved by iteration because the soil response is a nonlinear function of pile deflection. The computer code begins with an arbitrary set of values for E_{py} and solves for a set of deflections along the length of the pile such as shown in Figure 14.26. The *p-y* curves are employed and new values of E_{py} are obtained and a new set of *y*-values is computed. The new values of *y* are compared with the previous set and the solution continues until convergence occurs where all values of *y* for the present computation are within a given tolerance of the values of *y* in the just previous solution. The tolerance is set at a very small value of deflection to ensure an accurate solution of the difference equations.

14.5.1 Selection of the Length of the Increment

As shown in Figure 14.24 and in Eq. 14.68, the length of the increment between points along the pile where difference equations are written, as noted earlier, is an important parameter affecting the accuracy of a solution. A study is shown here to illustrate the importance of selecting an appropriate length for *h*. A steel-pipe pile with a diameter of 380 mm, used in a computation in Chapter 12, was selected for the study. The moment of inertia *I* was $4.414 \times 10^{-4}\ m^4$, the modulus of elasticity *E* of the steel was 2.0×10^8 kPa, and the yield strength of the steel was 250,000 kPa. Loading was selected for the pile to cause a significant deflection while keeping the bending stress below the yield point of the steel.

The computations are made for the case of a constant value of E_{py} for which a closed-form solution (Hetenyi, 1946) exists that will yield a precise result to compare with the output from the computer code. Comparison will be made of the pile-head deflection from the "exact" solution and the computer solution where the value of the increment length is changed through a wide range. The equations for the closed-form solution for deflection along the pile are

$$y = \frac{e^{-\beta x}}{2E_p I_p \beta^2}\left[\frac{P_t}{\beta}\cos \beta x + M_t(\cos \beta x - \sin \beta x)\right] \tag{14.77}$$

$$\beta = \left(\frac{E_{py}}{4E_p I_p}\right)^{1/4} \tag{14.78}$$

The equations are for a pile of infinite length, but Timoshenko (1941) pointed out that the solution is valid where the nondimensional length of the pile (βL) is equal to or greater than 4. For the solutions with the computer code, $P_t = 40$ kN, $M_t = 0$, and $E_{py} = 100$ kN/m^2, entered as constant along the length of a pile with a length of 40 m. In order to make computations to compare with the closed-form solution, the value of E_{py} was entered by implementing the numerical feature of the *p-y* curves with one linear curve at the top of the pile and the same curve at the bottom of the pile. The value of β was computed to be 0.129724, yielding a value of $\beta L = 5.2$.

The closed-form solution yielded a value of y_t of 0.103779 m. The computer code allowed the user to enter a number of increments from 40 to 300. The results from the computer were (no. of increments, y_t, m): 40, 0.103365; 60, 0.103607; 80, 0.103692; 100, 0.103731; 150, 0.103770; 200, 0.103783; 250, 0.103790; and 300, 0.103365. All of the computer values agreed with the closed-form solution for three places. The greatest error was the result from 40 increments, where the computed value was less than the closed-form value by 0.000414 m, an acceptable accuracy in view of the uncertainties in the input for most solutions. The greatest accuracy occurred for increments of 150 through 300, and the least accuracy occurred for an increment of 40.

However, if the pile length is increased from 40 m while maintaining the increments at 40, errors become significant. The results from the computer were (pile length m, y_tm): 60, 0.102801; 100, 0.101087; 150, 0.097831; 200, 0.093489; 250, 0.088301; and 300, 0.082571. None of the solutions is acceptable, and serious errors occur if the pile is subdivided too grossly. For a length of 300 m and for 40 increments, the length of an increment is 7.5 m, far too much to allow a correct depiction of the curved shape of the pile. Significant errors can occur, for example, if the length of the pile is relatively great because of axial loading and the same length is used for analysis of the response to lateral loading. The output should be examined, and the length for lateral analysis should be reduced if zero deflection is indicated more than three or four times.

A rule of thumb that seems to have some merit is to select an increment length of about one-half of the pile diameter. Using this approach, the number of increments for a pile length of 50 m in the first computations would be about 200. The specific data that were selected for the comparison between a closed-form solution show acceptable accuracy regardless of the number of increments selected, from 40 to 300, as allowed by the particular code. However, if the length of an increment becomes too great with respect to the diameter of a pile, significant errors will result. The user is urged to repeat the study shown here for piles with other dimensions.

14.5.2 Safe Penetration of Pile with No Axial Load

The design of a pile subjected only to lateral loading is required on occasion, and the engineer must determine the minimum penetration that is safe against excessive deflection. Problems will occur, particularly as the pile sustains cyclic loads, when there is one point of zero deflection and the tip of the pile deflects. The pile used in the above example will be used in the following study. The pile, loading, and resistance remain as before, and the length of the pile is reduced to examine its response. The length of the increment was made about one-half of the pile diameter.

The deflection at the top of the pile was unchanged, with a penetration of 30 m where there were just two points of zero deflection. Note that the computer runs were successfully completed as the penetration was decreased by increments of 5 m to a penetration of 10 m. However, the pile-head deflection increased by 60% and the tip of the pile had a deflection of 0.077 m, opposite in direction to that of the pile head. Thus, pile penetrations of less than 30 m for the example were unacceptable for the given data.

This example and the one directly above show that the results from the computer solution must be studied with care to ensure correctness.

14.5.3 Buckling of a Pipe Pile Extending Above the Groundline

The pile used in the above two studies will be used to study the prediction of buckling due to axial load. The pile was assumed to extend 5 m above the ground surface, and two loading conditions were investigated: a lateral load of 1.0 kN applied at the top of the pile and a lateral load of 10.0 kN at the top. Axial loading was increased in increments, and the pile-head deflection was noted in order to predict the load at which buckling would occur.

The pile was assumed to be installed in a stiff clay above the water table with the following properties: undrained shear strength, 75 kPa; total unit weight, 19.0 kN/m^3; K_{py}, 135 MN/m^3; and ε_{50}, 0.005. The results of the computations are shown in Figure 14.27. The amount of lateral load caused a significant effect on the load to cause buckling. While the application of such a pile under a large axial load is unusual, the application of the computer code is of interest and may be of practical use.

14.5.4 Steel Pile Supporting a Retaining Wall

The lateral load sustained by a retaining wall is due principally to active forces behind the wall, with possibly some amount being due to a moving vehicle operating close to the top of the wall. The forces have little cyclic component, so the formulation of *p-y* curves for static loading is appropriate. A single vertical pile is analyzed and is assumed to be far enough from other piles that no interaction occurs due to close spacing.

The pile being analyzed is an H-section with bending about the strong axis. Data from the AISC handbook for the section shows the width of the pile as 373 mm, the depth as 351 mm, the moment of inertia about the x-axis as 3.76×10^{-4} m^4, and the yield strength of the steel as 276 MPa. The tabulated modulus at yielding of the steel is 2.39×10^{-3} m^3, giving a value of the bending moment at which a plastic hinge would develop of 660 kN-m. The cross-sectional area of the section is 1.69×10^{-2} m^2, and the modulus of elasticity of the steel is 2×10^8 kPa. In the computer analysis, the pile was assumed to have a diameter of 373 mm. The penetration of the pile below the ground line was 15 m. The pile is assumed to extend about the ground surface and to be embedded in a concrete footing.

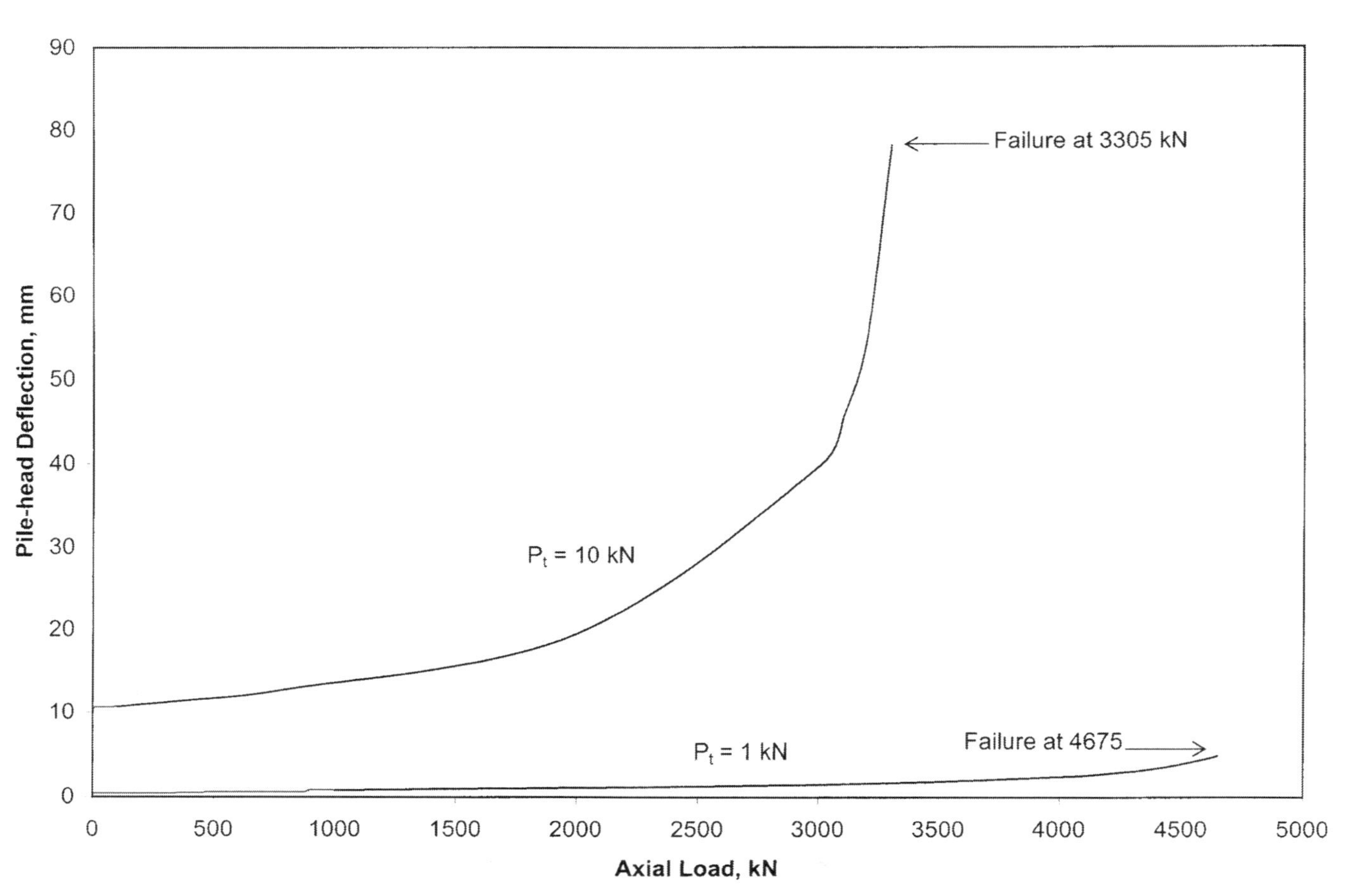

Figure 14.27 Buckling of a pile with 5 m of unsupported length and lateral load at the top.

The soil at the site is a stiff clay above the water table with the following properties: $\gamma = 18.7\ \text{kN/m}^3$, $c = 96.5$ kPa, and $\varepsilon_{50} = 0.007$. The data from a soil exploration invariably reveal scatter in the properties of the soil. In many cases, lower-bound and upper-bound solutions are desirable, taking into account the range in soil properties and possibly in the possible magnitude of the loadings. Under service conditions, the lateral load of the pile was 175 kN and the axial load was 90 kN.

Two conditions for the rotational restraint at the top of the pile were selected: the pile head was free to rotate (free head), and the pile head was entirely restrained against rotation (fixed head). The results of computations in Table 14.4 show the incrementing of the service load to find the loading at which a plastic hinge would develop. The pile-head deflection was also computed in the belief that the engineer could limit the loading if the deflection was considered excessive.

Table 14.4 shows that, for the free-head case, a plastic hinge developed at a lateral load of 426 kN, or for a load factor of 426/175 = 2.43. For the fixed-head case, a plastic hinge developed at a lateral load of 435 kN, or for a load factor of 435/175 = 2.49. Either of these load factors is acceptable for most designs. The pile-head deflection for the free-head case at the failure load was computed to be 98.3 mm, and for the fixed-head case it was computed to be only 23.4 mm. While the load factor at yield of the steel is about the same for both cases of pile-head rotation, the significant reduction in pile-head deflection for the fixed-head case could indicate special treatment in the design of the pile head support. Achieving full or partial fixity of a pile head is usually accomplished by extending the pile into a stiff pile cap for a distance of perhaps three pile diameters.

Table 14.4 shows the results for loading *above* what would cause a plastic hinge, which are incorrect because a constant *EI* was selected in computation. The improvements can be made if the non-linear *EI* vs. moment curve is

TABLE 14.4 Summary of the Ground Line Deflection and the Maximum Moment Under Different Lateral and Axial Loads for Both Cases of Pile-Head Fixity

	Lateral Load (kN)	Axial Load (kN)	Groundline Deflection (mm)	Maximum Moment (kN-m)
Free head	175	90	15.2	191.3
	350	180	65.1	503.5
	426	219	98.3	664.0
	525	270	157.0	907.1
Fixed head	175	90	3.6	188.0
	350	180	15.0	484.0
	435	224	23.4	652.0
	525	270	34.5	842.0

specified, and the code will indicate the development of a plastic hinge under excessive loadings. The erroneous results are shown here to alert the user to develop a full understanding of the input and output from computer solutions.

The bending-moment curves and pile-deflection curves, as a function of length along the pile, for both cases of pile-head fixity are shown in Figure 14.28. A number of interesting observations can be made. Considering the behavior of the pile under lateral loading, the bottom few meters of the pile are subjected to very little moment or deflection, suggesting that parametric studies could lead to a reduction in the length of the pile unless length is controlled by axial loading.

Also, in both cases, the portion of the pile where the bending moment is largest is relatively small compared to the length of the pile. If the pile consists of reinforced concrete or of a steel pile, the bending strength of the pile can be tailored to agree with the moment curve, perhaps leading to significant savings.

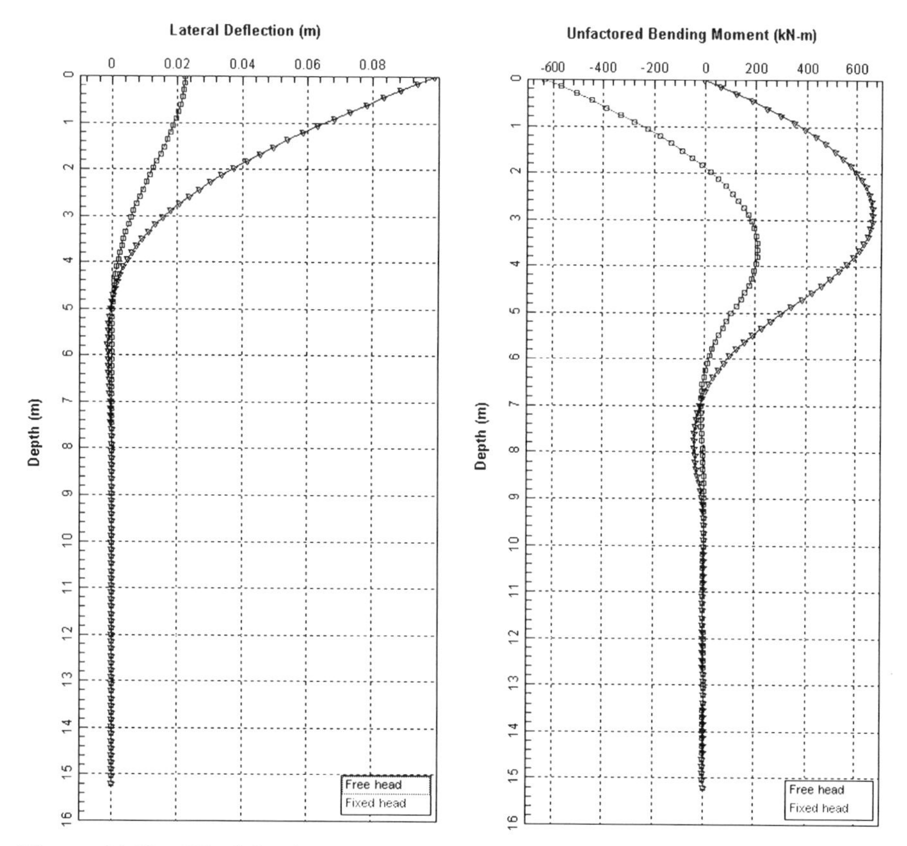

Figure 14.28 Pile-deflection curves and bending moment curves as a function of length along the pile for both cases of pile-head fixity.

The maximum negative moment for the fixed head occurs at the top of the pile and is over three times as large as the maximum positive moment at a depth of about 3.6 m. If the designer is able to allow some rotation of the pile head between fixed and free, the maximum negative moment and the maximum positive moment may be made close to equal, allowing the pile to sustain more lateral load before a plastic hinge develops.

14.5.5 Drilled Shaft Supporting an Overhead Structure

Many towers are being installed that transmit a shear and a moment to the groundline. Such towers are used for transmission lines, microwave supports, wind farms, advertising signs, and highway signs. Drilled shafts (bored piles) are frequently used for the foundation. Drilled shafts have the advantage that a wide variety of soils and rocks can be drilled. Further, using pile-drilling equipment, a geotechnical engineer at the site could arrange a program of rapid in situ testing. In addition, during the drilling of the production pile, the character and strength of the soil or rock could be logged by the engineer. The following paragraphs show how the engineer, with knowledge of the loadings on the overhead structure, could do preliminary computations for expected soil profiles, go to the site with a laptop, and complete the computations after obtaining information on the subsurface conditions.

While most designs are based on a soil boring, for structures requiring a single drilled shaft, a design can be based on soil properties from available geology and from information gained during the preliminary drilling. Loadings will be known from the nature of the structure and the environmental conditions, so a qualified engineer could go to a new site with a laptop and make the final design on site. Using the loadings, the maximum bending moment will occur within a diameter or two from the ground surface, regardless of the strength of the soil, so the required diameter of bending strength of the drilled shaft can be known fairly well at the outset.

A pile was selected for analysis with a diameter of 760 mm and reinforced with 12 vertical rebars with a diameter of 25 mm, spaced equally on a 610-mm-diameter circle that gave a clear space outside the steel of 76 mm. The ultimate strength of the reinforcing steel was 414 MPa, and that of the concrete was 27.6 MPa. The ultimate bending moment of the pile section was computed by the LPILE program as a function of the applied axial load, as shown in Figure 14.29. As may be seen, the axial load has a significant effect on the ultimate bending moment.

The axial load of the column and the superstructure were computed carefully and found to be 222 kN, yielding an ultimate bending moment of 734 m-kN, as shown in the figure. The resultant of the lateral load on the structure was due principally to forces from wind, assumed to be applied at 13 m above the groundline. The axial load will be maintained at 222 kN, which is conservative, as shown in the figure. Using that axial load, the bending stiffness for the section was computed as a function of the applied moment. The results

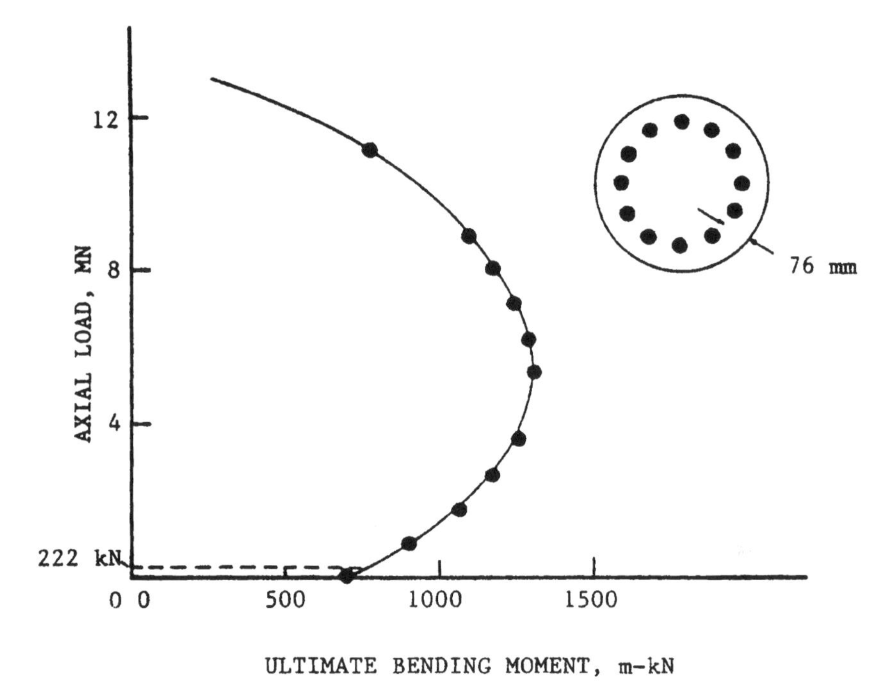

Figure 14.29 Interaction diagram for the drilled shaft of an example problem.

are shown in Figure 14.30. The figure shows a sharp decrease in E_pI_p when cracking of the concrete first occurs. For a range in bending moments, the E_pI_p value is relatively uniform until a significant decrease in E_pI_p occurs prior to the development of a plastic hinge. For the initial analyses, an E_pI_p value of 130,000 kN-m^2 was selected with the idea that the distribution of bending moments along the pile would be reviewed and adjustments made in E_pI_p if required. Alternatively, the computations could be made with a code that selected E_pI_p automatically as a function of the computed bending moment.

The problem is to compute the load required to cause a plastic hinge if the pile is founded above the water table in uniform sand or in overconsolidated clay. The sand has a friction angle of 36°, a total unit weight of 19.5 $\frac{kN}{m^3}$, and $E_{py\,\max}$ of 24 $\frac{MN}{m^3}$. The clay has a undrained shear strength of 75 kPa, a total unit weight of 19.0 $\frac{kN}{m^3}$, an ε_{50} of 0.005, and $E_{py\,\max}$ of 135 $\frac{MN}{m^3}$. Wind gusts will cause cyclic loading, and 200 was selected as the number of cycles to be used in the analysis with the clay. Cyclic loading with sand assumes that there are enough cycles to cause the limiting resistance.

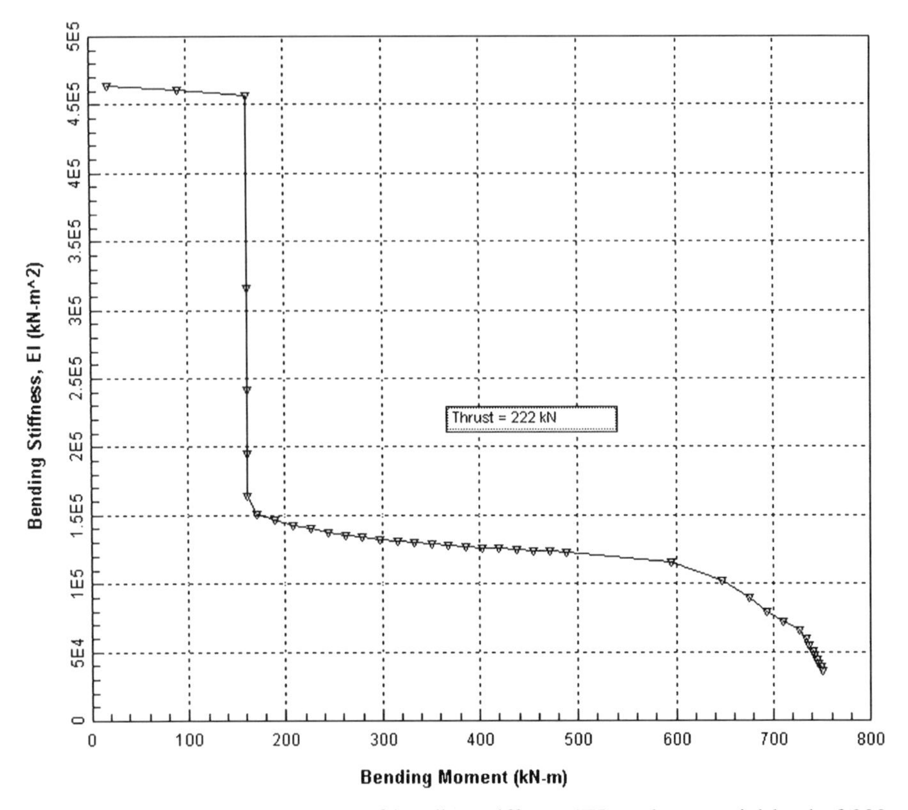

Figure 14.30 Computed values of bending stiffness (*EI*) under an axial load of 222 kN.

The first analysis was done assuming the pile to be installed in sand. The penetration of the pile was selected as 30 m to ensure *long-pile* behavior. The LPILE program was used, with a loading to cause a plastic hinge at the maximum bending moment of 734 m-kN. The lateral load at a height of 13.0 m was found to be 52.9 kN, yielding a moment at the groundline of 687.7 m-kN. The groundline deflection was 29.8 mm. The penetration of the pile was decreased in increments until the number of points of zero deflection was two or three. At a penetration of 10 m, the data showed almost three points of zero deflection. In regard to penetration, the engineer should err on the conservative side to ensure no significant deflection of the tip of the pile.

The bending moment over the length, using the above loading to cause pile failure, showed that about 2 m of the top of the pile showed a bending moment of more than 600 m-kN and about 3 m of the bottom of the pile showed a bending moment of less than 150 m-kN. No attempt was made to correct the E_pI_p values to improve the solution. However, the lateral load at 13 m was factored downward by 2.5, perhaps yielding the service load, and no moment along the pile exceeded 500 m-kN. Therefore, the use of 130,000 kN-m^2 for E_pI_p was judged to be satisfactory or somewhat conservative at the service load. The deflection at the groundline under this service load was computed to be 9 mm.

The analysis of the pile in clay showed that the maximum bending moment of 734 m-kN was developed by a lateral load of 55.1 kN at 13 m above the groundline, yielding a moment at the ground line of 716.3 m-kN. The deflection of the pile head was 30.9 mm. The penetration of the pile was reduced from 30 m to 8 m where more than 2 points of zero deflection were observed. The loads were reduced by a factor of 2.5 and the bending moment, as for the sand, was in the range to achieve some conservatism. The pile-head deflection at the service load was computed to be less than 10 mm.

The above analyses show that preliminary computation for a given pile with a given set of loadings can be done using a computer code, and a body of information can be developed that will provide guidance to a geotechnical engineer who may go with the drilling contractor to make final computations for the soil that exists at the site.

PROBLEMS

P14.1. Use the Student Version of the LPILE computer program to solve the following problem:

The pile is a steel pipe with an outside diameter of 30 in. (0.76 m) and a wall thickness of 1.0 in. (25.4 mm), cross-sectional area of 91.1 in.^{2} (0.0588 m^2), moment of inertia of 9589 in.^{4} (3.99×10^{-3} m^4), modulus of elasticity 29,000,000 psi (200,000,000 kPa), and yield strength of steel 36,000 psi (248,000 kPa).

The soil is a normally consolidated clay under water with 15 ft (4.6 m) of erosion. Properties at the ground surface: undrained shear strength of 173 psf (8.28 kPa), submerged unit weight of 39.1 lb/ft^3 (6.14 kN/m^3), ε_{50} of 0,02, and K_{py} of 90 lb/in.^{3} (24.4 MN/m^3). Properties at a depth of 85 ft (25.9 m): undrained shear strength 1100 psf (52.6 kPa), submerged unit weight 47.6 lb/ft^3 (7.48 kN/m^3), ε_{50} of 0.02, and K_{py} of 500 lb/in.^{3} (135.7 MN/m^3).

Apply a series of loads P_t at the top of the pile (M_t is zero, P_x is zero) with a pile length of 85 ft; the number of increments along the pile is 100 (the increment length is less than one-half of the pile diameter).

a. Use static loading, with loads increasing in increments of 10 kips to a total of 80 kips (the program will accept the entry of a number of inputs for the loading).

b. Use the Graphics feature of the program to print a plot of maximum bending moment versus applied load and explain the reason for the curvature in the plot. (Note: the entry for k in the input is K_{py}; use an average value over the depth of 85 ft.)

P14.2. Use the plot from Problem P14.1 to find the lateral load causing a bending stress of 15 ksi, noting that a series of runs can be made with the same entries around the correct value. (An answer within 1% of the precise value is acceptable.)

P14.3. Find a computer solution for the lateral load obtained in Problem P14.1b, print the tabulated output, and plot curves for deflection and soil reaction, both as a function of depth.

a. What penetration should the pile have so that it behaves as a "long" pile?

b. Make computations to check the agreement between deflection and soil resistance at ground surface and at a depth of 21.25 ft (255 in.).

P14.4. A free-standing pile is frequently used as a breasting dolphin to protect a dock or pier against damage from a docking vessel. The energy from the docking vessel may be a complex function of many factors. The following equation may be useful:

$$E = 0.5mv^2$$

where E = energy, m = mass of the vessel, and v = velocity on touching the breasting dolphin.

Because docking will occur many times, cyclic loading on the soil should be used. The problem is to compute the energy that can be sustained by the pile in Problem P14.1 if the vessel contacts the pile at 15 ft above the mudline. The maximum bending stress should not exceed 20 kips/in.^{2}.

a. Find the force that will act on the vessel if the given stress is not exceeded.

b. The breasting dolphins should be observed for performance. If locked-in deflection is observed, what steps can be taken to alleviate the problem?

c. Assuming that the vessel will contact two breasting dolphins and that the docking velocity will be 0.5 ft/sec, find the weight of a vessel that can dock without auxiliary fendering.

P14.5. Figure 14.31 shows a portion of an offshore platform. One leg of the jacket or template is shown. The jacket is fabricated on shore and placed in position by a derrick barge. Mud mats can be placed below the bottom braces to prevent the jacket from sinking below the mudline. Piles are stabbed through the jacket legs and driven with a swinging hammer. As shown, spacers welded inside the jacket serve to

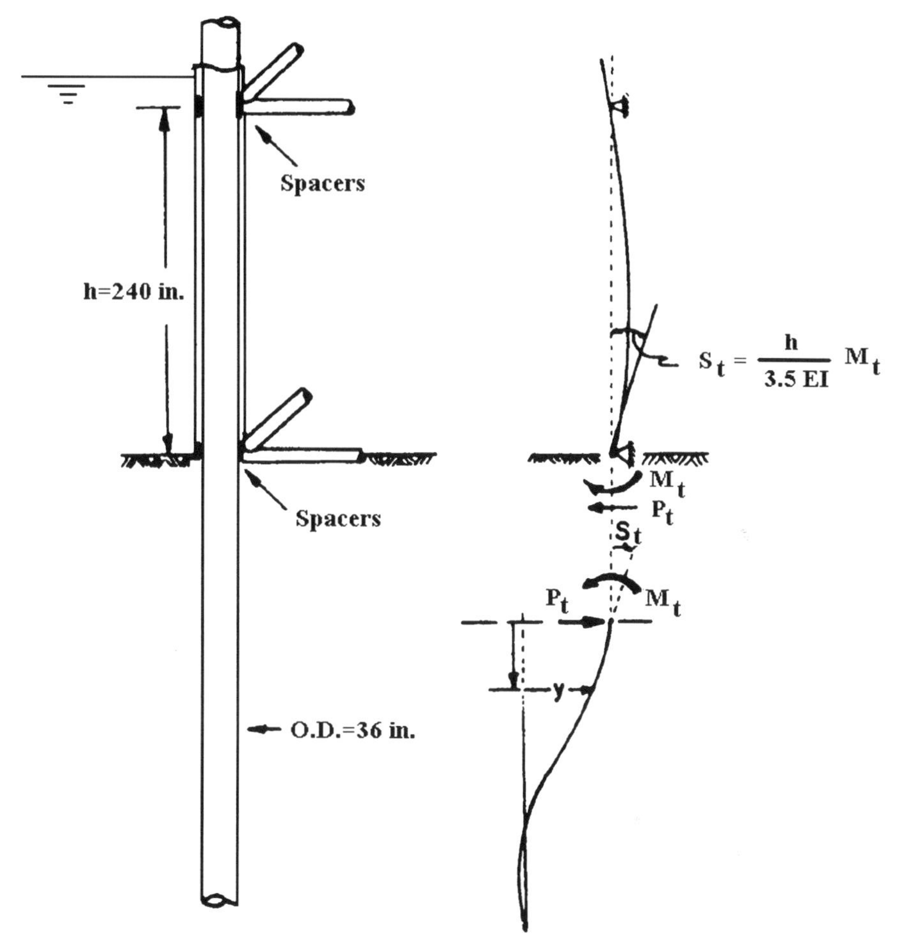

Figure 14.31

center the pile and allow it to rotate at the panel points. Alternatively, if the designer chooses, the annular space between the jacket and pile can be grouted. Both jacket and piles extend above the water surface. A deck section, with legs that extend through the zone of the maximum wave, is placed by a derrick barge and the joints are welded.

During a storm, wave forces, perhaps large, will act against the jacket and cause the structure to move laterally and rotate slightly; the piles must be analyzed by the GROUP computer program (see Chapter 15). A preliminary analysis, however, can be worthwhile even if the jacket rotation is temporarily ignored.

As shown in the free body in Figure 14.31, the pile is partially restrained against rotation. A moment at the pile head will cause a rotation that must be equal to the rotation of the pile within the jacket that acts as a continuous beam. In the absence of a detailed structural analysis, the end of the pile at the first panel point above the mudline is neither fixed nor free. Therefore, the ratio of moment to rotation can be given by the equation shown in the figure.

The pile has a diameter of 36 in., and the wall thickness in the zone of maximum bending moment is 1.5 in. To save money, the wall thickness can be reduced in steps to an amount that will not be damaged during pile driving. For analysis under lateral loading, a constant wall thickness can be assumed without error. The moment of inertia of the pile is 24,230 in.^{4} (0.01009 m^4), the cross-sectional area is 162.6 in.^{2} (0.105 m^4), the combined stress at failure is 36,000 lb/in.^{2} (248,000 kPa), and the modulus of elasticity is 29×10^6 lb/in.^{2} (2×10^8 kPa).

The pile is installed in a uniformly graded sand with a friction angle of 39°, a relative density of about 0.9, and a submerged unit weight of 66.3 lb/ft^3 (10.4 kN/m^3). The loading will be applied by wave forces that build up gradually with time. The maximum wave forces are usually used in design even though the number of such waves may be limited. Cyclic loading will be selected in developing the *p-y* curves for the sand.

The pile is assumed to penetrate 120 ft into the sand and to develop sufficient axial capacity so that any failure will be due to lateral loading. A load of 500 kips is assumed from the dead weight of the structure and will increase by a factor of 2 times the lateral load due to overturning forces.

a. Find the lateral load on this single pile at which the combined stress of the steel, 36,000 lb/in.^{2} (248,000 kPa), will be developed.

b. If the maximum negative moment and the maximum positive moment are considerably different, investigate the possibility of bringing the two moments closer to each other with increased capacity in loading.

c. Present a brief discussion of how a factor of safety should be developed for the pile.

CHAPTER 15

ANALYSIS OF PILE GROUPS

15.1 INTRODUCTION

Two general problems must be addressed in the analysis of pile groups: computation of the loads coming to each pile in the group and determination of the efficiency of a group of closely spaced piles. Both of these problems are important components in the area of soil–structure interaction. If piles are far apart in terms of multiples of pile diameter, pile–soil–pile interaction will not occur. As the piles become closer to each other, the stress in the soil from the distribution of axial load or lateral load to the soil will affect nearby piles. The simple way to consider the influence of the effect of the stresses in the soil is to think of the efficiency of closely spaced piles becoming less than unity. Methods of predicting efficiency will be discussed here.

15.2 DISTRIBUTION OF LOAD TO PILES IN A GROUP: THE TWO-DIMENSIONAL PROBLEM

The problem of solving for the distribution of axial and lateral loads to each pile in a group has long been of concern to geotechnical engineers, and various concepts have been proposed to find a solution. As the ideas of soil–structure interaction were developed, allowing the movements of pile heads to be computed for axial and lateral loading, it became possible to develop fully rational solutions to the problem of distribution of loading to piles in a group. The work of Hrennikoff (1950), based on linear analysis, provided an excellent guideline to the nonlinear problem. Useful information can be obtained by considering the piles as behaving in a linear fashion, but the actual behavior of a system can be found only by finding the loading that causes failure. This requires the analysis to be continued well into the nonlinear response of the piles to loading, either axial or lateral. Failure may occur in the analysis due to the computation of a plastic hinge in one of the piles in a group or due to excessive deflection. The ability to determine the movement of a group of piles if a nonlinear response of a single pile is considered, due to axial or lateral loading, allows the engineer to design a group of piles for superstructures with a wide range of tolerances to foundation movements.

The response of a group of piles is analyzed for two conditions; the first is where loading is symmetrical about the line of action of the lateral load. That is, no twisting of the pile group will occur, so no pile is subjected to torsion. Therefore, each pile in the group can undergo two translations and a rotation, the two-dimensional problem. However, the method can also be extended to the general case where each pile can undergo three translations and three rotations (Robertson, 1961; Reese et al., 1970; Bryant, 1977; O'Neill, et al., 1977).

The analyses presented in this section assume that the soil does not act against the pile cap. In many instances, of course, the pile cap is cast against the soil. With regard to lateral resistance, shrinkage of the soil could cause a gap between concrete and soil. A small settlement of the ground surface would eliminate most of the vertical resistance against the mat. A conservative assumption is that the piles under the pile cap support the total load on the structure. In some current designs, however, the vertical loading is resisted by both a mat (raft) and by piles, termed a *piled raft* (Franke, 1991; El-Mossallamy and Franke, 1997). The analysis of a piled raft requires the development of an appropriate model for the entire system of raft, piles, and supporting soil.

The derivation of the equations presented here is based on the assumption that the piles are spaced far enough apart that there is no loss of efficiency; thus, the distribution of stress and deformation from a given pile to other piles in the group need not be considered. The method that is derived can be used with a group of closely spaced piles, but another level of iteration will be required.

15.2.1 Model of the Problem

The problem to be solved is shown in Figure 15.1. Three piles supporting a pile cap are shown. The piles may be of any size and placed on any batter, and may have any penetration below the groundline. The bent may be supported by any number of piles, but the piles are assumed to be far enough apart that each is 100% efficient. The soil and loading may have any characteristics for which the response of a single pile may be computed.

The derivation of the necessary equations in general form proceeds conveniently from the consideration of a structure such as that shown in Figure 15.2 (Reese, 1966; Reese and Matlock, 1966). The sign conventions for the

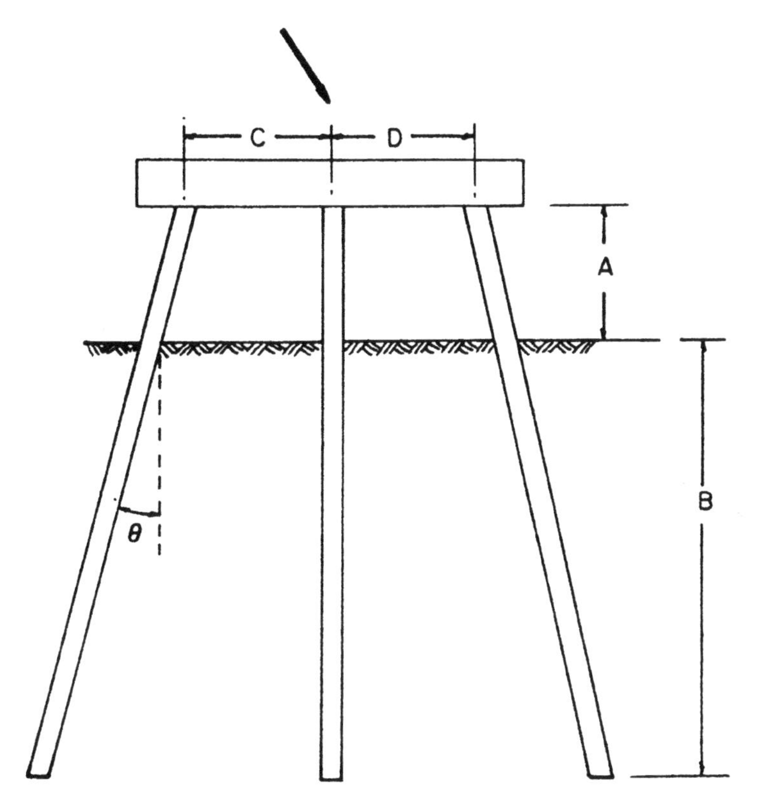

Figure 15.1 Example problem to be solved for the distribution of loads to piles in a group.

loading and the geometry are shown. A global coordinate system, *a-b*, is established with reference to the structure. A coordinate system, *x-y*, is established for each pile. For convenience in deriving the equilibrium equations, the *a-b* axes are located so that all of the coordinates of the pile heads are positive.

The soil is not shown, and the piles are replaced with a set of *springs* (mechanisms) that represent the interaction between the piles and the supporting soil (Figure 15.2). If the global coordinate system translates horizontally Δh and vertically Δv and rotates through the angle α_s, the movement of the head of each pile can be readily found. The angle α_s is assumed to be small in the derivation.

The movement of a pile head x_i in the direction of the axis of the pile is

$$x_t = (\Delta h + b\alpha_s)\sin\theta + (\Delta v + a\alpha_s)\cos\theta \tag{15.1}$$

The movement of a pile head y_t transverse to the direction of the axis of the pile (the lateral deflection) is

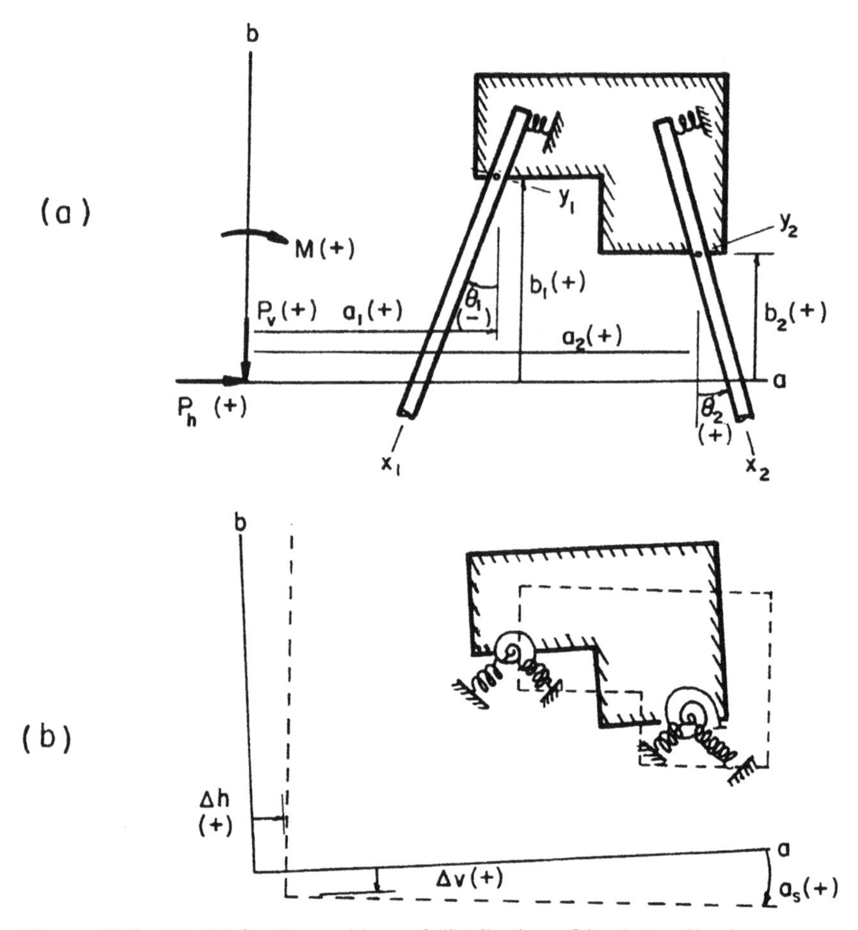

Figure 15.2 Model for the problem of distribution of loads to piles in a group.

$$y_t = (\Delta h + b\alpha_s)\cos\theta - (\Delta v + a\alpha_s)\sin\theta \tag{15.2}$$

The assumption made in deriving Eqs. 15.1 and 15.2 is that the pile heads have the same relative positions in space before and after loading. However, if the pile heads move relative to each other, an adjustment can be made in Eqs. 15.1 and 15.2 and a solution achieved by iteration.

The movements computed by Eqs. 15.1 and 15.2 will generate forces and moments at the pile head. The assumption is made that curves can be developed, usually nonlinear, that give the relationship between pile-head movement and pile-head forces. A secant to a curve is obtained and called the *modulus of pile-head resistance*. The values of the moduli, so obtained, can then be used, as shown below, to compute the components of movement of the structure. If the values of the moduli that were selected were incorrect, iterations are made until convergence is obtained.

Using sign conventions established for the single pile under lateral loading, the lateral force P_t at the pile head is defined as follows:

$$P_t = J_y y_t \tag{15.3}$$

If rotational restraint exists at the pile head, the moment is

$$M_s = -J_m y_t \tag{15.4}$$

The moduli J_y and J_m are not single-valued functions of pile-head translation but are also functions of the rotation α_s of the structure. Figures 15.1 and 15.2 indicate that some of the piles supporting a structure may be installed on a batter. The next section presents an empirical but effective procedure for making the required modifications if some of the piles in a structure are battered.

If it is assumed that a compressive load causes a positive deflection along the pile axis, the axial force P_x may be defined as follows:

$$P_x = J_x x_t \tag{15.5}$$

A pile under lateral loading will almost always experience a lateral deflection at the groundline that could cause some loss of axial capacity. However, the loss would be small, so P_x can be taken as a single-valued function of x_t.

A curve showing axial load versus deflection may be computed by one of the procedures recommended by several authors (Reese, 1964; Coyle and Reese, 1966; Coyle and Sulaiman, 1967; Kraft et al., 1981) in Chapter 13 or the results from a field load test may be used. A typical curve is shown in Figure 15.3a. The curve is shown in the first quadrant for convenience in plotting, but under some loadings, one or more of the piles in a structure may be subjected to uplift and a curve showing axial load in tension must be computed.

The methods for computing the response of a pile under lateral loading by computer have been presented in Chapter 14 and may easily be employed for computing the curves shown in Figures 15.3b and 15.3c. The method used to attach the piles to the superstructure must be taken into account because the pile-head rotation α_p will be affected. Also, if the pile heads are fully or partially restrained against rotation, the rotation of the structure α_s will affect the curves in Figures 15.3b and 15.3c. Alternatively, the nondimensional methods presented earlier in Chapter 12 may be used in obtaining the curves for the response of the pile to lateral loading.

The forces at the pile head defined in Eqs. 15.3 through 15.5 may now be resolved into vertical and horizontal components of force on the structure as follows:

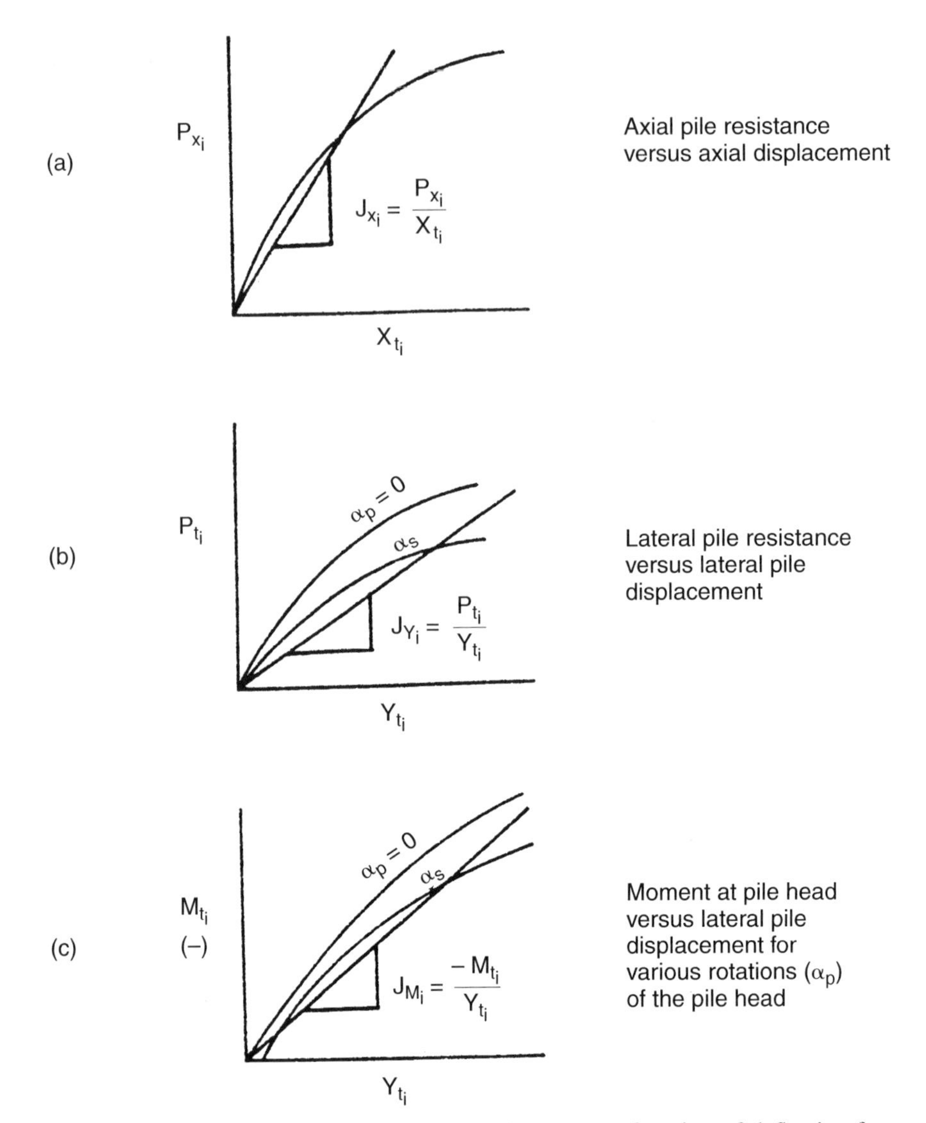

Figure 15.3 Typical curves showing pile resistance as a function of deflection for a pile in a group.

$$F_v = -(P_x \cos \theta - P_t \sin \theta) \tag{15.6}$$

$$F_h = -(P_x \sin \theta + P_t \cos \theta) \tag{15.7}$$

The moment on the structure is

$$M_s = J_m y_t \tag{15.8}$$

The equilibrium equations can now be written as follows:

$$P_v + \sum F_{v_i} = 0 \tag{15.9}$$

$$P_h + \sum F_{h_i} = 0 \tag{15.10}$$

$$M + \sum M_{s_i} + \sum a_i F_{v_i} + \sum b_i F_{h_i} = 0 \tag{15.11}$$

The subscript i refers to values of any "ith" pile. Using Eqs. 15.1 through 15.8, Eqs. 15.9 through 15.11 may be written in terms of the structural movements as shown in Eqs. 15.12 through 15.14.

$$\Delta v\left[\sum A_i\right] + \Delta h\left[\sum B_i\right] + \alpha_s\left[\sum a_i A_i + \sum b_i B_i\right] = P_v \tag{15.12}$$

$$\Delta v\left[\sum B_i\right] + \Delta h\left[\sum C_i\right] + \alpha_s\left[\sum a_i B_i + \sum b_i C_i\right] = P_h \tag{15.13}$$

$$\Delta v\left[\sum D_i + \sum a_i A_i + \sum b_i B_i\right] + \Delta h\left[\sum E_i + \sum a_i B_i + \sum b_i C_i\right] + \alpha_s\left[\sum a_i D_i + \sum a_i^2 A_i + \sum b_i E_i + \sum b_i^2 C_i + \sum 2a_i b_i B_i\right] = M \tag{15.14}$$

where

$$A_i = J_{x_i} \cos^2 \theta_i + J_{y_i} \sin^2 \theta_i \tag{15.15}$$

$$B_i = (J_{x_i} - J_{y_i}) \sin \theta_i \cos \theta_i \tag{15.16}$$

$$C_i = J_{x_i} \sin^2 \theta_i + J_{y_i} \cos^2 \theta_i \tag{15.17}$$

$$D_i = J_{m_i} \sin \theta_i \tag{15.18}$$

$$E_i = -J_{m_i} \cos \theta_i \tag{15.19}$$

The above equations are not as complex as they appear. For example, the origin of the coordinate system can usually be selected so that all of the b-values are zero. For vertical piles, the sine terms are zero and the cosine terms are unity. For small deflections, the J-values can all be taken as constants. Therefore, under many circumstances, it is possible to solve the above equations by hand. However, if the deflections of the group are such that the nonlinear portion of the curves in Figure 15.3 is reached, a computer solution is advantageous. Computer solutions are discussed later in this chapter.

15.2.2 Detailed Step-by-Step Solution Procedure

1. Study the foundation to be analyzed and select a two-dimensional bent the behavior of which is representative of the entire system.
2. Prepare a sketch such that the lateral loading comes from the left. Show all pertinent dimensions.
3. Select a coordinate center and find the horizontal component of load, the vertical component of load, and the moment through and about that point.
4. Compute a curve showing axial load versus axial deflection for each pile in the group or, preferably, use the results from a field load test. Computation of a curve showing axial load versus settlement is presented in detail in Chapter 13.
5. Compute curves showing lateral load as a function of lateral deflection and moment as a function of lateral deflection, taking into account the effect of structural rotation on the boundary conditions at each pile head.
6. Estimate trial values of J_x, J_y, and J_m for each pile in the structure.
7. Solve Eqs. 15.12 through 15.14 for values of Δv, Δh, and α_s.
8. Compute movements of pile heads, obtain loads at pile heads by use of the appropriate J-values, and check to ensure that equilibrium was established for the group.
9. If necessary, obtain new values of J_x, J_y, and J_m for each pile and solve Eqs. 15.12 through 15.14 again for the new values of Δv, Δh, and α_s.
10. Continue iteration until the computed values of the structural movements agree, within a given tolerance, with the values from the previous computation.
11. Compute the stresses along the length of each pile using the loads and moments at each pile head.

15.3 MODIFICATION OF *p-y* CURVES FOR BATTERED PILES

Kubo (1965) and Awoshika and Reese (1971) investigated the effect of batter on the behavior of laterally loaded piles. Kubo used model tests in sands and full-scale field experiments to obtain his results. Awoshika and Reese tested 2-in.-diameter piles in sand. The value of the constant showing the increase or decrease in soil resistance as a function of the angle of batter may be obtained for the line in Figure 15.4. The ratio of soil resistance was obtained

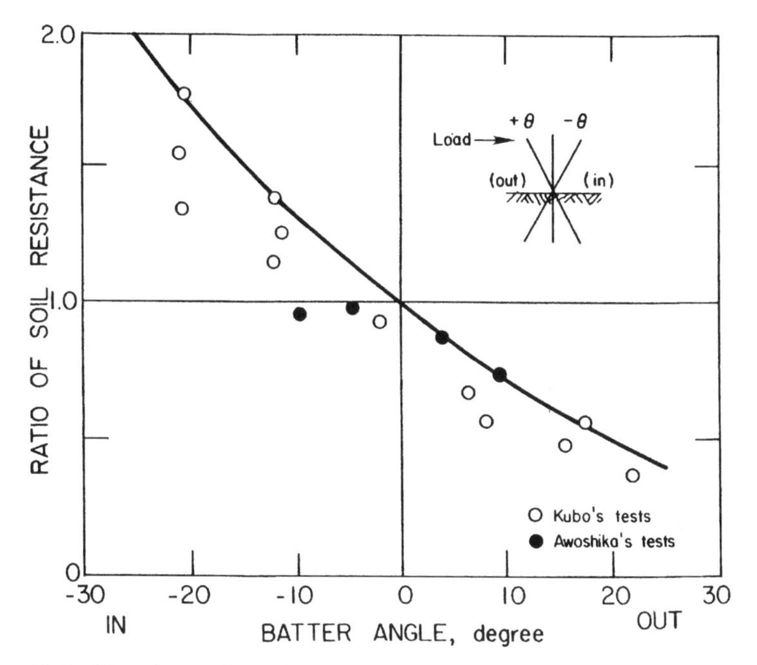

Figure 15.4 Experimental results showing loss of resistance as a function of batter angle for battered piles.

by comparing the groundline deflection for a battered pile with that of a vertical pile and is, of course, based purely on experiment.

The correction for batter is made as follows: (1) enter Figure 15.4 with the angle of batter, positive or negative, and obtain a value of the ratio; (2) compute groundline deflection as if the pile were vertical; (3) multiply the deflection found in (2) by the ratio found in (1); (4) vary the strength of the soil until the deflection found in (3) is obtained; and (5) use the modified strength found in (4) for the further computations of the behavior of the pile on a batter. The method outlined is obviously approximate and should be used with caution. If the project is large and expensive, field tests on piles that are battered and vertical could be justified.

15.4 EXAMPLE SOLUTION SHOWING DISTRIBUTION OF A LOAD TO PILES IN A TWO-DIMENSIONAL GROUP

15.4.1 Solution by Hand Computations

The detailed step-by-step procedure was presented earlier and is followed in the following example. Steps 1 through 3 were followed in preparing Figure 15.5. The pile-supported retaining wall has piles spaced 8 ft apart along the

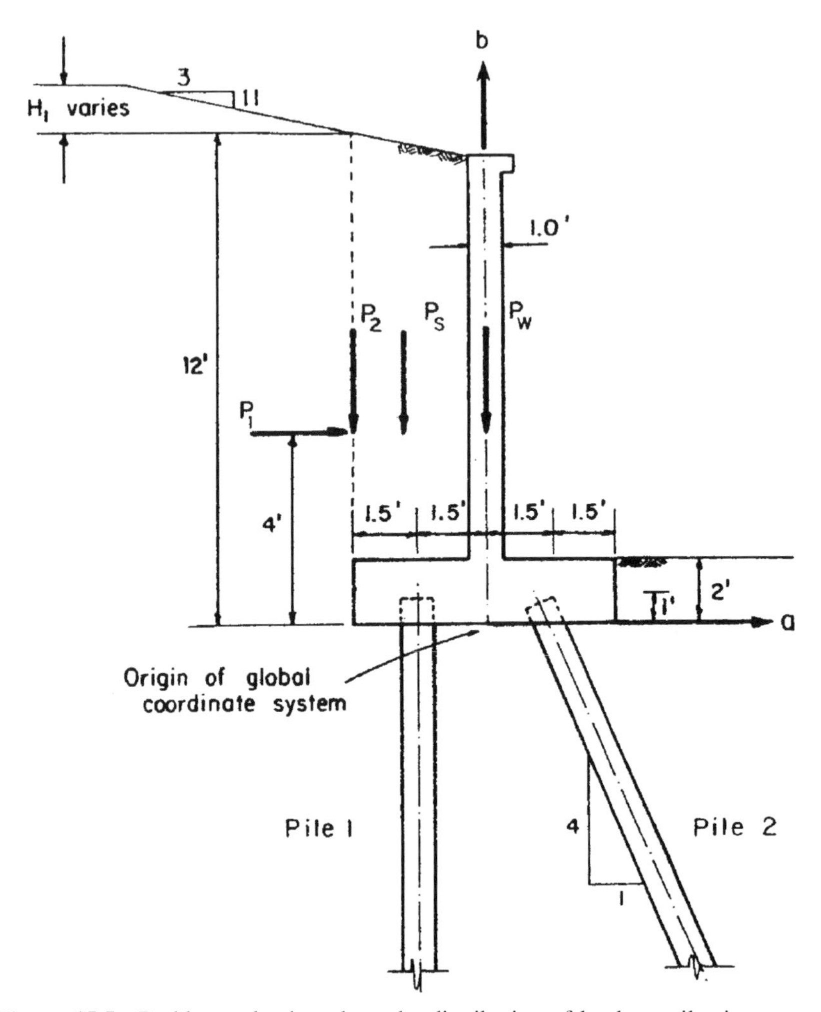

Figure 15.5 Problem solved to show the distribution of loads to piles in a group.

wall. The piles are steel pipes, 12 in. in outside diameter with a wall thickness of 0.5 in. The length of the piles is 40 ft. As shown in the figure, the piles are embedded only 1 ft. in the cap, and the assumption is made that the pile heads are unrestrained against rotation. This condition generally does not exist; if a pile is embedded a sufficient amount to sustain shear, some restraint against rotation must occur. Consideration of the effect of some rotational restraint shows that the assumption that the pile can rotate freely adds a small degree of safety to the solution.

The backfill is a free-draining granular soil without any fine particles. The surface of the backfill is treated to facilitate runoff, and weep holes are provided so that water will not collect behind the wall. The soil supporting the piles is a silty clay, with the water table reported as having a depth of 10 ft.

The water content averages 10% above and 20% below the water table. The undrained shear strength of the clay varies with depth, and a constant value of 2.0 kips/ft^2 was selected for the analyses. The unit weight of the clay is 118 lb/ft^3 above the water table, and a value of ε_{50} was estimated to be 0.005.

The pile to the right in the figure was battered at an angle of 14° with the vertical such that the soil resistance would be less than for a vertical pile. Figure 15.4 was employed, and a soil resistance ratio of 0.62 was found. Therefore, the soil strength around the battered pile was taken as (0.62)(2.0) = 1.24 kips/ft^2.

The forces P_1, P_2, P_s, and P_w (shown in Figure 15.5) were computed as to be: 21.4, 4.6, 18.4, and 22.5 kips, respectively. Resolution of the loads at the origin of the global coordinate system resulted in the following service loads: $P_v = 45.5$ kips, $P_h = 21.4$ kips, and $M = 44.2$ ft-kips. The value of the moment of inertia I of the pile is 299 in.^{4}. The yield strength of the steel is 36 kips/in.^{2}.

With regard to Step 4, a field load test was assumed at the site and the ultimate axial capacity of a pile was found to be 240 kips. The load-settlement curve is shown in Figure 15.6. Step 5 was accomplished by computing a set of *p-y* curves for Pile 1, the vertical pile, and for Pile 2, the battered pile, and shown in Figures 15.7 and 15.8. In complying with Step 5, with the sets of *p-y* curves, the curves showing lateral load versus deflection for the tops of Piles 1 and 2 may be developed by using the nondimensional curves demonstrated earlier. However, a computer solution was employed as described in Chapter 14, and the results are shown in Figures 15.9 and 15.10 for Piles 1 and 2, respectively. The figures do not indicate an influence of α_s, the rotation of the structure, because the heads of the piles are unrestrained against rotation.

Step 6 was accomplished by assuming that the movements of the pile heads would be small for the service loads indicated in Figure 15.5. The J-values were found by obtaining the slopes of the curves in Figures 15.6, 15.8, and 15.9 to the first points shown in the figures, with the following results: J_{x_1} and $J_{x_2} = 2800$ kips/in., $J_{y_1} = 333$ kips/in., and $J_{y_2} = 161$ kips/in.

Substitution of the J-values into Eqs. 15.15 through 15.19, with the value of θ as 14°, yields the following values: $A_1 = 2800.0$, $A_2 = 2645.5$, $B_1 = 0$, $B_2 = 619.46$, $C_1 = 333.00$, $C_2 = 315.45$, and the remainder of the terms in the equations have values of zero.

Step 7 is to substitute these values into Eqs. 15.12 through 15.14 to obtain equations for the movement of the global coordinate system. The results for the service loadings on the structure are as follows:

$$\Delta v[5445.5] + \Delta h[619.46] + \alpha_s[-2781] = 45.5 \text{ k}$$

$$\Delta v[619.46] + \Delta h[618.45] + \alpha_s[11150] = 21.4 \text{ k}$$

$$\Delta v[-2781] + \Delta h[11150] + \alpha_s[1{,}764{,}300] = 530 \text{ in.-k}$$

The three equations may be solved by hand with a little effort, but a spreadsheet solution is convenient, with the following results:

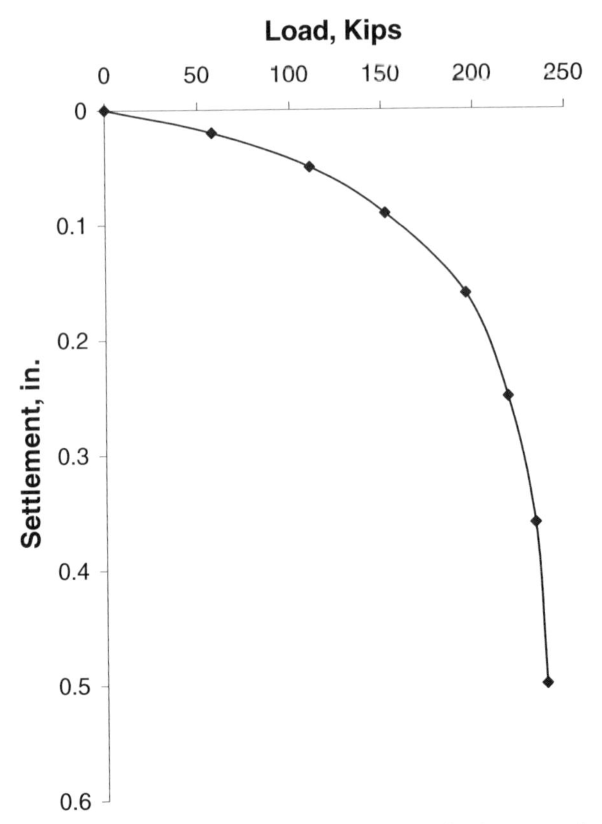

Figure 15.6 Axial load versus settlement for piles in the example problem.

$$\Delta v = 0.005579; \quad \Delta h = 0.025081; \quad \alpha_s = 0.000151$$

In Steps 8 through 10, the pile-head movements were computed with these movements of the global coordinate system, and the results from Eqs. 15.1 and 15.2 are as shown in the following table, along with the computed forces on the pile heads by use of the J-values assumed in the analysis.

Pile No.	x_t, in.	P_x, k	y_t, in.	P_t, k
1	0.002861	8.01	0.025081	8.35
2	0.014119	39.53	0.022329	3.59

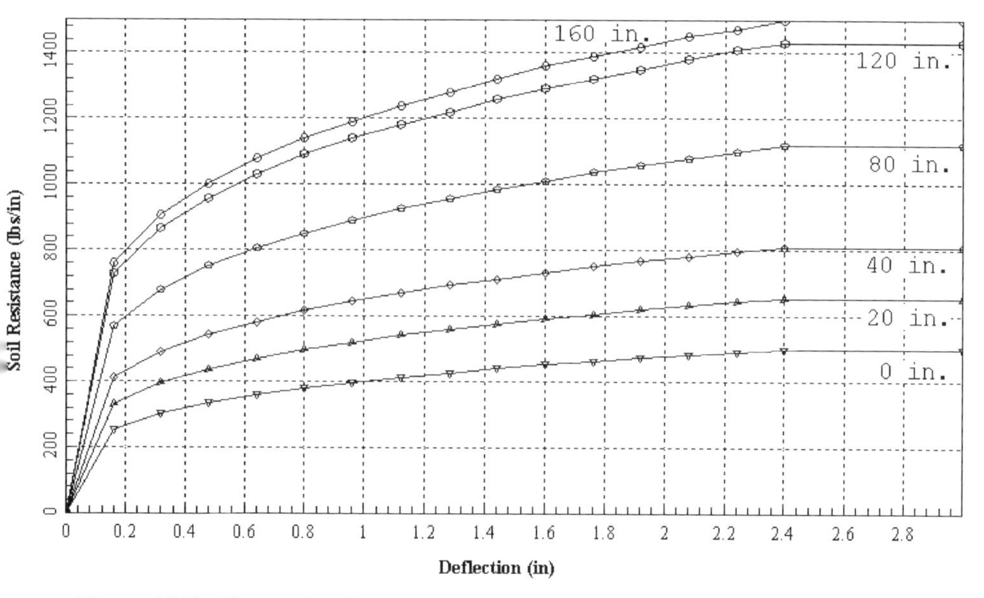

Figure 15.7 Curves showing soil resistance versus deflection (*p-y* curves) for vertical piles in the example problem.

The resisting loads on the pile cap generated by the pile-head movements show that the structure is in equilibrium; however, entering curves in Figures 15.6, 15.9, and 15.10 showed that the *J*-values had not been selected with sufficient accuracy. For example, entering Figure 15.9 with a pile-head deflection of 0.025081 yields a P_t of about 6 kips rather than 8.35 kips; thus, the *J*-values would need to be recomputed and a new solution obtained. This process or iteration requires the use of a computer. The GROUP program was employed, six iterations were required, and the following results were obtained:

Pile No.	x_t, in.	P_x, k	y_t, in.	P_t, k
1	0.00272	8.02	0.0321	7.05
2	0.0135	39.8	0.0298	4.83

As may be seen, the results were not very different, suggesting that useful solutions can be obtained by hand.

Step 11 will be demonstrated by using the results from a computer solution where the loadings were multiplied by a factor of 2.5, yielding P_v = 113,700

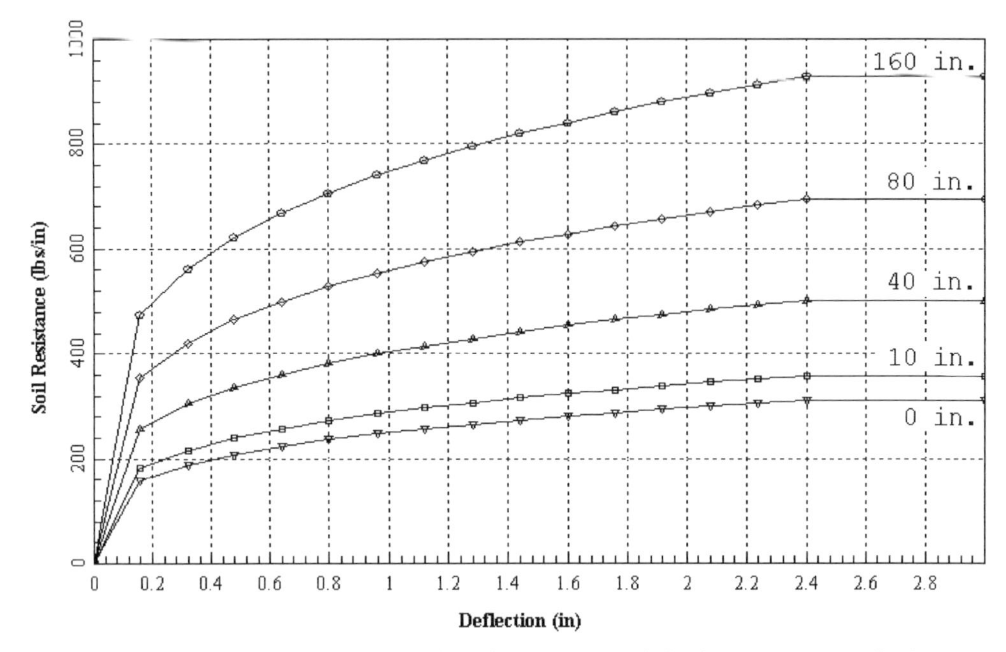

Figure 15.8 Curves showing soil resistance versus deflection (p-y curves) for batter piles in the example problem.

lb, P_h = 53,500 lb, and M = 1,325,000 in.-lb. The results for Piles 1 and 2 are as follows:

Pile No.	x_t, in.	P_x, k	y_t, in.	P_t, k	f_{max}, ksi
1	0.00679	20.0	0.289	18.2	10.9
2	0.0429	99.5	0.287	11.6	12.0

The pile-head movements and the loads with a load factor of 2.5 are shown in the above table, and the pile-head loads are presented in Figure 15.11. The table shows the lateral deflection with the load factor to be about 0.25 in. and the maximum stress to be far less than would cause a failure of the pile. The maximum axial load of 99.5 kips is far less than the load to cause failure of the pile (see Figure 15.6).

The above example may be viewed as part of an iteration to determine the pile size and spacing for the retaining-wall problem. The engineer may modify the pile diameter, wall thickness, and penetration and reconsider the distance between pile groups along the wall. The construction cost is increased for the driving of batter piles; two vertical piles could be investigated, along with other factors that influence the design.

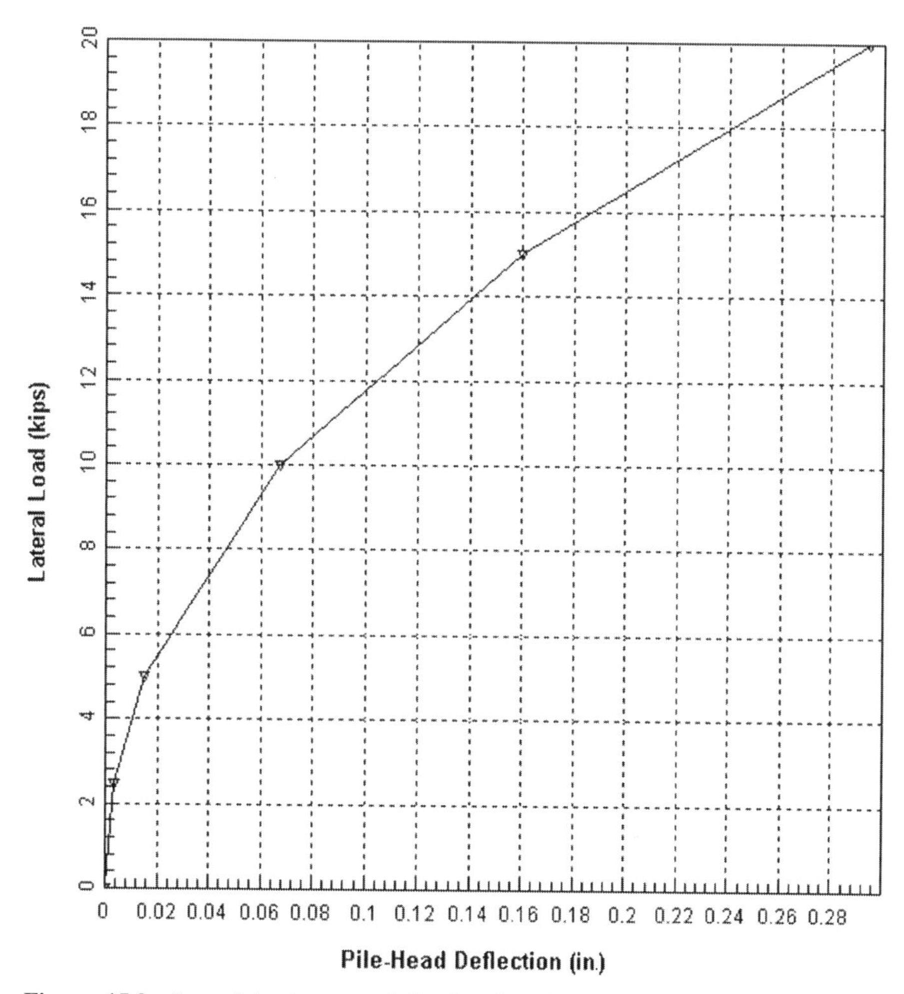

Figure 15.9 Lateral load versus deflection for pile head for vertical piles in the example problem.

15.5 EFFICIENCY OF PILES IN GROUPS UNDER LATERAL LOADING

15.5.1 Modifying Lateral Resistance of Closely Spaced Piles

O'Neill (1983) considered the response of a group of piles under both axial load and lateral load, and characterized the problem of closely spaced piles in a group as one of pile–soil–pile interaction. O'Neill listed a number of procedures that may be used in predicting the behavior of such groups. He states that none of the procedures should be expected to provide generally accurate predictions of the distribution of loads to piles in a group because none of the models account for installation effects. He concludes that more experimental data are needed. Sections 15.5.2 and 15.5.3 deal with lateral

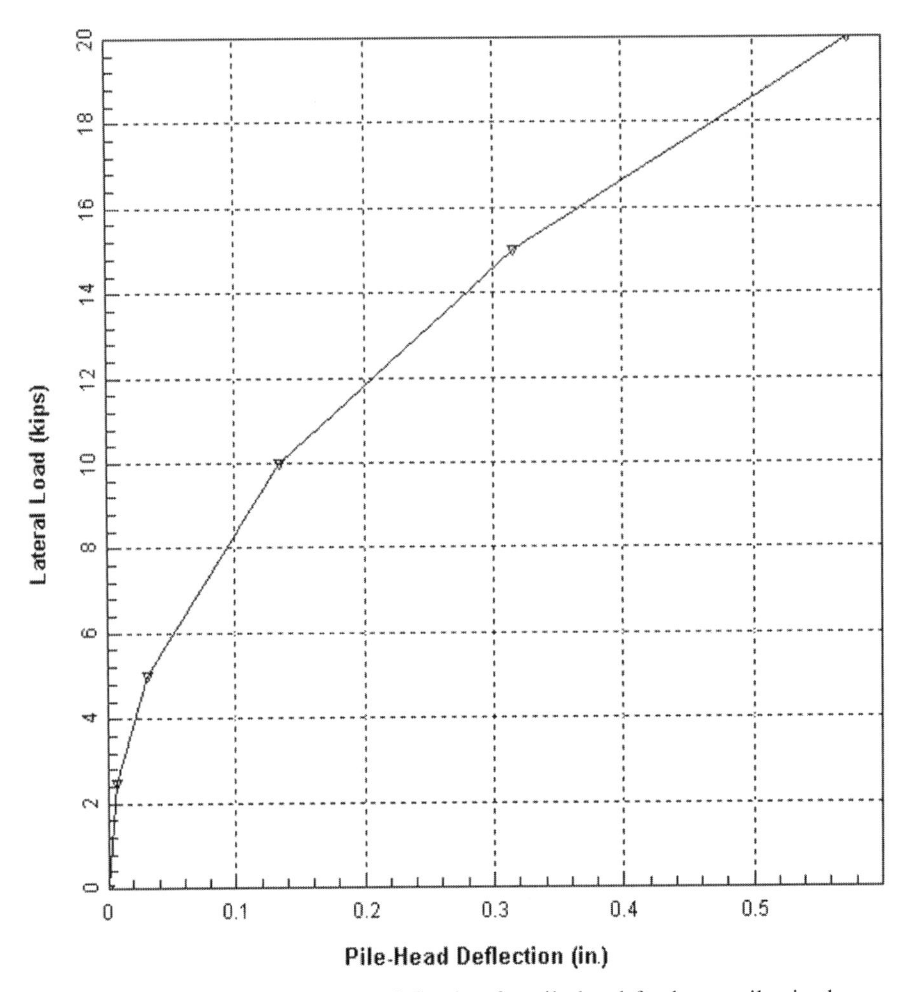

Figure 15.10 Lateral load versus deflection for pile head for batter piles in the example problem.

resistance. For the analysis of groups of piles, axial resistance must be modified for closely spaced piles. That topic was discussed in Section 15.6.

15.5.2 Customary Methods of Adjusting Lateral Resistance for Close Spacing

Pile–soil–pile interaction under close spacing is the reason that the piles in the group are less efficient than single piles. The theory of elasticity has been employed to take into account the effect of a single pile on other piles in the group. Solutions have been developed (Poulos, 1971; Banerjee and Davies,

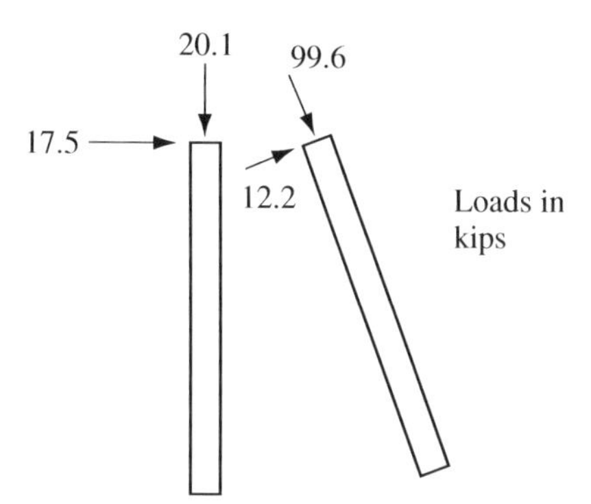

Figure 15.11 Loads at the pile heads for the example problem.

1979) that assume a linear response of the pile-soil system. While such methods are instructive, there is ample evidence to show that soils cannot generally be characterized as linear, homogeneous, elastic materials.

Two other approaches to the analysis of a group of closely spaced piles under lateral load are given in the following paragraphs. One method is to use a rather straightforward equation for the efficiency of piles in a group under lateral loading, such as Eq. 15.20; the other method is based on the assumption that the soil within the pile group moves laterally the same amount as the piles.

$$(Q_{\text{ult}})_G = En(Q_{\text{ult}})_p \tag{15.20}$$

where

$(Q_{\text{ult}})_G$ = ultimate lateral capacity of the group,
E = efficiency factor (1 or < 1),
n = number of piles in the group, and
$(Q_{\text{ult}})_p$ = ultimate lateral capacity of an individual pile.

Various proposals have been made for obtaining the value of E. For example, McClelland (1972) suggested that the value of E should be 1.0 for pile groups in cohesive soil with center-to-center spacing of eight diameters or more and that E should decrease linearly to 0.7 at a spacing of three diameters. McClelland based his recommendations on results from experiments in the field and in the laboratory. He did not differentiate between piles that are spaced front to back or side by side or spaced at some other angle between each other.

Unfortunately, experimental data on the behavior of pile groups under lateral load are limited. Furthermore, the mechanics of the behavior of a group of laterally loaded piles are more complex than those for a group of axially loaded piles. Thus, few recommendations have been made for efficiency formulas for laterally loaded groups.

The single-pile method of analysis is based on the assumption that the soil between the piles moves with the group. Thus, the pile group with the contained soil can be treated as a single pile of large diameter.

A step-by-step procedure for using the method is as follows:

1. The group to be analyzed is selected, and a plan view of the piles at the groundline is prepared.
2. The minimum length is found for a line that encloses the group. If a nine-pile (3 by 3) group consists of piles that are one foot square and three widths on center, the length of the line will be 28 ft.
3. The length found in Step 2 is considered to be the circumference of a pile of large diameter; thus, the length is divided by π to obtain the diameter of the imaginary pile.
4. The next step is to determine the stiffness of the group. For a lateral load passing through the tops of the piles, the stiffness of the group is taken as the sum of the stiffnesses of the individual piles. Thus, it is assumed that the deflection at the pile top is the same for each pile in the group and, further, that the deflected shape of each pile is identical. Some judgment must be used if the piles in the group have different lengths.
5. Then an analysis is made for the imaginary pile, taking into account the nature of the loading and the boundary conditions at the pile head. The shear and moment for the imaginary large-sized pile are shared by the individual piles according to the ratio of the lateral stiffness of each pile to that of the group.

The shear, moment, pile-head deflection, and pile-head rotation yield a unique solution for each pile in the group. As a final step, it is necessary to compare the single-pile solution to that of the group. The piles in the group may have an efficiency greater than 1; in this case, the single-pile solutions would control.

An example problem is presented in Figure 15.12. The steel piles are embedded in a reinforced-concrete mat in such a way that the pile heads do not rotate. The piles are 14HP89 by 40 ft long and placed so that bending is about the strong axis. The moment of inertia is 904 in.^{4} and the modulus of elasticity is 30×10^6 lb/in.^{2}. The width of the section is 14.7 in. and the depth is 13.83 in.

The soil is assumed to be sand with an angle of internal friction of 34°, and the unit weight is 114 lb/ft^3. The computer program presented in Chapter 14 may be utilized. Alternatively, the *p-y* curves for the imaginary large-diameter pile could be prepared as described by Reese et al. (1974) or API (1987) and a solution developed using nondimensional curves. For a pile with

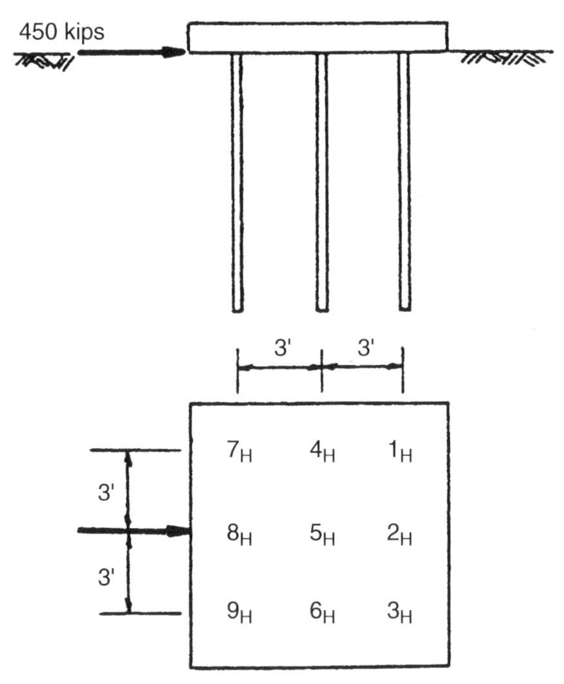

Figure 15.12 Example of a solution of the problem of closely spaced piles under lateral loading.

a diameter of 109.4 in. and a moment of inertia of 8136 in.^{4} (9 times 904), the results are as follows:

$y_t = 0.885$ in.
$M_t = M_{max} = 3.60 \times 10^7$ in.-lb for group
Bending stress $= 25.3$ kips/in.^{2}

The deflection and stress are for a single pile of large diameter.

If a single pile with a diameter of 14.7 in. is analyzed with a load of 50 kips, the groundline deflection is 0.355 in. and the bending stress is 23.1 kips/in.^{2}. Therefore, the solution with the imaginary large-diameter single pile is more critical.

15.5.3 Adjusting for Close Spacing Under Lateral Loading by Modified *p-y* Curves

Bogard and Matlock (1983) present a method in which the *p-y* curves for a single pile are modified to take into account the group effect. Excellent agreement was obtained between their computed results and results from field ıtlock et al., 1980). Much of the detailed presentation that on the work of Brown et al. (1987), who tested single piles lateral loading, all instrumented for the measurement of ...ent with depth. Two major experiments were performed, in overconsolidated clay and in sand, and analyses showed that the group effect could be taken into account most favorably by reducing the value of p for the p-y curve of the single pile to obtain p-y curves for the pile group. The curves in Figure 15.13 show that the factor f_m may be used to reduce the values of p_{sp} for the single pile to the value p_{gp} for the pile group. The proposals make use of other work in the technical literature, some based on the results of model tests of pile groups (Prakash, 1962; Schmidt, 1981, 1985; Cox et al., 1984; Dunnavant and O'Neill, 1986; Wang, 1986; Lieng, 1988).

Side-By-Side Reduction Factors The first pattern for the placement of piles in a group to be considered is for piles placed side by side. Values of β are found, termed β_a for this case, and may be summed as shown later to determine a composite factor of β for each pile in the group. The pattern for the placement of the side-by-side piles is shown in Figure 15.14, with the arrows showing the direction of loading. The values of β_a may be found from the curve or equations given in Figure 15.14. The plotted points in the figure are identified by the references previously cited. The value of β_a may be taken directly from the plot or may be found by using the equation in the figure. As may be seen, with s/b values of 3.28 or more, the value of β_a is unity. The smallest value of β_a for piles that touch is 0.5. A review of the

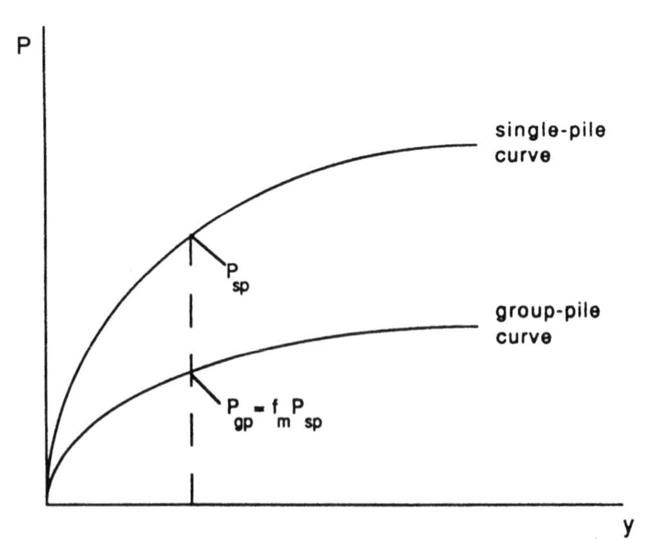

Figure 15.13 Employing a computed value of f_m to derive the p-y curve for piles in a group.

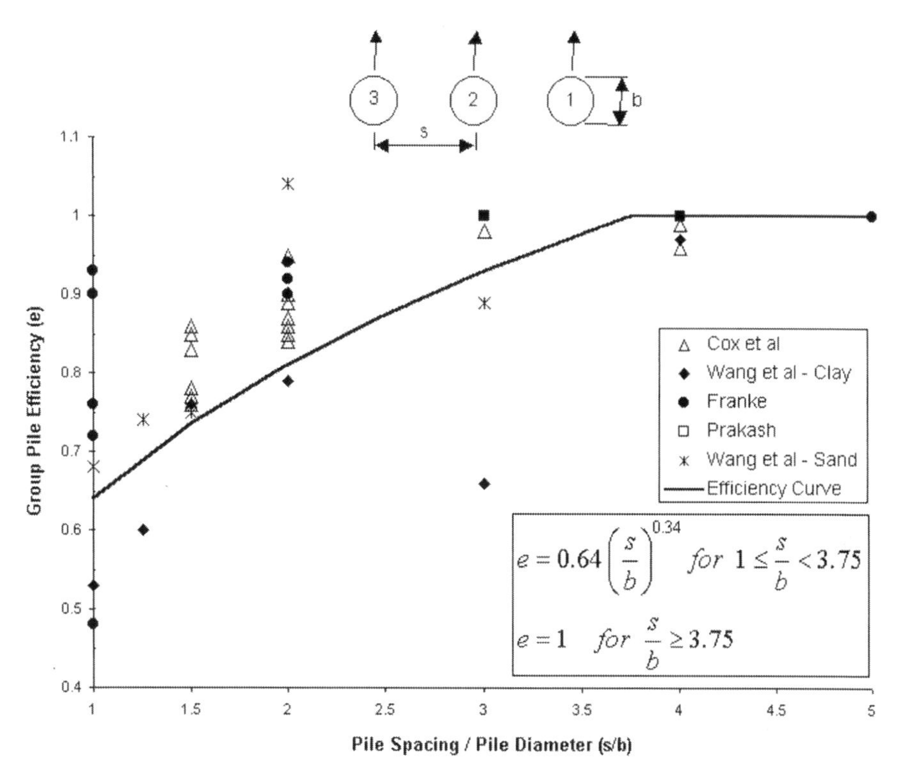

Figure 15.14 Curve giving reduction factors β_a for piles in a row.

plotted points reveals that the value of unity of β_a for s/b values of 3.28 or more is strongly supported. The value of 0.5 for piles that touch is found from mechanics. However, the first branch of the curve in Figure 15.14 is subject to uncertainty; this is not surprising, considering the variety of experiments that were cited.

Line-By-Line Reduction Factors, Leading Piles The next pattern of placement of the piles in a group to be considered is for piles placed in a line, as shown in Figure 15.15, with the arrow showing the direction of loading. The values of β_{bL} may be found from the curve or equations in Figure 15.15. The plotted points in the figure are identified by the references previously cited. The suggested curve agrees well with the plotted points except for four points to the left of the curve indicating an efficiency of unity for close spacing.

Line-By-Line Reduction Factors, Trailing Piles The experimental results, along with a suggested curve and equations, are given in Figure 15.16. The scatter of the plotted points indicates that the computed efficiency for the

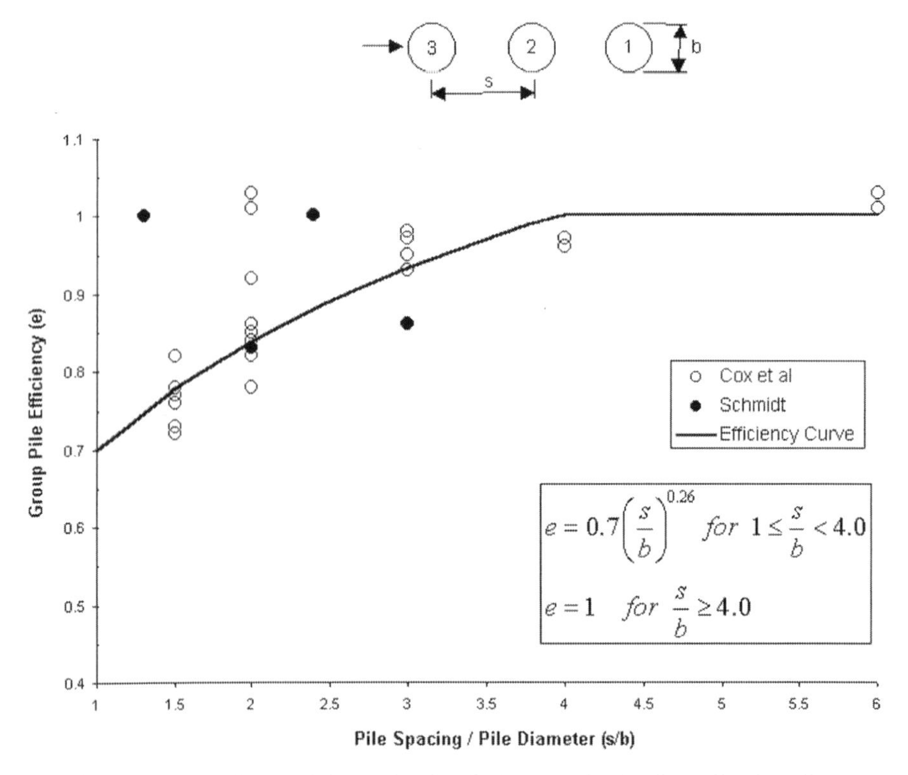

Figure 15.15 Curve giving reduction factors β_{bl} for leading piles in a line.

individual piles in a group is not likely to be precise. The scatter, as noted earlier, is not surprising in view of the many variables involved in the experiments. The authors maintain that the use of values suggested in the curves and equations in Figures 15.14 through 15.16 will yield a much better result than ignoring the effect of close spacing.

Skewed Piles The experiments cited above did not obtain data on skewed piles, but provision for skewed piles is necessary. A simple mathematical expression for the ellipse in polar coordinates was selected to obtain the reduction factor. The geometry of the two piles, A and B, is shown in Figure 15.17a. The side-by-side effect β_a may be found from Figure 15.14, where the spacing is r/b. The in-line effect β_b may be found from either Figure 15.16 or Figure 15.17, depending on whether Pile A or Pile B is being considered. The values of β_a and β_b are indicated in Figure 15.17b, and the value of β_s for the effect of skew may be found from the following equation:

$$\beta_s = (\beta_b^2 \cos^2 \phi + \beta_a^2 \sin^2 \phi)^{1/2} \tag{15.21}$$

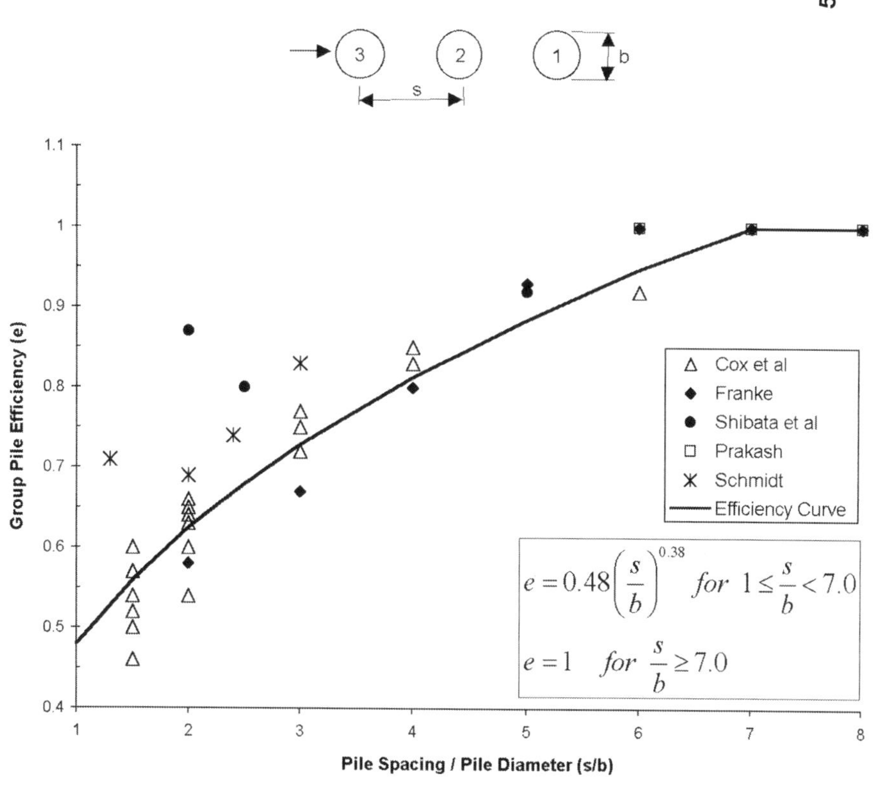

Figure 15.16 Curve giving reduction factors β_{bt} for trailing piles in a line.

Example Computation of Reduction Factors for Piles Under Lateral Loading The example selected for computation of the reduction factors for the group is presented in Figure 15.18. Pile 4 is selected as an example of the computation for each pile in the group. The value of f_m for Pile 4 to obtain the modified *p-y* curve for *that pile* in the group is obtained as follows:

$$
\begin{aligned}
f_{m4} = {} & \beta_{34} \quad \text{(side by side effect)} \\
& \times \beta_{24} \quad \text{(trailing effect)} \\
& \times \beta_{64} \quad \text{(leading effect)} \\
& \times \beta_{14} \quad \text{(skewed effect)} \\
& \times \beta_{54} \quad \text{(skewed effect)}
\end{aligned}
$$

Employing the information in Figures 15.14 through 15.16 and Eq. 15.21, the following values were computed:

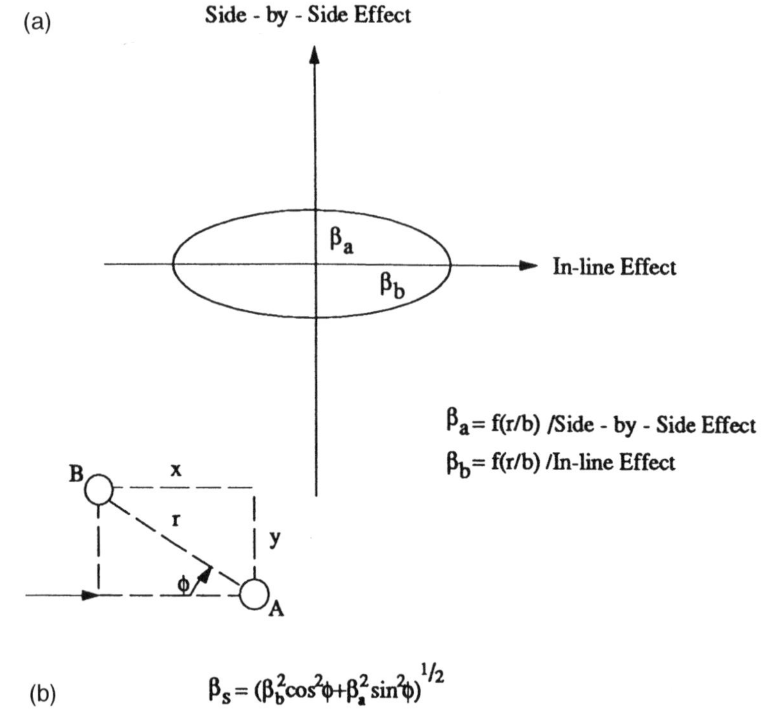

Figure 15.17 System for computing the reduction factor for skewed piles.

$$f_{m4} = (0.8101)(0.8509)(0.9314)(0.7287)(0.9809) = 0.46$$

For the pile selected in the group, f_{ma} is the reduction factor to the values of p for the p-y curve. Pile 4 in the group significantly reduces the values of p for a single pile.

For each pile i in the group, the group reduction factor may be computed by the following equation:

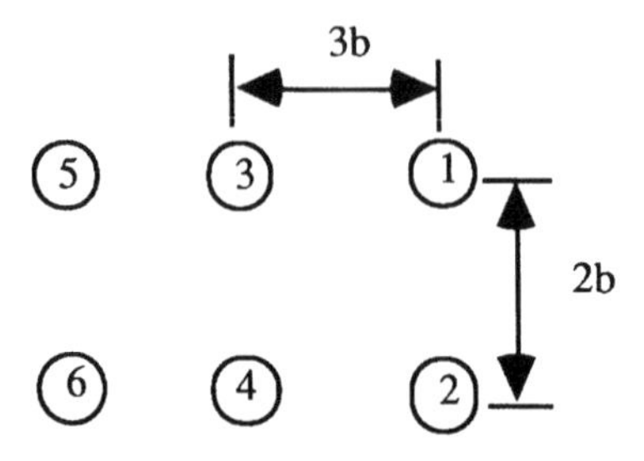

Figure 15.18 Pile group for example computations.

$$f_{mi} = \beta_{1i}\beta_{2i}\beta_{3i} \cdots \beta_{ji} \cdots m \neq i \tag{15.22}$$

Computation of the factor to reduce the values of p for the single pile to values for each pile in a group is tedious. Then the individual sets of p-y curves, perhaps all different, must be used in analyzing the behavior of the group under lateral loading. Such extensive computations require a properly written computer program.

15.6 EFFICIENCY OF PILES IN GROUPS UNDER AXIAL LOADING

15.6.1 Introduction

Most pile foundations consist not of a single pile, but of a group of piles for supporting superstructures. In a group of closely spaced piles, the axial capacity of the group is influenced by variations in the load-settlement behavior of individual piles because of pile–soil–pile interaction. The group effects of piles under axial loading, discussed in this chapter, will focus on proposals for determining the efficiency of the individual piles in the group.

The concept of group behavior is presented in Figure 15.19. Figure 15.19a shows a single pile and the possible downward movement of an imaginary surface at some distance below the groundline. As may be seen, that surface moves downward more at the wall of the pile than elsewhere, but movements do occur away from the wall.

Three piles spaced close together are shown in Figure 15.19b. The zones of influence overlap, so that the imaginary surface moves downward more for the group than for the single pile. The stresses in the soil around the center pile are larger than those for the single pile because of the transfer of stresses from the adjacent piles. Therefore, the designer must consider the ultimate capacity of a pile in a group as well as the settlement of the group.

The problem is complicated by the presence of the pile cap in two ways. First, if the cap is perfectly rigid and the axial loading is symmetrical, all of the piles will settle the same amount. However, if the cap is flexible, the settlement of the piles will be different. Second, if the cap rests on the ground surface, some of the axial load will be sustained by bearing pressure on the cap. Many authors have treated the problem of the distribution of the axial load to the piles and to the cap. However, conservatively, the assumption can be made that there can be settlement of the soil beneath the cap and that all of the load is taken by the piles. This above assumption is made in the discussion that follows.

Unlike the behavior of a group of piles under lateral loading, the behavior of a group under axial loading strongly depends on methods of installation, types of soils, and stress-induced settlement. The position of each pile in the group is less important than that of piles under lateral loading. A number of

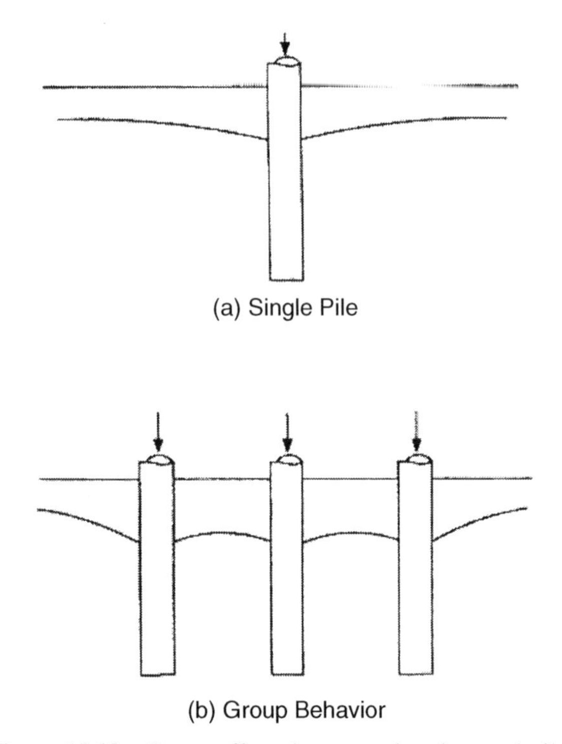

Figure 15.19 Group effects between closed-spaced piles.

investigators, such as Poulos and Davis (1980) and Focht and Koch (1973) have used the theory of elasticity to develop interaction recommendations. However, soils do not behave the same way in tension as in compression, and such theoretical results do not agree well with the results of experiments that have been conducted.

The concept of block failure (i.e., simultaneous failure of the piles and of the mass of soil within the pile group) is commonly used to calculate the ultimate capacity of a closed-spaced pile group. As shown in Figure 15.20, an imaginary block encompasses the pile group. The load carried by an imaginary block is the sum of the load carried by the base and friction on the perimeter of the block. The ultimate capacity developed by the block failure is compared with the sum of the ultimate capacity of individual piles in the group, and the smaller of these two values is selected as the load-carrying capacity of the pile group.

O'Neill (1983), in a prize-winning paper, presented a comprehensive summary of the efficiency of piles in a group. He reviewed the proposals that have been made for piles under axial loading and showed that piles in cohesive and cohesionless soils respond quite differently. Many investigations have been carried out to determine group efficiency under various soil conditions and pile spacings. A brief discussion of group efficiency in cohesionless and cohesive soils will now be presented for reference.

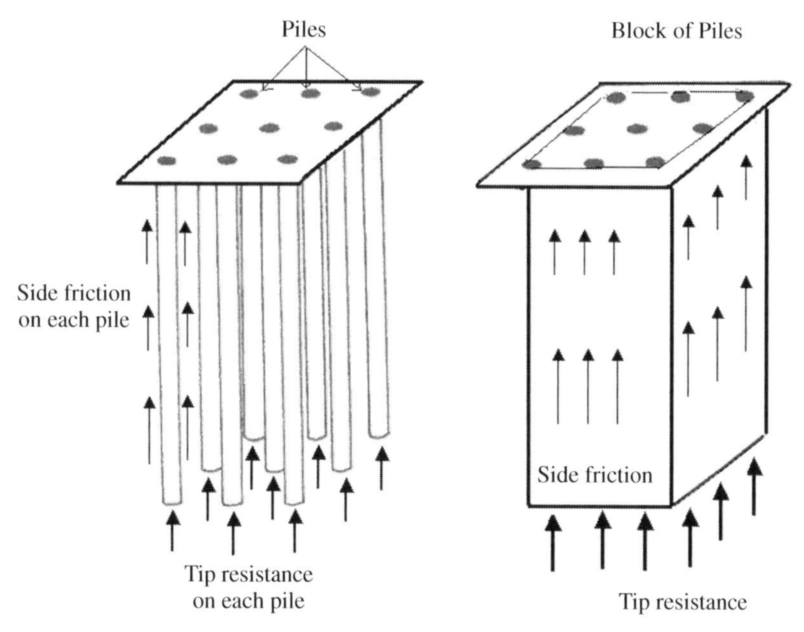

Figure 15.20 Block-failure model for closed-spaced piles.

15.6.2 Efficiency of Piles in a Group in Cohesionless Soils

The efficiency η of a pile in a group is defined formally by Eq. 15.23:

$$\eta = \frac{Q_g}{nQ_s} \tag{15.23}$$

where

η = number of piles,
Q_g = total capacity of the group, and
Q_s = capacity of a reference pile that is identical to a group pile but is isolated from the group.

Figure 15.21 shows the average efficiencies for driven piles. Theory predicts that the capacity of a pile in cohesionless soil is increased with an increase in the effective stress. Thus, the overlapping zones of stress at the base of a group of piles will cause an increase in end bearing. As shown in Figure 15.21, the average value of efficiency is slightly greater than unity. Also, the lateral compaction of the cohesionless soil during installation can cause an increase in effective stress along the sides of a driven pile (the shaft). Figure 15.21 shows that the efficiency of the shaft can range from slightly

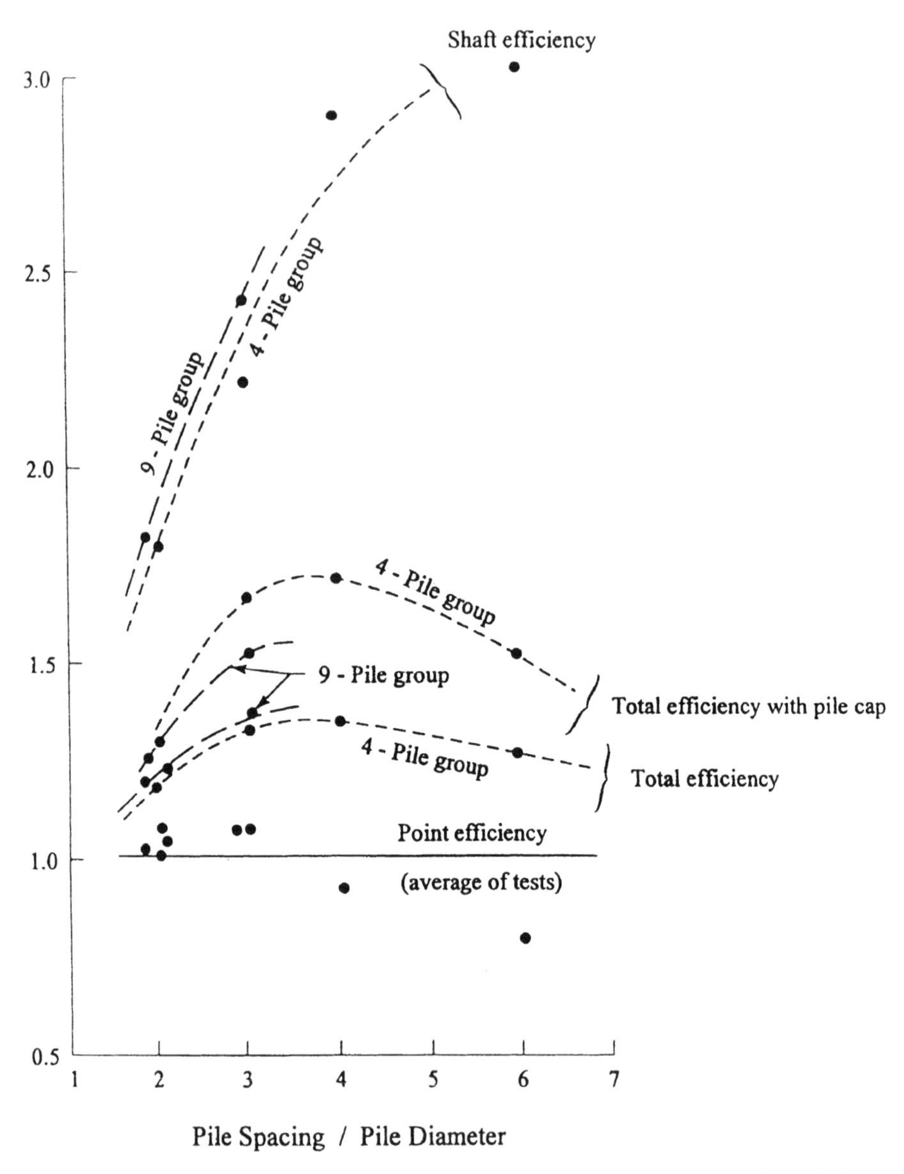

Figure 15.21 Group efficiency in cohesionless soils reported by Vesić (1969).

over 1.5 to almost 3.0. The figure shows that the overall efficiency of a group in cohesionless soils is well above 1.0.

While not shown in the figure, settlement frequently controls the capacity of piles in granular soils. Where settlement is critical, special study of the data from Vesić (1969) and from other authors is desirable.

O'Neill (1983) made a comprehensive study of the behavior of pile groups under axial loading. Figure 15.22 shows a compilation of the results of model tests for inserted piles (similar to those for driven piles). The trends in Figure 15.22 were described by O'Neill as follows: (1) η in loose sands always exceeds unity, with the highest values occurring at spacing-to-diameter ratios s/d of about 2, and in general, higher η occurs with increasing numbers of piles. Block failure (i.e., simultaneous failure of the piles and the mass of soil within the pile group) affects η in 4-pile groups only below $s/d = 1.5$ and in 9- to 16-pile groups only below $s/d = 2$. (2) η in dense sands may be either greater or less than unity, although the trend is toward $\eta > 1$ in groups of all sizes with s/d ranging from 2 to 4. Efficiency of less than unity is probably a result of dilatency and would not generally be expected in the field for other than bored or partially jetted piles, although theoretical studies of interference suggest η slightly below 1 at $s/d > 4$.

Conventional practice generally assigns a group efficiency value of 1 for driven piles in cohesionless soil unless the pile group is founded on dense soil of limited thickness underlain by a weak soil deposit. In such conditions, Meyerhof (1974) suggested that the efficiency value of the group should be based on the capacity of an equivalent base that punches through the dense sand in block failure.

15.6.3 Efficiency of Piles in a Group in Cohesive Soils

In general, piles in cohesive soils sustain load principally in side resistance or behave as "friction" piles. The load-carrying capacity of a group of friction piles in clay is the smaller of the following:

1. The sum of the failure loads of the individual piles or
2. The load carried by an imaginary block encompassing the group, where the loads is the sum of the load carried by the base and perimeter of the block.

In certain types of clay, particularly highly sensitive clay, the efficiency of closely spaced piles in a group is less than 1. However, there is insufficient data from testing of full-sized piles in the field to allow quantification of pile efficiency as a function of center-to-center spacing. Use of the imaginary block, described above, is the common way to investigate the efficiency of the group. The test results shown in Figure 15.23 were summarized by de Mello (1969) and show that failure of a pile group does not become pronounced until the pile spacing is less than two pile diameters.

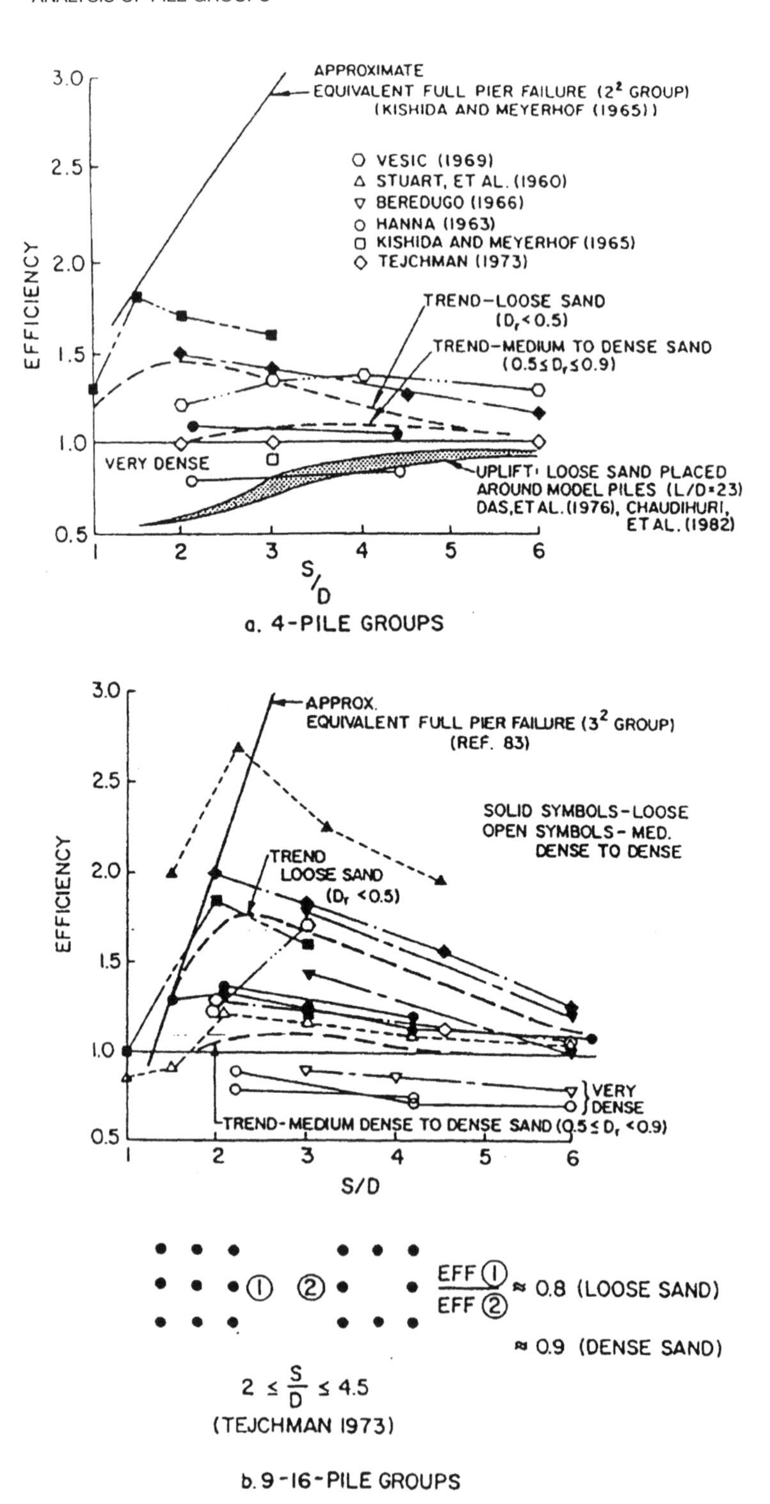

Figure 15.22 Group efficiency in cohesionless soils reported by O'Neill (1983).

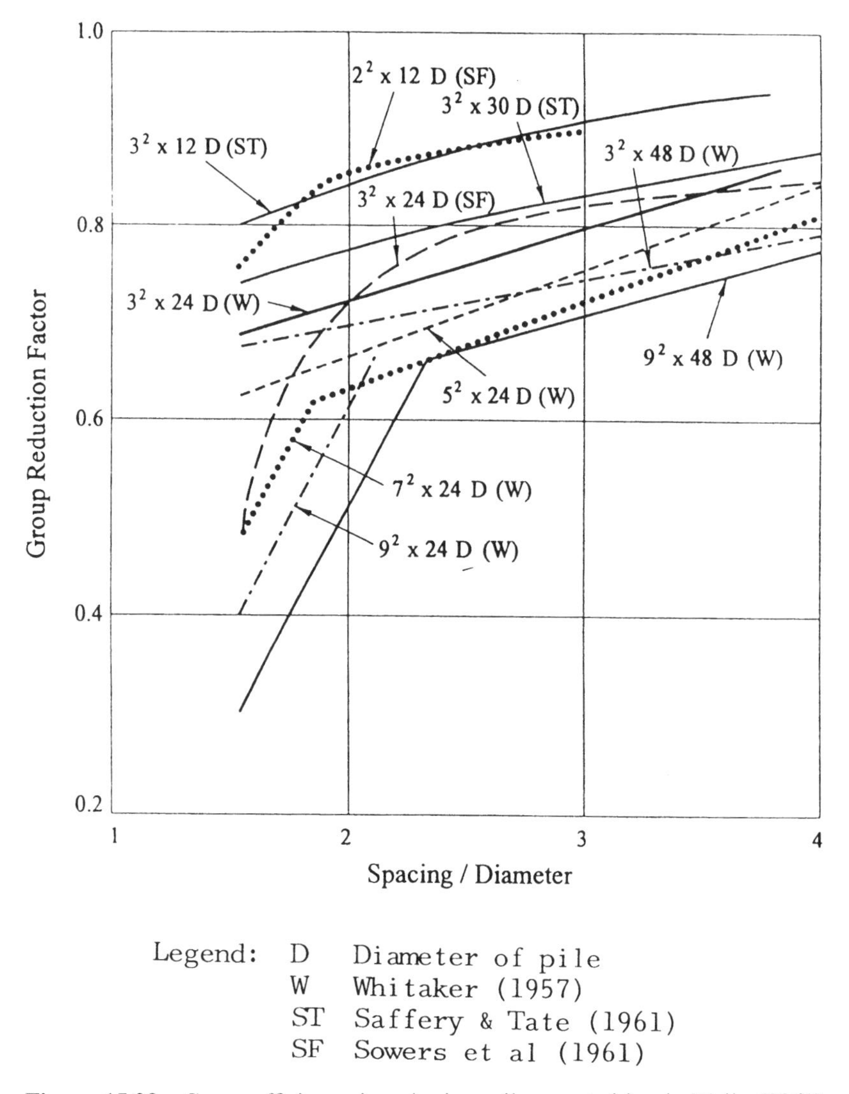

Figure 15.23 Group efficiency in cohesive soils reported by de Mello (1969).

O'Neill (1983) made studies of group efficiency in clay similar to those in sand. A compilation of the results of model tests of pile groups in clay is shown in Figure 15.24, except for the results reported by Matlock et al. (1980) where the test was in the field. Model tests in clay, in contrast to those in sand, always yield efficiencies less than 1, with a distinct trend toward block failure in square groups at values of s/d less than about 2. The group efficiency is higher in stiff clays than in soft clays.

O'Neill (1983) pointed out that the efficiencies are 1 or less for suspended caps and may be higher than 1 for caps in contact with the soil. However,

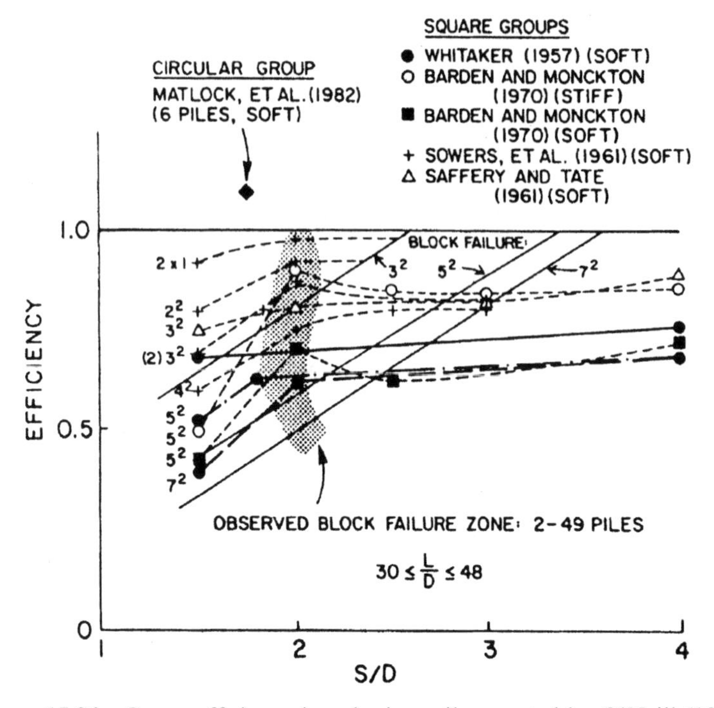

Figure 15.24 Group efficiency in cohesive soils reported by O'Neill (1983).

the possibility that the soil will settle eliminates the effectiveness of the cap. An important factor with respect to a group of piles in saturated clay is that the generation of porewater pressures during pile driving will create a large "bulb" of excess pore pressure around the group of piles. Thus, the dissipation of pore pressure around pile groups will be much slower than the dissipation around single piles, a fact that may not have been addressed in tests of full-scale piles that have been reported.

The settlement of a group of piles in saturated clay will be time-dependent and may be an overriding consideration with respect to the capacity of the group to sustain axial loading. The design of a group of piles in clay will depend on many factors, including site-specific conditions and the nature of the superstructure, as well as the center-to-center spacing of the piles.

15.6.4 Concluding Comments

The most reliable data on the efficiency of the piles in a group are derived from the results of full-scale tests. However, the behavior of pile groups under axial loading depends on many factors that can be investigated thoroughly only with a large number of carefully controlled tests, taking time into account. Such a program is beyond the capabilities of the agencies currently interested in the behavior of piles. As an example of the current limitations, guidelines are unavailable for estimating the efficiency of groups of piles in layered soils. The data presented in this chapter are useful and present an introduction to the efficiency of piles in groups under axial loading, but they are not meant to provide specific information for design, even for the most routine problem. The judgment of the engineer should take into account the type of piles, construction methods, distribution of frictional and tip resistance, and dominant soil layers. As a concluding comment, the design of piles in groups under lateral and axial loading should be based on the concept of soil–structure interaction. Curves showing the transfer of lateral load (p-y curves), in side resistance (t-z curves), and in end bearing (q-w curves) can potentially be modified to account for the efficiency of the individual piles. Then both capacity and settlement can be computed with a rational model. This approach must await the performance of the requisite research.

PROBLEMS

P15.1. Make a tour of your neighborhood and look for places of where piles have been used in groups. Hint: in numerous instances, bridges and overpasses have been put on piles and the upper portions of the piles are exposed.

- **a.** Examine the upper portion of the piles where they are fastened to the pile cap and estimate, if possible, whether the designer meant the pile heads to be fixed against rotation or free to rotate.
- **b.** Look closely at the piles, made of reinforced concrete in many instances, and observe any possible overstressing, such as cracks in the concrete.
- **c.** Look closely as the soil where the piles are penetrating and look for cracks or openings next to the piles that indicate past deflection.
- **d.** Make a sketch of the pile group showing a plan of the pile heads and two elevations. Distances may be estimated. From your knowledge of the performance of groups of piles, discuss the selection of pile placement. If batter piles were used, is their placement exact, as measured by eye, or do they deviate from the planned position? Are some of the piles close enough together to require use of the technology to account for pile–soil–pile interaction?

P15.2. Select one or more of the references, go to the engineering library, and request the librarian to assist in finding it. Prepare a one-page summary.

P15.3. The two-dimensional problem of the pile group under lateral load resulted in the following solution:

$$\Delta v = 0.005579; \quad \Delta h = 0.025081; \quad \alpha_s = 0.000151$$

Use appropriate equations to compute the pile-head movements from the above values, compute pile-head forces, and check to determine if the group is in static equilibrium.

P15.4. Provide input for the problem of the retaining wall to perform a push-over analysis and solve for the loading that will cause failure by excessive movement or by excessive bending stress. Assume that the axial capacity of the pile is 200 k and the maximum bending stress is 36 ksi (note that using a multiplier of the loads of 2.5 yielded a computed maximum axial load of 99.5 k and a bending stress of 12.0 ksi). Also, the maximum stress occurs down the length of the pile rather than at the top.

P15.5. Assume that you wish to redesign the system for failure at a load factor of 3.0. List some of the modifications you would consider.

APPENDIX

UNITS CONVERSION FACTORS

TABLE A1 Units of Distance

To convert	To	Multiply by
ft	in	12
ft	m	0.3048
in	ft	1/12
in	mm	25.40
m	ft	3.281
mm	in	0.03937

TABLE A2 Units of Force

To convert	To	Multiply by
k	kN	4.448
k	lb	1000
kg_f	lb	2.205
kg_f	N	9.807
kg_f	metric ton	0.001
kN	k	0.2248
lb	k	0.001
lb	kg_f	0.4536
lb	N	4.448
lb	ton (short)	1/2000
lb	ton (long)	1/2240
N	kg_f	0.1020
N	lb	0.2248
ton (short)	lb	2000
ton (long)	lb	2240
metric ton	kg_f	1000

TABLE A3 Units of Volume

To convert	To	Multiply by
ft^3	gal	7.481
gal	ft^3	0.1337

TABLE A4 Units of Stress and Pressure

To convert	To	Multiply by
Atmosphere	lb/ft^2	2117
Atmosphere	kPa	101.3
bar	kPa	100
kg_f/cm^2	kPa	98.07
kg_f/cm^2	lb/ft^2	2048
kPa	atmosphere	0.009869
kPa	bar	0.01
kPa	kg_f/cm^2	0.0102
kPa	lb/ft^2	20.89
kPa	lb/in^2	0.1450
kPa	metric ton/m^2	0.1020
lb/ft^2	atmosphere	4.725×10^{-4}
lb/ft^2	kPa	0.04787
lb/ft^2	lb/in^2	1/144
lb/in^2	kPa	6.895
lb/in^2	lb/ft^2	144
lb/in^2	MPa	6.895×10^{-3}
metric ton/m^2	kPa	9.807
MPa	lb/in^2	145.0

TABLE A5 Units of Unit Weight

To convert	To	Multiply by
kN/m^3	lb/ft^3	6.366
kN/m^3	metric ton/m^3	0.1020
kN/m^3	Mg_f/m^3	0.1020
lb/ft^3	Mg_f/m^3	0.1020
metric ton/m^3	kN/m^3	9.807
Mg_f/m^3	kN/m^3	9.807

REFERENCES

American Association of State Highway and Transportation Officials (AASHTO) (1983), "Standard Specifications for Highway Bridges."

American Concrete Institute (1993). ASTM Standards in ACI 301, 318, and 349, Publication SP-71.

American Petroleum Institute (1986). "API Recommended Practice for Planning, Designing, and Constructing Fixed Offshore Platforms," *Report RP-2A*.

American Petroleum Institute (1987). "API Recommended Practice for Planning, Designing, and Constructing Fixed Offshore Platforms," *Report RP-2A*.

American Petroleum Institute (1993). "API Recommended Practice for Planning, Designing, and Constructing Fixed Offshore Platforms," *Report RP-2A*.

American Society of Civil Engineers (1993). *Design of Pile Foundations Handbook*.

American Society of Civil Engineers (2000). *Standards of Professional Conduct for Civil Engineers*.

American Society for Testing and Materials (1992). *1992 Annual Book of ASTM Standards,* Vol. 04.08: "Standard Method for Penetration Test and Split-Barrel Sampling of Soils," ASTM D 1586; "Standard Practice for Thin-Walled Tube Sampling of Soils," ASTM D 1587; "Standard Practice for Diamond Core Drilling for Site Investigation," ASTM D 2113; "Standard Test Method for Field Vane Shear Test in Cohesive Soil," ASTM D 2573; "Standard Test Method for Triaxial Compressive Strength of Undrained Rock Core Specimens Without Pore Pressure Measurements," ASTM D 2664; "Standard Test Method for Unconfined Compressive Strength of Intact Rock Core Specimens," ASTM D 2938; "Standard Test Method for Deep, Quasi-Static, Cone and Friction-Cone Penetration Tests of Soil," ASTM D 3441; "Standard Practice for Ring-Lined Barrel Sampling of Soils," ASTM D 3550; "Standard Practice for Preserving and Transporting Soil Samples," ASTM D 4220; "Standard Test Method for Pressuremeter Testing of Soils," ASTM D 4719.

Andresen, A., Berre, T., Kleven, A., and Lunne, T. (1978). "Procedures Used to Obtain Soil Parameters for Foundation Engineering in the North Sea," Report 52107-4, Norges Geotekniske Institutt.

Andresen, A., and Simons, N. E. (1960). "Norwegian Triaxial Equipment and Techniques," *Proceedings,* Research Conference on the Shear Strength of Cohesive Soil, ASCE, Boulder, CO, pp. 695–710.

Apthorpe, M., Garstone, J., and Turner, G. D. (1988). "Depositional Setting and Regional Geology of North Rankin 'A' Foundation Sediments," *Proceedings,* International Conference on Calcareous Sediments, Perth, Western Australia. Balkema, Rotterdam, pp. 357–366.

Atterberg, A. (1911). "Über die physikalishe Bodeuntersuchung und über die Plastizität der Tone" (One the Investigation of the Physical Properties of Soils and on the Plasticity of Clays), Int. Mitt. für Bodenkunde, Vol. 1, pp. 10–43.

Aurora, R., and Reese, L. C. (1977). "Field Tests of Drilled Shafts in Clay-Shales," *9th International Conference on Soil Mechanics and Foundation Engineering,* ISSMFE, Tokyo, July.

Awoshika, K., and Reese, L. C. (1971a). "Analysis of Foundations with Widely Spaced Batter Piles," *Proceedings,* The International Symposium on the Engineering Properties of Sea-Floor Soils and Their Geophysical Identification, University of Washington, Seattle, July 25, pp. 200–211.

Awoshika, K., and Reese, L. C. (1971b). "Analysis of Foundation with Widely Spaced Batter Piles," Research Report 117-3F, Project 3-5-68-117, Center for Highway Research, University of Texas at Austin, February.

Banerjee, P. K., and Davies, T. G. (1979). "Analysis of Some Reported Case Histories of Laterally Loaded Pile Groups, *Proceedings,* Numerical Methods in Offshore Piling, Institution of Civil Engineers, London, May, pp. 101–108.

Baguelin, F., Jézéquel, J. F., and Shields, D. H. (1978). *The Pressuremeter and Foundation Engineering,* Trans Tech Publications, Clausthal, Germany.

Bedenis, T. H., Thelen, M. J., and Maranowski, S. (2004). "Micropiles to the Rescue," *Civil Engineering,* ASCE, Vol. 74, No. 3, pp. 66–71.

Berghinz, C. (1971), "Venice in Sinking into the Sea," *Civil Engineering,* ASCE, March, pp. 67–71.

Bernal, J. B., and Reese, L. C. (1983). "Study of the Lateral Pressure of Fresh Concrete as Related to the Design of Drilled Shafts," Research Report 308-1F, Center for Transportation Research, University of Texas at Austin.

Bieniawski, Z. T. (1984). *Rock Mechanics Design in Mining and Tunneling.* Balkema, Rotterdam/Boston.

Bishop, A. W. (1961). "The Measurement of Pore Pressure in the Triaxial Test," *Proceedings,* Conference on Pore Pressure and Suction in Soils, ICE, London, pp. 38–46.

Bishop, A. W., and Eldin, G. (1950). "Undrained Triaxial Tests on Sands and Their Significance in the General Theory of Shear Strength," *Geotechnique,* Vol. 2, pp. 13–32.

Bishop, A. W., and Henkel, D. J. (1957). *The Triaxial Test.* Edward Arnold Ltd., London.

Bjerrum, L. (1954). "Geotechnical Properties of Norwegian Marine Clays," *Geotechnique,* Vol. 4, pp. 49-69.

Bjerrum, L. (1960). "Some Notes on Terzaghi's Method of Working," in *From Theory to Practice in Soil Mechanics.* Wiley, New York, p. 22.

Bjerrum, L., Kringstad, S., and Kummeneje, O. (1961). "The Shear Strength of a Fine Sand," *Proceedings,* 5th International Conference on Soil Mechanics and Foundation Engineering, ISSMFE, Vol. 1, pp. 29–38.

Bjerrum, L., and Overland, A. (1957), "Foundation Failure of an Oil Tank in Fredrikstad, Norway," *Proceedings,* 4th International Conference on Soil Mechanics and Foundation Engineering, ICSMFE.

Bogard, D., and Matlock, H. M. (1983). "Procedures for Analysis of Laterally Loaded Pile Groups in Soft Clay," *Proceedings,* Geotechnical Practice in Offshore Engineering, American Society of Civil Engineers, pp. 499–535.

Boussinesq, J. (1885). *Applicication des Potentiels à l'Étude de l'Équilibre et du Mouvement des Solides Élastiques.* Gauthier-Villard, Paris.

Bowman, E. R. (1959). "Investigation of the Lateral Resistance to Movement of a Plate in Cohesionless Soil," Thesis, Master of Science in Civil Engineering, University of Texas at Austin.

Bozozuk, M., and Penner, E. (1971). "Land Subsidence in Built-up Marshland," *Geotechnical Engineering Journal,* Vol. 8, No. 4, November, pp. 592–596.

Brown, D. A., Reese, L. C., and O'Neill, M. W. (1987). "Cyclic Lateral Loading of a Large-Scale Pile Group," *Journal of the Geotechnical Engineering Division,* ASCE, Vol. 113, No. 11, November, pp. 1326–1343.

Brown, D. A., Shie, C.-F., and Kumar, M. (1989). "*p-y* Curves for Laterally Loaded Piles Derived from Three-Dimensional Finite Element Model," *Proceedings,* Third International Symposium, Numerical Models in Geomechanics, May. Elsivier Applied Science, New York, May, pp. 683–690.

Brown, L. L. (2003). "Rail Yard Complicated Design of South African Crossing," *Civil Engineering,* ASCE, June, Vol. 73, p. 26.

Browning, W. F., Jr. (1951). "Mapping of Geologic Formations by the Application of Aerial Photography," *Bulletin No. 46, Engineering Soil Survey Mapping,* Highway Research Board, December, pp. 67–86.

Brumund, W. F., and Leonards, G. A. (1972). "Subsidence of Sand Due to Surface Vibration," *Proceedings,* ASCE, Vol. 98, No. SM1, January, pp. 27–41.

Bryant, L. M. (1977) "Three-Dimensional Analysis of Framed Structures with Nonlinear Pile Foundations," Unpublished dissertation, University of Texas at Austin.

Bucher, F. (1997). "Design of Axially Loaded Piles—Swiss Practice," *Design of Axially Loaded Piles—European Practice,* Eds. F. De Cock and C. Legrand. Balkama, Rotterdam, pp. 343–351.

Building Research Advisory Board (1968). "National Research Council Criteria for Selection and Design of Residential Slabs on Ground," Publication 1571. U.S. National Academy of Sciences, Washington, D.C.

Bustamanta, M., and Frank, R. (1997). "Design of Axially Loaded Piles—French Practice," *Design of Axially Loaded Piles—European Practice,* Eds. F. De Cock and C. Legrand. Balkama, Rotterdam, pp. 161–175.

Cambefort, H. (1965). "Curiosités des Massifs Alluvionnaires et des Nappes d'eau," *Revue de la Fedération Internationale du Bâtiment et des Travaux Publics,* No. 45.

Canadian Geotechnical Society (1978). *Canadian Foundation Engineering Manual, Part 3, Deep Foundations.* 1978, p. 98.

Canadian Geotechnical Society (1992). *Canadian Foundation Engineering Manual,* third edition.

Caquot, A., and Kérisel, J. (1953). "Sur le terme de surface dans le calcul des foundations en milieu pulvérulent," *Proceedings,* Third International Conference on Soil Mechanics and Foundation Engineering, Zürich, Vol. I, pp. 336–337.

Carslaw, H. S., and Jaeger, J. C. (1947). *Conduction of Heat in Solids.* Oxford University Press, Oxford.

Carter, J. P. and Kulhawy, F. H. (1988), "Analysis and Design of Drilled Shaft Foundations Socketed Into Rock," *Final Report,* Project 1493-4, EPRI EL-5918, Geotechnical Group, Cornell University, Ithaca, NY, August.

Carter, J. P., and Kulhawy, F. H. (1983). "Analysis and Design of Drilled Shaft Foundations Socketed into Rock," *Final Report,* Project 1493-4, EPRI EL-5918, Cornell University, Ithaca, N.Y.

Casagrande, A. (1964). "The Role of the 'Calculated Risk' in Earthwork and Foundation Engineering," *Preprint,* Second Terzaghi Lecture, annual meeting of the ASCE, New York, October 21.

Casagrande, A. (1936), "The Determination of the Pre-consolidation Load and Its Practical Significance," *Proceedings,* 1st International Conference on Soil Mechanics and Foundation Engineering, Cambridge, Massachusetts., Vol. 3, pp. 60–64.

Casagrande, A. (1940), "Notes on Soil Testing for Engineering Purposes," Harvard University Graduate School of Engineering Publication No. 8, 74 pp.

Casagrande, A. (1960), "An unsolved problem embankment stability on soft ground," Proc. 1st Panamerican Conf. Soil Mech. and Found. Eng., Mexico, 2, pp. 721–746.

Casagrande, A., and R. E. Fadum (1940), "Notes on Soil Testing for Engineering Purposes," Harvard Univ. Grad. School of Engineering Publ. No. 8, 74 pp.

Chen, F. H. (1988). *Foundations on Expansive Soils.* Elsevier, New York.

Cox, W. R., Dixon, D. A., and Murphy, B. S. (1984). "Lateral Load Tests of 25.4 mm Diameter Piles in Very Soft Clay in Side-by-Side and In-line Groups," *Laterally Loaded Deep Foundations: Analysis and Performance,* SPT 835. American Society for Testing and Materials.

Cox, W. R., Reese, L. C., and Grubbs, B. R. (1974). "Field Testing of Laterally Loaded Piles in Sand," *Proceedings,* Offshore Technology Conference, Houston, Texas, Paper No. 2079, pp. 459–472.

Coyle, H. M., and Reese, L. C. (1966). "Load Transfer for Axially Loaded Piles in Clay," *Journal of the Soil Mechanics and Foundations Division,* ASCE, Vol. 92, SM2, Paper No. 4702, March, pp. 1–26.

Coyle, H. M., and Sulaiman, I. H. (1967). "Skin Friction for Steel Piles in Sand," *Journal of the Soil Mechanics and Foundations Division,* ASCE, Vol. 93, SM6, Paper No. 5590, November, pp. 261–278.

Cummings, A. E. (1941). *The Foundation Problem in Mexico City,* (unpublished).

Cummings, A. E. (1947). "The Foundation Problem in Mexico City," *Proceedings,* Seventh Texas Conference on Soil Mechanics and Foundation Engineering (Ed. R. F. Dawson). Bureau of Engineering Research, College of Engineering, University of Texas at Austin, January.

Cunha, R. P., Poulos, H. G., and Small, J. C. (2001). "Investigation of Design Alternatives for a Piled Raft Case History," *Journal of Geotechnical and Geoenvironmental Engineering,* ASCE, Vol. 127, No. 8, August, pp. 635–641.

D'Appolonia, E., and Romualdi, J. P. (1963). "Load Transfer in End Bearing Steel H-Piles," *Journal of the Soil Mechanics and Foundations Division,* ASCE, Vol. 89, SM2, Paper No. 3450, March, pp. 1–25.

D'Appolonia, E., and Spanovich, M. (1964). "Large Settlement of an Ore Dock Supported on End-Bearing Piles," *Preprint,* Conference on Deep Foundations, Mexican Society of Soil Mechanics, Mexico City.

Davisson, M. T. (1972). "High Capacity Piles," *Proceedings,* Lecture Series Innovations in Foundation Construction, ASCE, Illinois Section, Chicago.

Dawson, R. F. (1963). "A Review of Land Subsidence Problems in the Texas Gulf Coast Area," *Journal of the Surveying and Mapping Division,* ASCE, No. SU2, 1963.

de Mello, V. F. B. (1969), "Foundations of Buildings on Clay," State of the Art Report, Proceedings, 7th International Congress on Soil Mechanics and Foundation Engineering, Vol. 1, pp. 49–136.

De Nicola, A. (1996). "The Performance of Piles in Sand," Ph.D. Dissertation, Department of Civil Engineering, University of Western Australia.

Deere, D. V. (1968). "Geological Considerations," *Rock Mechanics in Engineering Practice,* Eds. K. G. Stagg and O. C. Zienkiewicz. Wiley, New York, pp. 1–20.

DNV//Risø (2001). *Guidelines for Design of Win Turbines.* Det Norske Veritas, Copenhagen, and Risø National Laboratory, Roskilde, Denmark.

Donald, I. B., Sloan, S. W., and Chiu, H. K. (1980). "Theoretical Analyses of Rock-Socketed Piles," *Proceedings,* International Conference on Structural Foundations on Rock, Sydney. Balkema, Rotterdam.

Dunn, I. S., Anderson, L. R., and Kiefer, F. W. (1980). *Fundamentals of Geotechnical Analysis.* Wiley, New York.

Dunnavant, T. W., and O'Neill, M. W. (1986). "Evaluation of Design-Oriented Methods for Analysis of Vertical Pile Groups Subjected to Lateral Load," *Numerical Methods in Offshore Piling.* Institut Français du Petrole, le Laboratoire Central des Ponts et Chaussees, pp. 303–316.

El-Mossallamy, Y., and Franke, E. (1997). "Pile Rafts—Numerical Modeling to Simulate the Behavior of Pile-Raft Foundations." Published by the authors, Darmstadt, Germany.

EM 1110-2-2906 (1991). *Design of Pile Foundations.* U.S. Army Corps of Engineers, Washington, DC.

EM 1110-2-2104 (1992). *Strength Design for Reinforced Concrete Hydraulic Structures.* U.S. Army Corps of Engineers, Washington, DC.

EM 1110-1-1904 (1994). *Settlement Analysis.* U.S. Army Corps of Engineers, Washington, DC.

EM 1110-1-1905 (1994). *Bearing Capacity of Soils.* U.S. Army Corps of Engineers, Washington, DC.

Emrich, W. J. (1971). "Performance Study of Soil Sampling for Deep Penetration Marine Borings," *Sampling of Soil and Rock,* ASTM, STP 483, pp. 30–50.

Engineering News Record (1962). "Engineers Erred on Bridge Failure," September 6, p. 24.

Engineering News Record (1963a). "Subsidence Suit, Trenching and Dewatering Damages are Charged," June 6, p. 24.

Engineering News Record (1963b). "$59 Million Job Shaken to a Stop?," November 28, pp. 20–21.

European Foundations. (2004). "Working Platform Guidelines Set Out to Cut Site Accidents," p. 6

Everts, H. J., and Luger, H. J. (1997). "Dutch National Codes for Pile Design," *Design of Axially Loaded Piles—European Practice,* Eds. F. De Cock and C. Legrand. Balkama, Rotterdam, pp. 243–265.

Feda, J., Krásný, J., Juranka, P., Pachta, V., and Masopust, J. (1997). "Design of Axially Loaded Piles—Czech Practice," *Design of Axially Loaded Piles—European Practice,* Eds. F. De Cock and C. Legrand. Balkama, Rotterdam, pp. 83–99.

Federal Highway Administration (1993). *Soils and Foundations, Workshop Manual,* second edition, Publication No. FHWA HI-88-009, Washington, D.C.

Feld, J. (1968). *Construction Failures,* Wiley, New York.

Fields, K. E., and Wells, W. L. (1944). "Pendleton Levee Failure," *Transactions,* ASCE, pp. 1400–1429.

Findlay, J. D., Brooks, N. J., Mure, J. N., and Heron, W. (1997). "Design of Axially Loaded Piles—United Kingdom Practice," *Design of Axially Loaded Piles—European Practice,* Eds. F. De Cock and C. Legrand. Balkama, Rotterdam, pp. 353–376.

Fisk, H. N. (1956). "Nearsurface Sediments of the Continental Shelf Off Louisiana," *Proceedings,* Eighth Texas Conference on Soil Mechanics and Foundation Engineering, Bureau of Engineering Research, University of Texas at Austin, September 14–15.

Focht, J. A. and Koch, K. J. (1973), "Rational Analysis of the Lateral Performance of Offshore Pile Groups," Proceedings, 5th Offshore Technology Conference, Houston, Vol. 2, pp. 701–708.

Franke, E. (1991). "Measurements beneath Piled Rafts," Keynote Lecture, ENPC Conference on Deep Foundations, Paris, March.

Franke, E., Lutz, B., and El-Mossallamy, Y., (1994). "Measurement and Numerical Modeling of High-Rise Building Foundations on Frankfurt Clay," Proceedings, ASCE Specialty Conference on Vertical and Horizontal Deformation of Foundations and Embankments, ASCE, New York, pp. 1335–1336.

Fredlund, D. G., and Rahardjo, H. (1993). *Soil Mechanics for Unsaturated Soils.* Wiley, New York.

Gaich, A., Fasching, A., Fuchs, R., and Schubert, W. (2003). "Structural Rock Mass Parameters Recorded by a Computer Vision System," *Proceedings,* Soil and Rock 2003, 12th Panamerican Conference on Soil Mechanics and Geotechnical Engineering, Cambridge, MA, Vol. 1, pp. 87–94.

Gazetas, G. (1983). "Analysis of Machine Foundation Vibrations: State of the Art," Soil Dynamics and Earthquake Engineering, Vol. 2(1), pp. 2–42.

Gazetas, G. (1991). "Formulas and Charts for Impedance of Surface and Embedded Foundation," J. Geotech. Engrg., ASCE, Vol. 117(9).

Georgiadis, M. (1983). "Development of *p-y* Curves for Layered Soils," *Proceedings,* Geotechnical Practice in Offshore Engineering, ASCE, April, pp. 536–545.

Gibbs, H. J., and Holtz, W. G. (1957). "Research on Determining the Density of Sands by Spoon Penetration Testing." *Proceedings,* 4th International Conference on Soil Mechanics, London, Vol. 1, pp. 35–39.

Glanville, W. H., Grime. G., Fox E. N. and Davies W. W. (1938), "An Investigation of the Stresses in Reinforced Concrete Piles During Driving," British Building Research Board Technical Paper No. 20, Dept. of Scientific and Industrial Research, His Majesty's Stationery Office, London.

Gleser, S. M. (1953). "Lateral Load Tests on Vertical Fixed Head and Free Head Piles," Symposium on Lateral Load Tests on Piles, ASTM Special Technical Publication 154, pp. 75–101.

Goble, G. G., Walker, F. K., and Rausche, F. (1972), "Pile Bearing Capacity—Prediction vs. Performance of Earth and Earth-Supported Structures," ASCE Special Conference, Purdue Univ. Vol. 1, Part 2, pp. 1243–1258.

Goble, G. G. and Rausche, F. (1986), "Wave Equation Analysis of Pile Driving—WEAP Program, U.S. Department of Transportation, Federal Highway Administration, Implementation Division, McLean, Volumes I-IV.

Gomez, J., Cadden, A., and Bruce, D. A. (2003), "Micropiles Founded in Rock. Development and Evolution of Bond Stresses Under Repeated Loading," *Proceedings,* 12th Panamerican Conference on Soil Mechanics and Geotechnical Engineering, Verlag Gluckauf, GMBH, Essen, Germany, Vol. 2, pp. 1911–1916.

Grand, B. A. (1970). "Types of Piles: Their Characteristics and General Use," *Highway Research Record, Number 333, Pile Founfations.* Highway Research Board, Washington, DC, pp. 3–15.

Gray, R. E., and Meyers, J. F. (1970). "Mine Subsidence and Support Methods in Pittsburgh Area," *Proceedings,* ASCE, Vol. 96, Paper No. SM4, July, pp. 1267–1287.

Gwizdala, K. (1997). "Polish Design Methods for Axially Loaded Piles," *Design of Axially Loaded Piles—European Practice,* Eds. F. De Cock and C. Legrand. Balkama, Rotterdam, pp. 291–306.

Hansen, B., and Christensen, N. H. (1969). Discussion of "Theoretical Bearing Capacity of Very Shallow Foundations," by A. L. Larkins, *Journal of the Soil Mechanics and Foundations Division,* ASCE, Vol. 95, Paper No. SM6, pp. 1568–1572.

Hansen, J. B. (1970). "A Revised and Extended Formula for Bearing Capacity," *Geoteknisk Instutut* (Danish Geotechnical Institute), Bulletin No. 28, pp. 5–11.

Heinonen, J., Hartikainen, J., and Kiiskilä, K. (1997). "Design of Axially Loaded Piles—Finnish Practice," *Design of Axially Loaded Piles—European Practice,* Eds. F. De Cock and C. Legrand. Balkama, Rotterdam, pp. 133–160.

Henkel, D. J. (1956), "The Effect of Overconsolidation on the Behavior of Clays During Shear," Geotechnique, Vol. 6, P. 139.

Hetenyi, M. (1946). *Beams on Elastic Foundation.* University of Michigan Press, Ann Arbor.

Hilf, J. W. (1956). "An Investigation of Pore Water Pressure in Compacted Cohesive Soils," *Technical Memorandum No. 654,* Bureau of Reclamation, U.S. Department of the Interior, Denver.

Hirsch, T. J., Carr, L. and Lowery, L. L. (1976), Pile Driving Analysis. TTI Program, U.S. Department of Transportation, Federal Highway Administration, Offices of Research and Development, IP-76-13, Washington D.C., Volumes I-IV.

Holeyman, A., Bauduin, C., Bottiau, M., Debacker, P., De Cock, F., Dupont, E., Hilde, J. L., Legrand, C., Huybrechts, N., Mengé, P., Miller, J. P., and Simon, G. (1997).

"Design of Axially Loaded Piles—Belgian Practice," *Design of Axially Loaded Piles—European Practice,* Eds. F. De Cock and C. Legrand. Balkama, Rotterdam, pp. 57–82.

Holloway, D. M. and Dover, A. R. (1978), "Recent Advances in Predicting Pile Drivability," Proceedings, Offshore Technology Conference, Houston, Texas, Oct.

Holtz, W. G., and Gibbs, H. J. (1954). "Engineering Properties of Expansive Clays," *Journal of the Soil Mechanics and Foundations Division,* ASCE, Vol. 80, pp. 641–663.

Horvath, R. G., and Kenney, T. C. (1979). "Shaft Resistance of Rock-Socketed Drilled Piers," *Proceedings,* Symposium on Deep Foundations, ASCE, Atlanta, pp. 182–214.

Hrennikoff, A. (1950). "Analysis of Pile Foundations with Batter Piles," *Transactions,* ASCE, 1950, Vol. 115, Paper No. 2401, pp. 351–374.

Hvorslev, M. J. (1949). *Subsurface Exploration and Sampling for Civil Engineering Purposes.* Waterways Experiment Station, U.S. Army Corps of Engineers, (reprinted 1962 and 1965), Vicksburg, Mississippi.

Ingra, T. S., and Baecher, G. B. (1983). "Uncertainty in Bearing Capacity of Sands," *Journal of Geotechnical Engineering,* ASCE, Vol. 109, No. 7, July, pp. 899–914.

Iowa Engineering Experiment Station (1959). *Screenings from the Soil Research Lab,* Vol. 3, No. 3, May-June, Iowa State College, Ames.

Isenhower, W. M., and Long, J. H. (1997). "Reliability Evaluation of AASHTO Design Equations for Drilled Shafts," *Transportation Research Record No. 1582: Centrifuge Modeling, Intelligent Geotechnical Systems, and Reliability-based Design.* pp. 60–67.

Japanese Road Association (1976). "Road Bridge Substructure Design Guide and Explanatory Notes, Designing of Pile Foundations." Tokyo, p. 67.

Jennings, J. E., and Burland, J. B. (1962). "Limitations to the Use of Effective Stresses in Partly Saturated Soils," *Geotechnique,* Vol. 12, pp. 125–144.

Jewell, R. J., and Andrews, D. C. (Eds.) (1988). "Engineering for Calcareous Sediments," *Proceedings,* International Conference on Calcareous Sediments, Perth, Vols. 1 and 2. Balkema, Rotterdam.

Johnson, A. M. (1951). "Engineering Significance of Sand Areas Interpreted from Airphotos," *Bulletin No. 46, Engineering Soil Survey Mapping,* Highway Research Board, December, pp. 28–47.

Johnson, G. W., (1982). "Use of the Self Boring Pressuremeter in Obtaining In Situ Shear Moduli of Clay," Thesis, Master of Science in Engineering, University of Texas at Austin.

Johnston, I. W., Donald, I. B., Bennet, A. B., and Edwards, J. W. (1980a). "The Testing of Large Diameter Pile Rock Sockets with a Retrievable Test Rig," *Proceedings,* Third Australian–New Zealand Conference on Geomechanics, Wellington.

Johnston, I. W., Williams, A. F., and Chiu, H. K. (1980b). "Properties of Soft Rock Relevant to Socketed Pile Design," *Proceedings,* International Conference on Structural Foundations on Rock, Sydney. Balkema: Rotterdam.

Katzenbach, R., and Moormann, C. (1997). "Design of Axially Loaded Piles and Pile Groups—German Practice," *Design of Axially Loaded Piles—European Practice,* Eds. F. De Cock and C. Legrand. Balkama, Rotterdam, pp. 177–201.

Kenney, T. C. (1959), Discussion, Proceedings, ASCE, Vol.85, No. SM3, pp. 67–69.

King, R. and Lodge, M. (15-18 March 1988), "North West Shelf Development—The Foundation Engineering Challenge," Engineering for Calcareous Sediments, Jewell & Khorshid (eds), Balkema, Rotterdam, Vol. 2, 333 pp.

Kitchens, Lance (2003). Russo Corporation, personal communication.

Kjellman, W., Kallstenius, T, and Wager, O. (1950), "Soil Sampler with Metal Foils, Device for Taking Undisturbed Samples of Very Great Length," *Proceedings,* Royal Swedish Geotechnical Institute, No. 1.

Kooijman, A. P. (1989). "Comparison of an Elastoplastic Quais Three-Dimensional Model for Laterally Loaded Piles with Field Tests," *Proceedings,* Third International Symposium, Numerical Models in Geomechanics. Elsevier Applied Science, New York, pp. 675–682.

Kraft, L. M., Jr., Focht, J. A., Jr., and Amerasinghe, S. F. (1981a). "Friction Capacity of Piles Driven into Clay," *Journal of the Geotechnical Engineering Division,* ASCE, Vol. 107, No. 11, November, pp. 1521–1541.

Kraft, L. M., Jr., Ray, R. P., and Kagawa. T. (1981b). "Theoretical *t-z* Curves," *Proceedings,* American Society of Civil Engineers, Vol. 107, No. GT11, November, pp. 1543–1561.

Kubo, K. (1965). "Experimental Study of the Behavior of Laterally Loaded Piles," *Proceedings,* Sixth International Conference on Soil Mechanics and Foundation Engineering, Vol. II, Montreal, pp. 275–279.

Kulhawy, F. H. (1983). "Transmission Line Structures Foundations for Uplift-Compression Loading," Geotechnical Group, Cornell University, *Report No. EL-2870,* Report to Electrical Power Research Institute, Geotechnical Group, Cornell University, Ithaca, NY.

Lacy, H. S. (1998). "Protecting Existing Structures Using Bored Piles," *Proceedings,* 6th Annual Great Lakes Geotechnical and Geoenvironmental Conference on the Design and Construction of Drilled Deep Foundations, Indianapolis, Indiana.

Lacy, H. S., and Gould, J. P. (1985). "Settlement from Pile Driving in Sands," *Vibration Problems in Geotechnical Engineering, Proceedings,* Symposium of the Geotechnical Engineering Division, ASCE, Detroit, October 22, pp. 152–173.

Lacy, H. S. and Moskowitz, J. (1993). "The Use of Augered Cast-In-Place Piles to Limit Damage of Adjacent Structures," Seminar on Innovation in Construction, Metropolitan Section, American Society of Civil Engineers, New York, NY.

Lacy, H. S., Moskowitz, J., and Merjan, S. (1994). "Reduced Impact on Adjacent Structures Using Augered Cast-In-Place Piles," *Transportation Research Record 1447.*

Ladd, C. C. (2003). "Recommended Practice for Soft Ground Site Characterization: Arthur Casagrande Lecture," *Proceedings,* Soil and Rock 2003, 12th Panamerican Conference on Soil Mechanics and Geotechnical Engineering, Cambridge, MA, Vol. 1, pp. 3–57.

Ladd, C. C., and Foott, R. (1974). "New Design Procedure for Stability of Soft Clays," *Journal of the Geotechnical Engineering Division,* ASCE, Vol. 100, No. GT7, July, pp. 763–766.

Lamb, T. W. and Whitman, R. V., (1969), Soil Mechanics, John Wiley & Sons, Inc., New York.

Le Laboratoire Central des Ponts et Chaussees (LCPC) (1986). "Bored Piles," English translation of *Les Pieux Flores* (FHWA TSS-86-206), March.

Lehane, B. M. (1997). "Design of Axially Loaded Piles—Irish Practice," *Design of Axially Loaded Piles—European Practice,* Eds. F. De Cock and C. Legrand. Balkama, Rotterdam, pp. 203–218.

Lemy, F., and Hadjigeorgiou, J. (2003). "Rock Mass Characterization Using Image Analysis," *Proceedings,* Soil and Rock 2003, 12th Panamerican Conference on Soil Mechanics and Geotechnical Engineering, Cambridge, MA, Vol. 1, pp. 95–100.

Leshchinsky, D., and Marcozzi, G. F. (1990). "Bearing Capacity of Shallow Foundations: Rigid versus Flexible Models," *Journal of Geotechnical Engineering,* ASCE, Vol. 116, No. 11, November, pp. 1750–1756.

Lieng, J. T. (1988). "Behavior of Laterally Loaded Piles in Sand-Large Scale Model Tests," Ph.D. Thesis, Department of Civil Engineering, Norwegian Institute of Technology.

Long, J. H. (1984). "The Behavior of Vertical Piles in Cohesive Soil Subjected to Repeated Horizontal Loading," Ph.D. Dissertation, University of Texas at Austin.

Long, W. R., Yi, D., and Jiang, S. Z. (1983). "Damage of a Pile-Supported Pier Due to Foundation Deformation and Its Repair," *Preprint,* Shanghai Symposium on Marine Geotechnology and Nearshore/Offshore Structures, Tong-Ji University, Shanghai, November 1–4.

Lysmer, J. (1978), "Analysis Procedures in Soil Dynamics," Earthquake Engrg. And Soil Dynamics, ASCE, Vol. 3.

Mandolini, A. (1997). "Design of Axially Loaded Piles—Italian Practice," *Design of Axially Loaded Piles—European Practice,* Eds. F. De Cock and C. Legrand. Balkama, Rotterdam, pp. 219–242.

Manoliu, I. (1997). "Design of Axially Loaded Piles—Romanian Practice," *Design of Axially Loaded Piles—European Practice,* Eds. F. De Cock and C. Legrand. Balkama, Rotterdam, pp. 307–320.

Matlock, H. (1970). "Correlations for Design of Laterally Loaded Piles in Soft Clay," *Proceedings,* Offshore Technology Conference, Houston, Texas, Paper No. 1204, pp. 577–594.

Matlock, H., Ingram, W. B., Kelley, A. E., and Bogard, D. (1980). "Field Tests of Lateral Load Behavior of Pile Groups in Soft Clay," *Proceedings,* Twelfth Annual Offshore Technology Conference, OTC 3871, Houston, Texas, pp. 163–179.

Matlock, H., and Reese, L. C. (1962). "Generalized Solutions for Laterally Loaded Piles, *Transactions,* ASCE, Vol. 127, Part 1, pp. 1220–1251.

Matlock, H., and Ripperger, E. A. (1956). "Procedures and Instrumentation for Tests on a Laterally Loaded Pile," *Proceedings,* Eighth Texas Conference on Soil Mechanics and Foundation Engineering, Special Publication 29, Bureau of Engineering Research, University of Texas at Austin.

Mattes, N. S., and Poulos, H. G. (1969). "Settlement of Single Compressible Pile," *Journal of the Soil Mechanics and Foundation Division,* ASCE, No. SM1, January, pp. 189–207.

May, R. W., and Thomson, S. (1978). "The Geology and Geotechnical Properties of Till and Related Deposits in the Edmonton, Alberta, Area," *Canadian Geotechnical Journal,* Vol. 15, No. 3, pp. 362–370.

Mayne, P. W., and Harris, D. E. (1993). "Axial Load-Displacement Behavior of Drilled Shaft Foundations in Piedmont Residuum," *FHWA Reference No. 41-30-2175.* Georgia Tech Research Corporation, Geotechnical Engineering Division, Georgia Institute of Technology, School of Civil Engineering, Atlanta.

McClelland, B. (1972). "Design and Performance of Deep Foundations," *Proceedings,* Specialty Conference on Earth and Earth Supported Structures, Purdue University, Soil Mechanics and Foundations Division, American Society of Civil Engineers.

McClelland, B., and Focht, J. A., Jr. (1958). "Soil Modulus for Laterally Loaded Piles," *Transactions,* ASCE, Vol. 123, pp. 1049–1086.

Mesri, G., Feng, T. W., and Benak, J. M. (1990). "Postdensification Penetration Resistance of Clean Sands," *Journal of Geotechnical Engineering,* ASCE, Vol. 116, No. 7, July, pp. 1095–1115.

Mets, M. (1997). "The Bearing Capacity of a Single Pile—Experience in Estonia," *Design of Axially Loaded Piles—European Practice,* Eds. F. De Cock and C. Legrand. Balkama, Rotterdam, pp. 115–132.

Meyerhof, G. G. (1963). "Some Recent Research on the Bearing Capacity of Foundations," *Canadian Geotechnical Journal,* Vol. 1, No. 1, pp. 16–26.

Meyerhof, G. G. (1976). "Bearing Capacity and Settlement of Pile Foundations," *Journal of the Soil Mechanics and Foundation Engineering, Proceedings,* ASCE, Vol. 102, No. GT3, March, pp. 197–228.

Meyerhof, G. G. (1978). "Ultimate Bearing Capacity of Footings on Sand Layer Overlying Clay," *Canadian Geotechnical Journal,* Vol. 15, No. 4, pp. 565–572.

Meyerhof, G, G., and Hanna, A. M. (1974). "Ultimate Bearing Capacity of Foundations on Layered Soil Under Inclined Loads," *Canadian Geotechnical Journal,* Vol. 11, No. 2, pp. 224–229.

Middlebrooks, T. A. (1940). "Fort Peck Slide," *Proceedings,* ASCE, December, pp. 1729–1749.

Miller, A. D. (2003). Personal communication.

Minami, K. (1983). "Comparison of Calculated and Experimental Load-Settlement Curves for Axially Loaded Piles," unpublished report, Geotechnical Engineering Center, University of Texas at Austin.

Mindlin, R. D. (1936). "Force at a Point in the Interior of a Semi-Infinite Solid," *Physics,* Vol. 7, No. 5, May, pp. 195–202.

Mitchell, J. K., and Solymar, Z. V. (1984). "Time-Dependent Strength Gain in Freshly Deposited or Densified Sand," *Journal of Geotechnical Engineering,* ASCE, Vol. 110, No. 11, November, pp. 1559–1576.

Mohan, D., Murthy, V. S. N., and Sen Gupta, D. P. (1970). "Structural Damages to Existing Buildings Due to Pile Driving in Calcutta Region," *Journal of the Indian Institute of Engineering,* Vol. 50, March, pp. 153–159.

Mokwa, R. L., and Duncan, J. M. (2001). "Laterally Loaded Pile Group Effects and *p-y* Multipliers," *Proceedings,* Conference on Foundations and Ground Improvement, Geotechnical Special Publication No. 113, ASCE, pp. 728–742.

Moore, W. L., and Masch, F. D. (1962). "Experiments on the Scour Resistance of Cohesive Sediments," *Journal of Geophysical Research,* Vol. 67, No. 4, pp. 1437–1446.

Moran, Proctor, Mueser, and Rutledge (1958), "Study of Deep Soil Stabilization by Vertical Sand Drains," U.S. Dept. of Commerce, Office Tech. Serv., Washington, DC, 192 pp.

Morgenstern, N. R., and Eigenbrod, K. D. (1974). "Classification of Argillaceous Soils and Rocks," *Journal of the Geotechnical Engineering Division,* ASCE, Vol. 100, No. GT10, October, pp. 1137–11156.

Mosher, R. L. (1984). "Load Transfer Criteria for Numerical Analysis of Axially Loaded Piles in Sand," U.S. Army Engineering Waterways Experimental Station Automatic Data Processing Center, Vicksburg, MS.

Moum, J., and Rosenqvist, I. T. (1957). "On the Weathering of Young Marine Clay," *Proceedings,* 4th International Conference on Soil Mechanics and Foundation Engineering, Vol. 1, pp. 77–79.

Muhs, H., and Weib, K. (1969). "The Influence of the Load Inclination on the Bearing Capacity of Shallow Footings," *Proceedings,* Seventh International Conference on Soil Mechanics and Foundation Engineering, Mexico, Vol. 2, pp. 187–194.

Nelson, J. D., and Miller, D. J. (1992). *Expansive Soils—Problems and Practice in Foundation and Pavement Engineering.* John Wiley & Sons, New York.

Newmark, N. M. (1942). "Influence Charts for Computing Stresses in Elastic Medium," *University of Illinois Engineering Experiment Station Bulletin 338.*

Novak, M. (1987), "State of the Art in Analysis and Design of Machine Foundations," Soil-structure Interaction, Elsevier and Computational Mechanics Ltd., New York, N.Y.

Novak, L. J., Reese, L. C., Wang, S-T. (2005), "Analysis of Pile-Raft Foundation with 3-D Finite Element Method," *Proceedings,* 2005 Structures Congress, New York, NY.

O'Neill, M. W. (1970). "Analysis of Axially Loaded Drilled Shafts in Beaumont Clay," Dissertation, University of Texas at Austin.

O'Neill, M. W. (1983). "Group Action in Offshore Piles," *Proceedings,* Conference on Geotechnical Practice in Offshore Engineering, ASCE, University of Texas at Austin, pp. 25–64.

O'Neill, M. W., and Dunnavant, T. W. (1984). "A Study of the Effects of Scale, Velocity, and Cyclic Degradability on Laterally Loaded Single Piles in Overconsolidated Clay," Report No. UHCE 84-7. Department of Civil Engineering, University of Houston, University Park, Houston, TX.

O'Neill, M. W., Ghazzaly, O. I. and Ha, H. B. (1977). "Analysis of Three-Dimensional Pile Groups with Nonlinear Soil Response and Pile–Soil–Pile Interaction," *Proceedings,* Offshore Technology Conference, Houston, TX, Vol. II, Paper No. 2838, pp. 245–256.

O'Neill, M. W., and Reese, L. C. (1970). "Behavior of Axially Loaded Drilled Shafts in Beaumont Clay," *Research Report No. 89-8,* Project No3-5-65-89, conducted for the Texas Highway Department, Federal Highway Administration, Center for Highway Research, University of Texas at Austin.

O'Neill, M. W., and Reese, L. C. (1999). "Drilled Shafts: Construction Procedures and Design Methods," *Report No. FHWA-IF-99-025,* prepared for the U.S. Department of Transportation, Federal Highway Administration, Office of Implementation, McLean, VA, in cooperation with ADSC: The International Association of Foundation Drilling.

O'Neill, M. W., and Sheikh, S. A. (1985). "Geotechnical Behavior of Underreams in Pleistocene Clay," *Drilled Piers and Caissons II,* Ed. C. N. Baker, Jr. ASCE, Denver, pp. 57–75.

O'Neill, M. W., Townsend, F. C., Hassan, K. H., Buller, A., and Chan, P. S. (1996), "Load Transfer for Drilled Shafts in Intermediate Geomaterials," Publication No. FHWA-RD-98-117, FHWA, November, 184 pages.

Olson, R. E. (1963). "Effective Stress Theory of Soil Compaction," *Journal of the Soil Mechanics and Foundations Division,* ASCE, Vol. 89, Paper No. SM2, Feb., pp. 27–45.

Olson, R. E. (1974). "Shearing Strengths of Kaolinite, Illite, and Montmorillonite," *Journal of the Geotechnical Engineering Division,* ASCE, Vol. 100, No. GT11, Nov., pp. 1215–1230.

Olson, R. E., and Dennis, N. D. (1983). "Axial Capacity of Steel Pipe Piles in Sand," *Proceedings,* Conference on Geotechnical Practice in Offshore Engineering, ASCE, pp. 389–401.

Olson, R. E., and Langfelder, L. J. (1965). "Pore Water Pressures in Unsaturated Soils," *Journal of the Soil Mechanics and Foundations Division,* ASCE, Vol. 91, No. SM4, April, pp. 127–150.

Otto, G. H. (1963). "Engineering Geology of the Chicago Area," *Proceedings of Lecture Series,* Foundation Engineering in the Chicago Area, January to June 1963, Department of Civil Engineering, Illinois Institute of Technology, Chicago, September.

Owens, M. J., and Reese, L. C. (1982). "The Influence of a Steel Casing on the Axial Capacity of a Drilled Shaft," *Research Report 255-1F,* Report to the Texas State Department of Highways and Public Transportation Center for Transportation Research, Bureau of Engineering Research, University of Texas at Austin, July.

Palmer, L. A., and Thompson, J. B. (1948). "The Earth Pressure and Deflection Along Embedded Lengths of Piles Subjected to Lateral Thrust," *Proceedings,* Second International Conference on Soil Mechanics and Foundation Engineering, Rotterdam, the Netherlands, Vol. 5, pp. 156–161.

Peck, R. B. (1944). "Discussion of Pendleton Levee Failure," *Transactions,* Vol. 109, ASCE, pp. 1414–1416.

Peck, R. B. (1967). "Bearing Capacity and Settlement: Certainties and Uncertainties," *Bearing Capacity and Settlement of Foundations,* Ed. A. S. Vesic. Department of Civil Engineering, Duke University, Durham, NC, pp. 3–8.

Peck, R. B. (1976). "Rock Foundations for Structures," *Proceedings,* Specialty Conference on Rock Engineering for Foundations and Slopes, Boulder, CO, ASCE, pp. 1–21.

Peck, R. B., and Bryant, F. G. (1953). "The Bearing-Capacity Failure of the Transcona Elevator," *Geotechnique,* Vol. 3, No. 5, pp. 201–208.

Peck, R. B., and W. C. Reed (1954), "Engineering Properties of Chicago Subsoils," Univ. of Ill. Eng. Exp. Sta. Bull. 423, 62 pp.

Peck, R. B., Hanson, W. E., and Thornburn, T. H. (1974). *Foundation Engineering,* second edition. Wiley, New York.

Peterson, R. (1954). "Studies of Bearpaw Shale at a Damsite in Saskatchewan," *Proceedings,* ASCE, Vol. 80, Separate No. 476, August.

Pile Dynamics, Inc. (2005), "GRLWEAP for Windows—Software for Dynamic Pile Analysis," Cleveland, Ohio.

Posey, C. J. (1971). "Protection of Offshore Structures Against Underscour," *Journal of the Hydraulics Division,* ASCE, Vol. 97 (HY7), pp. 1011–1016.

Poulos, H. G. (1968), "Analysis of the Settlement of Pile Groups," *Geotechnique,* Vol. 18, pp. 449–471.

Poulos, H. G. (1971), "Behavior of Laterally Loaded Piles: II—Pile Groups," *Proceedings,* American Society of Civil Engineers, Vol. 97, Paper No. SM5, May, pp. 733–751.

Poulos, H. G. (1994). "Alternate Design Strategies for Piled-Raft Foundations," *Proceedings,* Third International Conference on Deep Foundations Practice, Singapore.

Poulos, H. G., and Davis, E. H. (1968). "Settlement Behavior of Single Axially Loaded Incompressible Piles and Piers," *Géotechnique,* Vol. 18, No. 3, September, pp. 351–371.

Poulos, H. G., and Davis, E. H. (1980). *Pile Foundation Analysis and Design.* Wiley, New York.

Poulos, H. G., and Mattes, N. S. (1969). "Behavior of Axially Loaded End Bearing Piles," *Géotechnique,* Vol. 19, No. 7, June, pp. 285–300.

Prakash, S. (1962). "Behavior of Pile Groups Subjected to Lateral Load," Ph.D. Dissertation, Department of Civil Engineering, University of Illinois.

Prandtl, L. (1920). "Über die Härte Plasticher Körper," *Nachr. kgl. Ges. Wiss. Göttingen, Math. phys. Klasse.*

Quiros, G. W., and Reese, L. C. (1977). "Design Procedures for Axially Loaded Drilled Shafts," *Research Report No. 3-5-72-176,* conducted for the Texas Highway Department in cooperation with the U.S. Department of Transportation, Federal Highway Administration, by the Center for Highway Research, University of Texas at Austin, December.

Randolph, M. F., and Wroth, C. P. (1978). "Analysis of Deformation of Vertically Loaded Piles," *Journal of the Geotechnical Engineering Division,* ASCE, Vol. 104, No. 12, Dec., pp. 1465–1488.

Rankine, W. J. M. (1857). "On the Stability of Loose Earth," *Philosophical Transactions of the Royal Society of London,* Vol. 147.

Reavis, G. T., Reese, L. C., and Wang, S. T. (1995). "Engineering Characteristics of the GeoJet Foundation System," *Proceedings,* Meeting of the Texas Section, American Society of Civil Engineers, Waco, TX, April 24.

Reese, L. C. (1958). Discussion of "Soil Modulus of Laterally Loaded Piles," by B. McClelland and J. A. Focht, Jr., *Transactions,* ASCE, Vol. 123, pp. 1071–1074.

Reese, L. C. (1964). "Load versus Settlement for an Axially Loaded Pile," *Proceedings,* Part II, Symposium on Bearing Capacity of Piles, Central Building Research Institute, Roorkee, India, February, pp. 18–38.

Reese, L. C. (1966). "Analysis of a Bridge Foundation Supported by Batter Piles," *Proceedings,* Fourth Annual Symposium on Geology and Soil Engineering, Moscow, IA, April, pp. 61–73.

Reese, L. C. (1968). "Interpretation of Data from Load Test of a Drilled Shaft, Ship Channel Crossing, State Highway 146, Harris County Texas," Letter Report to McClelland Engineers, Houston, TX, April (unpublished).

Reese, L. C. (1983). *Behavior of Pile Groups Under Lateral Loading.* U.S. Department of Transportation, Federal Highway Administration, Washington, D.C.

Reese, L. C. (1984). *Handbook on Design of Piles and Drilled Shafts Under Lateral Load.* U.S. Department of Transportation, Federal Highway Administration.

Reese, L. C. (1997). "Analysis of Laterally Loaded Piles in Weak Rock," *Journal of Geotechnical and Geoenvironmental Engineering,* ASCE, Vol. 123, No. 11, pp. 1010–1017.

Reese, L. C., and Bowman, T. (1975). Unpublished report on load tests performed in Orange, TX.

Reese, L. C., Brown, D. A., and Morrison, C. (1988). "Lateral Load Behavior of a Pile Group in Sand," *Journal of Geotechnical Engineering,* ASCE, Vol. 114, November, pp. 1261–1276.

Reese, L. C., and Cox, W. R. (1968). "Soil Behavior from Analysis of Uninstrumented Piles Under Lateral Loading," *Performance of Deep Foundations,* ASTM, SPT 444, pp. 161–176.

Reese, L. C., Cox, W. R., and Koch, F. D. (1974). "Analysis of Laterally Loaded Piles in Sand," *Proceedings,* Offshore Technology Conference, Houston, TX, Vol. II, Paper No. 2080, pp. 473–484.

Reese, L. C., Cox, W. R., and Koch, F. D. (1975). "Field Testing and Analysis of Laterally Loaded Piles in Stiff Clay," *Proceedings,* VII Annual Offshore Technology Conference, Houston, TX, Vol. 2, Paper No. 2312, pp. 672–690.

Reese, L. C., and Matlock, H. (1956). "Non-dimensional Solutions for Laterally Loaded Piles with Soil Modulus Assumed Proportional to Depth," *Proceedings,* VIII Texas Conference on Soil Mechanics and Foundation Engineering, Bureau of Engineering Research, University of Texas at Austin, September.

Reese, L. C., and Nyman, K. J. (1978). "Field Load Tests of Instrumented Drilled Shafts at Islamorada, Florida," Report to the Girdler Foundation and Exploration Corporation, Clearwater, FL, February.

Reese, L. C., and O'Neill, M. W. (1988). *Drilled Shafts: Construction Procedures and Design Methods,* U.S. Department of Transportation, Federal Highway Administration, Office of Implementation, McLean, VA.

Reese, L. C., O'Neill, M. W., and Smith, R. E. (1970). "Generalized Solutions of Pile Foundations," *Proceedings,* American Society of Civil Engineers, Vol. 96, No. SM1, January, pp. 235–250.

Reese, L. C., and Van Impe, W. F. (2001). *Single Piles and Pile Groups Under Lateral Loading.* Balkema, Rotterdam.

Reese, L. C., and Wang, S.-T. (1983). "Evaluation of Method of Analysis of Piles Under Lateral Loading," *Preprint,* Shanghai Symposium on Marine Geotechnology and Nearshore/Offshore Structures, Shanghai, November 1–4.

Reese, L. C., and Welch, R. C. (1975). "Lateral Loading of Deep Foundations in Stiff Clay," *Proceedings,* ASCE, Vol. 101, No. GT7, February, pp. 633–649.

Reissner, H. (1924). "Zum Erddruckproblem," *Proceedings,* 1st International Congress on Applied Mechanics, Deflt, the Netherlands.

Reuss, R., Wang, S. T., and Reese, L. C. (1992). "Tests of Piles Under Lateral Loading at the Pyramid Building, Memphis, Tennessee," *Geotechnical News,* Vol. 10, Part 4, pp. 44–49.

Richards, A. F., and Zuidberg, H. M. (1983). "Sampling and In-Situ Geotechnical Investigations Offshore," *Marine Geotechnology and Nearshore/Offshore Structures,* ASTM Special Technical Publication 923, Eds. R. C. Chaney and H. Y. Fang, Symposium at Tongji University, Shanghai, November 1–4, pp. 51–73.

Richart, F. E. Jr., Woods, R. D., and Hall, J. R. (1970), *Vibration of Soils and Foundations,* Prentice-Hall, Inc., Englewood Cliffs, N.J.

Robertson, P. K., and Campanella, R. G. (1983). "Interpretation of Cone Penetration Data—Part I (Sand)," *Canadian Geotechnical Journal,* Vol. 20, No. 4, pp. 718–733.

Robertson, P. K., and Campanella, R. G. (1988). "Guidelines for Geotechnical Design Using CPT and CPTU Data" Volume II (with four appendices). Civil Engineering Department, University of British Columbia, Vancouver, B.C., February.

Robertson, R. N. (1961). "The Analysis of a Bridge Foundation with Batter Piles," unpublished thesis, Department of Civil Engineering, University of Texas at Austin.

Robinson, C. S., and Lee, F. T. (1964). "Geologic Research at the Straight Creek Tunnel Site, Colorado," *Highway Research Record, Number 57,* Highway Research Board, pp. 18–34.

Roesset, J. M. (1980), "Stiffness and Damping Coefficients of Foundations," *Proceedings Apec. Session on Dynamic Response of Pile Foundation: Analytical Aspects,* ASCE, pp. 1–30.

Rosenqvist, I. T. (1953). "Considerations on the Sensitivity of Norwegian Quick Clays," *Geotechnique,* Vol. 3, pp. 195–200.

Samson, C. H., Hirsch, T. J., and Lowery, L. L. (1963), "Computer Study of Dynamic Behavior of Piling," Proceedings, ASCE, Journal of Structure Division, Vol. 89, ST4.

Schmertmann, J. H. (1955), "The Undisturbed Consolidation Behavior of Clay," Trans. ASCE, 120, 1201–1227.

Schmertmann, J. H. (1970). "Static Cone to Compute Static Settlement Over Sand," *Proceedings,* ASCE, Vol. 96, Paper No. SM3, May, pp. 1011, 1043.

Schmertmann, J. H. (1977). "Report on Development of a Keys Limerock Shear Test for Drilled Shaft Design," Report to the Girdler Foundation and Exploration Company, Clearwater, FL, December.

Schmertmann, J. H. (1978). "Improved Strain Influence Factor Diagrams," *Proceedings,* ASCE, Vol. 104, Paper No. GT8, August, pp. 1131–1135.

Schmertmann, J. H., Hartman, J. P., and Brown, J. P. (1978), "Improved Strain Influence Factor Diagrams," Proceedings, ASCE, Vol. 104, No. GT8, August, pp. 1131–1135.

Schmidt, H. G. (1981). "Group Action of Laterally Loaded Bored Piles," *Proceedings,* Tenth International Conference, Soil Mechanics and Foundation Engineering, Stockholm, pp. 833–837.

Schmidt, H. G. (1985). "Horizontal Load Tests on Piles of Large Diameter Bored Piles," *Proceedings,* Eleventh International Conference, Soil Mechanics and Foundation Engineering, San Francisco, pp. 1569–1573.

Schram Simonsen, A., and Athanasiu, C. (1997). "Design of Axially Loaded Piles—Norwegian Practice," *Design of Axially Loaded Piles—European Practice,* Eds. F. De Cock and C. Legrand. Balkama, Rotterdam, pp. 267–289.

Seed, H. B., and Reese, L. C. (1957). "The Action of Soft Clay Along Friction Piles," *Transactions,* ASCE, Vol. 122, Paper No 2882, pp. 731–754.

Selig, E. T., and McKee, K. E. (1961). "Static and Dynamic Behavior of Small Footings," *Journal of the Soil Mechanics and Foundations Division,* ASCE, Vol. 87, Paper No. SM6, December, pp. 29–47.

Semple, R. M. and Ridgen, W. J. (1984), "Shaft Capacity of Driven Pile in Clay," Analysis and Design of Pile Foundations, ASCE, J. R. Meyer, Ed., October, pp. 59–79.

Sheikh, S. A., O'Neill, M. W., and Kapasi, K. (1985). "Behavior of 45-Degree Underream Footing in Eagle Ford Shale," *Research Report No. 85-12,* University of Houston, University Park, Houston, TX, December.

Shuster, L. A. (2004). "Redundancy Pays Off in Positioning Caissons for New Tacoma Narrows Bridge, *Civil Engineering,* ASCE, Vol. 74, No. 3, pp. 30–31.

Siegel, T. C., and Mackiewitz, S. M. (2003). "Failure of Axially Loaded Augered Cast-in-Place Piles in Coastal South Carolina," *Proceedings,* Soil and Rock America 2003, 12th Panamerican Conference on Soil Mechanics and Geotechnical Engineering, Cambridge, MA, Vol. 2, pp. 1839–1844.

Skempton, A. W. (1944), "Notes on the Compressibility of Clays," *Quart. J. Geol. Soc.,* London, *C,* pp. 119–135.

Skempton, A. W. (1948). "The $\phi = 0$ Analysis of Stability and Its Theoretical Basis," *Proceedings,* 2nd International Conference on Soil Mechanics and Foundation Engineering, Vol. 1, pp. 72–77.

Skempton, A. W. (1951). "The Bearing Capacity of Clays," *Proceedings,* Building Research Congress, Division I, London, pp. 180–189.

Skempton, A. W. (1954), "The Pore-pressure Coefficients *A* and *B,*" *Geotechnique,* Vol. 4, pp 143–147.

Skempton, A. W., and Bjerrum, L. (1957). "A Contribution to the Settlement Analysis of Foundations on Clay," *Géotechnique,* Vol. 7, pp. 168–178.

Skempton, A. W., and Northey, R. D. (1952). "The Sensitivity of Clays," *Geotechnique,* Vol. 3, pp. 40–51.

Skov, R. (1997). "Pile Foundation—Danish Design Methods and Piling Practice," *Design of Axially Loaded Piles—European Practice,* Eds. F. De Cock and C. Legrand. Balkama, Rotterdam, pp. 101–113.

Smith, E. A. L. (1960), "Pile Driving Analysis by the Wave Equation," ASCE, Journal of the Soil Mechanics and Foundations Division, 86(4), pp. 35–61.

Sowers, G. F. (1963). "Engineering Properties of Residual Soils Derived from Igneous and Metamorphic Rocks," *Proceedings,* Second Pan American Conference on Soil Mechanics and Foundation Engineering.

Sowers, G. B., and Sowers, G. F. (1970). *Introductory Soil Mechanics and Foundations,* third edition. Macmillan, New York, p. 338.

Spangler M. G. and Handy, R. L. (1982), Soil Engineering, 4th Edition, Harper & Row, Publishers, New York.

Spear, D., Reese, L. C., Reavis, G. T., and Wang, S. T. (1994). "Testing of GeoJet Units Under Lateral Loading," *Proceedings,* U.S. Federal Highway Administration, International Conference on Design and Construction of Deep Foundations, Orlando, FL, December 6–8.

Speer, D. (1992). "Shaft Lateral Load Test Terminal Separation," California Department of Transportation (unpublished).

Stevens, J. C. (1951). "Piedmont Soils Identified by Aerial Photographs," *Bulletin No. 46, Engineering Soil Survey Mapping,* Highway Research Board, December, pp. 48–66.

Stokoe, K. H. (1989). Personal communication.

Svensson, L., Olsson, C., and Grävare, C.-J. (1997). "Design of Axially Loaded Piles—Swedish Practice," *Design of Axially Loaded Piles—European Practice,* Eds. F. De Cock and C. Legrand. Balkama, Rotterdam, pp. 337–342.

Szecky, C. (1961). *Foundation Failures.* Concrete Publications, Ltd., London.

Taylor, D. W. (1948). *Fundamentals of Soil Mechanics.* Wiley, New York.

Terzaghi, K. (1913). "Beitrag zur Hydrographie und Morphologie des Kroatischen Karstes. Königlliche Ungarissche Geologische Reichsanstalt," *Mitteilungen aus dem Jahrbuche,* Vol. 20, No. 6, pp. 256–374.

Terzaghi, K. (1929). "Effect of Minor Geologic Details on the Safety of Dams," *American Institute of Mining and Metallurgical Engineers, Technical Publication 215,* pp. 31–44.

Terzaghi, K. (1943). *Theoretical Soil Mechanics.* Wiley, New York.

Terzaghi, K. (1944). "Discussion of Pendleton Levee Failure," *Transactions,* ASCE, Vol. 109, pp. 1416–1421.

Terzaghi, K. (1951). "The Influence of Modern Soil Studies on the Design and Construction of Foundations," *Building Research Congress,* Division 1, Part III, pp. 68–145.

Terzaghi, K. (1955). "Evaluation of Coefficients of Subgrade Reaction," *Geotechnique,* Vol. 5, pp. 297–326.

Terzaghi, K., and Peck, R. B. (1948). *Soil Mechanics in Engineering Practice.* Wiley, New York.

Terzaghi, K. and Peck, R. B. (1967), "Soil Mechanics in Engineering Practice," New York, Wiley.

Thompson, G. R. (1977). "Application of the Finite Element Method to the Development of *p-y* Curves for Saturated Clays," Thesis, Master of Science in Civil Engineering, University of Texas at Austin.

Thurman, A. G., and D'Appolonia, E. (1965). "Computed Movement of Friction and End Bearing Piles Embedded in Uniform and Stratified Soils," *Proceedings,* Sixth International Conference on Soil Mechanics and Foundation Engineering, Vol. 2, pp. 323–327.

Timoshenko, S. (1948). *Strength of Materials, Part II, Advanced Theory and Problems,* second edition. Van Nostrand, New York.

Timoshenko, S., and Goodier, J. N. (1951). *Theory of Elasticity,* McGraw-Hill, New York.

Tomlinson, M. J. (1980), Foundation Design and Construction, 4th Edition, Pitman Advanced Publishing Program.

Touma, F. T., and Reese, L. C. (1972). "Load Tests of Instrumented Drilled Shafts Constructed by the Slurry Displacement Method," Research Report conducted under Interagency Contract 108 for the Texas Highway Department, Center for Highway Research, University of Texas at Austin, January.

Trask, P. D., and Rolston, J. W. (1951). "Engineering Geology of San Francisco Bay, Califorina," *Bulletin of the Geological Society of America,* Vol. 62, pp. 1079–1109.

Ukritchon, B., Whittle, A. J., and Klangvijit, C. (2003). "Calculations of Bearing Capacity Factor Using Numerical Limit Analysis," *Journal of Geotechnical and Geoenvironmental Engineering,* ASCE, Vol. 129, No. 6, June, pp. 468–474.

United States Geological Survey (1953). "Interpreting Geologic Maps for Engineering Purposes," Hollidaysburg Quadrangle, Pennsylvania.

Vesić, A. S. (1961). "Bending of Beams Resting on Isotropic Elastic Solids," *Journal of the Engineering Mechanics Division,* ASCE, Vol. 87, Paper No. EN2, April, pp. 35–63.

Vesić, A. S. (1969), "Experiments with Instrumented Pile Groups in Sand," Performance of deep Foundations, ASTM, Special Technical Publication, No. 444, pp. 172–222.

Vesić, A. S. (1970). "Tests on Instrumented Piles, Ogeechee River Site," *Journal of the Soil Mechanics and Foundations Division,* ASCE, Vol. 96, Paper No. SM2, March, pp. 561–584.

Vesić, A. S. (1973). "Analysis of Ultimate Loads on Shallow Foundations," *Journal of the Soil Mechanics and Foundations Division,* ASCE, Vol. 99, Paper No. SM1, Jan., pp. 45–73.

Viggiani, C. (2000). "Analysis and Design of Piled Foundations," First Arrigo Croce Lecture, Naples, December.

Vijayvergiya, V. N. (1977). "Load-Movement Characteristics of Piles," *4th Symposium of Waterways, Port, Coastal and Ocean Division,* American Society of Civil Engineers, Long Beach, CA, Vol. 2, pp. 561–584.

Vijayvergiya, V. N., and Focht, J. A., Jr. (1972). "A New Way to Predict the Capacity of Piles in Clay," *Proceedings,* Offshore Technology Conference, Houston, TX, May.

Walker, F. C., and Irwin, W. H. (1954). "Engineering Problems in Columbia Basin Varved Clay," *Proceedings,* Soil Mechanics and Foundations Division, ASCE, Vol. 80, Separate No. 515, October.

Wang, S.-T. (1982). "Development of a Laboratory Test to Identify the Scour Potential of Soils at Piles Supporting Offshore Structures," Thesis, Master of Science in Civil Engineering, University of Texas at Austin.

Wang, S.-T. (1986). "Analysis of Drilled Shafts Employed in Earth-Retaining Structures," Ph.D. Dissertation, Department of Civil Engineering, University of Texas at Austin.

Welch, R. C., and Reese, L. C. (1972). "Laterally Loaded Behavior of Drilled Shafts," *Research Report No. 3-5-65-89,* Center for Highway Research, University of Texas at Austin, May.

Whitaker, T. and Cooke, R. W. (1966), "An Investigation of the Shaft and Base Resistance of Large Bored Piles in London Clay," *Proceedings,* Conference on Large Bored Piles, Institute of Civil Engineers, London, pp. 7–49.

Williams, A. F. (1980). "Principles of Side Resistance Development in Rock-Socketed Piles," *Proceedings,* Third Australian–New Zealand Conference on Geomechanics, Wellington.

Williams, A. F., Donald, I. B., and Chiu, H. K. (1980a). "Stress Distributions in Rock Socketed Piles," *Proceedings,* International Conference on Structural Foundations on Rock, Sydney. Balkema: Rotterdam.

Williams, A. F., and Erwin, M. C. (1980). "The Design and Performance of Cast-in-Situ Piles in Extensively Jointed Silurian Mudstone," *Proceedings,* Third Australian–New Zealand Conference on Geomechanics, Wellington.

Williams, A. F., Johnston, I. W., and Donald, I. B. (1980b). "The Design of Socketed Piles in Weak Rock," *Proceedings,* International Conference on Structural Foundations on Rock, Sydney, Balkema: Rotterdam, pp. 327–347.

Wilson, G., and Grace, H. (1939). "The Settlement of London Due to the Underdrainage of the London Clay," *Journal,* Institution of Civil Engineers, Vol. 12, pp. 100–127.

Womack, K. C., Whitaker, A. M., Caliendo, J. A., Goble, G. G., and Halling, M. W. (2003). "Physical Properties of Plywood and Waterboard as Pile Cushion Materials, Technical Note," *Journal of Geotechnical and Geoenvironmental Engineering,* ASCE, Vol. 129, No. 4, April, pp. 379–382.

Wright, S. J. (1977). "Limit States Design of Drilled Shafts," unpublished master's thesis, University of Texas at Austin.

Yamashita, K., Kakurai, M., and Yamada, T. (1994). "Investigation of a Piled Raft Foundation on Stiff Clay," *Proceedings,* Thirteenth International Conference on Soil Mechanics and Foundation Engineering, ISSMFE, India, Vol. 2, pp. 543–546.

Yamashita, K.,Yamada, T. and Kakurai, M. (1998). "Simplified Method for Analyzing Piled Raft Foundations," *Proceedings,* Deep Foundations on Bored and Auger Piles BAP III. Balkema, Rotterdam, pp. 457–464.

Yegian, M., and Wright, S. G. (1973). "Lateral Soil Resistance–Displacement Relationships for Pile Foundations in Soft Clays," *Proceedings,* Offshore Technology Conference, Houston, TX, Paper No. 1893, pp. 663–676.

Zeevaert, L. (1972), *Foundation Engineering for Difficult Soil Conditions.* Van Nostrand Reinhold, New York.

Zeevaert, L. (1975). Personal communication.

Zeevaert, L. (1980). "Deep Foundation Design Problems Related with Ground Surface Subsidence," *Proceedings,* 6th Southeast Asian Conference on Soil Engineering, Taipei, May 19–23, pp. 71–110.

SUBJECT INDEX

AUTHOR INDEX